NANOPARTICULATE DRUG DELIVERY SYSTEMS

NANOPARTICULATE DRUG DELIVERY SYSTEMS

Edited by
Raj K. Keservani, MPharm
Anil K. Sharma, MPharm, PhD

Apple Academic Press Inc.
3333 Mistwell Crescent
Oakville, ON L6L 0A2
Canada

Apple Academic Press Inc.
9 Spinnaker Way
Waretown, NJ 08758
USA

ISBN 13: 978-1-77463-388-5 (pbk)
ISBN 13: 978-1-77188-695-6 (hbk)

Library and Archives Canada Cataloguing in Publication

Nanoparticulate drug delivery systems / edited by Raj K. Keservani, MPharm, Anil K. Sharma, PhD, MPharm.

Includes bibliographical references and index.
Issued in print and electronic formats.
ISBN 978-1-77188-695-6 (hardcover).--ISBN 978-1-351-13726-3 (PDF)

1. Nanoparticles. 2. Drug delivery systems. I. Keservani, Raj K., 1981-, editor
II. Sharma, Anil K., 1980-, editor

| RS201.N35N36 2018 | 615.1'9 | C2018-905107-8 | C2018-905108-6 |

Library of Congress Cataloging-in-Publication Data

Names: Keservani, Raj K., 1981- editor. | Sharma, Anil K., 1980- editor

Title: Nanoparticulate drug delivery systems / editors, Raj K. Keservani, Anil K. Sharma.
Other titles: Nanoparticulate drug delivery systems (Keservani)

Description: Toronto ; New Jersey : Apple Academic Press, 2019. |
 Includes bibliographical references and index.

Identifiers: LCCN 2018041828 (print) | LCCN 2018042944 (ebook) | ISBN 9781351137263 (ebook) |
 ISBN 9781771886956 (hardcover : alk. paper)

Subjects: | MESH: Drug Delivery Systems | Nanoparticles--therapeutic use | Drug Carriers

Classification: LCC RS199.5 (ebook) | LCC RS199.5 (print) | NLM QV 785 | DDC 615.1/9--dc23

LC record available at https://lccn.loc.gov/2018041828

The Present Book is Dedicated to
Our Beloved
Aashna, Anika, Atharva, and Vihan

ABOUT THE EDITORS

Raj K. Keservani

Raj K. Keservani, MPharm, is associated with the Faculty of B. Pharmacy, CSM Group of Institutions, Allahabad, India. He has more than 10 years of academic experience in various institutes of India imparting pharmaceutical education. He has published 35 peer-reviewed papers in the field of pharmaceutical sciences in national and international reputed journals, 16 book chapters, 2 coauthored books, and 10 edited books. He is also active as a reviewer for several scientific international journals. His research interests encompass nutraceutical and functional foods, novel drug delivery systems (NDDS), transdermal drug delivery/drug delivery, health science/life science, and biology/cancer biology/neurobiology. Mr. Keservani graduated (B. Pharmacy) from the Department of Pharmacy, Kumaun University, Nainital, Uttarakhand, India in 2005. Afterwards, he qualified GATE in same year conducted by IIT Mumbai. He received his Master of Pharmacy (M. Pharmacy) from the School of Pharmaceutical Sciences, Rajiv Gandhi Proudyogiki Vishwavidyalaya, Bhopal, Madhya Pradesh, India, in 2008 with a specialization in pharmaceutics.

Anil K. Sharma

Anil K. Sharma, MPharm, PhD, is currently working as an assistant professor at the Department of Pharmacy, School of Medical and Allied Sciences, GD Goenka University, Gurugram, India. He has more than nine years of academic experience in pharmaceutical sciences. He graduated (B. Pharmacy) from the University of Rajasthan, Jaipur, India, in 2005. Afterwards, he qualified GATE in same year, conducted by IIT Mumbai. He received his Master of Pharmacy (M. Pharmacy) from the School of Pharmaceutical Sciences, Rajiv Gandhi Proudyogiki Vishwavidyalaya, Bhopal, Madhya Pradesh, India, in 2007, with a specialization in pharmaceutics. He has earned his PhD from the University of Delhi. In addition, he has published 29 peer-reviewed papers in the field of pharmaceutical sciences in national and international reputed journals, 15 book chapters, and 10 edited books. His research interests encompass nutraceutical and functional foods, novel drug delivery systems (NDDS), drug delivery, nanotechnology, health science/life science, and biology/cancer biology/neurobiology.

CONTENTS

CONTRIBUTORS

Hani Nasser Abdelhamid
Department of Chemistry, Assuit University, Assuit 71515, EgyptJuliana Palma Abriata
School of Pharmaceutical Sciences of RIbeirao Preto-University of Sao Paulo, Ribeirao Preto,
Sao Paulo, Brazil

Marcos Luciano Bruschi
Department of Pharmacy, Laboratory of Research and Development of Drug Delivery Systems,
Postgraduate Program in Pharmaceutical Sciences, State University of Maringá, Maringá,
Paraná, Brazil

Lakshmi Kiran Chelluri
Transplant Biology, Immunology & Stem Cell Unit, Global Hospitals, Hyderabad, India

Dariane Jornada Clerici
Laboratory of Nanotechnology, Post-Graduate Program of Nanosciences,
Centro Universitário Franciscano, Santa Maria, Brazil

Lizziane Maria Belloto de Francisco
Department of Pharmacy, Laboratory of Research and Development of Drug Delivery Systems,
Postgraduate Program in Pharmaceutical Sciences, State University of Maringá, Maringá,
Paraná, Brazil

Ljiljana Djekic
University of Belgrade-Faculty of Pharmacy, Belgrade, Serbia

Josimar O. Eloy
School of Pharmaceutical Sciences-Sao Paulo State University, Araraquara, Sao Paulo, Brazil

Surya Prakash Gautam
CT Institute of Pharmaceutical Sciences, Shahpur Campus, Jalandhar, India

Josef Jampílek
Department of Pharmaceutical Chemistry, Faculty of Pharmacy, Comenius University,
Odbojárov 10, 83232 Bratislava, Slovakia

Bhupinder Kaur
CT Institute of Pharmaceutical Sciences, Shahpur Campus, Jalandhar, India

Khushwinder Kaur
Department of Chemistry and Centre of Advanced Studies in Chemistry, Panjab University,
Chandigarh, Punjab 160014, India

Deepak Kaushik
Department of Pharmaceutical Sciences, Maharishi Dayanand University, Rohtak,
Haryana 124001, India

Katarína Kráľová
Institute of Chemistry, Faculty of Natural Sciences, Comenius University,
Ilkovičova 6, 84215 Bratislava, Slovakia

Narinder Kumar
CT Institute of Pharmaceutical Sciences, Shahpur Campus, Jalandhar, India

Sunil Kumar
Department of Pharmaceutical Sciences, Guru Jambheshwar University of Science and Technology, Hisar, Haryana 125001, India

Robert Lee
Ohio State University, Columbus, Ohio, USA

Marcela Tavares Luiz
School of Pharmaceutical Sciences of RIbeirao Preto-University of Sao Paulo, Ribeirao Preto, Sao Paulo, Brazil

Sheefali Mahant
Department of Pharmaceutical Sciences, Maharishi Dayanand University, Rohtak,, Haryana 124001, India

Juliana Maldonado Marchetti
School of Pharmaceutical Sciences of RIbeirao Preto-University of Sao Paulo, Ribeirao Preto, Sao Paulo, Brazil

Sanju Nanda
Department of Pharmaceutical Sciences, Maharishi Dayanand University, Rohtak, Haryana 124001, India

Mônica Villa Nova
Department of Pharmacy, Laboratory of Research and Development of Drug Delivery Systems, Postgraduate Program in Pharmaceutical Sciences, State University of Maringá, Maringá, Paraná, Brazil

Rakesh Pahwa
University Institute of Pharmaceutical Sciences, Kurukshetra University, Kurukshetra, Haryana 136119, India

Raquel Petrilli
University of Western Sao Paulo (UNOESTE), Presidente Prudente, São Paulo, Brazil

Meenakshi Ponnana
Transplant Biology, Immunology & Stem Cell Unit, Global Hospitals, Hyderabad, India

Rekha Rao
Department of Pharmaceutical Sciences, Guru Jambheshwar University of Science and Technology, Hisar, Haryana 125001, India

Giovanni Loureiro Raspantini
School of Pharmaceutical Sciences of RIbeirao Preto-University of Sao Paulo, Ribeirao Preto, Sao Paulo, Brazil

Hélen Cássia Rosseto
State University of Maringá, Postgraduate Program in Pharmaceutical Sciences, Department of Pharmacy, Laboratory of Research and Development of Drug Delivery Systems, Maringá, Paraná, Brazil

Roberto Christ Vianna Santos
Laboratory of Oral Microbiology Research, Universidade Federal de Santa Maria, Santa Maria, RS, Brazil

Ranjit Singh
CT Institute of Pharmaceutical Sciences, Shahpur Campus, Jalandhar, India

Márcia Ebling de Souza
a. Laboratory of Microbiological Research, Centro Universitário Franciscano, Santa Maria, Brazil
b. Laboratory of Nanotechnology, Post-Graduate Program of Nanosciences,
   Centro Universitário Franciscano, Santa Maria, Brazil

Lucas de Alcântara Sica de Toledo
State University of Maringá, Postgraduate Program in Pharmaceutical Sciences,
Department of Pharmacy, Laboratory of Research and Development of Drug Delivery Systems,
Maringá, Paraná, Brazil

Md. Sahab Uddin
Department of Pharmacy, Southeast University, Dhaka, Bangladesh

Hui-Fen Wu
Department of Chemistry and Center for Nanoscience and Nanotechnology, National Sun Yat-Sen
University, 80424 Kaohsiung, Taiwan; School of Pharmacy, College of Pharmacy, Kaohsiung Medical
University, 807 Kaohsiung, Taiwan; Institue of Medical Science and Technology, National Sun Yat-Sen
University, 80424 Kaohsiung, Taiwan; Doctoral Degree Program in Marine Biotechnology, National
Sun Yat-Sen University and Academia Sinica, 80424 Kaohsiung, Taiwan

ABBREVIATIONS

αCD	α-cyclodextrin
5-FU	5-fluorouracil
6-OHDA	6-hydroxydopamine
Aβ	amyloid beta
Ab	antibody
ACF	aceclofenac
ACh	acetylcholine
AD	Alzheimer's disease
AFM	atomic force microscopy
AgNPs	silver nanoparticles
ALPZ	alprazolam
AMB	amphotericin B
AMT	amantadine
ANP	asenapine
API	active pharmaceutical ingredient
APM	apomorphine
APZ	aripiprazole
ASGP-R	asialoglycoprotein receptor
ATL	amitriptyline
AUC	area under the curve
BBB	blood–brain barrier
BPR	buspirone
BRC	bromocriptine
BSA	bovine serum albumin
CBZ	carbamazepine
CCM	curcumin
CDS	cyclodextrins
CLP	citalopram
ClTox	chlorotoxin
CMC	carboxymethyl chitosan
CNS	central nervous system
CNT	carbon nanotube
COMT	catechol-o-methyltransferase
CPM	clomipramine

CPZ	chlorpromazine
CS	chitosan
CTC	citicoline
CTX	chlorotoxin
CZP	clozapine
DA	dopamine
DAA	diacetyl apomorphine
DAP	diallyl phthalate
DIA	diisobutyryl apomorphine
DMP	domperidone
DNA	deoxyribonucleic acid
DNP	donepezil
DOX	doxorubicin
DPPG	dipalmitoylphosphatidylglycerol
DSC	differential scanning calorimetry
DXT	duloxetine
DXT–NLC	NLC-containing duloxetine
EC	ethylcellulose
ECM	extracellular matrix
EE	entrapment efficiency
EGF	epidermal growth factor
EGFR	epidermal growth factor receptor
EL	endosome/lysosome
EMA	European Medicines Agency
EPR	electron paramagnetic resonance
EPR	enhanced permeability and retention
ESR	electron spin resonance
ETC	entacapone
ETX	ethosuximide
FA	folic acid
FA–BSA	folate-conjugated bovine serum albumin
FALT	fixed aqueous layer thickness
FDA	Food and Drug Administration
FR	folate receptor
FXT	fluoxetine
GABA	γ-aminobutyric acid
GBP	gabapentin
GLT	galantamine
GLT–Ce-Hap	GLT ceria nanodots-containing hydroxyapatite
Glu-GNPs	glucose-bound GNPs

GM1-rHDL	monosialotetrahexosylganglioside-modified reconstituted high-density lipoprotein
GMS	glyceryl monostearate
GNPs	gold nanoparticles
GRAS	generally recognized as safe
GRAS	generally regarded as safe
HA	hyaluronic acid
HAS	human serum albumin
HMC	highly ordered mesoporous carbon
HPH	high-pressure homogenization
HPL	haloperidol
HPLC	high-pressure liquid chromatography
HSH	high-shear homogenization
HUVEC	human umbilical vein endothelial cells
HXZ	Hydroxyzine
i.n.	intranasal
i.p.	intraperitoneal
IC	intracoronary
Ig	immunoglobulins
IPD	Iloperidone
IPM	imipramine
LBDDS	lipid-based drug delivery systems
LCNs	lipid-core nanocapsules
LD	laser diffraction
LD	levodopa
LDH	lactate dehydrogenase
LE	lipid emulsions
LN	lipid nanoparticles
LNC	lipid nanocarriers
LRD	lurasidone
LTG	lamotrigine
MAO-B	monoamine oxidase B
MASSA	melatonin agonist and selective serotonin antagonist
MDR	multidrug resistance
ME	microemulsion
MFH	magnetic fluid hyperthermia
MIP	molecularly imprinted polymer
MLD	molecular layer deposition
MLNPs	magnetic luminescent nanoparticles
MMA	methyl methacrylate

MMPNs	multifunctional magneto-polymeric nanohybrids
MMT	montmorillonite
MMTN	memantine
MNEG	mucoadhesive oil/water nanoemulgel
MNPs	magnetic nanoparticles
MPS	mononuclear phagocyte system
MR	magnetic resonance
MRI	magnetic resonance imaging
MSCs	mesenchymal stem cells
MSN	mesoporous silica NPs
MTX	methotrexate
MVs	microvesicles
MW	molecular weight
MWCNTs	multiwalled CNTs
NAP	neuroprotective peptide
NAs	nucleic acids
NC	nanocapsules
NCLs	nanostructured lipid carriers
NEs	nanoemulsions
NGs	nanogels
NIPAM	N-isopropylacrylamide
NIR	near-infrared
NMDA	N-methyl-d-aspartate
NMR	nuclear magnetic resonance
NPLC	nanostructured polymeric lipid carriers
o/w	oil-in-water
OBZ	oxcarbazepine
OVA	ovalbumin
OZP	olanzapine
P4VP	poly(4-vinylpyridine)
PACA	polyalkyl cyanoacrylates
PAMAM	polyamidoamine
PBCA	polybutyl cyanoacrylate
PCL	poly-ε-caprolactone
PCS	photon correlation spectroscopy
PD	Parkinson's disease
PDA	polidopamine
PDI	polydispersity index
PECA	polyethyl cyanoacrylate
PECs	polyelectrolyte complexes

PEG	polyethylene glycol
PEI	polyethyleneimine
PEO	poly(ethylene oxide)
PEO–PPO–PEO	poly(ethylene oxide)–poly(propylene oxide)–poly(ethylene oxide)
PET	positron emission tomography
PG	poly-L-glutamic
P-gp	P-glycoprotein
PHB	polyhydroxybutyrate
PHP	perphenazine
PHT	phenytoin (PHT
PiBCA	polyisobutyl cyanoacrylate
PLA	polylactic acid
PLGA	poly(lactic-co-glycolic acid)
PMAA	poly(methacrylic acid)
PMMA	polymethyl methacrylate
PMS	PEG monostearate
PNBs	perfluorocarbon nanobubbles
PNIPAM	poly(N-isopropylacrylamide)
PNPs	polymeric NPs
PPD	paliperidone
PPSu	poly(propylene succinate)
PSMA	prostate-specific membrane antigen
PTT	photothermal therapy
PVA	poly(vinyl alcohol)
PVP	polyvinylpyrrolidone
PXT	paroxetine
QDs	quantum dots
QTP	quetiapine
RES	reticuloendothelial system
RGL	rasagiline
RHAMM	receptor for hyaluronate-mediated motility
RM	reverse micelles
RNA	ribonucleic acid
ROS	reactive oxygen species
RPD	risperidone
RPN	ropinirole
RTG	rotigotine
RVG	rivastigmine
SC	stratum corneum

SELEX	systematic evolution of ligands by exponential enrichment
SEM	scanning electron microscopy
SGL	selegiline
siRNA	small interfering ribonucleic acid
SLNPs	solid lipid nanoparticles
SLNs	solid lipid nanoparticles
SNEDDS	self-nanoemulsifying drug delivery system
SPD	sulpiride
SPIONs	superparamagnetic iron oxide nanoparticles
SSRIs	serotonin-specific reuptake inhibitors
STR	sertraline
SWCNTs	single-wall carbon nanotubes
SWNT	single-wall nanotube
TBARS	thiobarbituric acid reactive substances
TCR	tacrine
TEM	transmission electron microscopy
TfR	transferrin receptor
TGA	thioglycolic acid
TPP	tripolyphosphate
TRZ	trazodone
UV	ultraviolet
VEGF	vascular endothelial growth factor
VLF	venlafaxine
VPA	valproic acid
WHO	World Health Organization
XRD	X-ray diffractometry
ZNS	zonisamide
ZP	zeta potential
ZPD	ziprasidone

PREFACE

The existence of diseases/disorders his commenced from the beginning of civilization on the Earth. In the ancient world, the occurrences of morbidity and mortality used to be higher, and this was ascribed to be resulting from the wrath of deities, owing to ignorance among the society. Contemporary scientists have faced a lot of trouble while attempting to demonstrate the treatment potential of medicines. However, as science evolved further, the significance of drugs to cure and mitigate certain ailments has recieved acceptance from people to whom these have done wonders. Substantially, it was recognized that the active ingredients warranted to be dispensed as dosage forms, which usually comprised drugs and excipients.

There has been an overwhelming growth of nanotechnology in a variety of aspects that we routinely encounter. The uses of nanotechnology embrace materials science, engineering, medicine, dentistry, drug delivery, and so forth. The development relevant to the delivery of pharmaceutical active ingredients are of paramount concern for researchers working in domains of academia and industry.

In line with the above, the objective of this present book is to provide its readers updated knowledge about several nanoparticulate carriers. To accomplish this, the content of this book is written by adept, qualified, and well-known scientists and researchers from all over the world. The audience of the present book encompasses postgraduate students, researchers, academicians, scientists, and industrialists.

This book, **Nanoparticulate Nanocarriers Approaches in Drug Delivery**, is comprised of 12 chapters divided into four sections, which entail an introduction of nanoparticulate, nanocarriers, physicochemical features, generalized, and specific applications dealing with drug delivery in particular. The materials used, as well as formulation and characterization, have been discussed in detail. The emphasis of certain chapters is to provide the authors' specific input regarding treatment of a disease/disorder causing high mortality.

PART I: APPLICATIONS OF NANOPARTICLES

Chapter 1: Polymeric Nanoparticles: General Features, Polymers, and Formulation Aspects, written by Marcos Luciano Bruschi and colleagues, introduces the terminology prevailing in nanoscience. Further, it provides an overview of various polymers explored for preparation of nanoparticles. Subsequent to this, techniques/procedures to formulate nanoparticles have been discussed.

The details of general principles of nanoparticles have been presented in Chapter 2: Nanoparticles in Drug Delivery: General Characteristics, Applications, and Challenges, written by Khushwinder Kaur. The chapter deliberates over the potential of nanoparticles in the treatment of diseases such as cancer, Alzheimer's, and so forth. Further, the challenges faced while attempting delivery through these nanocarriers are also described.

Chapter 3: Nanoparticles as Nanopharmaceuticals: Smart Drug Delivery Systems, written by Md. Sahab Uddin, gives an elaborated discussion of nanoparticles, covering a general introduction, characterization techniques, manufacturing, and classification of different nanoparticles. Future opportunities as well as challenges have been mentioned in the last section of the chapter.

A customized approach with respect to applications of nanoparticles in cancer is described by Chapter 4: Nanoparticles Advance Drug Delivery for Cancer Cells, written by Hani Nasser Abdelhamid and Hui-Fen Wu. The authors claim that this book chapter is a valuable reference source for those scientists working in the field of pharmaceutical sciences, medicine, bionanotechnology, materials science, biomedical sciences, and related areas of life sciences. They have focussed on the most relevant references and concisely summarize the findings with illustrated examples.

Chapter 5: Nanotechnology-Based Formulations for Drug Targeting to the Central Nervous Systems, written by Josef Jampílek and Katarína Králová, deals with specific targeting to organs/systems in the CNS. This contribution is focused on the effects and CNS targeting of nanoscale formulations containing drugs such as antiepileptics, antipsychotics, anxiolytics, antidepressants as well as drugs applied for treatment of schizophrenia and Parkinson's and Alzheimer's diseases. The benefits connected with their application (e.g., reduction of required drug dose at bioavailability increase, reduced side effects due to decreased toxicity against nontarget cells, prolongation of time in circulation) are highlighted as well.

PART II: METALLIC NANOPARTICLES

The description of various magnetic nanoparticles is given in Chapter 6: Magnetic Nanoparticles for Drug Delivery, written by Meenakshi Ponnana and Lakshmi Kiran Chelluri. This chapter covers the physical and chemical properties, applications, their fabrication, enhanced targeting through passive, active, and physical magnetic targeting mechanisms, route of their administration through oral, parenteral, intranasal, dermal and transdermal routes, imaging capabilities, innovative approaches for using stem cells and their *in vivo* tracking. In addition, future perspectives and challenges faced by magnetic nanoparticles in the drug delivery and their scope have also been addressed.

Chapter 7: Overview of Applications of Gold Nanoparticles in Therapeutics, written by Juliana Palma Abriata and colleagues, provides an overview of specialized gold nanoparticles. The objective is to present and discuss some aspects of gold nanoparticles (GNP) development and its use in the drug delivery field. In the beginning, different types of GNP, its production methods, and characterization have been described. Then, the different functionalization options, using small molecules or macromolecules, are discussed. Finally, the stimuli-responsive GNPs are described. Interestingly, ongoing clinical trials have been presented in tabular form.

The uses of iron oxide nanoparticles for treatment of cancer have been discussed in Chapter 8: Superparamagnetic Iron Oxide Nanoparticles: Application in Diagnosis and Therapy of Cancer, written by Ljiljana Djekic. This chapter provides an overview of the main research aspects with respect to superparamagnetic iron oxide nanoparticles (SPIONs), including common approaches for magnetite core synthesis; main strategies for coating and functionalization of SPIONs surface with targeting ligands, imaging or therapeutic moieties, drug release mechanisms; the principles of their usage in magnetic fluid hyperthermia (MFH) and magnetic resonance imaging (MRI); and safety considerations. The potential for enhancement of diagnostic and therapy of cancer is described in detail.

PART III: LIPID-BASED NANOPARTICULATES

Chapter 9: Solid Lipid Nanoparticles: General Aspects, Preparation Methods, and Applications in Drug Delivery, written by Marcos Luciano Bruschi and colleagues, addresses widespread applications of lipid nanoparticles as carriers for different drugs. This chapter brings a combination of information from the beginning of the development of these systems, in the 1990s decade until modern studies, approaching the main preparation methods and also the lipid nanoparticle evaluation as drug delivery systems through different application routes. In addition, toxicity issues have also been mentioned in the last section of this chapter.

The focused information of solid lipid nanoparticles in the treatment of skin lesions is given by Chapter 10: Solid Lipid Nanoparticles in Drug Delivery for Skincare, written by Sheefali Mahant and associates. The present chapter provides a detailed account of solid lipid nanoparticles (SLNs) as dermal carriers, covering their composition, production, characterization, release profile, cosmetic benefits, and the studies carried out. A list of recent patents on SLNs for drug delivery in skin care has also been included.

PART IV: NEWER NANOARCHITECTURES

Chapter 11: Dendrimers in Gene Delivery, written by Bhupinder Kaur and colleagues, describes the applications of dendrimers in gene delivery. This chapter focuses on the basic mechanisms of gene delivery and dendrimer-based gene delivery. The variety of approaches for gene delivery are described in the beginning, then uses of branched polymeric structures for transfer of genetic information are provided.

An overview of nanoscience-based preparations of carbon nanotubes for the treatment of infections has been presented in Chapter 12: Carbon Nanotubes for Drug Delivery: Focus on Antimicrobial Activity, written by Márcia Ebling de Souza and associates. The authors have provided a general introduction of CNTs in the first half of chapter. The latter half deals with diverse uses of these carriers in control of various microorganisms. The conclusion summarizes the chapter with mention of future perspectives.

PART I
Applications of Nanoparticles

CHAPTER 1

POLYMERIC NANOPARTICLES: GENERAL FEATURES, POLYMERS, AND FORMULATION ASPECTS

MARCOS LUCIANO BRUSCHI*,
LIZZIANE MARIA BELLOTO DE FRANCISCO, and
MÔNICA VILLA NOVA

Department of Pharmacy, Laboratory of Research and Development of Drug Delivery Systems, State University of Maringá, Maringá, PR, Brazil

Corresponding author. E-mail: mlbruschi@uem.br; mlbruschi@gmail.com

ABSTRACT

Nanotechnology has been extensively applied in pharmaceutical research to improve the drug delivery and the treatment of many diseases. Nanoparticle (NP) is a general term for colloidal particles smaller than 1000 nm in diameter in which a therapeutic agent can be encapsulated, adsorbed, or chemically coupled. Polymeric NPs represent one of the most investigated nanocarriers due to the great number of materials and methods available. In this sense, we propose an approach about polymeric NPs as delivery systems, including the general concepts, applications and advantages, and the natural and synthetic polymers and preparation methods commonly found throughout the literature to prepare them. Besides the increased efficacy and decreased toxicity in relation to the conventional therapy, it is expected that the development of methods that allow large-scale production as well as the use of low-cost materials can facilitate the reach of NPs to the market.

1.1 INTRODUCTION

Only nearly 20 years after the development of liposomes as drug carriers, a nanoparticle (NP)-based drug delivery system was approved by the FDA (Bruynesteyn et al., 2007; Schütz et al., 2013a; Swierczewska, 2016). Currently, less than 10 nanoparticulate systems are available on the market. Even so, few of them have therapeutic purposes, such as the liposomal transporter containing Amphotericin B (AmBisome®), solid lipid NPs to treat hepatitis C (Nanobase®), or virosomal vaccines (Inflexal® V and Epaxal®); the majority is used for diagnostic or as medical devices (Verigene®, Silverline®, Acticoat™, or Endorem™ superparamagnetic iron oxide nanoparticles [SPIONS]) (Beal et al., 2013; Bozzuto and Molinari, 2015; De Jong et al., 2005; Duncan, 2003; ESF, 2005; Ferrari, 2005; Marcato and Duran, 2008; Weissig et al., 2014; Wojewoda et al., 2013).

One reason for this small commercial portfolio of NPs may be related to the scale-up. The results of some studies cannot always be compared since they may use different cell lines, types of systems, formulations, or methods. Some clinically important parameters, such as the safe dose range or percentage of specific targeting in comparison with the amount found in nonspecific target tissues are not always reported or determined correctly (Bertrand and Leroux, 2012). For a good clinical translation, preclinical studies should primarily consider the therapeutic efficacy parameters used in clinical trials, followed by safety, either due to the concentration of material or drug in each tissue.

Some problems also arise in the magnification and appropriate cost-effective characterization of each batch of NPs (Bozzuto and Molinari, 2015). Consequently, the financial aspect is another important barrier during the marketing process. Once in the market, they must be cost-effective to be included in the common therapeutic arsenal. Currently, the price of nano-systems is very high, which obligate many countries to choose conventional treatments (Al-Badriyeh et al., 2009; Bamrungsap et al., 2012; Bertrand and Leroux, 2012; Bruynesteyn et al., 2007; Duncan and Gaspar, 2011; Naik et al., 2011; Zazo, 2016).

Most of the researches and patents in the nanotechnology area have been carried out at the universities by using government funding programs (Andrade et al., 2013; Schütz et al., 2013a,b). However, without the support of industry to achieve the clinical trials and without national and international regulations, the commercial success of nanomedicines become challenging. This can explain why there are only 276 clinical trials involving NPs to treat infectious diseases, in contrast to the thousands of published works (Chang and Yeh, 2012). Moreover, the interdisciplinary collaboration is essential to achieve a successful nanomedicine (Andrade et al., 2013; Schütz et al., 2013a; Zazo et al., 2016).

NPs can offer promising treatments of diseases in which specific targeting is beneficial. At the same time, they can optimize the physicochemical properties of drugs, allowing the clinical use of new active pharmaceutical ingredient (API) or the administration through more convenient routes. In addition, NPs may decrease drug dose and toxicity, as well as improve delivery schedules and bioavailability. However, little is known about the metabolism, clearance and toxicity of NPs, the preferential targets for certain infections, and the drug levels for therapeutic activity on the target site, improving the quality of life of patients due to less frequent and fewer serious adverse effects (Bosetti et al., 2013; Zazo, 2016).

1.1.1 TERMINOLOGY

Nanoencapsulation involves the development of drug-loaded particles with diameters ranging from 1 to less than 1000 nm (Reis et al., 2006; Schaffazick and Gutterres, 2003). Some authors consider that the NPs used as drug delivery vehicles are generally <100 nm in at least one dimension (Singh and Lillard, 2009).

Polymeric NPs can be defined as solid drug carriers that may or may not be biodegradable (Couvreur, 1988; Couvreur et al., 1995; Reis et al., 2006). The API can be found dispersed, linked, entrapped, adsorbed onto, and/or encapsulated into a polymer matrix (Singh and Lillard, 2009). NP is a general term for particulate systems in the nanoscale, which depending on the structure can be classified as nanospheres or nanocapsules (Fig. 1.1). These types of NPs may differ according to the properties, composition, and release characteristics (Singh and Lillard, 2009).

Nanocapsules are reservoir systems in which the active agent is restricted in a cavity consisting of an inner liquid core surrounded by a polymeric membrane (Couvreur et al., 1995). In this case, the active substances are usually dissolved in the inner core but may also be adsorbed to the capsule surface (Allémann et al., 1993; Reis et al., 2006). On the other hand, nanospheres are formed by a polymer matrix where the drug can be internally distributed and/or adsorbed onto the surface (Schaffazick and Gutterres, 2003).

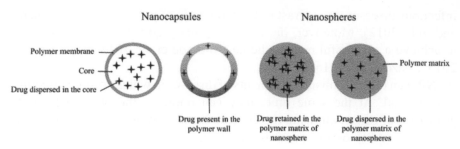

FIGURE 1.1   Schematic representation of the different types of nanoparticles.

In comparison with microparticles, NPs can be better absorbed by cells, and therefore, act as effective delivery systems (Brannon-Peppase and Blanchette, 2004; Pison et al., 2006; Schatzlein, 2006; Stylios et al., 2005; Yokoyama 2005).

1.1.2   APPLICATIONS AND ADVANTAGES OF NANOPARTICLE (NP) DRUG DELIVERY

The polymeric NPs can be obtained from natural and/or synthetic materials and have received the greatest attention because of their stability and ease of surface modification (Herrero-Vanrell et al., 2005; Vauthier et al., 2003). These features contribute to achieving both the controlled drug release and the specific targeting of the disease by adjusting the characteristics of the polymer and surface chemistry (Moghimi et al., 2001; Panyam et al., 2003a; Schaffazick and Gutterres, 2003). Moreover, these systems have been developed for therapeutic interventions through parenteral, oral, or ophthalmic administration.

Nanocarriers may preferentially concentrate on tumors and inflammatory sites because of the enhanced permeability and retention effect, providing a continuous supply of the API in the disease site. Therefore, specific (cell or tissue) drug delivery and improved bioavailability can be reached in addition to protection of the therapeutic agents against degradation (Ge et al., 2002).

The literature has shown the advantages of NPs over microparticles (Linhardt, 1989; McClean, 1998; Schaffazick and Gutterres, 2003) and liposomes (Benoit et al., 1986; Reis et al., 2006; Soppimath et al., 2001), which is mainly associated with the size and the use of biodegradable materials. The small size of NPs can facilitate the passage through the endothelium of inflammatory tissues, epithelia of the gastrointestinal tract, tumors, or microcapillaries in a wide variety of cell types (Desai et al., 1997; Panyam et al., 2003a).

NPs represent the most suitable systems for intravenous administration. The smaller capillaries are 5–6 µm in diameter, which means that the size of the particles should be less than 5 µm to be distributed through the bloodstream. In addition, the aggregation of the NPs must be avoided, thereby ensuring no embryos are formed.

The use of biodegradable polymers can prolong the drug release within the target site over a period of days or even weeks (Redhead et al., 2001; Singh and Lillard, 2009). Generally, NPs prepared with negatively charged hydrophobic polymers show a low affinity for the intestinal cells; contrariwise, uncharged and positively charged particles have a greater affinity for follicle-associated epithelia as well as absorbent enterocytes. In contrast, negatively charged hydrophilic polymer-based NPs demonstrate a strong increase in bioadhesive properties and are absorbed by M cells and absorbent enterocytes. The combination of both the surface charges and the increased hydrophilicity of the matrix material appears to affect gastrointestinal uptake in a positive manner (Reis et al., 2006).

1.1.3   IMPORTANT CHARACTERISTICS OF NPS FOR DRUG DELIVERY

1.1.3.1   PARTICLE SIZE

Cellular and intracellular targets are able to be reached by NPs. Additionally, they can cross the blood–brain barrier by using hyperosmotic mannitol or coating with Tween 80, which facilitate the opening of tight endothelial junctions (Kreuter et al., 2003; Kroll et al., 1998; Singh and Lillard, 2009). For example, a study demonstrated that nanostructures with a particle size of 100 nm had an uptake rate 2.5-fold higher than structures with 1 µm and six times greater compared to 10 µm particles by Caco-2 cells (Desai et al., 1997).

In another study, microparticles and NPs displayed different distribution in the rat intestine (Redhead et al., 2001; Singh and Lillard, 2009). Moreover, the size can also affect the release of the API. The larger surface area-to-volume ratio of smaller particles leads to a faster release; on the other hand, the large nuclei of bigger particles allow the encapsulation of a higher amount of drug resulting in a slower release (Panyam and Labhasetwar, 2003; Redhead et al., 2001; Singh and Lillard, 2009). Furthermore, a greater risk of aggregation is related to smaller particles during storage.

The particle size may also have an effect on the polymer degradation (Dunne et al., 2000). For example, degradation rate of poly(lactic-co-glycolic acid) (PLGA) decreases in larger particles supposedly due to the difficult diffusion of degradation products of the polymer through longer distances of the polymer matrix in comparison with smaller particles, which prolong the drug release and may also cause autocatalytic degradation of the polymer (Panyam et al., 2003a; Singh and Lillard, 2009).

Photon-correlation spectroscopy and dynamic light scattering are examples of methods used to measure particle size (Swarbrick and Boylan, 2002) and, usually, the results from both techniques are compared with the photomicrographs obtained by scanning or transmission electron microscopy. Toxicity, biodistribution, drug targeting, drug loading and release, and stability may be influenced by the size and size distribution.

The toxicity of NPs is very dependent on their size (Chowdhury et al., 2016; Swarbrick and Boylan, 2002). In general, NPs with 3–10 nm in diameter can access the entire human body (Egusquiaguirre et al., 2012). Different sizes are adequate to direct them to different organs (Choi et al., 2007; Gaitanis and Staal, 2010). Thus, NPs with 11–30 nm in diameter are good for the liver, tumor, and brain tissue (Souris et al., 2010), whereas those with a diameter of 31–80 nm are good for the lungs, tumors, and inflamed tissue (Chowdhury et al., 2016; Souris et al., 2010).

In addition, the elimination of NPs from the human body is often achieved through filtration and urinary excretion, which can filter structures with 8 nm diameter and positive charge. Alternatively, the elimination by the hepatobiliary system is also possible but it is less efficient than the renal filtration. Dialysis can be used as well, but it is expensive and time-consuming. Thus, besides the drug delivery, the toxicity of NPs related to the size and charge should also be considered in the development of these carriers (Chowdhury et al., 2016).

1.1.3.2   SURFACE PROPERTIES OF NPS

The association between conventional drugs and nanocarriers can modify the biodistribution. NPs can be recognized and eliminated by circulating phagocytes after intravenous injection, especially by the mononuclear phagocyte system (MPS) (Muller et al., 1996; Singh and Lillard, 2009). A direct relation between hydrophobicity of the nanocarriers and the number of blood components binding to their surface with consequent recognition of the systems as foreign elements has been established (Brigger et al., 2002; Muller et al., 1996).

Unmodified surface NPs (i.e., conventional systems) are rapidly opso-nized and eliminated by the MPS (Grislain et al., 1983). In this sense, for a prolonged blood circulation and enhanced possibility to achieve the desired tissue, it is necessary to minimize the opsonization of the NPs. This aim can be achieved by coating NPs with hydrophilic polymers and/or surfactants such as polysorbate 80, poloxamer, and polyethylene glycol (PEG) (Bhadra et al., 2002; Olivier, 2005; Villa Nova et al., 2015).

The zeta potential represents the surface charge of the particles (Couvreur et al., 2002) and is mostly influenced by the particle composition. In general, a zeta potential of ± 30 mV contributes to form a more stable NP suspension since the surface charge results in mutual repelling and prevent aggregation. This parameter can also indicate if a charged API is encapsulated or adsorbed.

1.1.3.3 DRUG LOADING

A nanosystem must be able to carry a high amount of drug using a low proportion of matrix materials (Singh and Lillard, 2009). In this context, the drug content and the entrapment efficiency analyses are important to verify if the method and material were suitable in terms of imprisonment capacity (Herrmann and Bodmeier, 1998). It is important to achieve high encapsula-tion efficiencies but many parameters can affect this parameter during the development (e.g., the method, material, and drug characteristics) (Jyothi et al., 2010).

The interactions between the API and the polymer(s) have been posi-tively associated to better encapsulation efficiencies (Boury et al., 1997; Govender et al., 1999; Govender et al., 2000; Panyam et al., 2004). On the other hand, if the hydrophobic interaction is dominant between them, the presence of hydrophobic terminal groups is more valuable (Mehta et al., 1996). Furthermore, macromolecules, drugs or proteins are more efficiently encapsulated at or near their isoelectric point (Calvo et al., 1997) and ionic interactions between small molecules and matrix materials can increase the drug loading (Chen et al., 1994, 2003).

1.1.3.4 DRUG RELEASE

The evaluation of in vitro drug release profile is an important step during the development of pharmaceutical forms to identify critical variables and to optimize the conditions of production. Moreover, the results determine the drug release kinetics and are useful to predict the performance of the system in vivo. Nevertheless, this test does not substitute preclinical and clinical tests (Bruschi et al., 2004; Villa Nova et al., 2013).

The drug release rate can be affected by drug solubility, surface drug desorption, drug diffusion through the matrix and matrix erosion/degradation (Singh and Lillard, 2009). In addition, the preparation method can also influence the release profile (Fresta et al., 1995).

Regarding the nanospheres, drug release may occur by diffusion or erosion of the matrix, being the prevalent event the responsible for the release mechanism. An initial fast release (burst release) is mostly related to drug adsorbed onto the surface of the nanocarriers (Magenheim et al., 1993).

Concerning the coated NPs, the polymer membrane represents a release barrier; consequently, both the drug solubility and its diffusion through the membrane control the release. Additionally, as previously stated, drug-excipient ionic interactions may also have an effect on the release (Singh and Lillard, 2009), especially if more hydrophobic complexes are formed, which may slow down the drug release with almost no burst effect (Chen et al., 1994).

Several methods described in the literature for in vitro release analysis include side-by-side diffusion cells, dialysis bag, reverse dialysis bag, shaking followed by ultracentrifugation/centrifugation and ultrafiltration (Singh and Lillard, 2009). Controlled shaking followed by centrifugation is the most used method but since it is laborious and difficult to separate from the release media, the use of dialysis technique has increased. Nevertheless, these methods find difficult to be used on the industrial scale (Singh and Lillard, 2009).

1.1.3.5   TARGETED DRUG DELIVERY

Targeted delivery can be actively or passively achieved. In the first case, the API or the nanosystem must be conjugated to a cell-specific ligand (Lamprecht et al., 2001). The passive target occurs due to the reduced size of NPs which can passively enter the target organ. Alternatively, catheters may be used to administrate the NPs to the desired site (Singh and Lillard, 2009).

1.2   POLYMERS

In the last decades, an increase in the polymer science was observed after an enhanced understanding of polymer structures and possible modifications to achieve the desired attribute (Fonte et al., 2015). The selection of an appropriate polymer is critical when formulating a drug delivery system. Polymer composition, stability, as well as water solubility are important factors with influence in some characteristics of the NPs, such as in the mechanism and rate of drug release, charge, encapsulation efficiency, bioavailability, targeting, and biodistribution (Naahidi et al., 2013; Prabhu et al., 2015). Moreover, biocompatibility and biodegradability are essential to develop NPs with low cytotoxicity (Naahidi et al., 2013).

Based on their source, polymers are mainly classified into natural or synthetic materials and can be used separately or together (Gaikwad and Bhatia, 2013; Sosnik et al., 2014). The most commonly used polymers for NPs production will be addressed in this section.

1.2.1   NATURAL POLYMERS

1.2.1.1   GELATIN

Gelatin is a natural protein derived from the hydrolysis of collagen, commercially available as both cationic (gelatin type A) or anionic (gelatin type B) protein (Elzoghby, 2013). It is water soluble, highly biocompatible, and biodegradable, being considered as Generally Regarded As Safe (GRAS) material by the FDA (Santoro et al., 2014). Furthermore, gelatin has mucoadhesive properties and a high number of functional groups susceptible to multiple modifications (Elzoghby, 2013).

In the field of drug delivery, by properly selecting either type A or B, gelatin has enabled the efficient loading of different drug molecules. By varying molecular weight or extent of cross-linking, gelatin degradation and thereby the release kinetics can be modulated. Additionally, modification of gelatin has three principal purposes: to create stealth carriers for prolonged blood circulation (e.g., PEGylated gelatin), to enhance drug entrapment (e.g., cationic gelatin for improved RNA and DNA delivery) and to target a specific tissue for reduced systemic toxicity (e.g., epidermal growth factor receptor-modified gelatin) (Santoro et al., 2014). All these characteristics contribute to a wide range of applications for gelatin in drug delivery (Elzoghby, 2013).

1.2.1.2   CHITOSAN

Chitosan (CS) is a linear polysaccharide derived from the deacetylation of the exoskeleton of crustaceans (chitin) (Zhang et al., 2016). It is biodegradable, biocompatible, nontoxic, and low immunogenic. In acidic to neutral environment, the amino group of CS is protonated resulting in a net positive charge, which can interact with negatively charged mucoproteins and thus act as a bioadhesive polymer (Mogoşanu et al., 2016).

CS contains three types of reactive functional groups, an amino group and primary and secondary hydroxyl groups, which are available for modification or grafting to produce different types of NPs. It is insoluble in water and most organic solvents, being soluble only in acidic solutions (pH below 6.5) since this condition allows protonation of the amino groups (Zhang et al., 2016). In basic or neutral conditions, CS is less water soluble. In order to enhance the solubility, modifications with hydrophilic molecules can be carried out, such as in the CS derivatives carboxymethyl-CS, glycol-CS, and PEG-CS (Swierczewska, 2016; Zhang et al., 2016). Thiolated CSs have better mucoadhesive properties since forms disulfide bonds with the cysteine-rich domains of mucus glycoproteins (Akhlaghi et al., 2010).

CS is normally cross-linked with glutaraldehyde, genipin, or tripolyphosphate (TPP). These compounds are particularly well accepted because they enable the preparation of particles in aqueous media under mild conditions, excluding the use of organic solvents or heat and avoiding cytotoxicity concerns (Fonte et al., 2015). In addition, CS–polyelectrolyte complexes (PECs) prepared with oppositely charged polysaccharides such as hyaluronic acid (HA), alginate, and carboxymethyl cellulose have also been reported (Swierczewska, 2016).

It is possible to control the drug release by varying the molecular weight and degree of deacetylation of CS as well as by the combination of CS with other polymers. In addition, specific targeting of the CS-NPs can be reached by attaching specific molecules at their surface.

1.2.1.3   ALGINATE

Alginate is a hydrophilic anionic carbohydrate-based polymer, derived mainly from alkaline extraction of brown sea algae (Goh et al., 2012). Alginate is biodegradable, biocompatible, nontoxic, nonimmunogenic, and has shown mucoadhesive properties (Paques et al., 2014). Alginate can easily be gelled under gentle conditions and is susceptible to chemical modification, making it applicable for encapsulation (Goyal et al., 2016; Paques et al., 2014).

As a negatively charged polymer, alginate can be ionically cross-linked by using calcium ions which induce alginate aggregation. Alternatively, cationic polymers such as poly-L-lysine, CS, and Eudragit E100 can be added to produce a PEC (Paques et al., 2014; Swierczewska et al., 2016). The NPs have a wide size variation and have been used to encapsulate molecules such as insulin, cisplatin, doxorubicin, DNA, and gliclazide (Paques et al., 2014).

Alginate NPs can also be prepared by water-in-oil reverse microemulsion method (Swierczewska et al., 2016). Water-in-oil or oil-in-water emulsions can be used, where alginate is present in the inner or outer aqueous phase, respectively, and cross-linker molecules (e.g., calcium, CS) are also employed for alginate gelation. The addition of surfactants or the use of sonication can enable the reduction of particle size (Paques et al., 2014).

1.2.1.4 ALBUMIN

Albumin is a protein obtained mainly from egg white (ovalbumin, OVA), bovine serum (BSA), and human serum (HSA). It is hydrophilic, biodegradable, nontoxic, nonimmunogenic, and easy to purify. Albumin is attractive for drug delivery due to the possibility of incorporating a significant amount of drug by electrostatic adsorption of charged molecules to its different drug binding sites.

Concerning the three types of albumin, OVA can form gel networks and stabilize emulsions and foams. Moreover, OVA has pH- and temperature-sensitive properties, which are interesting properties for controlled drug release. BSA is easy to purificate and widely accepted in the pharmaceutical industry. In order to avoid a possible immunologic response in vivo, HSA is a good alternative for BSA. It is the most abundant plasma protein, very soluble, stable in the pH range of 4–9, under temperature up to 60°C for about 10 h and in organic solvents. Moreover, HSA has a great noncovalent reversible binding capacity, providing a depot for a wide variety of compounds (Elzoghby et al., 2012).

Hydrophilic drugs can be encapsulated by incubation with the preformed albumin NPs, incorporation into the albumin solution before particles formation and cross-linking, or addition to the cross-linker solution prior to the particles formation. For water-insoluble drugs, electrostatic adsorption of charged molecules may occur, which depends upon the content of charged amino acids in the albumin molecule (Elzoghby et al., 2012).

In addition, albumin-based NPs are prone to surface modification in order to modify pharmacokinetic parameters (e.g., surfactants), enhance the stability (e.g., poly-L-lysine), prolong the blood circulation (e.g., PEG), slow down the drug release (e.g., cationic polymers) or target a specific site (e.g., folate, thermosensitive polymers, transferrin, apolipoproteins, and monoclonal antibodies).

1.2.1.5   CYCLODEXTRINS

Cyclodextrins (CDs) are a class of cyclic oligosaccharides derived from starch with low toxicity and immunogenicity and good biocompatibility. The most common natural CD consists of six (α-CD), seven (β-CD), and eight (γ-CD) glucopyranose units which give to CD a shape of cone with a hydrophilic outer surface and a lipophilic central cavity (Zhang and Ma, 2013). Among the natural CD, the β-CD has low water solubility, while α- and γ-CD are more hydrophilic. CD derivatives have been synthesized by processes such as amination, etherification, and esterification of hydroxyl groups modifying features such as cavity volume, solubility, photostability, and stability against oxygen (Zafar et al., 2014). Some examples of CD derivatives with pharmaceutical interest include hydroxypropyl derivatives of β- and γ-CD, randomly methylated β-CD, and sulfobutyl β-CD (Loftsson et al., 2005; Vyas et al., 2008).

Regarding drug delivery, both natural and chemically modified CDs are used to develop drug delivery systems. The hydrophobic cavity of CD permits the formation of inclusion complexes with drugs, proteins/peptides, and oligonucleotides (Loftsson et al., 2005). These complexes have been extensively used to increase drug solubility and stability, enhance drug absorption, mask organoleptic features, control drug release, and improve drug permeability across biological barriers (Zhang and Ma, 2013). In the preparation of CD NPs, these inclusion complexes can be formed before (e.g., by freeze-drying) applying the suitable method. CD can be used alone or in association with other polymers (e.g., CS, Eudragit, methacrylates, cyanoacrylates, PEG) to prepare nanocarriers by nanoprecipitation, emulsification solvent evaporation, emulsion solvent diffusion, ionic gelation, and polymerization (Zafar et al., 2014). Since CDs are fermented by colonic microflora in the large intestine, they offer a possibility for colon-specific drug delivery (Vyas, 2008).

1.2.1.6   DEXTRAN

Dextrans are polysaccharides produced from sucrose on the cell surface of bacteria or by chemical synthesis. The molecular weight and the degree of branching are dependent on the source and determine some physicochemical properties (e.g., a low degree of branching increases the water solubility). In fact, most of the commercially available dextrans are highly water soluble since they present a low degree of branching. Additionally, they have low cost, are biodegradable, biocompatible, nontoxic, show stability under mild acidic and basic conditions, and because a large number of hydroxyl groups are susceptible to conjugation (Mehvar, 2000).

In drug delivery, dextran has been used as the form of dextran–drug conjugates or nanosystems; normally, it suffers chemical modification to decrease its solubility (e.g., dextran esters, dextran sulfate) or is conjugated with other polymers (e.g., CS, poly-ε-caprolactone [PCL], polyvinylamine) (Hussain et al., 2017). Modified dextran-based NPs can be prepared by nanoprecipitation (Aumelas et al., 2007; Kaewprapan et al., 2012).

Moreover, dextran systems can decrease uptake by the MPS, which may prolong the blood circulation time and increase the accumulation in some cells (Aumelas et al., 2007).

1.2.1.7 HYALURONIC ACID

HA is a negatively charged, naturally occurring polysaccharide widely distributed in the human body, mainly in the epithelial and connective tissues, eyes, and synovial fluid (Choi et al., 2010; Choi et al., 2011). HA acts as a signaling molecule in cell motility, proliferation and regulates cell–cell adhesion, in addition to wound healing and inflammation (Swierczewska, 2016). It is biocompatible, biodegradable, nontoxic, and has mucoadhesive properties (Choi et al. 2010; Oyarzun-Ampuero et al., 2009).

HA has a specific receptor in cells such as hyaluronate-mediated motility (RHAMM), stabilin-2, and CD44. Particularly, the latter is a transmembrane glycoprotein overexpressed in many types of cancer, which has encouraged the use of HA as a cancer-targeting moiety for drug delivery systems (Jeong et al., 2008; Swierczewska, 2016). Another advantage is related to its rapid degradation by hyaluronidases (abundant in tumor cells), which contribute to a faster drug release (Yoon et al., 2012).

HA-based NPs are commonly prepared after conjugation of HA to hydrophobic biomoieties such as cholanic acid or ceramide. For example, amphiphilic HA–5β-cholanic acid (HACA) conjugates are able to self-assemble in aqueous conditions to form NPs, where the core of 5β-cholanic acid acts as the drug reservoir and the HA forms the shell for active tumor targeting (Swierczewska, 2016).

1.2.1.8   ETHYLCELLULOSE

Ethylcellulose (EC) is a semisynthetic polymer derived from cellulose in which some of the hydroxyl groups on the repeating anhydroglucose units are modified into ethyl ether groups (Murtaza, 2012). It is hydrophobic, nonbiodegradable, biocompatible, nontoxic, nonirritant, and listed as GRAS by the FDA and European Medicines Agency (EMA) (Eltayeb et al., 2015; Ghaderi et al., 2014; Lokhande et al., 2013). EC is available in different chain length, degree of polymerization, or number of anhydroglucose units, which generates different viscosity grades for the polymer (Murtaza, 2012; Wachsman and Lamprecht, 2014). In addition, aqueous dispersions of EC can be found in the market (e.g., Aquacoat® and Surelease®).

EC has been widely used for coating and nano/microencapsulation applications. Some methods for NP preparation are emulsification/solvent evaporation, coacervation, spray drying, nanoprecipitation, and emulsification/solvent diffusion (Ghaderi et al., 2014; Murtaza, 2012).

1.2.2   SYNTHETIC POLYMERS

1.2.2.1   POLY(LACTIC-CO-GLYCOLIC ACID)

PLGA, the copolymer of polylactic acid and polyglycolic acid, is a biocompatible and biodegradable polymer, approved by the FDA and EMA (Chereddy et al., 2016; Loureiro et al., 2016). It is extensively used in pharmaceutical formulations due to its low toxicity, which is related to the production of biodegradable metabolite monomers (lactic acid and glycolic acid) that are natural intermediate in carbohydrate metabolism (Kumari et al., 2010). It shows capability to encapsulate both hydrophobic and lipophilic drugs and allows surface functionalization for specific targeting or to prolong blood circulation time (e.g., PLGA–PEG). Furthermore, PLGA is available with different copolymer ratios and molecular weights, which have major importance on polymer degradation rate and drug release (Xu et al., 2016).

Several methods that can be used to formulate PLGA NPs include the solvent displacement, solvent diffusion, emulsification–evaporation, and phase inversion (Goyal et al., 2016).

1.2.2.2 POLY(E-CAPROLACTONE)

PCL is a hydrophobic, biocompatible, biodegradable, and nontoxic polymer, which is also approved by FDA (Dash and Konkimalla, 2012). Due to its semicrystalline structure, PCL degradation is slower in comparison to other polyesters (e.g., PLGA) permitting drug release for a longer time period (Goyal, 2016; Kumari et al., 2010). By copolymerization or blending with other polymers (e.g., PLGA, CS, PEG, polyethylene oxide, polyvinyl alcohol), the crystallinity, solubility, and biodegradability can be modified to achieve desired properties (Dash and Konkimalla, 2012; Wei et al., 2009).

The most used methods to prepare PCL NPs are emulsion solvent evaporation, solvent displacement, interfacial polymer disposition, and emulsion solvent diffusion. Normally, the encapsulation efficiency is influenced by the PCL concentration and particle size, while surface properties and drug release rate are influenced by the surfactant (e.g., PVA, poloxamer, sodium cholate) (Dash and Konkimalla, 2012).

1.2.2.3 POLYMETHYL METHACRYLATE

Polymethyl methacrylate (PMMA) is a synthetic homopolymer of methyl methacrylate (MMA) monomer approved by FDA. It is hydrophobic, biocompatible, has a low cost, and is easily available. Even PMMA is nonbiodegradable and is considered as a nontoxic material (Ali et al., 2015). PMMA NPs have been studied to deliver many classes of drugs, showing effective to stabilize chemically unstable drugs like antioxidants as well as to solubilize poorly soluble drugs (Bettencourt and Almeida, 2012). These particles can be synthesized by emulsification solvent evaporation, nanoprecipitation, and polymerization techniques, including emulsion, microemulsion, nonconventionally initiated emulsion polymerization, solvent shifting, and surfactant-free emulsion polymerization (Camli et al., 2010).

Due to the hydrophobic nature, PMMA does not swell in the presence of water. In this context, the formation of PMMA composites with hydrophilic polymers has been reported to increase the polymer water solubility and drug release (e.g., PMMA–polyethyleneimine, PMMA–polyvinylpyrrolidone, and PMMA–CS). Furthermore, the low mobility of PMMA chains also contributes to a slow drug release; thus, the use of compounds with the ability to decrease the polymer glass transition temperature (e.g., plasticizers) and increase the mobility of polymer chains can be investigated (Bettencourt and Almeida, 2012).

1.2.2.4   METHACRYLATES COPOLYMERS (EUDRAGIT®)

Eudragit is the trademark for synthetic anionic, cationic, and neutral copolymers based on methacrylic acid and methacrylic or acrylic esters or their derivatives in varying proportions. They are nonbiodegradable but nontoxic polymers. Basically, Eudragit is available in four main classes: cationic Eudragit E, anionic Eudragit L and S, neutral Eudragit RL and RS (pH-independent and soluble), neutral Eudragit NE and NM (swellable and permeable). Eudragit L dissolves at pH >6 and is used for enteric coating; Eudragit S dissolves at pH >7 and is used for colon targeting. For sustained drug release, more often neutral Eudragit has been investigated (Thakral et al., 2013).

Eudragit NPs have been prepared by emulsification solvent evaporation, nanoprecipitation, and emulsion solvent diffusion methods (Adibkia, 2011; Das, 2010; Katara et al., 2013). Eudragit RL and RS have different permeability (high and low, respectively) which interfere with dissolution and diffusion of the drug and result in different release profiles; thus, they can be mixed in varied proportions to achieve the desired release requirements (Thakral et al., 2013). Moreover, the combination of Eudragit with other polymers has also been reported (e.g., PLGA, CS, PCL).

1.2.2.5   POLYHYDROXYBUTYRATE

Polyhydroxybutyrate (PHB) is a hydrophobic and crystalline biopolymer with biocompatible and biodegradable properties. It is produced by bacteria from renewable raw materials such as sugarcane and also from certain genetically engineered plants (Melo et al., 2012). It is degraded by nonspecific lipases and esterases in nature-producing monomers and by-products with good tolerance in the human body. However, this polymer has a relatively high cost (Gumel et al., 2013).

PHB properties can be varied by changing processing and molecular weight. Due to the hydrophobic nature, PHB can result in an extended drug release over a period of months. In addition, the combination with other polymers has been reported (e.g., copolymer poly(3-hydroxybutyrate-co-3-hydroxyvalerate—PHBV); poly(3-hydroxybutyrate-co-3-hydroxy-hexanoate—PHBHHx) to produce carriers for drugs such as anesthetics, antibiotics, anti-inflammatory, anticancer, hormones, steroids, and vaccines have been encapsulated (Melo et al., 2012; Shrivastav et al., 2013).

1.2.2.6 POLYALKYL CYANOACRYLATES

Polyalkyl cyanoacrylates (PACA) comprises a class of polymers obtained by a chemical reaction between an alkyl cyanoacetate with formaldehyde to form PACA oligomers. Then, these oligomers undergo thermal depolymerization to achieve the pure alkyl cyanoacrylate monomer. Homologous with less than four carbons are well tolerated in vivo and the longer the alkyl side chains, the lower the toxicity (Nicolas and Couvreur, 2008). The monomers butyl cyanoacrylate and octyl cyanoacrylate are approved by FDA for human use.

PACA NPs are attractive systems because of biocompatibility and biode-gradability properties. The particles can be prepared by polymerization of monomers or from preformed PACA polymers (e.g., by nanoprecipitation or emulsification solvent evaporation) (Vauthier et al., 2007). The use of short alkyl chain homologous results in faster polymerization and degrada-tion rate since a less strong polymer network is formed (Graf et al., 2009). Polybutyl cyanoacrylate (PBCA), polyisobutyl cyanoacrylate (PiBCA), and polyethyl cyanoacrylate (PECA) as well as their copolymers (e.g., PiBCA–PEG diblock) are commonly used. Moreover, different stabilizing agents can be used to stabilize the NPs such as Pluronic, CS, and dextran (Vauthier et al., 2007).

1.3 PREPARATION METHODS

Many methods have been described in the literature for the preparation of polymeric NPs. These methods can be resultant of a polymerization reac-tion (emulsion or interfacial polymerization) or achieved directly from a preformed macromolecule or polymer (synthetic or natural materials) (Reis et al., 2006; Schütz et al., 2013b).

Moreover, in many of the preparation methods, two main steps are involved: the first one results in an emulsified mixture, while in the second one, the NPs are formed (Vauthier and Bouchemal, 2009). In some cases, the NPs are produced in the initial emulsification phase, forming emulsions, nanoemulsions, and microemulsions. On the other hand, some methods are not based on an emulsification process, but on the polymer precipitation (Vauthier and Bouchemal, 2009).

For other authors, the preparation methods are modifications of the following basic techniques: emulsification and solvent evaporation, coacervation and spray drying (Freitas et al., 2005), being many of them based on physical phenomena, chemical reactions, and others combine physical and chemical phenomena (Bock et al., 2011; Bruschi et al., 2002; Jyothi et al., 2010). Physicochemical methods include coacervation, emulsification and solvent evaporation, spray-on cross-linking agent, and liposomal wrapping; the chemical methods include the interfacial polymerization and the polymerization in situ; the most employed physical methods are spray drying, spray congealing, and fluidized bed (Cheng et al., 2009; Jyothi et al., 2010). Some of these techniques are presented below.

1.3.1   CROSS-LINKING TECHNIQUES

1.3.1.1   IONIC CROSS-LINKING

The ionic cross-linking method is simple and does not involve the use of organic solvent, high temperature, or chemical interaction (Ahmed and Aljaeid, 2016; Mao et al., 2006; Xu et al., 2003). Therefore, it is widely used and considered as an efficient and safe method to prepare NPs for encapsulation of proteins, peptides, hormones, and vaccines.

The particles are obtained from an acidic solution of the polymer where the ionic cross-linking agent is added dropwise under stirring and/or sonication. When anionic cross-linking agents are used (e.g. sodium sulfate or tripolyphosphate [TPP]), the process is named ionic gelation; however, the use of negatively charged polyelectrolyte macromolecules, such as CD derivatives, dextran sulfate, and polyglutamic acid result in the production of electrostatic PECs by a process called ionic cross-linking (Ahmed and Aljaeid, 2016).

This method is commonly used to prepare CS-NPs, which have shown effective to encapsulate proteins and antigens such as insulin, tetanus toxoid, albumin, and influenza subunit antigen (Amidi et al., 2006; Amidi et al.,2007). Interleukin-2 (IL-2) microparticles loaded with CS were prepared using sodium sulfate as an anionic cross-linking agent (Ahmed and Aljaeid, 2016; Ozbas-Turan, et al., 2002).

1.3.1.2 COMPLEX COACERVATION

Complex coacervation is a liquid–liquid-phase separation process that occurs when two ionic solutions with opposite charges are mixed, resulting in the formation of an ionic complex. Afterward, the precipitated NPs are separated by filtration or centrifugation and washed with hot and cold water. A cross-linking agent can be added to increase the hardness of the particles and prolong the drug release (Ahmed and Aljaeid, 2016).

Bozkir and Saka (2004) demonstrated that plasmid DNA was successfully loaded into CS-NPs ranging in size from 450 to 820 nm, with trap efficiency higher than 90% and prolonged release for 24 h. Since CS is cationic, it is soluble in acid solution and can be precipitated by the addition of alkali solutions, which is the theoretical principle of the precipitation/coacervation method.

1.3.2 CHEMICAL CROSS-LINKING

In this method, the nanostructures are obtained by the chemical interaction between the chemical groups of the polymer with the cross-linking agents, such as glutaraldehyde, p-phthaldehyde, ascorbyl palmitate, and dehydro-ascorbyl palmitate. The process may involve the formation of a water–oil emulsion, wherein the polymer and the API are in the aqueous phase, which is then emulsified in an external immiscible solvent (Ahmed and Aljaeid, 2016). The cross-linking agent is slowly incorporated and the particles are separated and washed (Jameela et al., 1998).

Other excipients may be added to increase the stability and the encapsulation efficiency (Ahmed and Aljaeid, 2016). Wang et al. (2006) demonstrated that the addition of gelatin to the aqueous phase increased the stability and the entrapment of insulin. Recently, CS-NPs containing rabeprazole within a water-in-oil nanoemulsion were formed by emulsifying the aqueous phase in paraffin oil containing a mixture of surfactants, where the optimized NPs showed a nanosize range of 120 ± 32 nm (Ahmed and Aljaeid, 2016).

1.3.3 DRYING TECHNIQUES

Drying is a process of removal of water/solvent from liquids, solids, or semi-solids by evaporation. Hot air, microwave, natural air-drying, freeze-drying, spray drying, and supercritical drying represent the general drying methods (Ahmed and Aljaeid, 2016).

1.3.3.1   SPRAY DRYING

Spray drying is widely used in the pharmaceutical industry to produce medicine and excipients as well as for encapsulation processes (Billon et al., 2000; Broadhead et al., 1992). In this technique, solid particles can be obtained from a complex liquid mixture, which contains a dissolved or dispersed API and excipients in an organic or aqueous polymer solution (Ahmed and Aljaeid, 2016; Benita, 2006). During the process, four simultaneous steps take place: atomization of the liquid into a spray nozzle, contact of the liquid with the hot air, drying of the droplets, and production of the dried powder (Barras et al., 2000; Broadhead et al., 1992; Bruschi et al., 2003).

Due to the rapid evaporation of the solvent, the droplet temperature can be kept below the drying air temperature, allowing the use of this technique for thermosensitive materials (Broadhead et al., 1992). In addition, spray drying is suitable for both hydrophilic and hydrophobic materials (Barras et al., 2000). Several biodegradable polymers have been successfully used to produce NPs such as PLA, PLGA, PCL, Eudragit, gelatin, polysaccharides, or related biopolymers (Benita, 2006).

Contrariwise to coacervation and emulsification methods, the spray drying method is a one-step continuous process, easy-to-scale expansion, inexpensive, and with the possibility of being free of organic solvent (Billon et al., 2000; Bruschi et al., 2002). However, the spray drying technique also has some drawbacks, such as variable yields, adherence of the product to the inner walls, agglomeration, high moisture content, and wide particle size distribution (Billon et al., 2000; Broadhead et al., 1992; Raffin et al., 2006). Even by using organic solvents, the residual amount of solvent in the product is often less than that achieved with the solvent emulsification/evaporation technique (Benita, 2006).

Spray drying has been reported for the manufacture of granules and dry powders from drug-excipient blends either in a solution or suspension (Chawla et al., 1994). The convenience of this method has been extended to include preparation of microparticles of different polymeric materials containing protein (Gander et al., 1995) and vaccine antigens (Lee et al., 1997). More recently, smooth-surface vildagliptin nanospheres were prepared by this technique, where nanospheres with a high drug content were obtained (Ahmed and Aljaeid, 2016).

1.3.3.2   SUPERCRITICAL DRYING

This process consists of drying the liquid mixture containing the drug and excipients by using a supercritical fluid. Substances above its critical point have intermediary properties of a gas and a liquid, that is, they are able to penetrate substances as a gas and to dissolve materials as a liquid (Reverchon et al., 1999). The examples of these substances include carbon dioxide, Freon, and nitrous oxide (Ahmed and Aljaeid, 2016; Reverchon et al., 1999).

Supercritical carbon dioxide has unique properties, such as high dissolving capacity, high diffusivity, and low viscosity. Many attractive techniques using supercritical carbon dioxide have been developed, for example, rapid expansion of supercritical solution, supercritical anti-solvent (Esfandiari, 2015), impregnation of supercritical solution, and supercritical drying (Champeau et al., 2015; Dowson et al., 2012; Walters et al., 2014). The use of carbon dioxide allows the solvent removal without affecting the nanostructured system network. Regarding the supercritical carbon dioxide/ organic solvent binary system, the homogeneous phase can be formed under pressure on the critical point of this system. Therefore, it is possible to conduct the drying process without interfacial tension.

1.3.4 REVERSE MICELLAR METHOD

When a surfactant is dispersed in a nonpolar solvent under specific conditions, they are arranged to form aggregate reverse micelles (RM) arranging in size between 1 and 10 nm, which can dissolve a polar solvent in its polar nucleus (Agazzi et al., 2016; Melo et al., 2001). Due to the large variety of substances, many conditions for the preparation of NPs can be investigated (Agazzi et al., 2016; Ahmed and Aljaeid, 2016).

The characteristics of the encapsulated solvents permit to obtain aqueous and nonaqueous RM (Agazzi et al., 2011; Correa et al., 2012). Moreover, the charge of the main group in the surfactant molecule determines if it is nonionic, anionic, cationic, or amphoteric.

1.3.5 EMULSIFICATION SOLVENT EVAPORATION

In this method, a volatile solvent containing the polymer and the drug is emulsified in an aqueous or nonaqueous phase. The system is kept under stirring until complete evaporation of the solvent, and the particles formed are filtered, washed, and finally dried (Ahmed and Aljaeid, 2016).

In general, the entrapment efficiency and particle size are affected by the polymer (concentration and molecular weight), drug solubility, the type and concentration of the surfactant, and speed of agitation.

Volatile solvents are used in the preparation of the polymer solution (e.g., dichloromethane, ethyl acetate), which after being evaporated, results in the precipitation of the polymer and the conversion of the emulsion into an NP suspension (Allémann et al., 1993; Anton et al., 2008; Vauthier and Bouchemal, 2009; Vauthier et al., 2004). Mechanical or magnetic stirring is necessary for the emulsification and droplet size reduction. In order to avoid the coalescence of the droplets during the evaporation, the interface oil–water is generally stabilized by using a surfactant. Normally, NPs greater than 250 nm are formed with this method (Quintanar-Guerrero et al., 1999).

This method was the first to be used in the preparation of nanocarriers from a preformed polymer (Gurny et al. 1981; Vanderhoff et al., 1979; Vauthier and Bouchemal, 2009) and was extensively applied to prepare composite NPs of PLA, PLGA, and PCL using pluronic F68 as a stabilizing agent (Anton et al., 2008; Avgoustakis, 2004; Mundargi et al., 2008; Vauthier et al., 2004). It can also be applied to formulate NPs with amphiphilic copolymers, including PEG and polysaccharides (Avgoustakis, 2002; Brigger et al., 2000). In this case, a surfactant is not necessary to ensure the formation of the emulsion and the stability of the final suspension of NPs.

1.3.6 NANOPRECIPITATION

Nanoprecipitation is one of the most accepted and reproducible techniques for laboratory encapsulation of drugs. It is based on the utilization of a solvent, where polymer and drug are soluble, and a non-solvent, which must be miscible with the solvent, but insoluble for drug and polymer. Once both solutions come into contact, a spontaneous diffusion of the polymer–drug solution occurs in direction to the non-solvent leading to simultaneous precipitation of the polymer and encapsulation of the API (Bilati et al., 2005; Chorny et al., 2002; Legrand et al., 2007; Villa Nova et al., 2015). NPs ranging from 100 to 300 nm and with low polydispersity index are normally obtained by nanoprecipitation (Bilati et al., 2005). The most common solvents are ethanol, acetone, and tetrahydrofuran, while water, containing or not containing a stabilizer, is used as the non-solvent (Lepeltier et al., 2014). For process optimization, solvent type, polymer concentration, solvent/non-solvent ratio, non-solvent addition flow rate, and stirring rate are commonly investigated (Beyer et al., 2015).

Nanoprecipitation is one of the most accepted methods to prepare PLGA NPs (Lepeltier et al., 2014; Natarajan et al., 2014). Although it is quite reproducible at the laboratory scale, the efficiency of the mixture and the precipitation can difficult the scale-up (Karnik et al., 2008; Wacker et al., 2011). Thus, microreactors have been used to assist the process since they permit better control of the parameters that involve precipitation processes and can produce the formulations at a medium scale (Roberge et al., 2009; Zhao et al., 2011).

1.4 CONCLUSIONS

Polymeric NPs have shown good prospects for diagnosis and therapy for many diseases since many drugs and macromolecules (e.g., proteins, peptides, vaccines, hormones, and genes) are able to be delivered by these systems. Increased bioavailability, stability, solubility, cellular permeability, absorption, biodistribution, and targeting to specific sites have been associated with the use of NPs. Moreover, they can provide the administration of the formulation through more convenient routes, increasing the patient compliance. In this context, the present chapter has presented the basics of the most common preparation methods described in the literature as well as the natural and synthetic polymers that are available to prepare polymeric NPs. It was demonstrated that particle size, size distribution, surface charge, drug loading, and drug release are important aspects to be considered during the development and may be influenced by the method and materials used to produce the NPs. In the last decades, progress has focused mainly on improving existing methods and on the use of methodologies that allow the reproducible production of large quantities of particles in order to facilitate the reach of these formulations to the market. It is expected that polymeric NPs meet the requirements of suitable drug delivery systems to bring consistent improvements to therapeutics and diagnosis.

KEYWORDS

- nanoencapsulation
- polymeric nanoparticles
- synthetic/natural polymers
- preparation methods
- drug targeting
- nanotechnology

REFERENCES

Adibkia, K.; Javadzadeh, Y.; Dastmalchi, S.; Mohammadi, G.; Niri, F. K.; Alaei-Beirami, M. Naproxen-Eudragit RS100 Nanoparticles: Preparation and Physicochemical Characterization. *Colloids Surf. B* **2011,** *83*(1), 155–159.

Agazzi, F. M.; Falcone, R. D.; Silber, J. J.; Correa, N. M. Solvent Blends can Control Cationic Reversed Micellar Interdroplet Interactions. The Effect of N-Heptane:Benzene Mixture on BHDC Interfacial Properties: Droplet Sizes and Micropolarity. *J. Phys. Chem. B* **2011,** *115*(42), 12076–12084.

Agazzi, F. M.; Falcone, R. D.; Silber, J. J.; Correa, N. M. Non-Aqueous Reverse Micelles Created with a Cationic Surfactant: Encapsulating Ethylene Glycol in BHDC/Non-Polar Solvent Blends. *Colloids Surf. A* **2016,** 509(20), 467–473.

Ahmed, T. A.; Aljaeid, B. M. Preparation, Characterization, and Potential Application of Chitosan, Chitosan Derivatives, and Chitosan Metal Nanoparticles in Pharmaceutical Drug Delivery. *Drug Des. Devel. Ther.* **2016,** *10,* 483–507.

Akhlaghi, S. P.; Saremi, S.; Ostad, S. N.; Dinarvand, R.; Atyabi, F. Discriminated Effects of Thiolated Chitosan-Coated pMMA Paclitaxel-Loaded Nanoparticles on Different Normal and Cancer Cell Lines. *Nanomedicine* **2010,** *6*(5), 689–697.

Al-Badriyeh, D.; Liew, D.; Stewart, K.; Kong, D. C. Economic Impact of Caspofungin as Compared with Liposomal Amphotericin B For Empirical Therapy in Febrile Neutropenia in Australia. *J. Antimicrob. Chemother.* **2009,** *63,* 1276–1285.

Ali, U.; Karim, K. J. Bt. A.; Buang, N. A Review of the Properties and Applications of Poly (Methyl Methacrylate) (PMMA). *Polym. Rev.* **2015,** *55*(4), 1–28.

Allémann, E.; Gurny, R.; Doelker, E. Drug-Loaded Nanoparticles Preparation Methods and Drug Targeting Issues. *Eur. J. Pharm. Biopharm.* **1993,** *39*(5), 173–191.

Amidi, M.; Romeijn, S. G.; Borchard, G.; Junginger, H. E.; Hennink, W. E.; Jiskoot, W. Preparation and Characterization of Protein-loaded N-trimethyl Chitosan Nanoparticles as Nasal Delivery System. *J. Controlled Release.* **2006,** *111,* 107–116.

Amidi, M.; Romeijn, S. G.; Verhoef, J. C.; Junginger, H. E.; Bungener, L.; Huckriede, A.; Crommelin, D. J.; Jiskoot, W. N-Trimethyl Chitosan (TMC) Nanoparticles Loaded with Influenza Subunit Antigen for Intranasal Vaccination: Biological Properties and Immunogenicity in a Mouse Model. *Vaccine* **2007,** *25,* 144–153.

Andrade, F.; Rafael, D.; Videira, M.; Ferreira, D.; Sosnik, A.; Sarmento, B. Nanotechnology and Pulmonary Delivery to Overcome Resistance in Infectious Diseases. *Adv. Drug Delivery Rev.* **2013,** *65,* 1816–1827.

Anton, N.; Benoit, J. P.; Saulnier, P. Design and Production of Nanoparticles Formulated from Nano-Emulsion Templates—A Review. *J. Controlled Release* **2008,** *128*(3), 185–199.

Aumelas, A.; Serrero, A.; Durand, A.; Dellacherie, E.; Leonard, M. Nanoparticles of Hydrophobically Modified Dextrans as Potential Drug Carrier Systems. *Colloids Surf. B* **2007,** *59,* 74–80.

Avgoustakis, K.; Beletsi, A.; Panagi, Z.; Klepetsanis, P.; Karydas, A. G.; Ithakissios, D. S. PLGA-mPEG Nanoparticles of Cisplatin: In Vitro Nanoparticle Degradation, in Vitro Drug Release and in Vivo Drug Residence in Blood Properties. *J. Controlled Release* **2002,** *79,* 123–135.

Avgoustakis, K. Pegylated Poly(Lactide) and Poly(Lactide-Coglycolide) Nanoparticles: Preparation, Properties and Possible Applications in Drug Delivery. *Curr. Drug Deliv.* **2004,** *1,* 321–333.

Choi, K. Y.; Min, K. H.; Yoon, H. Y.; Kim, K.; Park, J. H.; Kwon, I. C.; Choi, K.; Jeong, S. Y. PEGylation of Hyaluronic Acid Nanoparticles Improves Tumor Targetability in Vivo. *Biomaterials* **2011**, *32*(7), 1880–1889.

Chorny, M.; Fishbein, I.; Danenberg, H. D.; Golomb, G. Lipophilic Drug Loaded Nanospheres Prepared by Nanoprecipitation: Effect of Formulation Variables on Size, Drug Recovery and Release Kinetics. *J. Controlled Release* **2002**, *83*, 389–400.

Chowdhury, S.; Yusof, F.; Salim, W. W.; Sulaiman, N.; Faruck, M. O. An Overview of Drug Delivery Vehicles for Cancer Treatment: Nanocarriers and Nanoparticles Including Photovoltaic Nanoparticles. *J. Photochem. Photobiol. B* **2016**, *164*, 151–159.

Correa, N. M.; Silber, J. J.; Riter, R. E.; Levinger, N. E. Nonaqueous Polar Solvents in Reverse Micelle Systems. *Chem. Rev.* **2012**, *112*, 4569–4602.

Couvreur, P. Polyalkylcyanoacrylates as Colloidal Drug Carriers. *Crit. Rev. Ther. Drug Carrier Syst.* **1988**, *5*, 1–20.

Couvreur, P.; Dubernet, C.; Puisieux, F. Controlled Drug Delivery with Nanoparticles: Current Possibilities and Future Trends. *Eur. J. Pharm. Biopharm.* **1995**, *41*, 2–13.

Couvreur, P.; Barratt, G.; Fattal, E.; Legrand, P.; Vauthier, C. Nanocapsule Technology: A Review. *Crit. Rev. Ther. Drug Carrier Syst.* **2002**, *19*, 99–134.

Das, S.; Suresh, P. K.; Desmukh, R. Design of Eudragit RL 100 Nanoparticles by Nanoprecipitation Method for Ocular Drug Delivery. *Nanomedicine* **2010**, *6*, 318–323.

Dash, T. K.; Konkimalla, B. Poly-Є-Caprolactone Based Formulations for Drug Delivery and Tissue Engineering: A Review. *J. Controlled Release* **2012**, *158*, 15–33.

Davda, J.; Labhasetwar, V. Characterization of Nanoparticle Uptake by Endothelial Cells. *Int. J. Pharm.* **2002**, *233*, 51–59.

De Jong, W. H.; Geertsma, R. E.; Roszek, B. *Nanotechnology in Medical Applications: Possible Risks for Human Health;* Report 265001002/2005. National Institute for Public Health and the Environment (RIVM). Bilthoven: The Netherlands, **2005**.

Desai, M. P.; Labhasetwar, V.; Walter, E.; Levy, R. J.; Amidon, G. L. The Mechanism of Uptake of Biodegradable Microparticles in Caco-2 Cells is Size Dependent. *Pharm. Res.* **1997**, *14*, 1568–1573.

Dowson, M.; Grogan, M.; Birks, T.; Harrison, D.; Craig, S. Streamlined Life Cycle Assessment of Transparent Silica Aerogel Made by Supercritical Drying. *Appl. Energy* **2012**, *97*, 396–404.

Duncan, R. The Dawning Era of Polymer Therapeutics. *Nat. Rev. Drug Discov.* **2003**, 2:347–360.

Duncan, R.; Gaspar, R. Nanomedicine(s) Under The Microscope. *Mol. Pharm.* **2011**, *8*, 2101–2141.

Dunne, M.; Corrigan, I.; Ramtoola, Z. Influence of Particle Size and Dissolution Conditions on the Degradation Properties of Polylactide-Co-Glycolide Particles. *Biomaterials* **2000**, *21*, 1659–1668.

Egusquiaguirre, S. P.; Igartua, M.; Hernández, R. M.; Pedraz, J. L. Nanoparticle Delivery Systems for Cancer Therapy: Advances in Clinical and Preclinical Research. *Clin. Transl. Oncol.* **2012**, *14*, 83–93.

Eltayeb, M.; Stride, E.; Edirisinghe, M. Preparation, characterization and Release Kinetics of Ethylcellulose Nanoparticles Encapsulating Ethylvanillin as a Model Functional Component. *J. Funct. Foods* **2015**, *14*, 726–735.

Elzoghby, A. O. Gelatin-Based Nanoparticles as Drug and Gene Delivery Systems: Reviewing Three Decades of Research. *J. Controlled Release* **2013**, *172*, 1075–1091.

Elzoghby, A. O.; Samy, W. M.; Elgindy, N. A. Albumin-Based Nanoparticles as Potential Controlled Release Drug Delivery Systems. *J. Controlled Release* **2012**, *157*, 168–182.

Esfandiari, N. Production of Micro and Nano Particles of Pharmaceutical by supercritical Carbon Dioxide. *J. Supercrit. Fluids* **2015**, *100*, 129–141.

European Science Foundation. *Policy Briefing (ESF), ESF Scientific Forward Look on Nanomedicine.* IREG Strasbourg, France, 2005.

Ferrari, M. Cancer Nanotechnology: Opportunities and Challenges. *Nat. Rev. Cancer* **2005**, *5*, 161–171.

Fonte, P.; Araújo, F.; Silva, C.; Pereira, C.; Reis, S.; Santos, H. A.; Sarmento, B. Polymer-Based Nanoparticles for Oral Insulin Delivery: Revisited Approaches. *Biotechnol. Adv.* **2015**, *1*, 1342–1354.

Freitas, S.; Merkle, H. P.; Gander, B. Microencapsulation by Solvent Extraction/Evaporation: Reviewing the State of the Art of Microsphere Preparation Process Technology. *J. Controlled Release* **2005**, *102*, 313–332.

Fresta, M.; Puglisi, G.; Giammona, G.; Cavallaro, G.; Micali, N.; Furneri, P.M. Pefloxacine mesilate- and ofloxacin-loaded polyethylcyanoacrylate nanoparticles: characterization of the colloidal drug carrier formulation. *J Pharm Sci.* **1995**, *84*(7), 895-902.

Gaikwad, V. L.; Bhatia, M. S. Polymers Influencing Transportability Profile of Drug. *Saudi Pharm. J.* **2013**, *21*, 327–335.

Gaitanis, A.; Staal, S. Liposomal Doxorubicin and Nab-Paclitaxel: Nanoparticle Cancer Chemotherapy in Current Clinical Use. *Methods Mol. Biol.* **2010**, *624*, 385–392.

Gander, B.; Wehrli, E.; Alder, R.; Merkle, H. P. Quality Improvement of Spray-Dried, Protein-Loaded D, L-Pla Microspheres by Appropriate Polymer Solvent Selection. *J. Microencapsulation* **1995**, *12*, 83–97.

Ge, H.; Hu, Y.; Jiang, X.; Cheng, D.; Yuan, Y.; Bi, H.; Yang, C. Preparation, Characterization, and Drug Release Behaviors of Drug Nimodipine-Loaded Poly (Epsilon-Caprolactone)-Poly(Ethylene Oxide)-Poly(Epsilon-Caprolactone) Amphiphilic Triblock Copolymer Micelles. *J. Pharm. Sci.* **2002**, *91*, 1463–1473.

Ghaderi, S.; Ghanbarzadeh, S.; Mohammadhassani, Z.; Hamishehkar, H. Formulation of Gammaoryzanol-Loaded Nanoparticles for Potential Application in Fortifying Food Products. *Adv. Pharm. Bull.* **2014**, *4*, 549–554.

Goh, C. H.; Heng, P. W. S.; Chan, L. W. Alginates as a Useful Natural Polymer for Microencapsulation and Therapeutic Applications. *Carbohydr. Polym.* **2012**, *88*, 1–12.

Govender, T.; Stolnik, S.; Garnett, M. C.; Illum, L.; Davis, S. S. PLGA Nanoparticles Prepared by Nanoprecipitation: Drug Loading and Release Studies of a Water Soluble Drug. *J. Controlled Release* **1999**, *57*, 171–185.

Govender, T.; Riley, T.; Ehtezazi, T.; Garnett, M. C.; Stolnik, S.; Illum, L.; Davis, S. S. Defining the Drug Incorporation Properties of PLA-PEG Nanoparticles. *Int. J. Pharm.* **2000**, *199*, 95–110.

Goyal, R.; Macri, L. K.; Kaplan, H. M.; Kohn, J. Nanoparticles and Nanofibers for Topical Drug Delivery. *J. Controlled Release* **2016**, *240*, 77–92.

Graf, A.; McDowell, A.; Rades, T. Poly(Alkycyanoacrylate) Nanoparticles for Enhanced Delivery of Therapeutics – is there Real Potential? *Expert Opin. Drug Deliv.* **2009**, *6*, 371–387.

Grislain, L.; Couvreur, P.; Lenaerts, V.; Roland, M.; Deprez-Decampeneere, D.; Speiser, P. Pharmacokinetics and Distribution of a Biodegradable Drug Carrier. *Int. J. Pharm.* **1983**, *15*, 335–345.

Gumel, A. M.; Annuar, M. S. M.; Chist, Y. Recent Advances in the Production, Recovery and Applications of Polyhydroxyalkanoates. *J. Polym. Environ.* **2013,** *21,* 580–605.

Gurny, R.; Peppas, N. A.; Harrington, D. D.; Banker, G. S. Development of Biodegradable and Injectable Lattices for Controlled Release Potent Drugs. *Drug. Dev. Ind. Pharm.* **1981,** *7,* 1–25.

Herrero-Vanrell, R.; Rincón, A. C.; Alonso, M.; Reboto, V.; Molina-Martinez, I. T.; Rodríguez-Cabello, J. C. Self-Assembled Particles of an Elastin-Like Polymer as Vehicles for Controlled Drug Release. *J. Controlled Release* **2005,** *102,* 113–122.

Herrmann, J.; Bodmeier, R. Biodegradable, Somatostatin Acetate Containing Microspheres Prepared by Various Aqueous and Non-Aqueous Solvent Vaporation Methods. *Eur. J. Pharm. Biopharm.* **1998,** *45,* 75–82.

Hussain, A.; Zia, K. M.; Tabasum, S.; Noreen, A.; Ali, M.; Iqbal, R.; Zuber, M. Blends and Composites of Exopolysaccharides; Properties and Applications: A Review. *Int. J. Biol. Macromol.* **2017,** *94,* 10–27.

Jameela, S. R.; Kumary, T. V.; Lal, A. V.; Jayakrishnan, A. Progesterone-Loaded Chitosan Microspheres: A Long Acting Biodegradable Controlled Delivery System. *J. Controlled Release* **1998,** *52,* 17–24.

Jeong, Y. I.; Kim, S. T.; Jin, S. G.; Ryu, H. H.; Jin, Y. H.; Jung, T. Y.; Kim, I. Y.; Jung, S. Cisplatin-Incorporated Hyaluronic Acid Nanoparticles Based on Ion-Complex Formation. *J. Pharm. Sci.* **2008,** *97*(3), 1268–1276.

Jyothi, N. V. N.; Prasanna, P. M.; Sakarkar, S. N.; Prabha, K. S.; Ramaiah, P. S.; Srawan, G. Y. Microencapsulation Techniques, Factors Influencing Encapsulation Efficiency. *J. Microencapsulation* **2010,** *27,* 187–197.

Kaewprapana, K.; Inprakhona, P.; Mariec, E.; Durand, A. Enzymatically Degradable Nanoparticles of Dextran Esters as Potential Drug Delivery Systems. *Carbohydr. Polym.* **2012,** *88,* 875–881.

Karnik, R.; Gu, F.; Basto, P.; Cannizzaro, C.; Dean, L.; Kyei-Manu, W.; Langer, R.; Farokhzad, O. C. Microfluidic Platform for Controlled Synthesis of Polymeric Nanoparticles. *Nano Lett.* **2008,** *8,* 2906–2912.

Katara, R.; Majumdar, D. Eudragit RL 100-Based Nanoparticulate System of Aceclofenac for Ocular Delivery. *Colloids Surf. B* **2013,** *103,* 455–462.

Kreuter, J. Large-Scale Production Problems and Manufacturing of Nanoparticles. In *Specialized Drug Delivery System;* Tyle, P., Ed.; Marcel Dekker: New York, **1990;** 257–266.

Kreuter, J.; Ramge, P.; Petrov, V.; Hamm, S.; Gelperina, S. E.; Engelhardt, B.; Alyautdin, R. Von Briesen, H.; Begley, D. J. Direct Evidence that Polysorbate-80-Coated Poly (Butylcyanoacrylate) Nanoparticles Deliver Drugs to the CNS via Specific Mechanisms Requiring Prior Binding of Drug to the Nanoparticles. *Pharm. Res.* **2003,** *20,* 409–416.

Kroll, R. A.; Pagel, M. A.; Muldoon, L. L.; Roman-Goldstein, S.; Fiamengo, S. A.; Neuwelt, E. A. Improving Drug Delivery to Intracerebral Tumor and Surrounding Brain in a Rodent Model: A Comparison of Osmotic Versus Bradykinin Modification of the Blood–Brain and/or Blood–Tumor Barriers. *Neurosurgery* **1998,** *43*(4), 879–886 (discussion 886–889).

Kumari, A.; Yadav, S. K.; Yadav, S. C. Biodegradable Polymeric Nanoparticles Based Drug Delivery Systems. *Colloids Surf. B* **2010,** *75*(1), 1–18.

Lamprecht, A.; Ubrich, N.; Yamamoto, H.; Schäfer, U.; Takeuchi, H.; Maincent, P.; Kawashima, Y.; Lehr, C. M. Biodegradable Nanoparticles for Targeted Drug Delivery in Treatment of Inflammatory Bowel Disease. *J. Pharmacol. Exp. Ther.* **2001,** *299*(2), 775–781.

Lee, H. K.; Park, J. H.; Kwon, K. C. Double-Walled Microparticles for Single Shot Vaccine. *J. Controlled Release* **1997**, *44*(2–3), 283–293.

Legrand, P.; Lesieur, S.; Bochot, A.; Gref, R.; Raatjes, W.; Barratt, G.; Vauthier, C. Influence of Polymer 407 Behavior in Organic Solution on the Production of Polylactide Nanoparticles by Nanoprecipitation. *Int. J. Pharm.* **2007**, *344*(1–2), 33–43.

Lepeltier, E.; Bourgaux, C.; Couvreur, P. Nanoprecipitation and the "Ouzo Effect": Application to Drug Delivery Devices. *Adv. Drug Delivery Rev.* **2014**, *71*, 86–97.

Linhardt, R. J. Biodegradable Polymers for Controlled Release of Drugs. In *Controll Release Drugs;* Rosoff, M., Ed., VCH Publishers: New York, 1989; pp 53–95.

Loftsson, T.; Jarho, P.; Másson, M.; Järvinen, T. Cyclodextrins in Drug Delivery. Expert Opin. *Drug Deliv.* **2005**, *2*(2), 335–351.

Lokhande, A. B.; Mishra, S.; Kulkarni, R. D.; Naik, J. B. Influence of Different Viscosity Grade Ethylcellulose Polymers on Encapsulation and in Vitro Release Study of Drug Loaded Nanoparticles. *J. Pharm. Res.* **2013**, *7*(5), 414–420.

Loureiro, J. A.; Gomes, B.; Fricker, G.; Coelho, M. A.; Rocha, S.; Pereira, M. C. Cellular Uptake of PLGA Nanoparticles Targeted with Anti-Amyloid and Anti-Transferrin Receptor Antibodies for Alzheimer's Disease Treatment. *Colloids Surf. B* **2016**, *1*, 8–13.

Magenheim, B.; Levy, M. Y.; Benita, S. A New in Vitro Technique for the Evaluation of Drug Release Profile from Colloidal Carriers—Ultrafiltration Technique At Low Pressure. *Int. J. Pharm.* **1993**, *94*(1–3), 115–123.

Mao, S.; Bakowsky, U.; Jintapattanakit, A.; Kissel, T. Self-Assembled Polyelectrolyte Nanocomplexes Between Chitosan Derivatives and Insulin. *J. Pharm. Sci.* **2006**, *95*(5), 1035–1048.

Marcato, P. D.; Duranm, N. New Aspects of Nanopharmaceutical Delivery Systems. *J. Nanosci. Nanotechnol.* **2008**, *8*(5), 2216–2229.

McClean, S.; Prosser, E.; Meehan, E.; O'Malley, D.; Clarke, N.; Ramtoola, Z.; Brayden, D. Binding and Uptake of Biodegradable Poly-Lactide Micro and Nanoparticles in Intestinal Epithelia. *Eur. J. Pharm. Sci.* **1998**, *6*(2), 153–163.

Mehta, R. C.; Jayanthi, R., Calis, S.; Thanoo, B.C.; Burton, K. W.; De Luca, P. P. Biodegradable Microspheres as Depot System for Parenteral Delivery of Peptide Drugs. *J. Controlled Release* **1994**, *29*(3), 375–384.

Mehta, R. C.; Pike, G. B.; Enzmann, D. R. Magnetization Transfer Magnetic Resonance Imaging: A Clinical Review. *Top Magn. Reson. Imaging* **1996**, *8*(4), 214–230.

Mehvar, R. Dextrans for Targeted and Sustained Delivery of Therapeutic and Imaging Agents. *J. Controlled Release* **2000**, *69*(1), 1–25.

Melo, E. P.; Aires-Barros, M. R.; Cabral, J. M. Reverse Micelles and Protein Biotechnology. *Biotechnol. Annu. Rev.* **2001**, *7*, 87–129.

Melo, J. D. D.; Carvalho, L. F. M.; Medeiros, A. M.; Souto, C. R. O.; Paskocimas, C. A. A Biodegradable Composite Material Based on Polyhydroxybutyrate (PHB) and Carnauba Fibers. *Composites, Part B* **2012**, *43*(7), 2827–2835.

Moghimi, S. M.; Hunter, A. C.; Murray, J. C. Long-Circulating and Target-Specific Nanoparticles: Theory to Practice. *Pharmacol. Rev.* **2001**, *53*(2), 283–318.

Mogoşanu, G. D.; Grumezescu, A. M.; Bejenaru, C.; Bejenaru, L. E. Polymeric Protective Agents for Nanoparticles in Drug Delivery and Targeting. *Int. J. Pharm.* **2016**, *510*(2), 419–429.

Muller, R. H.; Maassen, S.; Weyhers, H.; Mehnert, W. Phagocytic Uptake and Cytotoxicity of Solid Lipid Nanoparticles (SLN) Sterically Stabilized with Poloxamine 908 and Poloxamer 407. *J. Drug Target.* **1996**, *4*(3), 161–170.

Mundargi, R. C.; Babu, V. R.; Rangaswamy, V.; Patel, P.; Aminabhavi, T. M. Nano/Micro Technologies for Delivering Macromolecular Therapeutics Using Poly(D, L-Lactide-Co-Glycolide) and its Derivatives. *J. Controlled Release* **2008**, *125*(3), 193–209.

Murtaza, G. Ethylcellulose Microparticles: A Review. *Acta Pol. Pharm.* **2012**, *69*(1), 11–22.

Naahidi, S.; Jafari, M.; Edalat, F.; Raymond, K.; Khademhosseini, A.; Chen, P. Biocompatibility of Engineered Nanoparticles for Drug Delivery. *J. Controlled Release* **2013**, *166*(2), 182–194.

Naik, S.; Lundberg, J.; Kumar, R.; Sjolin, J.; Jansen, J. P. Economic Evaluation of Caspofungin Versus Liposomal Amphotericin B for Empirical Antifungal Therapy in Patients with Persistent Fever and Neutropenia in Sweden. *Scand. J. Infect. Dis.* **2011**, *43*(6–7), 504–514.

Natarajan, J. V.; Nugraha, C.; Ng, X. W.; Venkatraman, S. Sustained-Release from Nanocarriers: A Review. *J. Controlled Release* **2014**, *193,* 122–138.

Nicolas, J.; Couvreur, P. Synthesis of Poly(Alkylcyanoacrylate)-Based Colloidal Nanomedicines. *Adv. Rev.* **2008**, *1*(1), 111–127.

Olivier, J.-C. Drug Transport to Brain with Targeted Nanoparticles. *NeuroRx* **2005**, *2*(1), 108–119.

Oyarzun-Ampuero, F. A.; Brea, J.; Loza, M. I.; Torres, D.; Alonso, M. J. Chitosan-Hyaluronic Acid Nanoparticles Loaded with Heparin for the Treatment of Asthma. *Int. J. Pharm.* **2009**, *3*(381), 122–129.

Ozbas-Turan, S.; Akbuga, J.; Aral, C. Controlled Release of Interleukin-2 from Chitosan Microspheres. *J. Pharm. Sci.* **2002**, *91*(5), 1245–1251.

Panyam, J.; Labhasetwar, V. Biodegradable Nanoparticles for Drug and Gene Delivery to Cells and Tissue. *Adv. Drug Deliv. Rev.* **2003a**, *55*(3), 329–347.

Panyam, J.; Dali, M. M.; Sahoo, S. K.; Ma, W.; Chakravarthi, S. S.; Amidon, G. L.; Levy, R. J.; Labhasetwar, V. Polymer degradation and in Vitro Release of a Model Protein from Poly(D, L-lactide-co-glycolide) Nano- and Microparticles. *J. Controlled Release* **2003b**, *92*(1–2), 173–187.

Panyam, J.; Williams, D.; Dash, A.; Leslie-Pelecky, D.; Labhasetwar, V. Solid-State Solubility Influences Encapsulation and Release of Hydrophobic Drugs from PLGA/PLA Nanoparticles. *J. Pharm. Sci.* **2004**, *93*(7), 1804–1814.

Paques, J. P.; Van der Linden, E.; Van Rijin, C.; Sagis, L. M. C. Preparation Methods of Alginate Nanoparticles. *Adv. Colloid Interface Sci.* **2014**, *209*, 163–171.

Pison, U.; Welte, T.; Giersing, M.; Groneberg, D. A. Nanomedicine for Respiratory Diseases. *Eur. J. Pharmacol.* **2006**, *533*(1–3), 341–350.

Prabhu, R. H.; Patravale, V. B.; Joshi, M. D. Polymeric Nanoparticles for Targeted Treatment in Oncology: Current Insights. *Int. J. Nanomed.* **2015**, *10*, 1001–1101.

Quintanar-Guerrero, D.; Allémann, E.; Fessi, H.; Doelker, E. Pseudolatex Preparation Using a Novel Emulsion-Diffusion Process Involving Direct Displacement of Partially Water-Miscible Solvents by Distillation. *Int. J. Pharm.* **1999**, 155–164.

Raffin, R. P.; Jornada, D. S.; Ré, M. I.; Pohlmann, A. R.; Guterres, S. S. Sodium Pantoprazole-Loaded Enteric Microparticles Prepared by Spray Drying: Effect of the Scale of Production and Process Validation. *Int. J. Pharm.* **2006**, *324*(1), 10–18.

Redhead, H. M.; Davis, S. S.; Illum, L. Drug Delivery in Poly(Lactide-Coglycolide) Nanoparticles Surface Modified with Poloxamer 407 and Poloxamine 908: in Vitro Characterisation and in Vivo Evaluation. *J. Controlled Release* **2001**, *70*(3), 353–363.

Reis, C. P.; Neufeld, R. J.; Ribeiro, A. J.; Veiga, F. Nanoencapsulation I Methods for Preparation of Drug-Loaded Polymeric Nanoparticles. *Nanomed. Nanotech. Biol. Med.* **2006**, *2*(1), 8–21.

Reverchon, E.; Daghero, J.; Marrone, C.; Mattea, M.; Poletto, M. Supercritical Fractional Extraction of Fennel Seed Oil and Essential Oil: Experiments and Mathematical Modeling. *Ind. Eng. Chem. Res.* **1999**, *38*(8), 3069–3075.

Roberge, D. M.; Gottsponer, M.; Eyholzer, M.; Kockmann, N. Industrial design Scale-Up, and Use of Microreactors. *Chim. Oggi* **2009**, *27*(4), 8–11.

Santoro, M.; Tatara, A. M.; Mikos, A. G. Gelatin Carriers for Drug and Cell Delivery in Tissue Engineering. *J. Controlled Release* **2014**, *190*, 210–218.

Schaffazick, S. R.; Guterres, S. S. Physicochemical Characterization and Stability of the Polymeric Nanoparticle Systems for Drug Administration. *Quim. Nova* **2003**, *26*(5), 726–737.

Schatzlein, A. G. Delivering Cancer Stem Cell Therapies – a Role for Nanomedicines. *Eur. J. Cancer* **2006**, *42*(9), 1309–1315.

Schütz, C. A.; Juillerat-Jeanneret, L.; Mueller, H.; Lynch, I.; Riediker, M.; NanoImpactNet Consortium. Therapeutic Nanoparticles in Clinics and Under Clinical Evaluation. *Nanomedicine* **2013a**, *8*(3), 449–467.

Schütz, C. A.; Juillerat-Jeanneret, L.; Soltmann, C.; Mueller, H. Toxicity Data of Therapeutic Nanoparticles in Patent Documents. *World Patent Inf.* **2013b**, *35*(2), 110–114.

Shrivastav, A.; Kim, H.-Y.; Kim, Y.-R. Advances in the Applications of Polyhydroxyalkanoate Nanoparticles for Novel Drug Delivery System. *BioMed Res. Int.* **2013**, 1–13.

Singh, R.; Lillard, Jr., J. W. Nanoparticle-Based Targeted Drug Delivery. *Exp. Mol. Pathol.* **2009**, *86*(3), 215–223.

Soppimath, K. S.; Aminabhavi, T. M.; Kulkarni, A. R.; Rudzinski, W. E. Biodegradable Polymeric Nanoparticles as Drug Delivery Devices. *J. Controlled Release* **2001**, *70*(1–2), 1–20.

Sosnik, A.; Das Neves, J.; Sarmento, B. Mucoadhesive Polymers in the Design of Nano-Drug Delivery Systems for Administration by Non-Parenteral Routes: A Review. *Prog. Polym. Sci.* **2014**, *39*(12), 2030–2075.

Souris, J. S.; Lee, C. H.; Cheng, S. H.; Chen, C. T.; Yang, C. S.; Ho, J. A. A.; Mou, C. Y.; Lo, L. W. Surface Charge-Mediated Rapid Hepatobiliary Excretion of Mesoporous Silica Nanoparticles. *Biomaterials* **2010**, *31*(21), 5564–5574.

Stylios, G. K.; Giannoudis, P. V.; Wan, T. Applications of Nanotechnologies in Medical Practice. *Injury* **2005**, *36*(4), S6–S13.

Swarbrick, J.; Boylan, J. *Encyclopedia of Pharmaceutical Technology*, 2nd ed.; Marcel Dekker: New York, 2002.

Swierczewska, M.; Han, H. S.; Kim, K.; Park, J. H.; Lee, S. Polysaccharide-Based Nanoparticles for Theranostic Nanomedicine. *Adv. Drug Delivery Rev.* **2016**, *99*, 70–84.

Thakral, S.; Thakral, N. K.; Majumdar, D. K. Eudragit®: A Technology Evaluation. *Expert Opin. Drug Delivery* **2013**, *10*(1), 131–149.

Vanderhoff, J. W.; El Aasser, M. S.; Ugelstad, J. Polymer Emulsification Process. U.S. Patent 4,177,177, 1979.

Vauthier, C.; Bouchemal, K. Methods for the Preparation and Manufacture of Polymeric Nanoparticles. *Pharm. Res.* **2009**, *26*(5), 1025–1058.

Vauthier, C.; Dubernet, C.; Chauvierre, C.; Brigger, I.; Couvreur, P. Drug Delivery to Resistant Tumors: the Potential of Poly(Alkyl Cyanoacrylate) Nanoparticles. *J. Controlled Release* **2003**, *93*(2), 151–160.

Vauthier, C.; Fattal, E.; Labarre, D. From Polymer Chemistry and Physicochemistry to Nanoparticular Drug Carrier Design and Applications. In *Biomaterial Handbook-Advanced Applications of Basic Sciences and Bioengineering;* Yaszemski, M. J., Trantolo, D. J.,

Lewamdrowski, K. U., Hasirci, V., Altobelli, D. E., Wise, D. L., Eds.; Marcel Dekker Inc.: New York, 2004; pp 563–598.

Vauthier, C.; Labarre, D.; Ponchel, G. Design Aspects of Poly(Alkylcyanoacrylate) Nanoparticles for Drug Delivery. *J. Drug Targeting* **2007**, *15*(10), 641–663.

Villa Nova, M.; Gonçalves, M. C. P.; Nogueira, A. N.; Herculano, L. S.; Medina, A. N.; Bazzote, R. B.; Bruschi, M. L. Formulation and Characterization of Ethylcellulose Microparticles Containing L-Alanyl- L-Glutamine Peptide. *Drug Dev. Ind. Pharm.* **2013**, *40*(10), 1–10.

Villa Nova, M.; Janas, C.; Schmidt, M.; Ulshoefer, T.; Gräfe, S.; Schiffmann, S.; de Bruin, N.; Wiehe, A.; Albrecht, V.; Parnham, M. J.; Bruschi, M. L.; Wacker, M. G. Nanocarriers for Photodynamic Therapy—Rational Formulation Design and Medium-Scale Manufacture. *Int. J. Pharm.* **2015**, *491*(1–2), 250–260.

Vyas, A.; Saraf, S.; Saraf, S. Cyclodextrin Based Novel Drug Delivery Systems. *J. Inclusion Phenom. Macrocyclic Chem.* **2008**, *62*(1), 23–42.

Wachsman, P.; Lamprecht, A. Ethylcellulose nanoparticles with bimodal size distribution as precursors for the production of very small nanoparticles. *Drug Dev. Ind. Pharm.* **2014**, *41*(7), 1–7.

Wacker, M.; Zensi, A.; Kufleitner, J.; Ruff, A.; Schütz, J.; Stockburger, T.; Marstaller, T.; Vogel, V. A Toolbox for the Upscaling of Ethanolic Human Serum Albumin (has) Desolvation. *Int. J. Pharm.* **2011**, *414*(1–2), 225–232.

Walters, R. H.; Bhatnagar, B.; Tchessalov, S.; Izutsu, K.; Tsumoto, K.; Ohtake, S. Next Generation Drying Technologies for Pharmaceutical Applications. *J. Pharm. Sci.* **2014**, *103*(9),2673–2695.

Wang, L. Y.; Gu, Y. H.; Zhou, Q. Z.; Ma, G. H.; Wan, Y. H.; Su, Z. G. Preparation and Characterization of Uniform-Sized Chitosan Microspheres Containing Insulin by Membrane Emulsification and a Two-Step Solidification Process. *Colloids Surf. B* **2006**, *50*(2), 126–135.

Wei, X. W.; Gong, C. Y.; Gou, M. L.; Fu, S. Z.; Gou, Q. F.; Shi, S.; Luo, F.; Gou, G.; Qui, L. Y.; Qiana, Z. Y. Biodegradable Poly(-Caprolactone)–Poly(Ethylene glycol) Copolymers as Drug Delivery System. *Int. J. Pharm.* **2009**, *381*(1), 1–18.

Weissig, V.; Pettinger, T. K.; Murdock, N. Nanopharmaceuticals (Part 1): Products on the Market. *Int. J. Nanomedicine* **2014**, *9*, 4357–4373.

Wojewoda, C. M.; Sercia, L.; Navas, M.; Tuohy, M.; Wilson, D.; Hall, G. S.; Procop, G. W.; Richter, S. S. Evaluation of the Verigene Gram-Positive Blood Culture Nucleic Acid Test for Rapid Detection of Bacteria and Resistance Determinants. *J. Clin. Microbiol.* **2013**, *51*(7), 2072–2076.

Xu, Y.; Du, Y.; Huang, R.; Gao, L. Preparation and Modification of N-(2-Hydroxyl) Propyl-3-Trimethyl Ammonium Chitosan Chloride Nanoparticle as a Protein Carrier. *Biomaterials* **2003**, *24*(27), 5015–5022.

Xu, Y.; Kim, C.-S.; Saylor, D. M.; Koo, D. Polymer Degradation and Drug Delivery in PLGA-Based Drug–Polymer Applications: A Review of Experiments and Theories. *J. Biomed. Mater. Res. Part B* **2016**, 1–25.

Yokoyama, M. Drug Targeting with Nano-Sized Carrier Systems. *J. Artif. Organs* **2005**, *8*(2), 77–84.

Yoon, H. Y.; Koo, H.; Choi, K. Y.; Lee, S. J.; Kim, K.; Kwon, I. C.; Leary, J. F.; Park, K.; Yuk, S. H.; Park, J. H.; Choi, K. Tumor-Targeting Hyaluronic Acid Nanoparticles for Photodynamic Imaging and Therapy. *Biomaterials* **2012**, *33*(15), 3980–3989.

Zafar, N.; Fessi, H.; Elaissari, A. Cyclodextrin Containing Biodegradable Particles: From Preparation to Drug Delivery Applications. *Int. J. Pharm.* **2014,** *461*(1–2), 351–366.

Zhang, J.; Ma, P. X. Cyclodextrin-Based Supramolecular Systems for Drug Delivery: Recent Progress and Future Perspective. *Adv. Drug Delivery Rev.* **2013,** *65*(9), 1215–1233.

Zhang, X.; Yang, X.; Ji, J.; Liu, A.; Zhai, G. Tumor Targeting Strategies for Chitosan-Based Nanoparticles. *Colloids Surf. B* **2016,** *148,* 460–473.

Zhao, C. X.; He, L.; Qiao, S. Z.; Middelberga, A. P. J. Nanoparticle Synthesis in Microreactors. *Chem. Eng. Sci.* **2011,** *66*(7), 1463–1479.

CHAPTER 2

NANOPARTICLES IN DRUG DELIVERY: GENERAL CHARACTERISTICS, APPLICATIONS, AND CHALLENGES

KHUSHWINDER KAUR

Department of Chemistry and Centre of Advanced Studies in Chemistry, Panjab University, Chandigarh, Punjab 160014, India, email: makkarkhushi@gmail.com

ABSTRACT

The world has all sorts of challenges and the fate narrows down the challenges to those who are capable of handling them. True to it the crucial issue facing the scientific community in the treatment of deadly diseases such as cancer, Alzheimer's, and so forth is the delivery of drug to the target site. Therefore, there is a growing interest to use nanotechnology in medicine and for drug delivery. Nanoparticles-based drug delivery system helps to strike a consistency between the drug–dose concentration and the resultant therapeutic outcome or toxic effects. Cell-specific targeting can be achieved by attaching drugs to specially designed carriers such as liposomes, metal nanoparticles, hybrid nanoparticles, and so forth. By successively addressing the encountered barriers such as nonspecific distribution and inadequate accumulation of therapeutics, innovative design features can be rationally incorporated to create a new generation of nanotherapeutics for realizing the exemplary shift in NP-based drug delivery. Therefore, the chapter deliberates over the potential of nanoparticles in the treatment of diseases such as cancer, Alzheimer's, and so forth.

2.1 INTRODUCTION

Drug delivery is a practice used for cautiously delivering pharmaceutical compounds in a biological system to obtain the required therapeutic effects. The delivery of drug to the identified organ or tissue, therefore, assumes crucial significance. The quest to unveil secrecy surrounding the disease and the pursuit for the eradication of disease has always remained a major challenge before the scientific community (Leung, 2006; Park, 2013; Qin et al., 2017). Therefore, an effective drug delivery system is an essential protocol for curing medical illness. The advent of new technologies has brought about dramatic change in the method of drug delivery during the preceding few decades and is poised for greater change in the years to come. Accordingly, the search for new drug delivery application requires a multidisciplinary scientific approach so as to find solutions for improving the therapeutic index and bioavailability for the targeted delivery of the therapeutic agents (Hare et al., 2016; Martinho et al., 2011; Tiwari et al., 2012).

The drug movement through cells, tissues, and its carriage through the circulatory system poses physiological barriers in efficient drug delivery. Biomedical engineers have, however, made a significant contribution to this effect by developing new modes of drug delivery. However, saving drug from degradation, reducing harmful side effects and increasing the bioavailability of the drug at the targeted organ remains major challenge for any drug delivery system. Therefore, effective and noninvasive drug delivery systems remain the most preferred and ideal option. Perpetual efforts are, therefore, being made in this regard to develop noninvasive drug delivery systems and technologies that can achieve selective delivery of therapeutic agents. Noninvasive-targeted drug delivery system has the potential to minimize or even eliminate the toxic side effects of the drug on tissues undergoing therapy. Several targeted drug delivery technologies are available that use nanoscale drug carriers. National Institutes of Health has termed these technological innovations as "nanomedicines" which are in scientific parlance and are effective within the targeted area (Andreas et al., 2015; Hare et al., 2016; Jain et al., 2015).

The human species need a drug delivery system potentially capable of efficiently delivering a drug to the patient in a manner which is less unpleasant. It is here that the nanomedicine can come to our rescue because it has the important mechanistic parameters which help it to qualify to be used as a drug delivery system. It can lay measurable factors forming one of a set that defines a system or sets the conditions of its operation which include its definite character which prompts the active drug to bind or hold on to the receptor. It also ensures intracellular uptake by the targeted cells besides ensuring sustained release over long time the release of active molecule.

Protracted technological advances in developing engineered nanoparticles (NPs) for drug delivery system have started showing promising results. Nanoscale technologies are set to change the perspective of the scientific community on the way it perceives disease diagnosis, its treatment, and prevention. Nanomedicine can transform molecular discoveries arising from genomics and proteomics into widespread benefits for patients (Frank et al., 2014; Hood, 2004).

Nanotechnology is an approach where a constant dose of drugs is delivered directly to the diseased sites or cells for an extended period and result in alternative therapeutic option for patients. Infection, tissue engineering, de novo synthesis, and so forth are some of the biological processes which nanoassemblies can mimic or alter. These nanoassemblies include a whole host of NPs. They are not limited to NPs but also functionalized carbon nanotubes, nanomachines (e.g., constructed from interchangeable deoxyribonucleic acid [DNA] parts and DNA scaffolds), nanofibers, self-assembling polymeric nanoconstructs, nanomembranes, and nanosized silicon chips for drug, protein, nucleic acid, or peptide delivery and release, and biosensors and laboratory diagnostics. However, what is challenging is the design of NPs which can efficiently deliver their cargo over an extended period of time to the targeted cell and enhance its bioavailability to achieve the desired clinical response (Leung, 2006; Martinho et al., 2011; Park, 2013; Qin et al., 2017;Tiwari et al., 2012).

2.2 NANOPARTICLES (NPS)

NPs represent a class of assemblies which are highly diverse from small molecule. They are solid and spherical structures with their dimensions floating or fluctuating around10–200 nm. NPs drug carriers, of late, have hogged the limelight for their potential to encapsulate therapeutic mediators with inherent properties to coordinate their release to the diseased cells. They have high surface to volume ratio and more successful pharmacokinetic features. Drug-encapsulated NPs gather both active and passive systems to ensure the extended release of drug during the complete period of its circulation. They secure constant drug discharge kinetics and recovery of damaged tissues. What makes NPs even more suitable for their medical application is the possibility of their subsuming in them such characteristics such as high chemical and biological stability. They also have the tendency to integrate both hydrophilic and hydrophobic pharmaceuticals besides the ability to administer them by diverse routes (including oral, inhalational, and parenteral) (Avgoustakis et al., 2002; Wilczewska et al., 2012).

NPs can be characterized for their ability to conjugate covalently with various ligands (such as antibodies, proteins, or aptamers) to target-specific tissues. They have a large surface area-to-volume ratio. It helps them attach multiple copies of a ligand. Multivalent functionalization dramatically increases their binding affinity. This also enhances serum stability, biocompatibility, and in vivo circulation time. All these attributes enable them to minimize the adverse side effects. Bare NPs are, however, subjected to opsonization and resultantly get cleared from circulation by the reticuloendothelial system (RES) mainly located in spleen and liver. The commonest of approaches adopted for escaping RES can be summarized as follows: (i) neutral surface charge is used to formulate the particles (ii) different hydrophilic surfactants, such as polysorbates and polyethylene glycol (PEG) be used to quote their surface, and (iii) small-sized nanoparticles be used. (e.g., <80 nm). NPs with these characteristics are effective in avoiding RES and have extended circulation time. This demonstrates their stability in blood (Balasubramanian et al., 2015;Mohanraj et al., 2006).

The uses of NPs to deliver drugs to cancer cells remain probably the most unexplored system under development. In this system, the particles are engineered to stick to diseased cells. The engineered particles directly treat the diseased cells while reducing damage to healthy cells in the body (Haleyand Frenkel, 2008). But this is the first hint or revelation of something larger or more complex hiding in the folds of future. Potential options are available in nanotechnology to efficiently deliver the drug and which are significantly less unpleasant for the patient. Some of the techniques are at the conceptual stage or are still fantasies. There are others techniques also which are at various stages of testing or are in use. Therefore, over the past two decades, the scientific community has evinced considerable interest in the development of NPs for use as effective drug delivery vehicles. There are other nanoassemblies which find mention in literature. Therefore, it is necessary that they are clearly segregated before we get into the depth of this chapter.

The various nano-based delivery vehicles have been shown in Figure 2.1 (Guterres et al., 2007).

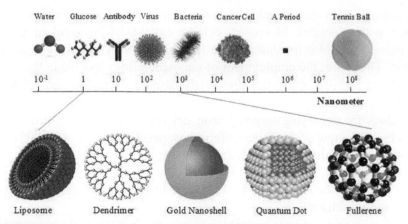

FIGURE 2.1 The various nano-based delivery vehicles.

2.3 NPS AS PREFERRED DELIVERY VEHICLES

Widely practized conventional drug administration methods have many problems. They may be possibly overcome by nanomethods (Singh et al., 2011; Vasir et al., 2005). NPs can offer significant advantages over the traditional delivery mechanisms in terms of high stability, high specificity, high drug carrying capacity, the ability for controlled release possibility to use in different types of drug administration, and the capability to transport both hydrophilic and hydrophobic molecules. Enhanced delivery or uptake by target cells and/ or reduction in the toxicity of the free drug to nontarget organs is the aims for NP entrapment of drugs. Both these situations will cause an uptick in the therapeutic index, increase the margin between the doses resulting in a therapeutic efficacy (e.g., cell death), and toxicity to other organ systems.

For all aforementioned aims what is required is the creation of enduring or resistant target-specific NPs. Conventionally, drugs are taken orally or through injections which circulate throughout the body which may cause harmful effects on the cells or tissues or organs. Protein and peptide drugs are poorly absorbed after oral administration because of their susceptibility across the intestinal epithelium. Conventional drug delivery needs high doses to make up the bioavailability. Nanotechnology can also help to improve oral absorption.

The efficacy of highly lipophilic drugs is much lower than the desirable level. Therefore, vital issues such as their poor solubility and reduced systemic exposure need to be addressed to enhance their pharmacological effects. Probucol is one of the lipophilic drugs. Zhang et al. (2013) reported that as compared with free probucol suspension, the blood concentration of probucol was considerably enhanced with the nanodelivery system loaded with probucol. Moreover, cellular uptake of probucol in Caco-2 cell monolayers also increased when the nanodelivery system was administered. Therefore, the employment of nanotechnology has made the use of such drugs possible that were previously discarded or that were tricky to administer. Distinct pathophysiological features of diseased tissues provide vital clues for achieving targeted drug delivery. Nanopaticles offer many advantages in view of the above-referred properties (Vasir et al., 2005).

Some of the advantages are listed below.

- Improve the stability of hydrophobic drugs, rendering them suitable for administration
- Release drug at sustained rate and lower the frequency of administration
- Improve biodistribution and pharmacokinetics resulting in improved efficacy
- Control and ensure sustained release of the drug to achieve increase in drug therapeutic efficiency and reduction in side effects during the transportation and at the site of localization altering organ distribution of the drug and the resultant clearance of the drug
- Reduce toxicity by using biocompatible nanomaterials
- Provide improvement in the therapeutic performance of the drug over conventional systems making treatment a better experience with reduced treatment expenses. They also provide comfort and compliance to the patient
- Provide ideal targeting system having long circulating time and ensure presence with appropriate concentrations at the target site and are not likely to lose its activity or therapeutic efficacy while in circulation
- Improve the solubility of poorly water-soluble drugs, prolong the shelf life of the drug, and systemic circulation by reducing immunogenicity

2.4   ESSENTIALS OF NPS FOR EFFICIENT DRUG DELIVERY

i.   Surface modification—An efficient drug delivery system necessitates that NPs used for delivery of drug to the affected site must have the ability to prevail in the bloodstream for a significant time without being destroyed. A RES, such as the liver and the spleen, usually traps bare NPs without surface modification. These get excreted out of the body. Therefore, the fate of injected NPs is controlled by their size and surface. Persual of the literature reveals that the destiny of

NPs as efficient delivery systems can be controlled because hydrophilic surface modification allows the NP to escape the macrophage capture (Liang et al., 2008; Rege et al., 2002). Besides the preceding, enhanced permeability and retention effects (EPR) are the other options by which duly sized nanocarriers can be accumulated ideally at the target site, that is, tumor, inflammatory, and infection sites (Yin et al., 2014).The EPR effect is not associated with organ or normal tissue but instead, it has definite attributes. It provides superior critical targeting.

ii. Particle size and size of distribution are the other important factors. The particle size determines the in vivo distribution, biological fate, toxicity, and targeting ability of these formulated delivery systems (Borkovec, 2002; Gaumet et al., 2008). Particle size is documented to influence drug loading, drug release, and stability of NPs. Smaller particles have a larger surface area-to-volume ratio. Resultantly, the greatest quantity/part of the drug associated with small particles would accumulate at or near the particle surface. It helps in the faster release of the drug. But smaller particles have enhanced the risk of aggregation during storage, transport, and dispersion. NPs with a hydrodynamic diameter of 10–100 nm have optimal pharmacokinetic properties. Therefore, they are generally accepted for in vivo applications. On the contrary, larger particles have a large core, which allows the additional drug to be encapsulated per particle which allows steady release (Redhead et al., 2001). The larger size of NPs impedes them from promptly leaking into blood capillaries. Therefore, the size of NPs to be used in a drug delivery system need be such as be able to help them escape from being captured by fixed macrophages.

iii. Biodegradability—particle size can also affect the biodegradability of NP (polymer/capping degradation). It can be inferred from above that as the size of particle made from the said polymer increased, the rate of poly(lactic-co-glycolic acid) (PLGA) degradation was also found to increase (Dunne et al., 2000). This phenomenon can be attributed to PLGA degradation products which in smaller NPs can just get dispensed through shorter distances. Conversely, the polymer matrix of larger particles increases the time of release due to the greater distance. It is also potentially capable of causing autocatalytic degradation of the polymer material (Panyam et al., 2004). Therefore, it has been hypothesized that larger particles will chip-in towards faster polymer degradation. Commercially available nanoformulations have been listed in Table 2.1 (Weissig et al., 2014).

TABLE 2.1 Available Marketed Drug Nanoformulations.

Name	Drug entrapped or linker	Mechanism of action	Current stage of approval
AmBisome® liposome	Amphotericin B encapsulated in liposomes (60–70 nm) composed of hydrogenated soy phosphatidylcholine, cholesterol, and distearoyl phosphatidylglycerol (2/0.8/1 M)	Mononuclear phagocyte system (MPS) targeting: liposomes preferentially accumulate in organs of the MPS. Negative charge contributes to MPS targeting. Selective transfer of the drug from lipid complex to target fungal cell with minimal uptake into human cells has been postulated	Food and Drug Administration (FDA) 1997 systemic fungal infections (IV)
DaunoXome®	Daunorubicin citrate encapsulated in liposomes (45 nm) composed of distearoyl phosphatidylcholine and cholesterol (2/1 M)	Passive targeting through enhanced permeability and retention effects (EPR) effect: concentration of available liposomal drug in tumors exceeds that of free drug. Liposomal daunorubicin persists at high levels for several days	FDA 1996 HIV-related KS (IV)
DepoCyt®	Cytarabine encapsulated in multivesicular liposomes (20 μm; classified as nanopharmaceutical based on its individual drug containing "chambers") made from dioleoyl lecithin, dipalmitoyl phosphatidylglycerol, cholesterol, and triolein	Sustained release: this formulation of cytarabine maintains cytotoxic concentrations of the drug in the cerebrospinal fluid for more than 14 days after a single 50 mg injection	FDA 1999/2007 Lymphomatous malignant meningitis (IV)
DepoDur®	Morphine sulfate encapsulated in multivesicular liposomes (17–23 μm; per se not a nanopharmaceutical—classified as such based only on its individual drug-containing "nano-sized chambers") made from dioleoyl lecithin cholesterol, dipalmitoyl phosphatidylglycerol, tricaprylin, and triolein	Sustained release: after the administration into the epidural space, morphine sulfate is released from the multivesicular liposomes over an extended period of time	FDA 2004 For treatment of chronic pain in patients requiring a long-term daily around-the-clock opioid analgesic (administered into the epidural space)

TABLE 2.1 *(Continued)*

Name	Drug entrapped or linker	Mechanism of action	Current stage of approval
Doxil®	Doxorubicin hydrochloride encapsulated in stealth liposomes (100 nm) composed of N-(carbonyl-methoxy polyethylene glycol 2000)-1,2-distearoyl-sn-glycero3-phosphoethanolamine sodium, fully hydrogenated soy phosphatidylcholine, and cholesterol	Passive targeting through EPR effect: extravasation of liposomes by the passage of the vesicles through endothelial cell gaps present in solid tumors. Enhanced accumulation of doxorubicin in lesions of acquired immune deficiency syndrome (AIDS)-associated KS after administration of polyethylene glycol (PEG)-liposomal doxorubicin	FDA 1995 AIDS-related KS, multiple myeloma, ovarian cancer (IV)
Inflexal®V	Influenza virus antigens (hemagglutinin, neuraminidase) on the surface of 150 nm liposomes	Mimicking native antigen presentation: liposomes mimic the native virus structure, thus allowing for the cellular entry and membrane fusion. Retention of the natural presentation of antigens on the liposomal surface provides for high immunogenicity	Switzerland 1997 Influenza vaccine
Marqibo®	Vincristine sulfate encapsulated in sphingomyelin/cholesterol (60/40 M) 100 nm liposomes	Passive targeting through EPR effect: extravasation of liposomes through fenestra in bone marrow endothelium	FDA 2012 Acute lymphoid leukemia, Philadelphia chromosome-negative, relapsed or progressed (IV)
Mepact™	Mifamurtide (synthetic muramyl tripeptide-phosphatidylethanolamine) incorporated into large multilamellar liposomes composed of 1-palmitoyl-2-oleoyl-sn-glycerol-3-phosphocholine and 1,2-dioleoyl-sn-glycero-3-phospho-L-serine	MPS targeting: the drug, an immune stimulant, is anchored in negatively charged liposomal bilayer membrane	Europe 2009 Non-metastasizing resectable osteosarcoma (IV)

TABLE 2.1 (Continued)

Name	Drug entrapped or linker	Mechanism of action	Current stage of approval
Myocet®	Doxorubicin encapsulated 180 nm oligolamellar liposomes composed of egg phosphatidylcholine/cholesterol (1/1 M)	MPS targeting: forms "MPS depot," slow release into blood circulation resembles prolonged infusion	Europe 2000 Metastatic breast cancer (IV)
Visudyne®	Verteporfin in liposomes made of dimyristoyl-phosphatidylcholine and egg phosphatidylglycerol (negatively charged); lyophilized cake for reconstitution	Drug solubilization: rendering drug biocompatible and enhancing ease of IV administration. No other apparent function of liposomes. Liposomal formulation unstable in the presence of serum. Fast transfer of verteporfin from visudyne to lipoproteins	FDA 2000 Photodynamic therapy of wet age-related macular degeneration, pathological myopia, ocular histoplasmosis syndrome (IV)
Lipid-based (nonliposomal) formulations			
Abelcet®	Amphotericin B complex 1:1 with DMPC and DMPG (7:3), >250 nm, ribbon-like structures of a bilayered membrane	MPS targeting: selective transfer of drug from lipid complex to fungal cell with minimal uptake into human cells has been postulated	FDA 1995 and 1996 Marketed outside the USA as Amphocil®systemic fungal infections (IV)
PEGylated proteins polypeptides and aptamers			
Adagen®	PEGylated adenosine deaminase one enzyme molecule is modified with up to 17 strands of PEG, molecular weight (MW) 5000,114 oxymethylene groups per strand	Increased circulation time and reduced immunogenicityPEGylation generally increases the hydrodynamic radius, prolongs circulation and retention time, decreases proteolysis, decreases renal excretion, and shields antigenic determinants from immune detection without obstructing the substrate-interaction site	FDA 1990 Adenosine deaminase deficiency—severe combined immunodeficiency disease

TABLE 2.1 *(Continued)*

Name	Drug entrapped or linker	Mechanism of action	Current stage of approval
Cimzia®	PEGylated antibody (Fab′ fragment of a humanized anti-tumor necrosis factor-alpha antibody)		FDA 2008 Crohn's disease, rheumatoid arthritis
Neulasta®	PEGylatedfilgrastim (granulocyte colony-stimulating factor)		FDA 2002 Febrile neutropenia, in patients with nonmyeloid malignancies; prophylaxis (SC)
Oncaspar®	PEGylated L-asparaginase		FDA 1994 Acute lymphoblastic leukemia
Pegasys®	PEGylated interferon alpha-2b		FDA 2002 Hepatitis B and C
PegIntron®	PEGylated interferon alpha-2b		FDA 2001 Hepatitis C
Somavert®	PEGylated human growth hormone receptor antagonist		FDA 2003 Acromegaly, second-line therapy
Macugen®	PEGylated anti-VEGF aptamer		FDA 2004 Intravitreal Neovascular age-related macular degeneration
Mircera®	PEGylatedepoetin beta (erythropoietin receptor activator)		FDA 2007 Anemia associated with chronic renal failure in adults

TABLE 2.1 *(Continued)*

Name	Drug entrapped or linker	Mechanism of action	Current stage of approval
		Nanocrystals	
Emend®	Aprepitant as nanocrystal	Increased bioavailability due to increased dissolution rate: below 1000 nm, the saturation solubility becomes a function of the particle size leading to an increased saturation solubility of nanocrystals, which in turn increases the concentration gradient between gut lumen and blood, and consequently the absorption by passive diffusion	FDA 2003 Emesis, antiemetic (oral)
Megace ES®	Megestrol acetate as nanocrystal		FDA 2005 Anorexia, cachexia (oral)
Rapamune®	Rapamycin (sirolimus) as nanocrystals formulated in tablets		FDA 2002 Immunosuppressant (oral)
Tricor® Triglide®	Fenofibrate as nanocrystals		FDA 2004
Triglide®	Fenofibrate as insoluble drug-delivery microparticles		Hypercholesterolemia, hypertriglyceridemia (oral)
		Polymer-based nanoformulations	
Copaxone®	Polypeptide (average MW 6.4 kDa) composed of four amino acids (glatiramer)	No mechanism attributable to nanosize. Based on its resemblance to myelin basic protein, glatiramer is thought to divert as a "decoy" an autoimmune response against myelin	FDA 1996/2014 multiple sclerosis (SC)

TABLE 2.1 *(Continued)*

Name	Drug entrapped or linker	Mechanism of action	Current stage of approval
Eligard®	Leuprolide acetate (synthetic gonadotropin-releasing hormone or or luteinizing hormone releasing hormone agonist (LH-RH analog) incorporated in nanoparticles (NPs) composed of PLGH copolymer (DL-lactide/glycolide; 1/1 M)	Sustained release	FDA 2002 Advanced prostate cancer (SC)
Genexol®	Paclitaxel in 20–50 nm micelles composed of block copolymer poly(ethylene glycol)-poly(D,L-lactide)	Passive targeting through EPR effect	South Korea 2001 Metastatic breast cancer, pancreatic cancer (IV)
Opaxio®	Paclitaxel covalently linked to solid NPs composed of polyglutamate	Passive targeting through EPR effect: drug release inside solid tumor through enzymatic hydrolysis of polyglutamate	FDA 2012 Glioblastoma
Renagel®	Cross-linked polyallylamine hydrochloride MW variable	No mechanism attributable to nano size. Phosphate binder	FDA 2000 Hyperphos-phatemia (oral)
Zinostatinstimal-amer®	Conjugate protein or copolymer of styrene—maleic acid and an antitumor protein neocarzinostatin (NCS). Synthesized by conjugation of one molecule of NCS and two molecules of poly(styrene-co-maleic acid)	Passive targeting via EPR effect	Japan 1994 Primary unresect-able hepatocellular carcinoma
Eligard®	Leuprolide acetate (synthetic gonadotropin-releasing hormone or LH-RH analog) incorporated in NPs composed of PLGH copolymer (DL-lactide/glycolide; 1/1 M)	Sustained release	FDA 2002 Advanced prostate cancer (SC)

TABLE 2.1 *(Continued)*

Name	Drug entrapped or linker	Mechanism of action	Current stage of approval
Genexol®	Paclitaxel in 20–50 nm micelles composed of block copolymer poly(ethylene glycol)-poly(D,L-lactide)	Passive targeting through EPR effect	South Korea 2001 Metastatic breast cancer, pancreatic cancer (IV)
Polymer based			
Abraxane®	NPs (130 nm) formed by albumin with conjugated paclitaxel	Passive targeting through EPR effect: dissociation into individual drug-bound albumin molecules, which may mediate endothelial transcytosis of paclitaxel through albumin-receptor-mediated pathway	FDA 2005 Metastatic breast cancer, non-small-cell lung cancer (IV)
Kadcyla®	Immunoconjugate. Monoclonal antibody (against human epidermal growth factor receptor-2)–drug (DM1, a cytotoxin acting on microtubule) conjugate, linked through thioether	No mechanism attributable to nanosize	FDA 2013 Metastatic breast cancer
Ontak®	Recombinant fusion protein of fragment A of diphtheria toxin and subunit binding to interleukin-2 receptor	Fusion protein binds to the interleukin-2 receptor, followed by receptor-mediated endocytosis; fragment A of diphtheria toxin then released into the cytosol where it inhibits protein synthesis	FDA 1994/2006 Primary cutaneous T-cell lymphoma, CD25-positive, persistent or recurrent disease
Fungizone® (also referred to as "conventional AMB")	Lyophilized powder of amphotericin B with added sodium deoxycholate. Forms upon reconstitution colloidal (micellar) dispersion	Drug solubilization: rendering drug biocompatible and enhancing ease of administration after IV injection no other apparent function of micelles, which dissociate into monomers following dilution in circulation	FDA 1966 Systemic fungal infections (IV)

TABLE 2.1 *(Continued)*

Name	Drug entrapped or linker	Mechanism of action	Current stage of approval
Diprivan®	Oil-in-water emulsion of propofol in soybean oil/glycerol/egg lecithin	Drug solubilization: rendering drug biocompatible and enhancing ease of administration after IV injection	FDA 1989 Sedative-hypnotic agent for induction and maintenance of anesthesia (IV)
Estrasorb™	Emulsion of estradiol in soybean oil, polysorbate 80, ethanol, and water	Drug solubilization	FDA 2003 Hormone replacement therapy during menopause (transdermal)
Fungizone® (also referred to as "conventional AMB")	Lyophilized powder of amphotericin B with added sodium deoxycholate. Forms upon reconstitution colloidal (micellar) dispersion	Drug solubilization: rendering drug biocompatible and enhancing ease of administration after IV injection no other apparent function of micelles, which dissociate into monomers following dilution in circulation	FDA 1966 Systemic fungal infections (IV)
Diprivan®	Oil-in-water emulsion of propofol in soybean oil/glycerol/egg lecithin	Drug solubilization: rendering drug biocompatible and enhancing ease of administration after IV injection	FDA 1989 Sedative-hypnotic agent for induction and maintenance of anesthesia (IV)
Estrasorb™	Emulsion of estradiol in soybean oil, polysorbate 80, ethanol, and water	Drug solubilization	FDA 2003 Hormone replacement therapy during menopause (transdermal)

TABLE 2.1   *(Continued)*

Name	Drug entrapped or linker	Mechanism of action	Current stage of approval
		Metal-based nanoformulations	
Feridex®	Superparamagnetic iron oxide NPs coated with dextran. Iron oxide core 4.8–5.6 nm, hydrodynamic diameter 80–150 nm	MPS targeting: 80% taken up by the liver and up to 10% by spleen within minutes of administration. Tumor tissues do not take up these particles and thus retain their native signal intensity	FDA 1996 Liver/spleen lesion magnetic resonance imaging (MRI) (IV) Manufacturing discontinued in 2008
Feraheme™ (Ferumoxytol)	Superparamagnetic iron oxide NPs coated with dextran. Hydrodynamic diameter >50 nm	MPS targeting: iron released inside macrophages, subsequently enters into intracellular storage iron pool, or is transferred to plasma transferrin	FDA 2009 Treatment of iron deficiency anemia in adults with chronic kidney disease
NanoTherm®	Aminosilane-coated superparamagnetic iron oxide 15 nm NPs	Thermal ablation: Injecting iron oxide NPs exposed to alternating magnetic field causing the NPs to oscillate, generating heat directly within the tumor tissue	Europe 2013 Local ablation in glioblastoma, prostate, and pancreatic cancer (intratumoral)
Feridex®	Superparamagnetic iron oxide NPs coated with dextran. Iron oxide core 4.8–5.6 nm, hydrodynamic diameter 80–150 nm	MPS targeting: 80% taken up by the liver and up to 10% by spleen within minutes of administration. Tumor tissues do not take up these particles and thus retain their native signal intensity	FDA 1996 Liver/ spleen lesion MRI (IV) Manufacturing discontinued in 2008
Feraheme™ (Ferumoxytol)	Superparamagnetic iron oxide NPs coated with dextran. Hydrodynamic diameter >50 nm	MPS targeting: Iron released inside macrophages, subsequently enters into intracellular storage iron pool or is transferred to plasma transferring	FDA 2009 Treatment of iron deficiency anemia in adults with chronic kidney disease

TABLE 2.1 *(Continued)*

Name	Drug entrapped or linker	Mechanism of action	Current stage of approval
Gendicine®	Recombinant adenovirus expressing wildtype-p53 (rAd-p53)	"The adenoviral particle infects tumor target cells and delivers the adenovirus genome carrying the therapeutic p53 gene to the nucleus. The expressed p53 gene appears to exert its antitumor activities."	People's Republic of China 2003 Head and neck squamous cell carcinoma
Rexin-G®	Gene for dominant-negative mutant form of human cyclin G1, which blocks endogenous cyclin-G1 protein and thus stops cell cycle, inserted into retroviral core (replication-incompetent retrovirus) devoid of viral genes. About 100 nm particle	Targeted gene therapy: this retrovirus-derived particle targets specifically exposed collagen, which is a common histopathological property of metastatic tumor formation	

2.5   SYNTHESIS OF NPS

There are different methods for synthesizing of NPs (Mohd et al., 2014). The outline of important methods has been mentioned in Table 2.2.

TABLE 2.2   Methods for the Fabrication of Nanoparticles.

	Method	Process
1.	Gas condensation	This technique is used to synthesize nanocrystalline metals and alloys. In this technique, a metallic or inorganic material is vaporized using thermal evaporation sources such as a Joule heated refractory crucibles, electron beam evaporation devices, in an atmosphere of 1–50 mbar.
2.	Plasma arcing	Synthesis is done through plasma which is achieved by making gas conduct electricity by providing a potential difference across two electrodes.
3.	Vacuum deposition and vaporization	Elements alloys or compounds are vaporized and deposited in a vacuum. The vaporization source is the one that vaporizes materials by thermal processes. The process is carried out at pressure of less than 0.1 Pa (1 m Torr) and in vacuum levels of 10–0.1 MPa. The substrate temperature ranges from ambient to 500°C.
4.	Chemical vapor deposition (CVD) and Chemical vapor condensation (CVC)	It involves the formation of nanomaterials from the gas phase at elevated temperatures—usually onto a solid substrate or catalyst. Solid is deposited on a heated surface through a chemical reaction from the vapor or gas phase. CVC reaction requires activation energy to proceed which can be provided by several methods.
4.	Electrodeposition	Molecules and atoms are separated by vaporization and then allowed to deposit in a carefully controlled and orderly manner to form NPs.
5.	Solgel synthesis	It involves the transition of a system from a liquid "sol" (mostly colloidal) into a solid "gel" phase.
6.	Solvent evaporation	Added water-immiscible organic solvent is removed by heating and or under reduced pressure, helpful for matrix bounding.
7.	High energy milling/ball milling	Macrocrystalline structures are broken down into nanocrystalline structures, but the original integrity of the material is retained.
8.	Use of natural NPs	Involve modification and further use of natural materials.
9.	Denaturation	The macromolecule is dissolved in an aqueous solution and the drug is entrapped in an oil and emulsification carried out through the homogenizer.
10	Chemical precipitation	This involves the synthesis and studies of the nanomaterials in situ, that is, in the same liquid medium avoiding the physical changes and aggregation of tiny crystallites.

2.6 ANALYSIS AND CHARACTERIZATION OF NP FORMULATIONS

From a technical as well as regulatory perspective identifying the appropriate analytical tests to fully characterize NPs, whether physical, chemical, or biological is one of the most challenging aspects of nanomedicine development (Ranjitand Baquee, 2013). Complex nature of nanomedicines, as compared to standard pharmaceuticals, requires a more sophisticated level of testing to fully characterize a nanoproduct. Each component of a nanomedicine serves a specific motive or function. Therefore, what is important is the imperative necessity of determining the quantity of each component. Besides it is also crucial that the relationships and interactions between these components with particular reference to both stoichiometry and their spatial orientation critical are also evaluated.

Several orthogonal characterization techniques are critical to ensuring that the NPs have all the desired properties for the purposive therapeutic use. There can be no compromise about the standard analytical tests such as quantification of active (API) and inactive ingredients, impurities, and so forth in pharmaceutical products. But still, characterization of physicochemical properties of NPs is resorted to purposely with the aid of various additional techniques.

These tests encompass a broad range of methods. Atomic force microscopy, transmission electron microscopy (TEM), and scanning electron microscopy are some of the methods which can be clubbed under the heading visualization of NPs by microscopy. Other methods include analytical ultracentrifugation, capillary electrophoresis, and field flow fractionation; measurement of particle size and size distribution with light scattering (static and dynamic), analysis of surface charge, or zeta potential; and examination of surface chemistry by X-ray photoelectron spectroscopy or Fourier transform infrared spectroscopy (Ballauff et al., 2007; Singh et al., 2009).

X-ray diffraction and differential scanning calorimetry can assess the crystalline state of drugs encapsulated in the NPs (Bunjes et al., 2007; Dorofeev et al., 2012). In the meantime, more original testing methods are in the process of being developed and applied to the analysis of NPs. To ensure their robust performance, it is paramount that they can deliver under accelerated conditions such as higher temperature. They need to be tested for all vital parameters including overall stability while being in the solid state, in suspension, and in biological medium. But these tests may not be sufficient to functionally segregate an "active" formulation from the one that is "inactive" or "less active." For instance, consider the case of a NP carrying a payload of an active drug and having at its surface moieties that allow stealth features, such as polyethylene glycol (PEG) and moieties that may allow targeting including peptides, nucleic acids, proteins, or antibody fragments that bind to certain receptors.

To analyze the individual component, quantitative techniques are essential, but they are not infallible on standalone basis. They can miss out on critical guidance about the diffusion of these moieties on the surface of the particles or whether indeed these moieties are buried or they exist on the surface. The spatial distribution of these moieties is critical to the intended behavior of the particular NP. To determine these aspects, a different chain of tests may be required. Such tests are likely to be nonconventional for use in pharmaceutical and may include surface analysis, as well as biological function tests to assure that the manufacturing process produces NPs that are active in cellular uptake, transcytosis, or binding to appropriate biological materials. These "structure-function" tests are critical at least up to the point where one can validate a highly reproducible manufacturing process. Bioassays to confirm activity of the product are often employed for testing biological drugs. Several proposed nanotherapeutics have complex components. These include proteins or nucleic acids forming an integral part of the nanomedicine. These may be sensitive to the conditions regulating manufacturing process. In some cases, manufacturing processes can bring about the changes in composition. These components may not be the "active" pharmaceutical ingredient in the nanomedicine. However, their presence may play a role in targeting specific cells/biological pathways or distribution of the active ingredient in the body. These components cannot be considered as inactive substances. They play an important role in the safety and efficacy of the product. Therefore, they need to be characterized after appropriate analytical investigation. Therefore, it is imperative that appropriate in-process tests during early manufacturing activity besides carrying out tests to assure that the product is consistently produced and controlled according to the designated standards are put in place. It will generate important information in the development of a robust manufacturing process.

These are the tests that are designed to give an early lead into how varying process conditions can affect the nanomedicine composition at intermediate stages of the process and can be conducted with relatively quick turnaround time.

2.7 NPS IN DRUG DELIVERY

As mentioned above, NPs for drug delivery can be prepared from a variety of materials such as proteins, polysaccharides, synthetic polymers, and the others mentioned above (Andreas et al., 2015; Hare et al., 2016; Jain et al., 2015; Leung, 2006; Martinho et al., 2011; Park, 2013; Qin et al., 2017;Tiwari et al., 2012;).

The selection of material depends on many factors such as the size of NPs, inherent properties of the drug, surface characteristics such as charge and permeability, the degree of biodegradability, and so forth. NPs containing encapsulated, dispersed, absorbed, or conjugated drugs have unique characteristics that can lead to enhanced performance in a variety of dosage forms. When formulated correctly, drug particles are resistant to settling and can have higher saturation solubility, rapid dissolution, and enhanced adhesion to biological surfaces thereby providing rapid onset of therapeutic action and improved bioavailability. Variety of NPs has been reported in the literature that can be used for drug delivery.

2.7.1 NOBLE METAL NPS DECORATED WITH POLYMERS

Noble metal NPs have been widely used for drug delivery. NPs, particularly noble metal NPs, normally use biological pathways to achieve drug delivery to cellular and intracellular targets, including transport through the blood–brain barrier (BBB). Owing to the smaller size, they can easily enter host cells and circulate through the body. They are suitable for site-specific delivery vehicles to carry large doses of chemotherapeutic agents/drugs/therapeutic genes to the target site. Surface modification has been used for the decoration of Ag, Au, Pt, and other metal NPs. The physical and chemical properties of the NPs influenced their pharmacokinetic behavior, which ultimately determined their affected site accumulation capacity (Rai et al., 2015; Sreeprasad and Pradeep, 2013).

2.7.2 GOLD NANOPARTICLES

Gold is basically inert and nontoxic. Monodisperse nanoparticles with core size ranging from 1 to 150 nm can be created easily. They have wide-ranging advantages in the matter of modification, functionalization, and bioconjugation (Sperling and Park, 2010; Yoo et al., 2011). Laced with biodegradable polymers and, protein, in particular, may be of help in controlling colloidal stability, assembly, and delivery of nanoparticles to the targeted sites. Enough information is available in literature to educate about drug delivery with Au NPs (Du et al., 2013; Rai et al., 2015; Sperling and Park, 2010; Sreeprasad and Pradeep, 2013; Yoo et al., 2011).

Au NPs are known to possess unique chemical and physical properties which enable them to upload and transport pharmaceuticals to the affected site. Chemisorption, electrostatic attraction, or hydrophobic interaction is often provided by a head group of the ligand molecule attached to the NP surfaces by some attractive interactions.

Thiol group is accepted to have the advantage of highest affinity to noble metal surfaces and the gold in particular (approximately 200 kJ/mol) (Sperling and Park, 2010). Therefore, it can be inferred from forging that generally thiol-containing molecules are firmly attached to the Au NP surface through covalent-like bonds and are favored for drug delivery. The effect of PEGylated (10–100 nm) gold NPs and surface chemistry on the passive targeting of tumors in vivo has been investigated scientifically by Perrault et al. (2009). They observed that the movement of NPs inside tumor is contingent upon the overall size of the NP, that is, while smaller nanoparticles rapidly disperse throughout the tumor matrix, the larger ones appear to stay near the vasculature. Therefore, the most important issue is to optimize through design the ability of the drug-carrying medium that can penetrate and pass through the tumor interstitial space without causing rupture or displacement or can remain at the perivascular space. Solvent evaporation method has been used to prepare fluorouracil (5-Fu), containing polylactic acid-co-ethyl cellulose nanocapsules, in the presence and absence of AuNPs. The distinctive nature of drug-entrapped nanocapsules was explained and the study was conducted on the controlled release of the anticancer drug-entrapped nanocapsules. From the study, it was found that the drug released by nanocapsules containing AuNPs was regulated and slower in comparison to 5-Fu incorporated polymeric nanocapsules without AuNPs (Sathishkumar, 2012). Selective delivery of folic acid FA-NPs by folate receptor (FR)-positive KB cells had been established by Andres and coworkers (Dixit et al., 2006). The study reveals that folic acid was conjugated onto gold nanoparticles (d = 10 nm) with PEG spacer and the particles were successfully embrace by KB cell. However, little cellular uptake was detected in WI cells which did not recognize the FR.

2.7.3 SILVER NANOPARTICLES

Ag is commonly used in biomedical applications and is one of the most commonly known metals for its antimicrobial properties (Benyettou et al., 2015). To synthesize silver NPs and load anticancer agents, (paclitaxel and doxorubicin) self-assembly of multilayer polymeric micelles of poly(ethylene oxide)-b-poly(n-butyl acrylate)-b-poly(acrylic acid) (PEO113-b-PnBA235-b-PAA14) triblockter polymer were exploited as template. Then the hydrophobic PnBA micellar cores were loaded with anticancerous drug curcumin to assess the combined effect of two anticancer agents spontaneously released from the micelles, on the vitality of acute myeloid leukemia (HL-60), its multidrug-resistant subline HL-60/DOX and human urinary bladder carcinoma (EJ) cells. In vitro experiments revealed enhanced efficacy in regard to the cytotoxic activity of silver NPs and curcumin (Petrov et al., 2016). AgNPs decorated with poly(glycidyl methacrylate) were prepared and evaluated for their antibacterial activity (Anandhakumar et al., 2012). The reported experiments elucidate the mechanism of formation of AgNP-coated polymer spheres. Besides it also confirmed the size- and concentration-dependent activity against *Escherichia coli*.

Yet another study refers to the development of AgNPs embedded NPs for the controlled release of drug from the layer-by-layer assembly of polyelectrolytes (Anandhakumar et al., 2012). It was engineered by the alternate assembly of polyallylamine hydrochloride and dextran sulphate on silica template. Antimicrobial potential of metallic silver and silver-based compounds along with its mechanism of action has also been studied. The effect of size and shape of silver-based NPs on their antimicrobial potential was also reviewed. The bactericidal effect of silver NPs against multidrug-resistant bacteria, for instance, *Pseudomonas aeruginosa*, ampicillin-resistant *E coli*, and erythromycin-resistant *Streptococcus pyogenes* have also been examined by Lara et al., 2010.

A panel of drug-resistant and drug-susceptible bacteria with MBC and MIC concentrations in the range of 30–100 mm, respectively can be inactivated with the help of silver NPs. This occurrence has been established by Luciferase assays through Kiby–Bauer tests. The tests revealed that bacteriostatic mechanisms of silver NPs arrested the cell wall. It also proved the protein synthesis and nucleic acid synthesis Co-dependent effects of silver NPs with antibiotics and other agents. In other words, silver NPs in amalgam with antibiotics such as penicillin G, amoxicillin, erythromycin, and vancomycin resulted in improved antimicrobial effects against different Gram-negative and Gram-positive bacteria (Fayaz et al., 2010). In a study by Panax ginseng, meyer fresh leaves have been used for (P.gAgNPs) (Castro-Aceituno et al., 2016). The study revealed that P.gAgNPs decreased the cell viability and induced reactive oxygen species production in A549, Michigan Cancer Foundation-7 (MCF7), and human hepatoma cell line 2. P.gAgNPs also decreased the cell migration and reduced mRNA-induced phosphorylation of epidermal growth factor receptors (EGFR) of A54. In addition, P.gAgNPs induces cell apoptosis and upregulated the p38 mitogen-activated protein kinase/p53-mitochondria caspase-3 pathway in A549 cells. It was discovered that in response to the changes of gene expression in inflammatory response, oxidative stress, and amyloid-β (Aβ) degradation, 3–5 nm AgNPs could enter in mouse neural cells to induce pro-inflammatory cytokine secretion and increase Aβ deposition. The study indicated that AgNPs-induced neuroinflammatory response and Aβ deposition. It might evolve to unveil the progress of neurodegenerative disorders.

2.7.4 PLATINUM NPS

Pt NPs are one among noble metals that have antioxidant properties. This called upon the scientific community to conduct extensive research on them which ultimately facilitated discovery of possible applications in different areas such as catalysis, medicine, biology, and preparation of nanocomposite materials with distinctive characteristics. PtNPs demonstrate dual functionality and when used in soluble forms they generate strand breaking in DNA.

These characteristics of PtNPs have made them helpful for effective anticancer drug delivery except that when circulating in bloodstream they have been found to be toxic to normal cells. The later disadvantage usually discourages the use of Pt NPs in drug delivery. Teow and Valiyaveettil (2010) synthesized PVP and folic acid-capped platinum NPs to assess their bioactivities in commercial cell lines. The capping agents while influencing the uptake and toxicity of NPs by live cells also had the effect on their stability, shape, and size distribution. After 24 h of incubation with Pt-FA, the viability of HeLa, MCF7, and IMR90 dropped significantly to 65, 44, and 43%, respectively at 25 μg/ml. The success rates were > 90% in the cases for Pt-PVP. Increased accumulation of Pt-FA in comparison to Pt-PVP revealed by dark field optical microscope images also indicated the effectiveness of folic acid in receptor-mediated endocytosis. Pt-NPs treatment caused an increase in proportions of MCF7 cells undergoing apoptosis, particularly in the case of Pt-FA. For IMR90 cells, Pt-PVP treatment caused cells to undergo apoptosis, while Pt-FA caused necrosis to take place. This suggests that different capping agents caused differing cellular component targeting. Toxicity is inherent to platinum NPs and their trait provides them with an additional advantage because they do not need to attach anticancer drug for destroying cancer cells.

2.7.5   HYBRID NPS

Both inorganic and organic components add up to form hybrid NPs that retain the beneficial features of both inorganic and organic nanomaterials. Besides they hold exclusive advantages over the preceding two types. They have the ability to combine a multitude of organic and inorganic components in a modular fashion thus allowing systematic tuning of the properties of the resultant hybrid nanomaterial. Silica-based nanomaterials and nanoscale metal-organic frameworks are the newly explored two major classes of hybrid nanomaterials which have therapeutic applications. Mesoporous silica NPs (MSNs) are popularly used in controlled and targeted drug delivery system. They are less toxic besides having higher drug loading capacity. Better biocompatibility is another trait of silica which makes it a preferred choice when compared with other metal oxides of the likes of titanium and iron oxides. Surface functionalization (hydrophobic/hydrophilic or positive/negative charged) is crucial for loading correct type of drug molecules. Further, functionalization through chemical links with other materials such as stimuli-responsive, luminescent or capping materials, MSNs with smart and multifunctional properties can be prepared. Researchers have attempted to conjugate a range of targeting ligands including folic acid, transferrin, aptamer, and herceptin to $mSiO_2$ NPs for in vitro or in vivo tumor cell targeting but have met with limited success.

Trewyn et al. (2007) were the first who developed MSN-based drug delivery systems. They capped the NP pores by a stimuli-responsive system that permitted a controlled cargo release. Figure 2.2 is a case in point to illustrate photolabile linker used to conjugate gold NPs to MSN materials. Exposure to ultraviolet (UV) irradiation will facilitate the release of Au NP pore caps by the hybrid system. This system was used to deliver paclitaxel to human liver and fibroblast cells. Marked reduction in cell viability was observed after UV irradiation. This approach was adopted for delivering other drug molecules and neurotransmitters, such as vancomycin and adenosine triphosphate (Giri et al., 2005; Lai et al., 2003; Vivero-Escoto et al., 2009).

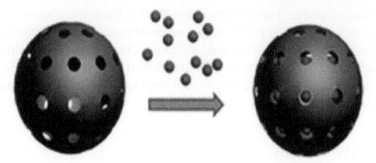

FIGURE 2.2   Release of cargo from mesoporous silica nanoparticles (NPs) material.

2.7.6   POLYMERIC NPS

Biodegradable polymeric NPs capped with hydrophilic, nontoxic, blood compatible polymers such as polyvinyl alcohol, PEG, monomethoxy PEG, polysorbate, and vitamin E. D-alpha tocopheryl PEG 1000 succinate have acquired significant recognition for being used as possible drug delivery devices (Pridgen et al., 2007; Zhang et al., 2014). They find variety of applications in the controlled release of drugs, targeting particular organs/tissues, as carriers of DNA in gene therapy, and in their ability to deliver proteins, peptides, and genes through an oral route of administration. These polymers circulate plasma proteins in the blood and cast a cloud of hydrophilic chains on the outer surface of NPs which prevents the process of phagocytosis and opsonization. Thus, the surface-modified amphiphilic drug delivery systems containing a drug within its hydrophobic core surrounded by a hydrophilic outer shell. Ultimately, they have increased systemic circulation time and persistence in the blood.

To functionalize drug-loaded polymeric NPs (PNPs)-specific ligands including small molecules, peptide, proteins, antibodies, engineered antibody fragments, and aptamers are used to festoon NPs. Particular receptor (membrane protein domains) expressed on cancer cells and/or tumor endothelial cells (vasculature of a tumor) is the picky target of these ligands. Most cancer cells usually have FRs, the (EGFRs), and the transferrin receptors (whereas vascular cell adhesion molecule-1, integrins, aminopeptidase N/CD13, vascular EGFR, and membrane type 1 matrix metalloproteinase are expressed on the vasculature of tumor cells.

Besides being excellent biocompatible the other great features of this system are its nontoxicity and its effectiveness in circumventing the resistance associated with the multiple drug dosing. Without affecting healthy cells, PNPs-based targeted drug delivery systems, by active and passive tumor targeting, has the additional advantage of boosting the tissue-specific distribution of anticancer substances. Defective, leaky structure of tumor vessels and the impaired lymphatic system of tumor cells allow PNPs easy access to tumor tissues during passive targeting. In the case of active tumor targeting, a ternary structure which is composed of a ligand as a targeting moiety, an active chemotherapeutic drug, and a polymer as a carrier enters the tumor cell or tumor vasculature through receptor-mediated endocytosis.

In the preclinical animal trials reportedly polymeric NPs and micelles also enhance insulin oral bioavailability (Alai et al., 2015). For the delivery of ocular drug Econazole nitrate with Chitosan NPs, sulfobutyl ether-cyclodextrin as polyanionic cross-linker has been used.

2.7.7 DENDRIMERS

Dendrimers (Fig. 2.3) have defined polymeric architectures (Madaan et al., 2014; Nanjwade et al., 2009). They have been studied to be ideal delivery vehicles. They are known for their versatility in delivering drugs and have high functional properties resembling biomolecules. They are biocompatible, highly water soluble, polyvalence with precise molecular weight. A dendrimer-based prodrug has been developed for paclitaxel (P-gp efflux substrate) that has focused on the enhancement of permeability and transportation of drug across cellular barriers. The highly functional lauryl-modified G3 PAMAM dendrimer-paclitaxel conjugates demonstrate good stability under physiological conditions. They exhibit 12-fold better permeability across Caco-2 cells and porcine brain endothelial monolayers. However, nanometric cellular components such as proteins, cell organelles, and cell membrane are not immune from interacting with dendrimers. Dendrimers with cationic surface groups tend to interact with the lipid bilayer. It has the effect of increasing their permeability and lessens the integrity of the biological membrane. Leaking cytoplasmic proteins, caused by the interaction of dendrimers with cell membranes such as luciferase and lactate dehydrogenase, provide them passage to move. This mechanism causes disruption and cell lysis. Therefore, they have a limited use in biological systems because of the toxicity issues associated with them.

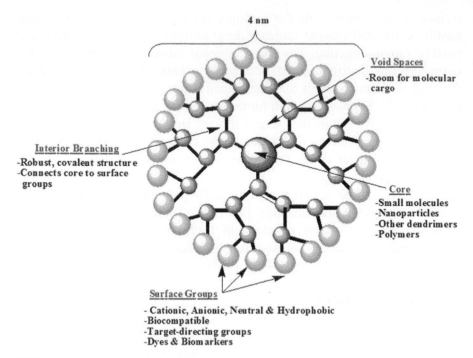

FIGURE 2.3   Three-dimensional, immensely branched, well-organized nanoscopic macromolecules: dendrimers.

2.8   DRUG RELEASE FROM NPS

Following the release of drug from NPs after encapsulation of drug into it, the kinetics of drug release from NPs is a crucial factor and it is this property that holds a central position for the quality control of NP formulations. Diffusion or erosion from the core across the coated membrane or matrix controls the release of drug from NPs usually coated with a polymer/peptide/protein. Membrane coating acts as a barrier for release of the drug. Therefore, solubility and diffusivity of the drug in polymer membrane is the decisive element in drug release. For establishing in vitro–in vivo correlations, dependable choice of in vitro release kinetics is also a prerequisite. It in turn characterizes the performance of a formulation in vivo. Ionic interaction between the drug and addition of auxiliary ingredients can also affect the release rate of the drug. Less water-soluble complex formation as the auxiliary ingredients interact with the drug, it decelerates the drug release with minimal burst release effect. (Chen et al., 2009).

Polymer biodegradation is one of the two principal considerations in the development of a successful nanoparticulate system—the other being the drug release. Usually, drug release rate depends on (1) solubility of drug, (2) desorption of the surface-bound/adsorbed drug, (3) drug diffusion through the NP matrix, (4) NP matrix erosion/degradation, and (5) combination of erosion/diffusion process (Mohanraj and Chen, 2006; Valo et al., 2013). Hence, the release process is governed by diffusion, biodegradation, and solubility of the matrix material. Different techniques have been used to determine the release rate of the drug from NPs which include sample-and-separate method but it suffers from a number of drawbacks which reveal erroneous results, for example, a premature release caused by the centrifugation which was generally used to separate the NPs resulting in the release of the free drug. When a sample is taken for measuring, there can also be occasion when the release of the drug may occur during the runtime of the separation process (McCarron et al., 2009). Besides, we may require special release medium with surfactant addition. It will, however, depend on the properties of the drug. But it can be tricky for high-pressure liquid chromatography (HPLC) or UV-visible spectrometer. Pre-treating, for instance, solid-phase extraction is necessary in this case (Kamberi and Tran, 2012). Solid-phase extraction coupled with HPLC or mass spectrometry, assures sensitive analysis of the released drug (Monteiro-Riviere et al., 2005), particularly in the case of samples obtained from dissolution media containing proteins or serum. Most importantly, when the drug formulation persists for a few days or even weeks of release, tedious and repetitive measurements during the long-lasting process are inevitable.

Dialysis membranes are usually employed to separate NPs from the released drug. A range of systems and setups are available which are based on this model such as rotating dialysis cell. The other methods are Franz diffusion (especially for skin permeability tests), the dialysis bag method, and reverse dialysis.

Dynamic dialysis is, however, one of the most commonly used techniques for the determination of release kinetics from NPs. Only recently, Zambito et al. (2012) revealed that most of the reports that appeared in literature tell about the use of dynamic dialysis to measure the release kinetics. It is popular over other methods, for example, ultracentrifugation and ultrafiltration. It eliminates additional steps required for separating NPs from the free drug at various time points during the kinetic study. The application of external pressure when applied for separation in other methods can disturb the equilibrium, and incomplete separation can lead to significant measurement errors.

Albeit dynamic dialysis is generally considered as a simple first-order process, but the diffusion from the NPs followed by diffusion across the dialysis membrane leads to the appearance of drug in the "sink" receiver compartment. Either of the two, that is, the disappearance of the drug from the donor compartment containing the NPs or appearance in the receiver compartment can, however, be measured by means of an experiment. The rate release is given by the net result of drug that gets transported across the barriers. Therefore, the rate constant obtained may not necessarily reflect the rate at which the drug gets released from the NPs.

It necessitates the study of both (i) the properties of the dialysis membrane and (ii) the driving force involved in transferring the drug across the membrane barrier. It necessitates the study of both, that is, the driving force for drug transport across that membrane barrier and the properties of the dialysis membrane. Accordingly, what is required is a careful analysis of the risks involved in interpreting apparent release data about the assessment of the reliability of rate constants determined by dynamic dialysis (Dash et al., 2010).

The drawbacks in dynamic drug release technique have been rectified by mathematical models to obtain the actual rate constant for NP release and to estimate its level of certainty. Noyes and Whitney in 1897 offered a fundamental principle for the evaluation of the kinetics of drug release. The same is given as Noyes-Whitney Rule (Noyes and Whittney, 1897).

$$dM/dt = KS\,(Cs - Ct)$$

where M = the mass of the transferred drug with respect to time, t, by dissolution from the NP of instantaneous surface S

$(Cs - Ct)$ = concentration driving force
Ct = concentration at time t
Cs = equilibrium solubility

The rate of dissolution dM/dt is the amount dissolved per unit area per unit time and most solids can be expressed in units of $\mathrm{gcm^{-2\,s^{-1}}}$.

2.8.1   RELEASE KINETIC MODELLING

A number of kinetic models which describe the overall release of drug from the dosage form exist. However, the release of behavior shows wide variations in vitro and in vivo developing tools. But what is important is that it facilitates the development of a product.

The most commonly used method based on mathematical models to describe release profiles include zero order, first order, Higuchi, Hixson–Crowell, Korsmeyer–Peppas, Baker–Lonsdale, Weibull, Hopfenberg, Gompertz, and regression models.

2.8.1.1 ZERO-ORDER MODEL

Drug dissolution from the dosage forms that do not disaggregate and release the drug slowly can be resented by the equation

$$Q_0 - Q_t = k_0 t \tag{2.1}$$

Q_t = amount of drug dissolved in time t
Q_0 = initial amount of drug in solution
K_0 = zero-order release constant

The data obtained from in vitro drug release is plotted as the cumulative amount of drug release against time.

2.8.1.2 FIRST-ORDER MODEL

This model has also been used to describe absorption and/or elimination of drugs although it's difficult to conceptualize the mechanism on theoretical basis. The release of drug can be expressed as

$$\frac{dC}{dt} = -K_c \tag{2.2}$$

where K is the first-order rate constant expressed in time^{-1}
The above equation can also be expressed as

$$\log C = \log C_0 - \frac{Kt}{2.303} \tag{2.3}$$

where C_0 = initial concentration of the drug

K = first-order release constant
t = time

The data obtained are plotted as log cumulative percentage of remaining drug versus time which yields a straight line with slope $-K/2.303$.

2.8.1.3   HIGUCHI MODEL

The attempt to describe the drug release from matrix system was proposed by Higuchi (Higuchi, 1963). This model is based on the hypothesis (i) initial drug concentration in the matrix is much higher than drug solubility (ii) drug diffusion takes place only in one dimension (iii) drug particles are much smaller than system thickness (iv) matrix swelling and dissolution are negligible (v) drug diffusivity is constant (v) perfect sink conditions are always obtained in the release environment.

$$f_t = Q = A\sqrt{D(2C - C_s)C_s t}$$ (2.4)

Q = amount of drug released in time t per unit area A, C is the initial concentration
C_s = drug solubility in the matrix media
D = diffusion coefficient

However, the relation is not valid when the total depletion of the drug in the therapeutic system is achieved.

To study the dissolution of a planer heterogeneous matrix system, where the drug concentration in the matrix is lower than its solubility and release occurs through pores in the matrix, the expression can be written as

$$f_t = Q = \sqrt{\frac{D\delta}{\tau}}(2C - \delta C_s)C_s t$$ (2.5)

D = diffusion co-efficient
δ = = porosity of the matrix
τ = tortuisity of the matrix

Q, A, C_s, t have their usual meaning as mentioned above.
However, it is possible to simplify the Higuchi model

$$f_t = Q = K_H * t^{\frac{1}{2}}$$

where K_H is the Higuchi dissolution constant. The obtained data are plotted as cumulative percentage drug versus square root of time. The relationship can be used to describe drug dissolution from several types of modified release pharmaceutical dosage forms, as in case of some transdermal systems and matrix tablets with water-soluble drugs.

2.8.1.4 HIXSON–CROWELL MODEL

Hixson and Crowell considered that the particles regular area is proportional to the square root of its volume equation. The equation describes release from system where there is change in surface area and diameter of particles or tablets.

$$W_0^{13} - W_t^{1/3} = \kappa t$$

W_0 = initial amount of drug in the pharmaceutical dosage form
$W\tau$ = the remaining amount of drug in the pharmaceutical dosage form at time t
κ = constant incorporating the surface volume relation.

The expression implies to pharmaceutical dosage forms such as tablets where dissolution occurs in planes that are parallel to the drug surface if the tablets dimension diminish proportionally, in such a manner that the initial geometric firm keeps constant all the time.

2.8.1.5 KORSMEYER–PEPPAS MODEL

Korsmeyer et al. (1983) derived a simple relationship which described drug release from a polymeric system equation. To elucidate the mechanism of drug release, drug release data were fitted in Korsmeyer–Peppas model (Korsmeyer et al., 1983)

$$\frac{M_t}{M_\infty} = Kt^n$$

where $\frac{M_t}{M_\infty}$ is the fraction of drug released at time t, K is the release constant and is the release exponent. The n value is used to characterize different release for cylindrical shaped matrices.

2.8.1.6 BAKER–LONSDALE MODEL

This model was developed by Baker and Lonsdale (1974) from the Higuchi model and described the drug release from spherical matrices according to the equation

$$f_1 = 3/2\left[1-\left(1-\frac{M_t}{M_\infty}\right)^{2/3}\right]\frac{M_t}{M_\infty} = k_t$$

where the release constant k corresponds to the slope.

To study the release kinetics, data obtained from in vitro release studies were plotted as $\dfrac{\left[d\left(\dfrac{M_t}{M_\infty}\right)\right]}{dt}$ / with respect to the root of time inverse.

2.8.1.7 HOPFENBERG MODEL

Hopfenberg developed a mathematical model to correlate the drug release from surface eroding polymers so long as the surface area remains constant during the degradation process. The cumulative fraction of drug released at time t was described as

$$M_t/M_\infty = 1-[1-{k_0 t}/{C_L}\,a]^n$$

k_0 = zero-order rate constant describing the polymer degradation (surface erosion) process

C_L = initial drug loading throughout the system

a = systems half thickness (i.e., radius of sphere or cylinder)

n = exponent varying with geometry

2.8.1.8 REGRESSION MODEL

Several types of regression models are used to optimize the formulation for in vitro release study.

2.8.2 LINEAR REGRESSION METHOD

Linear regression is a method for determining the parameters of a linear system. The empirical model relating the response variable to the independent variables are described by the following equation:

$$Y = \beta_0 + \beta_1 X_1 + \beta_2 X_2$$

where Y represents the response, X_1 and X_2 represent the two independent variables. The parameter β_0 signifies the intercept of the plane. β_1 and β_2, called partial regression coefficients, where β_1 measures the expected change in Y_1, the response, per unit change in X_1 when X_2 kept constant and vice versa for β_2. The equation can be rewritten in general form as

$$Y = \beta_0 + \beta_1 X_1 + \beta_2 X_2 + \beta_k X_k \qquad (2.6)$$

The model is a multiple linear regression model with "k" regression variables. The model describes a hyperplane in the k-dimensional space. Further complex model (Eq. 2.6) are often analyzed by multiple linear regression technique by adding interaction terms to the first-order linear model

$$Y = \beta_0 + \beta_1 X_1 + \beta_2 X_2 + \beta_{12} X_2$$

Quadratic model or second-order regression model

$$Y = \beta_0 + \beta_1 X_1 + \beta_2 X_2 + \beta_{11} X_1^2 + \beta_{22} X_2^2 + \beta_{12} X_1 X_2$$

If we put $X_{21} = X_3$ and $X_{22} = X_4, X_1 X_2 = X_5$ and $\beta_{11} = \beta_3, \beta_{22} = \beta_4, \beta_{12} = \beta_5$

$$Y = \beta_0 + \beta_1 X_1 + \beta_2 X_2 + \beta_3 X_3 + \beta_4 X_4 + \beta_5 X_5$$

Utilizing the above factors the equation is reduced to linear model. Any model is linear if $\beta_{\text{co-efficients}}$ are linear.

2.9 NPS IN TREATMENT OF DISEASES

Therapeutic drugs often lead to adverse effects and result in systemic toxicity. NPs hold the promise of being effective and improving the therapeutic efficacy of the drug. A variety of NPs have been fabricated which can be attached with targeting moiety. They can be used as theranostic agents because they can deliver the drug. They have imaging capability and can also be used in combination. These NPs have been explored for application in the treatment diseases like Alzheimer's, cancer, kidney diseases, and so forth besides for their use in identifying various pathological conditions.

2.9.1 CANCER

United States of America is still grappling with cancer; a disease identified with the dragon of unrestrained growth and spread of abnormal cells. It is still the second most general reason of death in the UnitedStates. The American Cancer Society puts the estimate of deaths it causes at 1500 deaths per day.

Presently cancer can be treated by surgery, radiation, hormone therapy and chemotherapy. Solid tumors can be treated with radiation therapy. These are conventional therapies. Though they improve survival rate yet these therapies have their shortcomings. For instance chemotherapy distributes cancer therapeutic agents in the human body loosely. It affects both cancerous and normal cells. While it limits the therapeutic dose for cancer cells, it also provides excessive toxicities to normal cells, tissues, and organs adversely affecting tissues and organs causing hair loss, weakness, and renders organs dysfunctional. All this leads to patient's poor quality of life. First clinical trial for NPs as anticancer drug delivery systems occurred in the mid-1980 s. However, it was only in 1995 when the first NP, for example, liposomal with encapsulated doxorubicin attained commercialization. With the passage of time numerous new NPs for cancer drug delivery have been approved or are under at various stages of development. They have numerous advantages which include enhanced solubility of hydrophobic drugs, prolonging circulation time, minimizing non-specific uptake, preventing undesirable off-target and side effects, improving intracellular penetration, and allowing for specific cancer-targeting. While securing normal cells, from toxicity, active targeting or passive targeting, NPs have been revealed to enhance the intracellular concentration of drugs/genes in cancer cells. The Figure 2.4 explains the phenomenon. In addition active targeting cannot be separated from the passive since the former occurs only after passive accumulation in tumors. Maeda and Matsumura (1989) first described how in passive targeting macromolecules including NPs primarily accumulate in the neoplastic tissues. This happens because of EPR behaviour of NPs (Chen et al., 1994; Matsumura et al., 1986).

The EPR is influenced by the nanometer size range of the NPs besides the two fundamental characteristics of the neoplastic tissues. These are the leaky vasculature and impaired lymphatic drainage. However, active targeting is based on biorecognition of molecules that have been attached to the surface of the nanovectors to target specific markers that are over expressed by the neoplastic cells.

In addition, the targeted NPs can also be designed as either pH-sensitive or temperature-sensitive carriers. The pH-sensitive drug delivery system can deliver and release drugs within the more acidic microenvironment of the cancer cells and/or components within cancer cells. The acidic microenvironment of tumor cells is caused by the accumulation of Lactic acid rapidly growing tumor cells owing to their elevated rates of glucose uptake but reduced rates of oxidative phosphorylation. NPs have been formulated for pH-dependent drug release by using polymers.

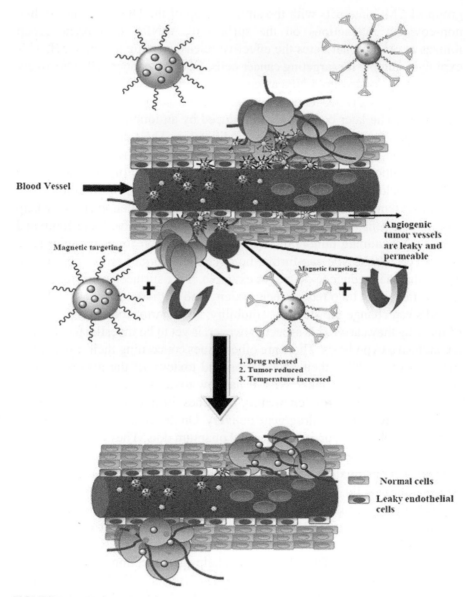

Blood Vessel

Magnetic targeting

Angiogenic
tumor vessels
are leaky and
permeable

Magnetic targeting

1. Drug released
2. Tumor reduced
3. Temperature increased

Normal cells

Leaky endothelial
cells

FIGURE 2.4 Various nano-based delivery methods active targeting, passive targeting, and magnetic targeting.

Capped with carboxymethyl chitosan (CMC), paralyzed or incapacitated on gold NPs (AuNPs), have been reported by Singh et al doxorubicin (DOX) for effective delivery to cancer cells. It has been reported that the carboxylic group of CMC interacts with the amino group of the DOX forming stable, non-covalent interactions on the surface of AuNPs. Carboxylic group ionizes at pH and facilitates the effective release of drug at acidic pH. This exploit is suitable for targeting cancer cells. Cervical cancer cells effectively sop up DOX-loaded gold NPs as compared to free DOX. Studies revealed that the intake of DOX-loaded gold NPs get further increased under acidic conditions. The later condition was induced by an ionophore, that is, nigericin. Nigericin causes intracellular acidification. The findings suggest that drug releasing properties of DOX-loaded gold NPs with pH is a pioneering nanotheraputic approach found to solve the problem of drug resistance. The magnetic field, ultrasound waves and the like can provide information about changes in temperature locally in the tumor region. This phenomenon can help temperature sensitive system carry and release drug and facilitate combined therapy. NPs aiming tumors through cancer-specific features/moieties have also revealed that NPs can further be doctored or organized to perform an action that may reduce toxicity. For example, if the surface chemistry of NPs is modified it can help reduce their toxicity and immunotoxicity.

NPs can change the stability, solubility, and pharmacokinetic properties of the drug they carry. This aspect, however, is yet to be investigated to reach a conclusive hypothesis. There are other issues concerning their use, namely the shelf life of NPs, their aggregation and toxicity of the materials used. As (PLGA) is used to make NPs. It has low toxicity but degrades quickly. This trait prevents it from circulating in tissues for a long enough period of time to ensure sustained drug/gene delivery. On the contrary, there are other materials such as carbon nanotubes and quantum dots. They are durable and can persist in the body for weeks, months, or even years. This trait of theirs makes them potentially toxic and forbids their repeated use in the treatment. Silicon/silica (solid, porous, and hollow silicon NPs), to name one; have been developed to make targeted NPs. However, their use to deliver the drug to cancer patients was yet to be established. The anticancer drug remains passive till the same enters the tumor cell to which it is targeted to by the monoclonal antibody. This phenomenon helps in protecting the normal cells intact while exclusively delivering the drug to cancer cells. But still, enough research remains to be done before a perfected technology can hit the market.

The anticancer drug remains passive till the same enters the tumor cell to which it is targeted to by the monoclonal antibody. This phenomenon helps in protecting the normal cells intact while exclusively delivering the drug to cancer cells. But still enough research remains to be done before a perfected technology can hit the market.

2.9.2 ALZHEIMER'S DISEASE

Alzheimer's disease (AD) is a devastating neurodegenerative disorder. It is the most common type of dementia prevalent amongst elderly of over 65 years of age. This neuropathological condition is characterized by a progressive loss of cognitive function and presents two established pathophysiological hallmarks in the brain. Pathophysiological conditions include extracellular accumulations which are mainly composed of Aβ peptide, a condition referred to as senile plaques, besides intracellular neurofibrillar tangles of hyperphosphorylated τ protein.

For proper functioning of neuronal circuits and synaptic transmission, it is important to maintain the chemical composition of the neuronal "milieu." The BBB (BBB) which exists at the level of the endothelial cells of the cerebral capillaries is a formidable gatekeeper in the body towards exogenous substances helps in maintaining the chemical composition of the neuronal "milieu" Fig. 2.5). BBB essentially causes the most important contact between the blood and the brain. BBB is, therefore, the linchpin in the development of new drugs and biologics for the central nervous system (CNS). Most small molecules and the pharmaceuticals, in general, do not negotiate BBB. During the several passed years focused research has been conducted on this crucial subject when different strategies were designed that may help the drug to successfully negotiate BBB. The most important among them are the nanotechnology-based strategies because some of them are capable of overcoming the limitations inherent to BBB passage. These strategies include various types of lipidic, polymeric inorganic, and other types of NPs. These help in controlled drug delivery and release which is of significance to various CNS conditions. Nanosystems which can act as "Trojan horses" can be created by clocking the physicochemical properties of the encapsulated therapeutic agents and provide a medium to get such molecules successfully negotiate BBB. Since these encapsulated molecules are masked they are protected against enzymatic degradation. Their life gets enhanced. Such nanocarriers can increase their bioavailability in the brain. Amid-enhanced efficacy and with lesser quantities of therapeutic agent required any likely detrimental side effect at sites other than the targeted site would also be minimized.

For mitigation of symptoms, in mild to moderate dementia cases, US Food and Drug Administration approved "Rivastigmine" is used. Clinically, the limited ability of the free compound to successfully negotiate the BBB and its spinoffs on peripheral organs, prevent it from reaching its full therapeutic potential. It is thus an ideal bet for NP-mediated brain drug delivery in which the role of the NP is to facilitate the delivery of drug to the target that it would not reach in its natural course.

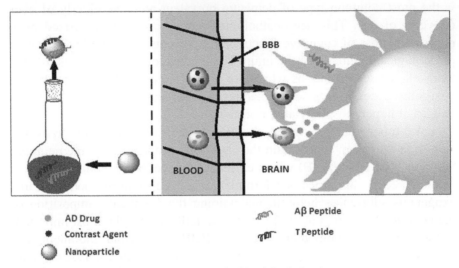

FIGURE 2.5   NPs delivering drugs across the blood–brain barrier.

Biodegradable polymeric NPs have very recently been employed for this function. Intravenously injected rivastigmine which was bound to PnBCA NPs coated with the chemical polysorbate 80 was found to have a significantly better uptake by the brain relative to the free drug (an enhancement of up to 3.82-fold was found). The superior delivery is attributable to a system which seeks to bind lipoproteins existing in the blood to the surface of the NP.

Kreuter et al. (2002) have also put across the same findings. A pharmacodynamic study was conducted on amnesic mice. For faster reversal of memory loss, PLGA and PnBCA NPs were used as carriers for rivastigmine. It was, however, not the case with free drug. The research result indicated that referred NPs were associated with speedy and effective delivery into the mouse brain (Joshi et al., 2010). Not long ago an approach based on the use of single-wall carbon nanotubes (SWCNTs) has been suggested to carry neurotransmitter aacetylcholine (ACh) to treat the disrupted cholinergic neurotransmission in the AD (Yang et al., 2010). In this process, the drug was carried to the brain through olfactory nerve axons instead of BBB. Previously, there were biosafety concerns about the use of SWCNTs particularly about their toxicity toward the mitochondria in cells. The biosafety concern can, however, be overcome by carefully regulating the release of drug dosages. A regulated release of drug can safely deliver Ach, especially to the target organelle. It, at the same time, avoids its mitochondrial toxicity.

The study revealed that to functionalize CNTs with biomolecules, for instance, branched polyethylene-glycol chain the later may provide way to augment their biocompatibility and suitability as vehicles for drug delivery. Another study discusses the improvement in the transportation of polymeric NPs to brain and amyloid plaque retention of 125I-clioquinol (5-chloro-7-iodo-8-hydroxyqunioline (I-CQ). For this route, polymeric NPs were prepared and encapsulated with radiolabelled ICQ).

Radioiodinated CQ-NPs have been discovered promising delivery vehicles for in vivo single-photon emission computed tomography. They can also be used as photon emission computed tomography amyloid imaging agents. Aβ peptides are used as targets in the development of biological markers for the diagnosis of the AD. The use of nanotechnology in the diagnostic tools is dependent upon the detection of amyloid peptides. The use of thioflavin-T entrapped in polymeric NPs has been described for use as a probe to detect Aβ in senile plaques. The photoconversion of fluorescent thioflavin-T as a model drug was achieved in tissues, within 3 days after injection, and thioflavin-T delivered from nanospheres was predominantly found in neurons and microglia. This describes the use of polymeric NPs entrapped thioflavin-T which are used as a probe to detect Aβ in the brain (Härtig et al., 2003).

Toxicity can also be reduced with the use of nanosystems. There are certain anticancer drugs which have platinum as its constituent. But it has its drawbacks, platinum is nephrotoxic and neurotoxic and the drug resistance caused by its use has led to its restricted use. On the contrary, nontarget toxicity is reduced if platinum nanocarrier-aided delivery is employed. In some cases, nanocarriers also check drug resistance against platinum. Besides, nanocarrier-aided delivery systems can be exploited in cases of multidrug resistance cancer. Commercially available Taxol plus XR9576 has been found to be less efficient as compared to paclitaxel-loaded NPs (Oberoi et al., 2013; Ma and Mumper, 2013). Nanocarrier systems can help increase drug solubility. A study has been conducted with water-insoluble drug simvastatin. However, when simvastatin was loaded, highly ordered mesoporous carbon (HMC) samples were synthesized by nanocasting techniques and the process yielded an enhanced dissolution rate. Simvastatin-loaded spherical HMC nanomatrix when matched with Zocor which is commercially marketed conventional tablet yielded exceptionally shortert_{max} and higher C_{max} and larger area under curve (0–24 h) (Bai et al.,

2013). Nanotechnology can help to improve oral absorption. The efficacy of highly lipophilic drugs is not of required level. Therefore, to improve their pharmacological properties it is of paramount significance that the key issues connected with their poor solubility and reduced systemic exposure are appropriately addressed. Zhang et al. (2013) reported that as compared to free probucol suspension, the blood concentration of lipophilic drug (probucol) got considerably improved with nanodelivery system-loaded probucol. When nanodelivery system was adopted, cellular uptake of probucol in Caco-2 cell monolayers also increased.

2.10 CHALLENGES FOR NPS-BASED DRUG DELIVERY SYSTEMS

The pace at which nanotechnology is growing is bound to have both toxicological and advantageous impacts and consequences on human health and environment. As is ought to be, new technology-driven targeted NPs for cancer therapy will also face numerous challenges. NPs can change the solubility, stability, and pharmacokinetic properties of the carried drug. But these traits of the NPs still remain to be investigated. Besides, there are other limitations in their use because of the concerns related to their shelf life, leakage, aggregation, and the toxicity of the material used to make NPs. For example, substances used to create as poly(lactic-co-glyoclic acid) NPs have low toxicity but on the contrary, it quickly degenerates. It prohibits its circulation in tissues for long enough for sustained drug/gene delivery. There are other materials such as carbon nanotubes and quantum dots also. These are resilient and can persist in the body for even for year together. Their resilience makes them toxic limiting their constant use in the treatment of cancer patients.

Though innovative materials such as silica/silicon (solid, porous, and hollow silicon NPs) to make targeted NPs have emerged yet their usage in drug delivery has not taken off with equilibrium and aplomb. This is because of the inherent health risk connected with them. Vital issues still need to be tackled before these new materials are put into the human body making. Some of the issues are summarized below

1. Occasions have emerged when researchers have developed the usage of NPs for industry including pharmacology. The unique properties of the NPs have helped it happen so. But they have toxic characteristics. Therefore, before they are fed into the human body it needs to be assured that they are purged of their toxic characteristics.

2. Approval from appropriate health authorities such as US FDA is required for the commercialization of NPs human use. It is, therefore, of utmost importance that the scientific community holds a meaningful and focused discussion on larger issue connected with their impacts on the environment and society as a whole.

3. Better understanding of bionanotechnologies is necessary before it finds application in humans. To achieve the objective, it is important to first understand the mechanism underlying intracellular uptake trafficking and the fate of the nanomaterials in the complex biological human system. The current delivery system suffers from various drawbacks. They have low targeting efficiency. They find it difficult to successfully negotiate the biological barriers. Besides they get rapidly cleared by the immune system. All these impediments in the way of their safe usage require a thorough understanding of the basic science connected with NP carrier before we can attain the capability to exploit and handle drug delivery.

4. To scale up the production through the laboratory to pilot-scale and commercialization is another issue connected with nanodrug delivery. Nanodrug delivery technologies or processes may not be compatible with bulk production because of production process adopted and expensive input used. The concentration of nanomaterial may be a challenge in bulk production. Aggregation and chemical process may be the other issues. For improved performance, it is far easier to manipulate, sustain, and maintain the size and composition of nanomaterial at laboratory scale. However, it may not be the case when produced at commercial levels.

5. Administration of nanomedicines may also encounter economic barriers and financial constraints. Relatively expensive new diagnostics test before administering nanodrug has emerged as major limitation in the development of personalized medicine in general, particularly, when the reimbursement from public and private health insurers is limited. To sum-up, nanoproducts are likely to encounter cost and complexity hurdles. Sound strategy aligned with the overall development of nanotechnology is the key to its efficient commercial implementation.

2.11 CONCLUSION

NP-based drug delivery holds great promise for helping to overcome some of the obstacles to efficiently target a number of diverse cell types. This causes scientific community to explore wonderful and exciting possibilities to overcome problems associated with drug resistance in target cells and to facilitate the movement of drugs across barriers (e.g., BBB). However, toxicity, clearance by the immune system, approvals from regulatory authorities such as FDA and production scale-up remains the major issues associated with NP-based drug delivery systems.

KEYWORDS

- nanoparticles
- drug delivery
- cancer
- alzheimer
- nanoformulations
- nanotherapeutics

REFERENCES

Alai, M. S.; Lin, W. J.; Pingale, S. S. Application of Polymeric Nanoparticles and Micelles in Insulin Oral Delivery. *J. Food Drug Anal.* **2015,** *23,* 351–358.

Anandhakumar, S.; Mahalakshmi, V.; Raichur, A. M. Silver Nanoparticles Modified Nanocapsules for Ultrasonically Activated Drug Delivery. *Mater. Sci. Eng. C.* **2012,** *32*(8), 2349–2355.

Andreas, W.; Witzigmann, D.; Vimalkumar, B.; Huwyler, J. Nanomedicinein Cancer Therapy: Challenges, Opportunities, and Clinical Applications. *J. Control. Release* **2015,** *200,* 138–157.

Avgoustakis, K.; Beletsi, A.; Panagi, Z.; Klepetsanis, P.; Karydas, A. G.; Ithakissios, D. S. PLGA-mPEG Nanoparticles of Cisplatin: In Vitro Nanoparticle Degradation, in Vitro Drug Release and in Vivo Drug Residence in Blood Properties. J. Control. Release **2002,** *79,* 123–135.

Bai, F.; Wang, C.; Lu, Q.; Zhao, M.; Ban, F.-Q.; Yu, D.-H. Nanoparticle-Mediated Drug Delivery to Tumor Neovasculature to Combat P-gp Expressing Multidrug Resistant Cancer. *Biomaterials* **2013,** *34,* 6163–6174.

Baker, R.W.; Lonsdale, H.K. Controlled release: mechanisms and release. In: *Controlled release of biologically active agents*, Tanquary, A.C.; Lacey, R.E., Eds.; Plenum Press: New York, 1974; p 15–71.

Balasubramanian, J.; Sravanthi, T.; Sujitha, V. Ruminative Announcement on Nanoparticles and Mononuclear Phagocytic System. *Int. J. Pharma. Sci. Drug Res.* **2015,** *7*(2), 129–137.

Ballauff, M.; Lu, Y. 'Smart' Nanoparticles: Preparation, Characterization and Applications. *Polymer* **2007,** *48,* 1815–1823.

Benyettou, F.; Rezgui, R.; Ravaux, F.; Jaber, T.; Blumer, K.; Jouiad, M.; Motte, L.; Olsen, J. C.; Platas-Iglesias, C.; Magzoub, M.; Trabolsi, A. Synthesis of Silver Nanoparticles for the Dual Delivery of Doxorubicin and Alendronate to Cancer Cells. *J. Mater. Chem. B.* **2015,** *3,* 7237–7245.

Borkovec, M. Measuring Particle Size by Light Scattering. In *Handbook of Applied Surface and Colloid Chemistry;* Holmberg, K., Ed.; Wiley: Hoboken, New Jersey 2002; pp 357–370.

Bunjes, H.; Unruh, T. Characterization of Lipid Nanoparticles by Differential Scanning Calorimetry, X-ray and Neutron Scattering. Adv. Drug. Delivery Rev. **2007,** *59,* 379–402.

Castro-Aceituno, V.; Ahn, S.; Simu, S. Y.; Singh, P.; Mathiyalagan, R.; Lee, H. A.; Yang, D. C. Anticancer Activity of Silver Nanoparticles from Panax Ginseng Fresh Leaves in Human Cancer Cells. *Biomed. Pharmacother.* **2016,***84,*158–165.

Chen, Y. S.; Hung, Y. C.; Liau, I.; Huang, G. S. Assessment of the In Vivo Toxicity of Gold Nanoparticles. *Nanoscale Res Lett.* **2009,** 4(8):858-864

Chen , Y.; McCulloch, R. K.; Gray, B. N. Synthesis of Albumin-Dextran Sulfate Microspheres Possessing Favourable Loading and Release Characteristics for the Anti-Cancer Drug Doxorubicin. *J. Controlled Release* **1994,** *31*(1), 49–54.

Chen, G.; Lu, J.; Lam, C.; Yu, Y. A Novel Green Synthesis Approach for Polymer Nanocomposites Decorated with Silver Nanoparticles and their Antibacterial Activity. Analyst **2014,** *139,* 5793–5799.

Dash, S.; Padala, N. M.; Nath, L.; Chowdhury, P. Kinetic Modeling on Drug Release from Controlled Drug Delivery Systems. *Acta. Pol. Pharm.* **2010,** *67*(3), 217–223.

Dixit, V.; Van den Bossche, J.; Sherman, D. M.; Thompson, D. H.; Andres, R. P. Synthesis and Grafting of Thioctic Acid-PEG-Folate Conjugates onto au Nanoparticles for Selective Targeting of Folate Receptor-Positive Tumor Cells. *Bioconjugate Chem.* **2006,** *17,* 603–609.

Dorofeev, G. A.; Streletskii A. N.; Povstugar, I. V.; Protasov, A. V.; Elsukov, E. P. Determination of Nanoparticle Sizes by X-Ray Diffraction. Colloid *J.* **2012,** *74,* 675–685.

Du, L.; Miao, X.; Jiang, Y.; Jia, H.; Tian, Q.; Shen, J.; Liu, Y. An Effective Strategy for the Synthesis of Biocompatible Gold Nanoparticles Using Danshensu Antioxidant: Prevention of Cytotoxicity via Attenuation of Free Radical Formation. *Nanotoxicology* **2013,** *7,* 94–300.

Dunne, M.; Corrigan, I.; Ramtoola, Z. Influence of Particle Size and Dissolution Conditions on the Degradation Properties of Polylactide-Co-Glycolide Particles. *Biomaterials* **2000,** *21,* 1659–1668.

Fayaz, A. M.; Balaji. K.; Girilal, M.; Yadav, R.; Kalaichelvan, P. T.; Venketesan, R. Biogenic Synthesis of Silver Nanoparticles and their Synergistic Effect with Antibiotics: A Study Against Gram-Positive and Gram-Negative Bacteria. Nanomed. *Nanotechnol. Biol. Med.* **2010,** *6,* 103–109.

Frank, D.; Tyagi, C.; Tomar, L.; Choonara, Y. E.; Toit, L. C.; Kumar, P.; Penny, C.; Viness. P. Overview of the Role of Nanotechnological Innovations in the Detection and Treatment of Solid Tumors. Int. J. Nanomed. **2014,** *9,* 589–613.

Gaumet, M.; Vargas, A.; Gurny, R.; Delie, F. Nanoparticles for Drug Delivery: The Need for Precision in Reporting Particle Size Parameters. *Eur. J. Pharm. Biopharm.* **2008,** *69,* 1–9.

Giri, S.; Trewyn, B. G.; Stellmaker, M. P.; Lin, V. S. Y., Stimuli-Responsive Controlled-Release Delivery System Based on Mesoporous Silica Nanorods Capped with Magnetic Nanoparticles. *Angew. Chem. Int. Ed. Engl.* **2005,** *44*(32), 5038–5044.

Guterres, S. S.; Alves, M. P.; Pohlmann, A. R. Polymeric Nanoparticles, Nanospheres and Nanocapsules, for Cutaneous Applications. Drug Target Insights **2007**, *2*, 147–157.

Haley, B.; Frenkel, E. Nanoparticles for Drug Delivery in Cancer Treatment. Urol Oncol. **2008**, *26*, 57–64.

Hare, J. I.; Lammers, T.; Ashford, M. B.; Puri, S.; Storm, G.; Barry, S. T. Challenges and Strategies in Anti-Cancer Nanomedicine Development: An Industry Perspective. Adv. Drug Deliv. Rev. (in press) **2016.** http://dx.doi.org/10.1016/j.addr.2016.04.025

Härtig, W.; Paulke, B. R.; Varga, C.; Seeger, J.; Harkany, T.; Kacza, J. Electron Microscopic Analysis of Nanoparticles Delivering Thioflavin-T After Intrahippocampal Injection in Mouse: Implications for Targeting Beta-Amyloid in Alzheimer's Disease. Neurosci. Lett. **2003**, *338*(2), 174–176.

Higuchi, T. Mechanism of Sustained-Action Medication-Theoretical Analysis of Rate of Release of Solid Drugs Dispersed in Solid Matrices. *J. Pharm. Sci.* **1963**, *84*, 1464.

Hood, E. Nanotechnology: Looking as we Leap. Environ. Health Perspect. **2004**, *112*, A740–A749.

Jain, V.; Jain, S.; Mahajan, S. C. Nanomedicines Based Drug Delivery Systems for Anti-Cancer Targeting and Treatment. Curr. Drug Deliv. **2015**, *12*, 177–191.

Joshi, S. A.; Chavhan, S. S.; Sawant, K. K. Rivastigmine-Loaded PLGA and PBCA Nanoparticles: Preparation, Optimization, Characterization, in vitro and Pharmacodynamic Studies. *Eur. J. Pharm. Biopharm.* **2010**, *76*, 189–199.

Kamberi, M.; Tran, T. N. UV-Visible Spectroscopy as an Alternative to Liquid Chromatography for Determination of Everolimusin Surfactant-Containing Dissolution Media: A Useful Approach Based on Solid-Phase Extraction. J. Pharm. Biomed. Anal. **2012**, *70*, 94–100.

Korsmeyer, R.; Gurny, R.; Doelker, E.; Buri, P.; Peppas, N. . Mechanisms of solute release from porous hydrophilic polymers. *Int J Pharm.* **1983**, *15*(1), 25-35.

Kreuter, J.; et al. Apolipoprotein-Mediated Transport of Nanoparticle-Bound Drugs Across the Blood-Brain Barrier. J. Drug Target. **2002**, *10*, 317–325.

Lai, C.-Y.; Trewyn, B. G.; Jeftinija, D. M.; Jeftinija, K.; Xu, S.; Jeftinija, S.; Lin, V. S. A Mesoporous Silica Nanosphere-Based Carrier System with Chemically Removable CdS Nanoparticle Caps for Stimuli-Responsive Controlled Release of Neurotransmitters and Drug Molecules. *J. Am. Chem. Soc.* **2003**, *125*(15), 4451–4459.

Lara, H. H.; Ayala-Núñez, N. V.; Turrent, L.; Padilla, C. R. Silver Nanoparticles are Broad-Spectrum Bactericidal and Virucidal Compounds. World J. Microb. Biot. **2010**, *26*, 615–621.

Leung, A. Y. Traditional Toxicity Documentation of Chinese Materia Medica—An Overview. Toxicol. Pathol. **2006**, *34*, 319–26.

Liang, X.; Chen, C.; Zhao, Y.; Jia, L.; Wang, P. C. Biopharmaceutics and Therapeutic Potential of Engineered Nanomaterials. Curr. Drug Metab. **2008**, *9*, 697–709.

Ma, P.; Mumper, R. J. Paclitaxel Nano-Delivery Systems: A Comprehensive Review. *J. Nanomed. Nanotechnol.* **2013**, *4*, 1000164.

Madaan, K.; Kumar, S.; Poonia, N.; Lather, V.; Pandita, D. Dendrimers in Drug Delivery and Targeting: Drug-dendrimer Interactions and Toxicity Issues. J. Pharm. *Bio Allied* Sci. **2014**, *6*(3), 139–150.

Maeda, H.; Matsumura, Y. Tumoritropic and Lymphotropic Principles of Macromolecular Drugs. Crit. Rev. Ther. Drug Carrier Syst. **1989**, *6*(3), 193–210.

Martinho, N.; Damgé, C.; Reis, C. P. Recent Advances in Drug Delivery Systems. *J. Biomater. Nanobiotechnol.* **2011**, *2*, 510–526.

Matsumura, Y.; Maeda, H. A New Concept for Macromolecular Therapeutics in Cancer Chemotherapy: Mechanism of Tumoritropic Accumulation of Proteins and the Antitumor Agent Smancs. Cancer Res. **1986**, *46,* 6387–6392.

McCarron, A.; David Woolfson, A.; Morrissey, A.; Juzenas, P.; et al. Microneedle Arrays Permit Enhanced Intradermal Delivery of a Preformed Photosensitizer. Photochem. Photobiol. **2009**, *85*(1), 195–204.

Mohd, A.; Amar, J. D.; Therapeutic Nanoparticles: State-of-the-Art of Nanomedicine. *Adv. Mater. Rev.* **2014**, *1,* 25-37.

Mohanraj, V.J. Nanoparticles – A Review. *Trop. J. Pharm. Res.* **2006**, *5*(1), 561–573.

Mohanraj, V. J.; Chen, Y. Nanoparticles – A Review. *Trop. J. Pharm. Res.* **2006**, *5*(1), 561–573.

Monteiro-Riviere, N. A.; Nemanich, R. J.; Inman, A. O.; et al. Multi-Walled Carbon Nanotube Interactions with Human Epidermal Keratinocytes. Toxicol. Lett. **2005**, *155,* 377–384.

Nanjwade, B. K.; Bechra, H. M.; Derkar, G. K.; Manvi, F. V.; Nanjwade, V. K. Dendrimers: Emerging Polymers for Drug-Delivery Systems. Eur. J. Pharm. Sci. **2009**, *38*(3), 185–196.

Noyes, A. A.; Whittney, W. R. The Rate of Solution of Solid Substances in their Own Solutions. *J. Am. Chem. Soc.* **1897**, *19,* 930–934.

Oberoi, H. S.; Nukolova, N. V.; Kabanov, A. V.; Bronich, T. K. Nanocarriers for Delivery of Platinum Anticancer Drugs. *Adv. Drug Delivery Rev.* **2013**, *65*(13–14), 1667–1685.

Panyam, J.; Williams, D.; Dash, A.; Leslie-Pelecky, D.; Labhasetwar, V.; Solid-State Solubility Influences Encapsulation and Release of Hydrophobic Drugs from Plga/Pla Nanoparticles. J. Pharm. Sci. **2004**, *93,* 1804–1814.

Park, K. Facing the Truth About Nanotechnology in Drug Delivery. ACS Nano. **2013**, *7,* 7442–7447.

Perrault, S. D.; Walkey, C.; Jennings, T.; Fischer, H. C.; Chan, W. C. Mediating Tumor Targeting Efficiency of Nanoparticles Through Design. Nano Lett. **2009**, *9*(5), 1909–1915.

Petrov, P. D.; Yoncheva, K.; Gancheva, V.; Konstantinov, S.; Trzebicka, B. Multifunctional Block Copolymer Nanocarriers for Co-Delivery of Silver Nanoparticles and Curcumin: Synthesis and Enhanced Efficacy Against Tumor Cells. *Eur. Polym. J.* **2016**, *81,* 24–33.

Pridgen, E. M.; Langer, R.; Farokhzad, O. C. Biodegradable, Polymeric Nanoparticle Delivery Systems for Cancer Therapy. *Nanomedicine* **2007**, *2*(5), 669–680.

Qin, S. Y.; Zhang, A. Q.; Cheng, S. X.; Rong, L.; Zhang, X. Z. Drug Self-Delivery Systems for Cancer Therapy. *Biomaterials* **2017**, *112,* 234–247.

Rai, M.; Ingle, A. P.; Gupta, I.; Brandelli, A. Bioactivity of Noble Metal Nanoparticles Decorated with Biopolymers and their Application in Drug Delivery. *Int. J. Pharm.* **2015**, *496,* 159–172.

Ranjit, K.; Baquee, A. A. Nanoparticle: An Overview of Preparation, Characterization and Application. *Int. Res. J. Pharm.* **2013**, *4*(4), 47–57.

Ravishankar Rai, V.; Jamuna Bai, A.; Nanoparticles and their Potential Application as Antimicrobials. In *Science Against Microbial Pathogens: Communicating Current Research and Technological Advances;* Méndez-Vilas, A., Ed.; Formatex Research Center: Badajoz, 2011; *3*(1), 197–209.

Rege, B. D.; Kao, J. P. Y.; Polli, J. E. Effects of Nonionic Surfactants on Membrane Transporters in Caco-2 Cell Monolayers. *Eur. J. Pharm. Sci.* **2002**, *16*(4–5), 237–246.

Sathishkumar, K. Gold Nanoparticles Decorated Polylactic Acid-Co-Ethyl Cellulose Nanocapsules for 5-Fluorouracil Drug Release. *Int. J. Nano Biomater.* **2012**, *4,* 12–20.

Singh, R.; Lillard, J. W. Nanoparticle-Based Targeted Drug Delivery. Exp. Mol. Pathol. **2009**, *86,* 215–223.

Singh, S.; Pandey, V. K.; Tewari, R. P.; Agarwal, V. Nanoparticle Based Drug Delivery System: Advantages and Applications. *Indian J. Sci. Technol.* **2011,** *4*(3), 177–180.

Sperling, R. A.; Parak, W. J. Surface Modification, Functionalization and Bioconjugation of Colloidal Inorganic Nanoparticles. *Philos. Trans. R. Soc. A* **2010,** *368,* 1333–1383.

Sreeprasad, T. S.; Pradeep, T. Noble Metal Nanoparticles. In *Springer Handbook of Nanomaterials;* **2013,** 303–388.

Teow, Y.; Valiyaveettil, S. Active Targeting of Cancer Cells Using Folic Acid-Conjugated Platinum Nanoparticles. *Nanoscale* **2010,** *2,* 2607–2613.

Tiwari, G.; Tiwari, R.; Sriwastawa, B.; Bhati, L.; Pandey, P.; Bannerjee, S. K. Drug Delivery Systems: An Updated Review. Int. J. Pharm. Invest. **2012,** *2,* 2–11.

Trewyn, B. G.; Giri, S.; Slowing, I. I.; Lin, V. Mesoporous Silica Nanoparticle Based Controlled Release, Drug Delivery, and Biosensor Systems. *Chem. Commun.* **2007,** *31,* 3236–3245.

Valo, H.; Arola, S.; Laaksonen, P.; Torkkeli, M.; Peltonen, L.; Linder, M. B.; Serimaa, R.; Kuga, S.; Hirvonen. J.; Laaksonen, T. Drug Release from Nanoparticles Embedded in Four Different Nanofibrillar Cellulose Aerogels. Eur. J. Pharm. Sci. **2013,** *50*(1), 69–77.

Vasir, J. K.; Reddy, M. K.; Labhasetwar, V. Nanosystemsin Drug Targeting: Opportunities and Challenges. *Curr. Nanosci.* **2005,** *1,* 47–64.

Vivero-Escoto, J. L.; Slowing, I. I.; Wu, C.-W.; Lin, V. S. Y. Photo Induced Intracellular Controlled Release Drug Delivery in Human Cells by Gold-Capped Mesoporous Silica Nanosphere. *J. Am. Chem. Soc.* **2009,** *131*(10), 3462–3463.

Weissig, V.; Pettinger, T. K.; Murdock, N. Nanopharmaceuticals (part 1): Products on the Market. *Int. J. Nanomed.* **2014,** *9,* 4357–4373.

Wilczewska, A. Z.; Niemirowicz, K.; Markiewicz, K. H.; Car, H. Nanoparticles as Drug Delivery Systems. Pharmacol. Rep. **2012,** *64,* 1020–1037.

Yang, Z.; et al. Pharmacological and Toxicological Target Organelles and Safe use of Single-Walled Carbon Nanotubes as Drug Carriers in Treating Alzheimer's Disease. Nanomed. Nanotechnol. Biol. Med. **2010,** *6,* 427–441.

Yin, H.; Liao, L.; Fang, J. Enhanced Permeability and Retention (EPR) Effect Based Tumor Targeting: The Concept, Application and Prospect. *JSM Clin. Oncol. Res.* **2014,** *2*(1), 1010.

Yoo, D.; Lee, J. H.; Shin, T. H.; Cheon, J. Theranostic Magnetic Nanoparticles. *Acc. Chem. Res.* **2011,** *44,* 863–874.

Zambito, Y.; Pedreschi, E.; Di Colo, G. Is Dialysis A Reliable Method for Studying Drug Release from Nanoparticulate Systems?—A Case Study. Int. J. Pharm. **2012,** *434*(1-2), 28–34.

Zhang, Z.; Huang, J.; Jiang, S.; Liu, Z.; Gu, W.; Yu, H.; Li, Y. Porous Starch Based Self Assembled Nano Delivery System Improves the Oral Absorption of Lipophilic Drug. Int. J. Pharm. **2013,** *444,* 162–168.

Zhang, S. Y.; Wu, Y.; He, B.; Luo, K.; Gu, Z. Biodegradable Polymeric Nanoparticles Based on Amphiphilic Principle: Construction and Application in Drug Delivery. Sci. China Chem. **2014,** *57,* 461–475.

CHAPTER 3

NANOPARTICLES AS NANOPHARMACEUTICALS: SMART DRUG DELIVERY SYSTEMS

MD. SAHAB UDDIN

Department of Pharmacy, Southeast University, Dhaka, Bangladesh

E-mail: msu-neuropharma@hotmail.com; msu_neuropharma@hotmail.com

ABSTRACT

Nanopharmaceuticals represent an emerging field and one of the utmost applications of nanotechnology using nanoparticles (NPs). The trouble of delivering the proper dose of therapeutic agents to specific disease site especially brain is a long-standing issue. Nanopharmaceuticals have tremendous potential intending to this failure of the traditional therapeutics. They offer site-specific targeting of active agents as well as reduce adverse/ side effects, subsequently superior patient compliance. Drug delivery to the brain is a promised research theme in the domain of nanomedicine. Furthermore, nanopharmaceuticals abate the cost of drug discovery, design, and development and boost the drug delivery process. Therefore, this chapter deals with the profile of NPs and uses of NPs as nanopharmaceuticals as well as future opportunities and challenges of nanomedicine.

3.1 INTRODUCTION

Nanotechnology is an emerging technology that deals with nanometer-sized objects and has crucial roles in numerous areas of research and applications. It is the engagements of knowledge from various disciplines such as physics, chemistry, biology, materials science, health sciences, and engineering

(Mazzola, 2003). It has immense applications in electronics, biology, and medicine. The prefix nano is derived from the Greek word "dwarf." In 1974, the term "nanotechnology" was first used by Norio Taniguchi, a scientist at the University of Tokyo, Japan, referring to materials in nanometers (Taniguchi, 1974). The structural and functional units of all kind of living organisms are called cells that are typically 10 μm across. However, the cellular parts are much smaller and in the size range of the submicron. The proteins are even smaller with a typical size of just 5 nm, which is comparable with the dimensions of smallest man-made nanoparticles (NPs). This high through thinking is the main theme of using NPs at the cellular and or subcellular apparatus (Taton, 2002). Milk is a nanoscale colloid with a natural nanoscale assembly. In nature, NPs have been in existence for thousands of years as the products of combustion and cooking food (Thassu, 2007).

NPs are solid, colloidal particles consisting of macromolecular substances that vary in size from 10 to 1000 nm (Kreuter, 1994). According to National Nanotechnology Initiative, the sizes range of the NPs is from 1 to 100 nm in at least one dimension. Figure 3.1 shows the nanometer in context. Conversely, the prefix "nano" is frequently used for particles that are up to several hundred nanometers in size (Singh and Lillard, 2009). Based on the method of formulation, NPs, nanospheres, or nanocapsules can be prepared. Actually, to obtain different release characteristics for the best delivery or encapsulation of the therapeutic agent, different manufacturing methods are used (Barratt, 2000). In case of nanospheres, the drug is physically and uniformly distributed to the matrix systems, whereas for nanocapsules, the drug is incorporated into a cavity enclosed by a polymer membrane. On the other hand, particles <200 nm (i.e., the width of microcapillaries) are called nanomedicine. Typically, the drug of interest is dissolved, entrapped, adsorbed, attached, and or encapsulated into or onto a nano-matrix (Athar and Das, 2014).

NPs can alter biological processes such as infection, tissue engineering, de novo synthesis, and so forth. These devices include, but are not limited to, functionalized carbon nanotubes; nanomachines (e.g., constructed from interchangeable deoxyribonucleic acid [DNA] parts and DNA scaffolds); nanofibers; self-assembling polymeric nanoconstructs; nanomembranes and nanosized silicon chips for drug, protein, nucleic acid, or peptide delivery and release; and biosensors and laboratory diagnostics (Singh and Lillard, 2009). In recent years, biodegradable polymers, mainly those coated with hydrophilic polymer, have been examined widely over the past few decades as potential drug delivery devices because of their ability to circulate for a prolonged period of time and target to a particular organ (e.g., as carrier of DNA in gene therapy) and their ability to deliver proteins, peptides, and genes those coated with hydrophilic polymer. Substantial research is accomplished for the development of biodegradable polymeric NPs for drug delivery and tissue engineering, owing to their applications in controlling the release of drugs, stabilizing labile molecules (e.g., proteins, peptides, or DNA) from degradation as well as site-specific drug targeting (Singh and Lillard, 2009). The purpose of development of biodegradable NPs from poly-D-L-lactide-co-glycolide and polylactic acid is intracellular target and sustained drug delivery (Davda and Labhasetwar, 2002; Panyam and Labhasetwar, 2003). In addition, several studies suggested the quick discharge of hydrophobic polycaprolactone-coated NPs from endolysosomes to the cytoplasm (Barrera et al., 1993; Woodward et al., 1985). Vascular smooth muscle cells that were treated with dexamethasone-loaded NPs exerted greater as well as sustained antiproliferative activity (Redhead et al., 2001). Hence, NPs can be effective in delivering their contents to intracellular targets. Assigning drugs to separately intended carriers is the way for cell-specific targeting. Abundant studies suggested that NPs have an inordinate possibility as drug carriers. The unique physicochemical and biological properties of NPs are due to their small size that makes them a promising material for biomedical applications.

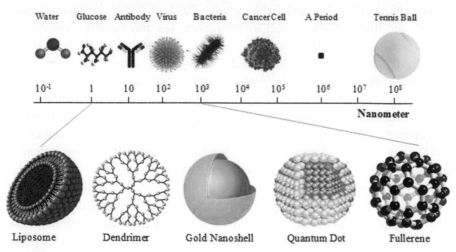

FIGURE 3.1   Nanoscale and nanostructures.

In the scientific knowledge and commercial applications, nanomaterials are playing greatest protagonist (Mirza et al., 2014). NP-based drug delivery system created a breakthrough in the pharmaceutical industry for the development of new specialized treatment strategy for exotic diseases to remodeling existing treatments. Gradually, the number of patents and products in this field is increasing markedly (Nikam et al., 2014). Therefore, the purpose of this chapter is to give an overview about NPs and their biomedical applications as well as nanopharmaceuticals.

3.2   CHARACTERISTICS OF NANOPARTICLES (NPS) IN DRUG DELIVERY

3.2.1   PARTICLE SIZE AND SIZE DISTRIBUTION

The most important characteristics of NPs are particle size and size distribution. They regulate various parameters such as in vivo distribution, biological fate, drug loading, drug release, stability, and toxicity of NPs (Panyam and Labhasetwar, 2003). NPs can easily cross the blood–brain barrier (BBB) that may offer sustained delivery of therapeutic agents for the treatment of numerous brain diseases such as brain tumors (Kroll et al., 1998). In fact hyperosmotic mannitol and polysorbate-80-coated NPs have been exposed to cross the BBB (Kreuter et al., 2003). NPs are more advantageous than microparticles (Panyam and Labhasetwar, 2003). Usually, cellular uptakes of NPs are higher with respect to microparticles and rapidly available in cellular and intracellular targets owing to small particle size (Kroll et al., 1998). Actually, 100 nm NPs had a 2.5-fold greater uptake rate than 1 μm microparticles and a 6-fold greater uptake than 10 μm microparticles by Caco-2 cells (Desai et al., 1997). In a parallel study, NPs were shown to penetrate the submucosal layers of a rat intestinal loop model, while microparticles were mostly confined in the epithelial lining (Redhead et al., 2001). This indicates that particle distribution can, in part at least, be tuned by controlling particle size. Drug release is strongly influenced by particle size. The surface-area-to-volume ratio is larger for smaller particles that lead to faster drug release and vice versa for larger particles (Redhead et al., 2001). However, the problems faced by the smaller particles are greater risk of aggregation during storage, transport, and dispersion (Dunne et al., 2000).

3.2.2   SURFACE PROPERTIES OF NPS

The connection of a drug to conventional carriers is responsible for the alteration of the distribution pattern of the drug in the biological system as it is generally transported to the mononuclear phagocyte system (MPS), for example, lungs, liver, spleen, and bone marrow. In addition, intravenously administered NPs can be identified by the immune system of the host and finally cleared by phagocytes (Muller et al., 1996b). The hydrophobicity of the NP controls the amount of blood components (e.g., opsonins) that bind with this surface and finally that impacts the in vivo fate of NPs (Brigger et al., 2002; Muller et al., 1996b). Indeed, once in the bloodstream, surface-nonmodified NPs (i.e., conventional NPs) are rapidly opsonized and massively cleared by the MPS (Grislain et al., 1983). In order to increase the probability of completion in drug targeting, it is required to lessen the opsonization and extend the circulation of NPs. The coating of NPs with hydrophilic polymers/surfactants is one of the effective ways. Furthermore, this can be obtained by formulating NPs with the help of biodegradable copolymers having hydrophilic characteristics, for example, polyethylene oxide, poloxamer, polyethylene glycol (PEG), poloxamine, and polysorbate 80. Several studies suggested that PEG-coated NP surfaces can stop opsonization by influencing complement and other serum factors. The configuration of PEG is most important for opsonization. The surfaces containing PEG with the brushlike and intermediate structure reduced phagocytosis and complement activation but PEG with mushroomlike structures exert opposite phenomena (Bhadra et al., 2002; Olivier, 2005). Usually, zeta potential is used to characterize the surface charge property of the NPs (Couvreur et al., 2002).

3.2.3 DRUG LOADING

The NP-based drug delivery system must have the higher drug-loading ability in order to reduce the amount of matrix materials. Drug loading can be obtained with the help of incorporation and adsorption or absorption methods. In case of incorporation method, drugs are incorporated during the formulation of NPs. The adsorption or absorption method requires the drug to be incorporated after the formulation of NPs with the help of adsorption or absorption by placing the nanocarrier in concentrated solution of drugs, which is linked to the matrix composition, molecular weight, drug–polymer interactions, and the existence of functional groups (i.e., ester or carboxyl) in either the drug or matrix (Govender et al., 2000; Panyam et al., 2004). In some cases, the suitable polymer is PEG, owing to its very little or no effect on drug loading and interactions (Peracchia et al., 1997). Moreover, at or near their isoelectric point, the macromolecules, drugs, or protein encapsulated in

NPs show the greatest loading efficiency (Calvo et al., 1997). However, for small molecules, ionic interaction is reported among the drugs plus matrix materials and this approach is very operative in increasing drug loading (Chen et al., 1994; Chen et al., 2003).

3.2.4   DRUG RELEASE

Generally, the drug release rate depends on the solubility of the drug, amount of adsorbed drug, diffusion of the drug in the NP matrix, erosion or degradation of NP matrix, and the combination of erosion and diffusion processes. Consequently, solubility, diffusion, and biodegradation of the particle matrix control the drug release. Diffusion or matrix erosion is the operative process of drug release if the particles are nanospheres where the drug is homogeneously distributed. If the diffusion of the drug is faster than matrix erosion, then the mechanism of release is largely controlled by a diffusion process. The rapid, initial release, or "burst," is mainly attributed to weakly bound or adsorbed drug to the relatively large surface of NPs (Magenheim et al., 1993). In some cases, the method of drug loading also influences the release profile. Sustained release of the drug is predictable if the drug is loaded by the incorporation method (Fresta et al., 1995). For NPs coated by the polymer, the release is regulated by diffusion of the drug from the polymeric membrane. Membrane coating is a causal factor in drug release since drug solubility and diffusion occur through the polymer membrane. Likewise, the ionic interactions among drugs and auxiliary ingredients also disturb the release rate. Actually, the interaction of drugs with auxiliary ingredients causes the formation of a less water-soluble complex, which slows the drug release (Chen et al., 1994). However, if the addition of auxiliary ingredients, for example, ethylene oxide–propylene oxide block copolymer (polyepsilon-polycaprolactone) to chitosan, reduces the interaction of the drug with the matrix material due to competitive electrostatic interaction of ethylene oxide–propylene oxide block copolymer with chitosan, then an increase in drug release could be achieved (Calvo et al., 1997).

3.2.5   TARGETED DRUG DELIVERY

Recently, various methods have been updated for targeting drug delivery system (Moghimi et al., 2001). Targeted delivery can be actively or passively achieved. The therapeutic agent for active targeting requires to be achieved by conjugating the therapeutic agent or carrier system to a tissue or cell-specific ligand (Lamprecht et al., 2001). Passive targeting is achieved by incorporating the therapeutic agent into a macromolecule or NP that passively reaches the target organ. On the other hand, catheters can be used to infuse NPs to the target organ or tissues (Sahoo et al., 2002). Liposomes have been used as pharmaceutical agents. These systems involve binding or interacting with the targeted cell membrane. This permits enhanced lipid–lipid exchange with the lipid monolayer of the NP, which accelerates the convective flux of lipophilic drugs to dissolve through the outer lipid membrane of the NPs to targeted cells (Guzman et al., 1996). For prolonged release mechanism, such nanosystems can be served at the target site. Several biological barriers used in NPs can be formulated to deliver drugs (Lockman et al., 2002). Antineoplastics, antiviral drugs, and several other types of drugs are markedly hindered because of the inability of these molecules to cross the BBB. The application of NPs to deliver across this barrier is extremely promising. Different research findings reported that NPs can cross the BBB following the opening of tight junctions by hyperosmotic mannitol (Avgoustakis et al., 2002). Polysorbate-80-coated NPs also have been shown to cross the BBB (Beletsi et al., 1999).

3.3 CLASSES OF NPS

The NPs of different sizes, shapes, and with various chemical and surface properties have been constructed by various laboratories. Nanotechnology field is under constant and rapid growth and new additions continue to supplement these laboratories. Some general and multifunctional types of NPs are listed below.

3.3.1 FULLERENES

Carbon is the main molecule of fullerene where it is in the form of a hollow sphere, ellipsoid, or tube. Spherical fullerenes are known as buckyballs and carbon nanotubes or buckytubes are called as cylindrical fullerenes. Fullerenes and graphite are similar in structure and they are composed of stacked graphene sheets of linked hexagonal rings; also, they may contain pentagonal or heptagonal ringlike structure to form porous molecule (Holister et al., 2003). Buckyball clusters or buckyballs are composed of endohedral fullerenes, which are composed of 300 carbon atoms among these fullerenes, buckminsterfullerene, C_{60} is the most common. Megatubes which are potentially used for the transport of a variety of molecules of different sizes are larger in diameter than nanotubes and prepared with walls of different thickness (Mitchell et al., 2001). Another spherical particle is nano "onions" formed by multiple carbon layers surrounding a buckyball core which are proposed for lubricants (Sano et al., 2001). These properties of fullerenes hold great promise in versatile biomedical applications.

3.3.2   SOLID LIPID NANOPARTICLES

Mainly, lipids form solid lipid nanoparticles (SLNs) that are in solid phase at the room temperature and surfactants for emulsification; the mean diameters is between 50 and 1000 nm for colloid drug delivery applications procedure (Zur Mühlen and Mehnert, 1998). Small size, large surface area, high drug loading, the interaction of phases at the interfaces of pharmaceuticals are some common parameters of SLNs (Cavalli et al., 1993). Common methods of preparing SLNs include spray drying (Freitas and Müller, 1998), high shear mixing (Domb, 1993), ultrasonication (Eldem et al., 1991), and high-pressure homogenization (Müller et al., 1996a). Fatty acids (e.g., palmitic acid, decanoic acid, and behenic acid), triglycerides (e.g., trilaurin, trimyristin, and tripalmitin), steroids (e.g., cholesterol), partial glycerides (e.g., glyceryl monostearate and glyceryl behenate), and waxes (e.g., cetyl palmitate) are used in the formulation of SLNs. Various types of surfactants are used as emulsifiers including soybean lecithin, phosphatidylcholine, sodium cholate, and sodium glycocholate for lipid dispersion (Zhang et al., 2010). SLNs improve the bioavailability of drugs and also guard the sensitive drug molecules from the environment such as water and light, and also control drug release (Müller et al., 2002). They possess a better stability than liposome, which increase production scale. This property is very important for targeting the drug. SLNs form the basis of colloidal drug delivery systems, which are biodegradable and capable of being stored for at least 1 year.

3.3.3   LIPOSOMES

Liposomes are amphiphilic lipid of spherical lipid vesicles with an aqueous core surrounded by a hydrophobic lipid bilayer. Phospholipids are generally recognized as safe ingredients, therefore minimizing the potential for adverse effects. Drugs in the core cannot pass through the hydrophobic bilayer; however; hydrophobic molecules can be absorbed into the bilayer, thus enabling the liposome to carry both hydrophilic and hydrophobic molecules. Other bilayers such as cell membrane can fuse lipid bilayer of liposomes, which promotes the release of its contents, making them useful for drug delivery and cosmetic delivery applications. Liposomes with the size of nanometers are also known as nanoliposomes (Cevc, 1996 and Zhang and Granick, 2006). Liposomes can vary in size, from 15 nm up to several μm. It can also be classified into small unilamellar vesicles and large unilamellar vesicles depending on their size range (Vemuri and Rhodes, 1995). In liposomes, a lipid membrane structure is surrounded by an aqueous cavity, enabling them to carry both hydrophobic and hydrophilic compounds. Liposome surface can be easily functionalized with other material to increase their in vivo property. They may also target ligands to enable preferential delivery of liposomes. These properties of liposomes make them to be used as potent carriers for various drugs such as antibacterials, antivirals, insulin, antineoplastics, and plasmid DNA.

3.3.4 NANOSTRUCTURED LIPID CARRIERS

Nanostructured lipid carriers (NLCs) are produced from a blend of solid and liquid lipids. Lipids are a special type of molecules that may form differently structured solid matrices, such as the NLCs and the lipid–drug conjugate NPs to improve drug-loading capacity (Wissing et al., 2004). The NLC production is based on solidified emulsion technologies. The release of drug from lipid particles occurs by diffusion and simultaneously by lipid particle in the body. In some cases, it might be desirable to have a controlled fast release going beyond diffusion and degradation. Ideally, this release should be triggered by an impulse when the particles are administered. NLCs accommodate the drug because of their highly unordered lipid structures. Drug release can be initiated by applying the trigger impulse to the matrix to convert in a more ordered structure. Few NLC structures function in this way (Radtke and Müller, 2001). NLCs can generally be applied where solid NPs are advantageous for the delivery of drugs. Major application areas in pharmaceutics are topical drug delivery, oral and parenteral (i.e., subcutaneous or intramuscular and intravenous) route. Lipid–drug conjugate NPs have proved particularly useful for targeting water-soluble drug administration. They also have been utilized in the delivery of anti-inflammatory compounds, cosmetic preparation, topical cortico therapy and they also increase bioavailability and drug-loading capacity.

3.3.5   NANOSHELLS

Nanoshells are spherical cores of a particular compound and they are surrounded by a shell or thin-layer coating of another material (Xia et al., 2000). Nanoshell particles are highly functional materials which show modified and improved properties than their single component counterparts or NPs of the same size. Their properties can be changed by modifying the outer part (Oldenberg et al., 1998). Nanoshell materials can be synthesized from semiconductors (dielectric materials such as silica and polystyrene), metals, and insulators. Silica and polystyrene are commonly used as core because they are highly stable (Kalele et al., 2006a and Kalele et al., 2006b). Metal nanoshells are a novel type of composite spherical NPs consisting of a dielectric core covered by a thin metallic shell which is typically gold. Nanoshells possess highly favorable optical and chemical properties for biomedical imaging and therapeutic applications. Nanoshells offer other advantages over conventional organic dyes including improved optical properties and reduced susceptibility to chemical or thermal denaturation. Conjugation process that is used to bind biomolecules to gold colloid is easily modified for nanoshells (Loo et al., 2004). When a nanoshell and polymer matrix is illuminated with resonant wavelength, nanoshells absorb heat and transfer it to the local environment. This causes the collapse of the network and release of the drug. The drug can be encapsulated or adsorbed onto the shell surface in core-shell-particle-based drug delivery systems (Sparnacci et al., 2002). The shell interacts with the drug through a specific functional group or by electrostatic stabilization method. When it comes in contact with the biological system, it directs the drug. In imaging applications, nanoshells can be tagged with specific antibodies for diseased tissues or tumors.

3.3.6   QUANTUM DOTS

Quantum dots (QDs) consist of semiconductor nanocrystals and core–shell nanocrystals containing interface between various semiconductor materials. The range of QDs can be from 2 to 10 nm, which, generally increases to 5–20 nm in diameter after polymer encapsulation. Small-size particles like 5 nm are quickly cleared by renal filtration (Choi et al., 2007a and Choi et al., 2007b). Semiconductor nanocrystals have unique and fascinating optical properties. These have become an indispensable tool in biomedical research, especially in quantitative and long-term fluorescence imaging and detection (Michalet et al., 2005 and Smith et al., 2006). QD core can serve as the structural scaffold and the imaging contrast agent and small-molecule hydrophobic drugs can be embedded between the inorganic core and the amphiphilic polymer coating layer. Hydrophilic therapeutic agents including small interfering ribonucleic acid (RNA) and antisense oligodeoxynucleotide and targeting biomolecules such as antibodies, peptides, and aptamers can be immobilized into the hydrophilic side of the amphiphilic polymer either through covalent or noncovalent bonds. This fully integrated nanostructure may behave like magic bullets that will not only identify but also bind to diseased cells and treat it. Its alternative unique property is that it may emit detectable signals for real-time monitoring of its trajectory (Qi and Gao, 2008). These benefits enable applications of QDs in medical imaging and disease detection.

3.3.7   SUPERPARAMAGNETIC NPS

Superparamagnetic molecules are those that are attracted to a magnetic field but do not retain residual magnetism after the field is removed. Techniques involve coating the particles with antibodies to cell-specific antigens for separation from the surrounding matrix. Superparamagnetic NPs can be visualized in magnetic resonance imaging due to their paramagnetic properties (Irving, 2007). Superparamagnetic NPs are iron oxide core coated by either inorganic (silica, gold) or organic (phospholipids, fatty acids, polysaccharides, peptides, or other surfactants and polymers) materials (Babic et al., 2008 and Gupta and Curtis, 2004). In contrast to other NPs, the inducible magnetic properties of superparamagnetic NPs allow them to be directed to a defined location or heated in the presence of an externally applied AC magnetic field. These characteristics make them attractive for many applications and magnetically assisted transfection of cells (Gupta and Gupta, 2005 and Horák, 2005). Already marketable products, so-called beads, are micron-sized polymer particles loaded with superparamagnetic iron oxide NPs (SPIONs). Such beads can be functionalized with molecules that allow a specific adsorption of proteins or other biomolecules and subsequent separation in a magnetic field gradient for diagnostic purposes. More interesting applications, such as imaging of single cells or tumors, delivery of drugs or genes, local heating and separation of peptides, signaling molecules or organelles from a single living cell or from a living body (i.e., human) are still subjects of intensive research. The transdisciplinarity of basic and translational research carried out in superparamagnetic NPs during the last decades lead to a broad field of novel applications for superparamagnetic

NPs. The issues such as the mechanisms utilized by cells to take up multi-functional SPIONs in human cells in culture; the involvement of membrane molecules; specific adsorption of SPIONs to targeted subcellular compo-nents after uptake, transport of drugs, plasmids, or other substances to specific cells followed by controlled release; separation of SPIONs from the cells after cell uptake and specific adsorption to subcellular components or to biomolecules like proteins without interfering with cell function; preven-tion of uncontrolled agglomeration of modified SPIONs in physiological liquids; short- and long-term impact on cell functions by loading cells of different phenotypes with such NPs are not yet fully understood (Hofmann-Amtenbrink et al., 2009).

3.3.8   DENDRIMERS

Dendrimers are unimolecular, monodisperse, micellar nanostructures, around 20 nm in size, with a well-defined, regularly branched symmetrical structure, and a high density of functional end groups at their periphery. The structure of dendrimers consists of three distinct architectural regions: a focal moiety or a core, layers of branched repeat units emerging from the core, and functional end groups on the outer layer of repeat units. Robust, covalently fixed, three-dimensional structures possessing both a solvent-filled interior core (i.e., nanoscale container) as well as a homogenous, mathematically defined, exterior surface functionality are some properties of dendrimers (Grayson and Frechet, 2001 and Svenson and Tomalia, 2005). The divergent method can be used to prepare dendrimers. In case of paren-teral injections, dendrimeric vectors are most commonly used (Tomalia et al. 2007). Dendrimers used in drug delivery studies typically incorporate one or more of the following polymers: polyamidoamine (PAMAM), melamine, poly-L-glutamic (PG) acid, polyethyleneimine, polypropyleneimine, PEG, and chitin. Dendrimers may be used in two major modalities for targeting vectors for diagnostic imaging, drug delivery, gene transfection, detec-tion and therapeutic treatment of cancer and other diseases, first, passive targeting-nanodimension mediated through enhanced permeability retention effect (Matsumura and Maeda, 1986) involving primary tumor vascular-ization or organ-specific targeting (Kobayashi and Brechbiel, 2003); and second, active targeting-receptor-mediated cell-specific targeting involving receptor-specific targeting groups (Hofmann-Amtenbrink et al., 2009). There are several potential applications of dendrimers in the field of imaging, drug delivery, gene transfection, and nonviral gene transfer.

3.4   MANUFACTURING OF NANOMATERIALS

There is a wide variety of techniques that are used for generating nanostructures with various degrees of quality, speed, and cost. These manufacturing methods of NPs fall under aforementioned two types.

3.4.1   BOTTOM-UP MANUFACTURING

Bottom-up manufacturing implies the assembling of nanostructures atom by atom or molecule by molecule. This can be performed in the aforementioned three ways: chemical synthesis, self-assembly, and positional assembly given in Figure 3.2.

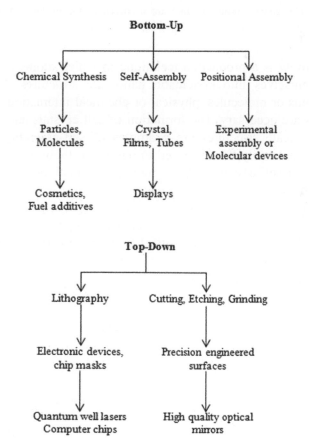

FIGURE 3.2   The use of bottom-up and top-down techniques in manufacturing nanoparticles (NPs).

Chemical synthesis is a production technique in which raw materials, such as molecules or particles are produced. The generated molecules or particles can then be used either directly in products in their bulk-disordered form or as the building blocks of more advanced ordered materials. In Figure 3.3, the general processes that are involved in the production of NPs are presented (The Royal Society, 2004).

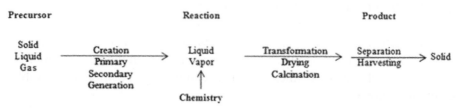

FIGURE 3.3   The general processes that are involved in the production of NPs using chemical synthesis.

Self-assembly is a production technique in which atoms or molecules assemble themselves into systematic nanoscale structures. In order to assemble atoms or molecules, physical or chemical interactions within the smaller units are necessary. The formation of salt crystals and snowflakes with their intricate structures are examples of self-assembly processes. Although self-assembly has happened in nature for thousands of years, till now, the use of self-assembly in the industry is relatively new and not a well-established process (The Royal Society, 2004).

Positional assembly is a production technique in which, atoms, molecules, or clusters are intentionally manipulated and positioned one by one. Scanning probe microscopy and optical tweezers are the techniques that are used for positional assembly. Positional assembly is very difficult and infrequently used in the industries.

3.4.2   TOP-DOWN MANUFACTURING

Top-down manufacturing implies the breaking down of a larger piece of material followed by cutting, etching, grinding to generate the required nanostructures (Fig. 3.2). This can be done by using techniques such as precision engineering and lithography and has been developed and refined by the semiconductor industry over the past 30 years. This method suggests consistency and device complexity. In this method, energy usage is higher and produce more waste than the bottom-up methods (Hoet et al., 2004 and Bandi Ramesh et al., 2011).

3.5 BENEFITS OF NPS OVER TRADITIONAL DRUG DELIVERY

Conventionally, drugs are taken orally or through parenteral routes that circulate them throughout the body, which may have harmful effects on the cells or tissues or organs. Protein and peptide drugs are poorly absorbed after oral administration because of their susceptibility across the intestinal epithelium. Conventional drug delivery needs high doses to make up the bioavailability (Jain et al., 2010 and Zhang et al., 2008). The advantages of using NPs as a drug delivery system include the following (Singh et al., 2011):

1. Controlled and sustained release of the drug during the transportation and at the site of localization, altering organ distribution of drug, and subsequent clearance of the drug so as to achieve an increase in drug therapeutic efficiency and reduction in side effects.
2. Drug can be incorporated into the system without any chemical reaction; it is an important factor for preserving the drug.
3. Controlled release and drug degradation characteristics can be readily modulated.
4. There is no wastage of drug and thus enhanced bioavailability of drug at specific site in right proportion for prolonged period of time.
5. It improves the solubility of poorly water-soluble drug.
6. It prolongs the biological half-life of the drug.
7. It releases drug at sustained rate and lowers the frequency of administration.
8. It provides comfort and compliance to the patient and yet improves the therapeutic performance of the drug over conventional systems.

3.6 BIOMEDICAL APPLICATIONS OF NPS

Nanotechnology is now extending the medical tools, knowledge, and therapies at present available to clinicians. The application of nanotechnology in medicine draws on the natural scale of biological phenomena to produce actual solutions for disease prevention, diagnosis, and treatment (Salata et al., 2004). A list of some of the applications of nanomaterials to biomedicine is specified in Table 3.1.

TABLE 3.1 Biomedical Application of nanoparticles.

Name of the study	Composition of the NP	Application
	Fullerenes	
Friedman et al., 1993	Fullerene	HIV proteases
Marchesan et al., 2005	Fulleropyrrolidines	HIV-1 and HIV-2
Schuster et al., 2000	Dendrofullerene 1	HIV-1 replication
Kotelnikova et al., 2003	Amino acid derivatives of fullerene	HIV and human cytomegalovirus replication
Kaesermann and Kempf, 1997	Buckminsterfullerene	Semliki forest virus or vesicular stomatitis virus
Mashino et al., 2005	Cationic, anionic, and amino-acid-type fullerene	HIV-reverse transcriptase and hepatitis C virus replication
Krusic et al., 1991	Fullerene 34 methyl radicals	Free radicals and oxidative stress
Slater et al., 1985	Fullerene	Liver toxicity and diminished lipid peroxidation
Dugan et al., 1997	C3-Fullero-tris-methanodicarboxylic acid	Apoptosis of neuronal cells
Huang et al.,1998	Carboxyfullerene	Apoptosis of hepatoma cells
Dugan et al., 1997	Carboxyfullerenes	Neurological disease including Parkinson's disease
Azzam and Domb, 2004	Fullerene with organic cationic compounds, viral carriers, recombinant proteins, and inorganic NPs	Gene transfer
Thrash et al., 1999	Metallofullerol	Leukemia and bone cancer
	Solid lipid nanoparticles	
Pandey and Khuller, 2005	Stearic acid	*Mycobacterium tuberculosis*
Jain and Banerjee, 2008	Stearic acid, soya phosphatidylcholine, and sodium taurocholate	Gram-negative bacteria, Gram-positive bacteria, and mycoplasma
Souto et al., 2004	Glyceryl tripalmitate and tyloxapol	Fungi (e.g., yeast, Aspergilli, dermatophytes)
Souto and Müller, 2005	Glyceryl behenate and sodium deoxycholate	Fungi

TABLE 3.1 (Continued)

Name of the study	Composition of the NP	Application
Bhalekar et al., 2009	Glyceryl behenate, propylene glycol, polysorbate 80, and glyceryl monostearate	Fungi
Sanna et al., 2007	Glycerol palmitostearate	Fungi
Sarmento et al., 2007	Cetyl palmitate	Type 1 diabetes
Jain et al., 2009a and Jain et al., 2009b	Lecithin, sodium taurocholate	Inflammation
Panga et al., 2009	Oleic acid	Inflammation
Wong et al., 2006	Soyabean oil	Breast cancer
Jain et al., 2009a and Jain et al., 2009b	Hyaluronic-acid-coupled chitosan	Colorectal cancer
Serpe et al., 2004	Cholesteryl butyrate	Colorectal cancer
Fontana et al., 2005	Solid lipid NP	Breast cancer
Ruckmani et al., 2006	Solid lipid NP	Carcinoma
Liposomes		
Takemoto et al., 2004	Hydrogenated soya, phosphatidylcholine, cholesterol, and distearoylphosphatidylglycerol	*Aspergillus fumigatus*
Omri et al., 2002	1,2-Dipalmitoyl-sn-glycero-3-phosphocholine and cholesterol	*Pseudomonas aeruginosa*
Schumacher and Margalit, 1997	Hydrogenated soya phosphatidylcholine and cholesterol	*Micrococcus luteus and Salmonella typhimurium*
Magallanes et al., 1993	Dipalmitoyl-phosphatidylcholine, dipalmitoyl-phosphatidylglycerol, and cholesterol	*Salmonella dublin*
Kim and Jones, 2004	Dipalmitoyl-phosphatidylcholine, cholesterol, and dimethylammonium ethane carbamoyl cholesterol	*Staphylococcus aureus*

TABLE 3.1 *(Continued)*

Name of the study	Composition of the NP	Application
Mimoso et al., 1997	Phosphatidylcholine, cholesterol, and phosphatidylinositol	*Bacillus subtilis* and *Escherichia coli*
Gangadharam et al., 1991	Phosphatidyl glycerol, phosphatidyl choline, and cholesterol	*Mycobacterium avium*
Fielding et al., 1998	Hydrogenated soy phosphatidylcholine, cholesterol, and distearoylphosphatidylglycerol	Gram-negative bacteria
Kaur et al., 2008	Stearylamine and dicetyl phosphate	HIV
Onyeji et al., 1994	Egg phosphatidylcholine, diacetylphosphate, and cholesterol	methicillin-resistant *S. aureus*
Park, 2002	Liposome	Breast cancer
Di Paolo et al., 2009	Liposome	Neuroblastoma
Spangler, 1990	Hepatically targeted liposomes	Diabetes mellitus
Nanostructured lipid carriers		
Bhaskar et al., 2009	Phosphatidylcholine, dynasan, and flurbiprofen	Sustained release of anti-inflammatory drug
Silva et al., 2009	Stearic acid, oleic acid, carbopol, and minoxidil	Pharmaceutical, cosmetic, and biochemical purposes
Doktorovová et al., 2010	Fluticasone propionate, glyceryl palmito-stearate, and PEG	Topical corticotherapy
Hentschel et al., 2008	Beta-carotene-loaded propylene glycol monostearate	Evaluate the feasibility
Hu et al., 2006	Monostearin, caprylic and capric triglycerides	Improved drug-loading capacity and controlled release properties
Nanoshells		
Ung et al., 1998	Silica coating of silver colloids	Stability of colloids
Thaxton et al., 2005	Gold nanoshell	Detection of DNA

TABLE 3.1 *(Continued)*

Name of the study	Composition of the NP	Application
Hirsch et al., 2003a	Gold nanoshell	Immunoassay to detect analytes
Loo et al., 2004	Nanoshell	Detection of cancer cells
Hirsch et al., 2003b	Nanoshell	Detection of tumors
Kalele et al., 2005	Silica–silver core–shell particles	Detection of antibodies
Kalele et al., 2006b	Silver nanoshell	Detection of microorganisms
Kalele et al., 2006a	Silver nanoshells	Detection of toxic metal ions such as Cd, Hg, and Pb present in water
QDs		
Hohng and Ha, 2005	QDs	For measuring protein conformational changes
Pathak et al., 2001	QDs-conjugated oligonucleotide sequences	Gene technology
Hasegawa et al., 2005	Conjugation of QD with Tat protein and by encapsulation in cholesterol-bearing pullulan modified with amine groups coating with a silica shell	Fluorescent labeling of cellular proteins and different intracellular structures
Dubertret et al., 2002	QDs encapsulated in phospholipid micelles	Cell tracking and color imaging of live cells
Yang and Li, 2006	Transferrin-bound QDs, wheat germ agglutinin, and transferrin-bound QDs, p53 conjugated with QDs	Pathogen and toxin detection such as *Cryptosporidium parvum* and *Giardia lamblia*, *E. coli* and *Salmonella typhi*, Hepatitis B and C viruses, and *Listeria monocytogenes*
Gao et al., 2004	Polyethylene-glycol-encapsulated QDs	In vivo animal imaging, lymph node mapping
Akerman et al., 2002	QDs	Barriers to use in vivo
Choi et al., 2007	Combination of QDs imaging with second-harmonic generation, CdTe-bound QDs	Tumor biology investigation, cell motility and metastatic potential, measurement of different cancer antigens

TABLE 3.1 *(Continued)*

Name of the study	Composition of the NP	Application
	Superparamagnetic NPs	
Smith et al., 2007	SPIONs coated with organic molecules	MRI contrast agents for detecting liver tumors
Zur Mühlen et al., 2007	SPIONs	Identify dangerous arteriosclerotic plaques by MRI
Yoo et al., 2007	SPIONs coated with polyvinylbenzyl-O-β-D-galactopyranosyl-D-gluconamide with galactose moieties	Liver-targeting MRI contrast agent
Meng et al., 2009	SPIONs conjugated to luteinizing hormone releasing hormone	Enhanced MRI contrast in breast cancer xeno-grafts and metastases in the lungs
Minard, 2009	SPIONs	Magnetic particle imaging
McIlwain, 2008	Combidex— an ultrasmall superparamagnetic iron oxide covered dextran	Molecular imaging agent during contrast-en-hanced MRI
Morishige et al., 2010	Monocrystalline iron oxide NPs-47	Measures macrophage burden in atherosclerosis
Jordan et al., 1999	Colloidal dispersions of SPIONs	Magnetic fluid hyperthermia in cancer treatment
Chiang et al., 2005	Nanosized superparamagnetic NPs coated with the multi-valent cationic agent, polyethylenimine	Purification of plasmid DNA from bacterial cells
	Dendrimers	
Wiener et al., 1994	PAMAM	Diagnose certain disorders of the heart, brain, and blood vessels
Thomas et al., 2004	PAMAM	Cell binding and internalization
Abeylath et al., 2008	PAMAM	Strep throat (*Streptococcus*), staph infection (*S. aureus*), and flu (*Haemophilus influenzae*)
Cheng et al., 2007	PAMAM	Various bacteria
D'Emanuele et al., 2004	PAMAM	Hypertension
Devarakonda et al., 2005	PAMAM dendrimers	Tapeworm

TABLE 3.1 *(Continued)*

Name of the study	Composition of the NP	Application
Bhadra et al., 2005	PEGylated lysine-based copolymeric dendrimer	*Plasmodium falciparum*
Tolia et al., 2008	PAMAM dendrimers with carboxylic or hydroxyl surface groups	Glaucoma
Bai et al., 2007	PAMAM	Pulmonary embolism
Cheng et al., 2007	PAMAM	Inflammation
Chauhan et al., 2003	PAMAM	Inflammation
Rupp et al., 2007	Polylysine dendrimer	HIV, HSV, and sexually transmitted infections
Schumann et al., 2003	Dendrimer	Diagnostic tool for arteriosclerotic vasculature, tumors, infarcts, kidneys, or efferent urinary tract
Culver, 1994	Dendrimer	Induce a systemic antitumor immune response against residual tumor cells
Zhuo et al., 1999	PAMAM	Tumor
Hawthorne, 1993	Dendrimer	Cancer

HIV: Human immunodeficiency virus, DNA: Deoxyribonucleic acid, Cd: Cadmium, Hg: Mercury, Pb: Lead, MRI: Magnetic resonance imaging, HSV: Herpes simplex virus, PEG: Polyethylene glycol, PAMAM: Polyamidoamine, SPIONs: Superparamagnetic iron oxide nanoparticles, QDs: Quantum dots.

3.7   NANOPHARMACEUTICALS

Nanopharmaceuticals epitomize an emergent field in which the sizes of the drug particle or a therapeutic delivery system work at the nanoscale. However, for the pharmaceutical industry, the main rate-limiting step is to deliver the appropriate dose of a particular active agent to specific disease site (Boisseau et al., 2011). Nanopharmaceuticals have vast potential in addressing the failure of traditional therapeutics by providing site-specific targeting of active agents to provide better patient compliance which reduces toxic systemic side effects (Nikalje et al., 2015). In Table 3.2, some commercially available nanopharmaceuticals are presented (Weissig et al., 2014).

3.8   FUTURE OPPORTUNITIES AND CHALLENGES

NPs provide massive advantages regarding drug targeting, delivery, and release, and with their additional potential to combine diagnosis and therapy, they emerge as one of the major tools in nanomedicine. The main goals are to improve their stability in the biological environment; mediate the biodistribution of active compounds; and improve drug loading, targeting, transport, release, and interaction with biological barriers. The cytotoxicity of NPs or their degradation products remains a major problem and improvements in biocompatibility obviously are a main concern of future research (Bawa et al.,2005; Pragati et al., 2009 and Wadher et al., 2009).

3.9   CONCLUSION

The exclusive size-dependent properties of NPs make them superior and indispensable in numerous parts of human activity. Nanodelivery systems can efficiently target numerous varied cell types without any difficulties associated with traditional therapy. The outstanding features of NPs enable them to remove the chance of drug resistance in target cells and facilitate the movement of drugs across biological barriers. However, the challenge remains the exact characterization of the drugs target and ensuring that they only affect targeted organs. Moreover, it is imperative to recognize the fate of the drugs once transported to the targets.

TABLE 3.2 Commercially Available Nanopharmaceuticals.

Brand name	Active ingredient	Form	Approval	Application
AmBisome®	Amphotericin B	Liposomes	FDA 1997	Systemic fungal infections
DaunoXome®	Daunorubicin citrate	Liposomes	FDA 1996	HIV-related Kaposi's sarcoma
DepoCyt®	Cytarabine	Liposomes	FDA 1999/2007	Lymphomatous malignant meningitis
DepoDur®	Morphine sulfate	Liposomes	FDA 2004	For treatment of chronic pain
Doxil®	Doxorubicin hydrochloride	Liposomes	FDA 1995	AIDS-related Kaposi's sarcoma, multiple myeloma, ovarian cancer
Inflexal® V	Influenza virus antigens (hemagglutinin, neuraminidase)	Liposomes	Switzerland 1997	Influenza vaccine
Marqibo®	Vincristine sulfate	Liposomes	FDA 2012	Acute lymphoid leukemia, Philadelphia chromosome-negative, relapsed or progressed
Mepact™	Mifamurtide	Liposomes	Europe 2009	Non-metastasizing respectable osteosarcoma
Myocet®	Doxorubicin	Liposomes	Europe 2000	Metastatic breast cancer
Visudyne®	Verteporfin	Liposomes	FDA 2000	Pathological myopia, ocular histoplasmosis syndrome
Abelcet®	Amphotericin B	Lipid-based (non-liposomal) formulations	FDA 1995 and 1996	Systemic fungal infections
Adagen®	PEGylated adenosine deaminase	PEGylated proteins, polypeptides, aptamers	FDA 1990	Immunodeficiency disease
Cimzia®	PEGylated antibody (Fab' fragment of a humanized anti-TNF-alpha antibody)	PEGylated proteins, polypeptides, aptamers	FDA 2008	Crohn's disease, rheumatoid arthritis
Neulasta®	PEGylated filgrastim (granulocyte colony-stimulating factor)	PEGylated proteins, polypeptides, aptamers	FDA 2002	Febrile neutropenia

TABLE 3.2 *(Continued)*

Brand name	Active ingredient	Form	Approval	Application
Oncaspar®	PEGylated L-asparaginase	PEGylated proteins, polypeptides, aptamers	FDA 1994	Acute lymphoblastic leukemia
Pegasys®	PEGylated interferon alfa-2b	PEGylated proteins, polypeptides, aptamers	FDA 2002	Hepatitis B and C
PegIntron®	PEGylated interferon alfa-2b	PEGylated proteins, polypeptides, aptamers	FDA 2001	Hepatitis C
Somavert®	PEGylated human growth hormone receptor antagonist	PEGylated proteins, polypeptides, aptamers	FDA 2003	Acromegaly
Macugen®	PEGylated anti-VEGF aptamer	PEGylated proteins, polypeptides, aptamers	FDA 2004	Neovascular age-related macular degeneration
Mircera®	PEGylated epoetin beta (erythropoietin receptor activator)	PEGylated proteins, polypeptides, aptamers	FDA 2007	Anemia associated with chronic renal failure in adults
Emend®	Aprepitant	Nanocrystals	FDA 2003	Emesis, antiemetic
Megace® ES	Megestrol acetate	Nanocrystals	FDA 2005	Anorexia, cachexia
Rapamune®	Rapamycin (sirolimus)	Nanocrystals	FDA 2002	Immunosuppressant
Tricor®, Triglide®	Fenofibrate	Nanocrystals	FDA 2004	Lipid-lowering agent
Copaxone®	Polypeptide of four amino acids (glatiramer)	Polymer-based nanoformulations	FDA 1996/2014	Multiple sclerosis
Eligard®	Leuprolide acetate	Polymer-based nanoformulations	FDA 2002	Advanced prostate cancer

TABLE 3.2 (*Continued*)

Brand name	Active ingredient	Form	Approval	Application
Genexol®	Paclitaxel	Polymer-based nanoformulations	South Korea 2001	Metastatic breast cancer, pancreatic cancer
Opaxio®	Paclitaxel	Polymer-based nanoformulations	FDA 2012	Glioblastoma
Renagel®	Polyallylamine hydrochloride	Polymer-based nanoformulations	FDA 2000	Hyperphosphatemia
Zinostatin stimalamer®	Conjugate protein or copolymer of styrene–maleic acid and an antitumor protein neocarzinostatin	Polymer-based nanoformulations	Japan 1994	Primary unresectable hepatocellular carcinoma
Abraxane®	Paclitaxel	Protein–drug conjugates	FDA 2005	Metastatic breast cancer, non-small cell lung cancer
Kadcyla®	Ado-trastuzumab emtansine	Protein–drug conjugates	FDA 2013	Metastatic breast cancer
Ontak®	Denileukin diftitox	Protein–drug conjugates	FDA 1994/2006	Primary cutaneous T-cell lymphoma, CD25-positive, persistent or recurrent disease
Fungizone®	Amphotericin B	Surfactant-based nanoformulations	FDA 1966	Systemic fungal infections
Diprivan®	Propofol	Surfactant-based nanoformulations	FDA 1989	Sedative–hypnotic agent for induction and maintenance of anesthesia
Estrasorb™	Estradiol	Surfactant-based nanoformulations	FDA 2003	Hormone replacement therapy during menopause
Feridex®*	Superparamagnetic iron oxide	Metal-based nanoformulations	FDA 1996	Liver/spleen lesion MRI

TABLE 3.2 *(Continued)*

Brand name	Active ingredient	Form	Approval	Application
Feraheme™	Superparamagnetic iron oxide	Metal-based nanoformulations	FDA 2009	Treatment of iron deficiency anemia in adults with chronic kidney disease
NanoTherm®	Superparamagnetic iron oxide	Metal-based nanoformulations	Europe 2013	Local ablation in glioblastoma, prostate, and pancreatic cancer
Gendicine®	Recombinant adenovirus expressing wildtype-p53	Virosomes	People's Republic of China 2003	Head and neck squamous cell carcinoma
Rexin-G®	Gene for dominant-negative mutant form of human cyclin G1	Virosomes	Philippines 2007	For all solid tumors

Note: *Manufacturing discontinued in 2008.

FDA: Food and Drug Administration, AIDS: *Acquired immune deficiency syndrome*, HIV: Human immunodeficiency virus, TNF: Tumor necrosis factor, VEGF: Vascular endothelial growth factor, PEG: Polyethylene glycol, MRI: Magnetic resonance imaging.

Source: Adapted from Weissig et al. (2014)

KEYWORDS

- **nanotechnology**
- **nanoparticles**
- **nanopharmaceuticals**
- **nanomedicine**
- **drug delivery system**

REFERENCES

Abeylath, S. C.; Turos, E.; Dickey, S.; Lim, D. V. Glyconanobiotics: Novel Carbohydrated Nanoparticle Antibiotics for MRSA and *Bacillus anthracis*. *Bio. Med. Chem.* **2008,** *16*(5), 2412–2418.

Akerman, M. E.; Chan, W. C.; Laakkonen, P.; Bhatia, S. N.; Ruoslahti, E. Nanocrystal Targeting in Vivo. *Proc. Natl. Acad. Sci. USA* **2002,** *99*(20), 12617–12621.

Athar, M.; Das, A. J. Therapeutic Nanoparticles: State-of-The-Art of Nanomedicine. *Adv. Mater. Rev.* **2014,** *1*(1), 25–37.

Avgoustakis, K.; Beletsi, A.; Panagi, Z.; Klepetsanis, P.; Karydas, A. G; Ithakissios, D. S. PLGA-mPEG Nanoparticles of Cisplatin: in Vitro Nanoparticle Degradation, in Vitro Drug Release and in Vivo Drug Residence in Blood Properties. *J. Control Release* **2002,** *79,* 123–135.

Azzam, T.; Domb. A. J. Current Developments in Gene Transfection Agents. *Curr. Drug Delivery* **2004,** *1*(2), 165–193.

Babic, M.; Horák, D.; Trchová, M.; Jendelová, P.; Glogarová, K.; Lesný, P.; Herynek, V.; Hájek, M.; Syková, E. Poly(L-Lysine)-Modified Iron Oxide Nanoparticles for Stem Cell Labeling. *Bioconjug. Chem.* **2008,** *19*(3), 740–750.

Bai, S.; Thomas, C.; Ahsan, F. Dendrimers as a Carrier for Pulmonary Delivery of Enoxaparin, a Low Molecular Weight Heparin. *J. Pharm. Sci.* **2007,** *96*(8), 2090–2106.

Bandi Ramesh; Ganesh, N. S.; Kotha N. S.; Lova R.; Kavitha, K. Nano Particulate Drug-Delivery Systems: an Overview. *J. Pharm. Res.* **2011,** *4*(6), 1688–1690.

Barratt, G. M. Therapeutic Applications of Colloidal Drug Carriers. *Pharm. Sci. Tech. Today* **2000,** *3,* 163–171.

Barrera, D. A. Synthesis and RGD Peptide Modification of a New Biodegradable Copolymer: Poly(Lactic Acid-Co-Lysine). *J. Am. Chem. Soc.* **1993,** *115,* 11010–11011.

Bhalekar, M. R.; Pokharkar, V.; Madgulkar, A.; Patil, N.; Patil, N. Preparation and Evaluation of Miconazole Nitrate-Loaded Solid Lipid Nanoparticles for Topical Delivery. *AAPS Pharm. Sci. Tech.* **2009,** *10*(1), 289–296.

Bawa, R.; Bawa, S. R.; Meibius, S. B. Protecting New Ideas and Invention in Nanomedicine with Patent. *Nanomedicine* **2005,** *1*(2), 150–158.

Beletsi, A.; Leontiadis, L.; Klepetsanis, P.; Ithakissios, D. S.; Avgoustakis, K, Effect of Preparative Variables on the Properties of Poly (Dl-Lactide-Co-Glycolide)-Methoxypoly

(Ethyleneglycol) Copolymers Related to Their Application in Controlled Drug Delivery. *Int. J. Pharm.* **1999**, *182,* 187–197.

Bhadra, D.; Bhadra, S.; Jain, N. K. Pegylated Lysine Based Copolymeric Dendritic Micelles for Solubilization and Delivery of Artemether. *J. Pharm. Pharm. Sci.* **2005**, *8*(3), 467–482.

Bhadra, D.; Bhadra, S.; Jain, P.; Jain, N. K. Pegnology: a Review of PEG-Ylated Systems. *Pharmazie* **2002**, *57,* 5–29.

Bhaskar, K.; Anbu, J.; Ravichandiran, V.; Venkateswarlu, V.; Rao, Y. M. Lipid Nanoparticles for Transdermal Delivery of Flurbiprofen: Formulation, in Vitro, Ex Vivo and in Vivo Studies. *Lipids Health Dis.* **2009**, *8,* 6.

Boisseau, P.; Loubaton, B. N. Nanotechnology in Medicine. *C. R. Phys.* **2011**, *12,* 620–636.

Brigger, I.; Dubernet, C.; Couvreur, P. Nanoparticles in Cancer Therapy and Diagnosis. *Adv. Drug Del. Rev.* **2002**, *54,* 631–651.

Calvo, P.; Remuñan-López, C.; Vila-Jato, J. L.; Alonso, M. J. Chitosan and Chitosan/Ethylene Oxide-Propylene Oxide Block Copolymer Nanoparticles as Novel Carriers for Proteins and Vaccines. *Pharm. Res.* **1997**, *14,* 1431–1436.

Cavalli, R.; Caputo, O.; Gasco, M. R. Solid Lipospheres of Doxorubicin and Idarubicin. *Int. J. Pharm.* **1993**, *89*(1), R9–R12.

Cevc, G. Transfersomes, Liposomes and Other Lipid Suspensions on the Skin: Permeation Enhancement, Vesicle Penetration, and Transdermal Drug Delivery. *Crit. Rev. Ther. Drug Car. Syst.* **1996**, *13*(3–4), 257–388.

Chauhan, A. S.; Sridevi, S.; Chalasani, K. B.; Jain, A. K.; Jain, S. K.; Jain, N. K.; Diwan, P. V. Dendrimer-Mediated Transdermal Delivery: Enhanced Bioavailability of Indomethacin. *J. Control Release* **2003**, *90*(3), 335–343.

Chen, Y.; McCulloch, R. K.; Gray, B. N. Synthesis of Albumin-Dextran Sulfate Microspheres Possessing Favourable Loading and Release Characteristics for the Anticancer Drug Doxorubicin. *J. Control Release* **1994**, *31,* 49–54.

Chen, Y.; Mohanraj, V. J.; Parkin, J. E. Chitosan-Dextran Sulfate Nanoparticles for Delivery of an Anti-Angiogenesis Peptide. *Int. J. Peptide Res. Ther.* **2003**, *10*(5), 621–629.

Cheng, Y.; Man, N.; Xu, T.; Fu, R.; Wang, X.; Wang, X.; Wen, L. Transdermal Delivery of Nonsteroidal Anti-Inflammatory Drugs Mediated by Polyamidoamine (PAMAM) Dendrimers. *J. Pharm. Sci.* **2007**, *96*(3), 595–602.

Chiang, C. L.; Sung, C. S.; Wu, T. F.; Chen, C. Y.; Hsu, C. Y. Application of Superparamagnetic Nanoparticles in Purification of Plasmid DNA from Bacterial Cells. *J. Chromatoger. B: Anal. Technol. Biomed. Life Sci.* **2005**, *822*(1–2), 54–60.

Choi, A. O.; Cho, S. J.; Desbarats, J.; Lovric J; Maysinger, D. Quantum Dot-Induced Cell Death Involves Fas Upregulation and Lipid Peroxidation in Human Neuroblastoma Cells. *J. Nanobiotechnol.* **2007a**, *5,* 1–4.

Choi, H. S.; Liu, W.; Misra, P.; Tanaka, E.; Zimmer, J. P.; Ipe, B. I.; Bawendi, M. G.; Frangion, J. V. Renal Clearance of Quantum Dots. *Nat. Biotechnol.* **2007b**, *25*(10), 1165–1170.

Couvreur, P.; Barratt, G.; Fattal, E.; Legrand, P.; Vauthier, C. Nanocapsule Technology: a Review. *Crit. Rev. Ther. Drug Carrier Syst.* **2002**, *19*(2), 99–134.

Culver, K. W. Clinical Applications of Gene Therapy for Cancer. *Clin. Chem.* **1994**, *40*(4), 510–512.

D'Emanuele, A.; Jevprasesphant, R.; Penny, J.; Atwood, D. The Use of a Dendrimer-Propranolol Prodrug to Bypass Efflux Transporters and Oral Bioavailability. *J. Control Release* **2004**, *95*(3), 447–453.

Davda, J.; Labhasetwar, V. Characterization of Nanoparticle Uptake by Endothelial Cells. *Int. J. Pharm.* **2002**, *233*(1–2), 51–59.

Desai, M. P.; Labhasetwar, V.; Walter, E.; Levy, R. J.; Amidon, G. L. The Mechanism of Uptake of Biodegradable Microparticles in Caco-2 Cells is Size Dependent. *Pharm. Res.* **1997,** *14*(11), 1568–1573.

Devarakonda, B.; Hill, R. A.; Liebenberg, W.; Brits, M.; de Villiers, M. M. Comparison of the Aqueous Solubilization of Practically Insoluble Niclosamide by Polyamidoamine (PAMAM) Dendrimers and Cyclodextrins. *Int. J. Pharm.* **2005,** *304*(1–2), 193–209.

Di Paolo, D.; Loi, M.; Pastorino, F.; Brignole, C.; Marimpietri, D.; Becherini, P.; Caffa, I.; Zorzoli, A.; Longhi, R.; Gagliani, C.; Tacchetti, C.; Corti, A.; Allen, T. M.; Ponzoni, M.; Pagnan, G. Liposome-Mediated Therapy of Neuroblastoma. *Meth. Enzymol.* **2009,** *465,* 225–249.

Doktorovová, S.; Araújo, J.; Garcia, M. L.; Rakovský, E.; Souto, E. B. Formulating Fluticasone Propionate in Novel PEG-Containing Nanostructured Lipid Carriers (PEG-NLC). *Colloids Surf. B.* **2010,** *75*(2), 538–542.

Domb, A. J. Liposphere Parenteral Delivery System. *Proc. Intl. Symp. Con. Rel. Bioact. Mater.* **1993,** *20,* 346–347.

Dubertret, B.; Skourides, P.; Norris, D. J.; Noireaux, V.; Brivanlou, A. H.; Libchaber, A. In Vivo Imaging of Quantum Dots Encapsulated in Phospholipid Micelles. *Science* **2002,** *298*(5599), 1759–1762.

Dugan, L. L.; Turetsky, D. M.; Du, C.; Lobner, D.; Wheeler, M.; Almli, C. R.; Shen, C. K.; Luh, T. Y.; Choi, D. W.; Lin, T. S. Carboxyfullerenes as Neuroprotective Agents. *Proc. Natl. Acad. Sci. USA.* **1997,** *94*(17), 9434–9439.

Dunne, M.; Corrigan, I.; Ramtoola, Z. Influence of Particle Size and Dissolution Conditions on the Degradation Properties of Polylactide-Co-Glycolide Particles. *Biomaterials* **2000,** *21,* 1659–1668.

Eldem, T.; Speiser, P.; Hincal, A. Optimization of Spray-Dried and Congealed Lipid Microparticles and Characterization of Their Surface Morphology by Scanning Electron Microscopy. *Pharm. Res.* **1991,** *8*(1), 47–54.

Fielding, R. M.; Lewis, R. O.; Moon-McDermott, L. Altered Tissue Distribution and Elimination of Amikacin Encapsulated in Unilamellar, Low-Clearance Liposomes (MiKasome). *Pharm. Res.* **1998,** *15*(11), 1775–1781.

Fontana, G.; Maniscalco, L.; Schillaci, D.; Cavallaro, G.; Giammona, G. Solid Lipid Nanoparticles Containing Tamoxifen Characterization and in Vitro Antitumoral Activity. *Drug. Deliv.* **2005,** *12*(6), 385–392.

Freitas, C.; Müller, R. H. Spray-Drying of Solid Lipid Nanoparticles (SLN™). *Eur. J. Pharm. Biopharm.* **1998,** *46*(2), 145–151.

Fresta, M.; Puglisi, G.; Giammona, G.; Cavallaro, G.; Micali, N.; Furneri, P. M. Pefloxacine Mesilate- and of loxacin-loaded Polyethylcyanoacrylate Nanoparticles: Characterization of the Colloidal Drug Carrier Formulation. *J. Pharm. Sci.* **1995,** *84*(7), 895–902.

Friedman, S. H.; DeCamp, D. L.; Sijbesma, R. P.; Srdanov, G.; Wudl, F.; Kenyon, G. L. Inhibition of the HIV-1 Protease by Fullerene Derivatives: Model Building Studies and Experimental Verification. *J. Am. Chem. Soc.* **1993,** *115*(15), 6506–6509.

Gangadharam, P. R.; Ashtekar, D. A.; Ghori, N.; Goldstein, J. A.; Debs, R. J.; Duzgunes, N. Chemotherapeutic Potential of Free and Liposome Encapsulated Streptomycin Against Experimental Mycobacterium Avium Complex Infections in Beige Mice. *J. Antimicrob. Chemother.* **1991,** *28*(3), 425–435.

Gao, X. H.; Cui, Y. Y.; Levenson, R. M.; Chung, L. W. K.; Nie, S. M. In Vivo Cancer Targeting and Imaging with Semiconductor Quantum Dots. *Nat. Biotechnol.* **2004,** *22,* 969–976.

Govender, T.; Riley, T.; Ehtezazi, T.; Garnett, M. C.; Stolnik, S.; Illum, L.; Davis, S. S. Defining the Drug Incorporation Properties of PLA-PEG Nanoparticles. *Int. J. Pharm.* **2000,** *199,* 95–110.

Grayson, S. M.; Frechet, J. M. Convergent Dendrons and Dendrimers: from Synthesis to Applications. *Chem. Rev.* **2001,** *101*(12), 3819–3868.

Grislain, L. Pharmacokinetics and Distribution of a Biodegradable Drug-Carrier. *Int. J. Pharm.* **1983,** *15,* 335–345.

Gupta, A. K.; Curtis, A. S. G. Lactoferrin and Ceruloplasmin Derivatized Superparamagnetic Iron Oxide Nanoparticles for Targeting Cell Surface Receptors. *Biomaterials* **2004,** *25*(15), 3029–3040.

Gupta, A. K.; Gupta, M. Synthesis and Surface Engineering of Iron Oxide Nanoparticles for Biomedical Applications. *Biomaterials* **2005,** *26*(18), 3995–4021.

Guzman, L. A.; Labhasetwar, V.; Song, C.; Jang, Y.; Lincoff, A. M.; Levy, R.; Topol, E. J. Local Intraluminal Infusion of Biodegradable Polymeric Nanoparticles. A Novel Approach for Prolonged Drug Delivery After Balloon Angioplasty. *Circulation* **1996,** *94,* 1441–1448.

Hasegawa, U.; Nomura, S. I. M.; Kaul, S. C.; Hirano, T.; Akiyoshi, K. Nanogel-Quantum Dot Hybrid Nanoparticles for Live Cell Imaging. *Biochem. Biophys. Res. Commun.* **2005,** *331*(4), 917–921.

Hawthorne, M. F. The Role of Chemistry in the Development of Boron Neutron Capture Therapy of Cancer. *Angew. Chem.* **1993,** *32*(7), 950–984.

Hentschel, A.; Gramdorf, S.; Müller, R. H.; Kurz, T. Beta-Carotene-loaded Nanostructured Lipid Carriers. *J. Food Sci.* **2008,** *73*(2), 1–6.

Hirsch, L. R.; Jackson, J. B.; Lee, A.; Halas, N. J.; West, J. L. A Whole Blood Immunoassay Using Gold Nanoshells. *Anal. Chem.* **2003a,** *75*(10), 2377–2381.

Hirsch, L. R.; Stafford, R. J.; Bankson, J. A.; Sreshen, S. R.; Rivera, B.; Price, R. E.; Hazle, J. D.; Halas, N. J.; West, J. L. Nanoshell-Mediated Near-Infrared Thermal Therapy of Tumours Under Magnetic Resonance Guide. *Proc. Natl. Acad. Sci. USA.* **2003b,** *100*(23), 13549–13554.

Hoet, P.; Hohlfeld, I. B.; Salata, O. V. Nanoparticles—Known and Unknown Health Risks. *J. Nanobiotechnol.* **2004,** *2,* 12.

Hofmann-Amtenbrink, M.; von Rechenberg, B.; Hofmann, H. Superparamagnetic Nanoparticles for Biomedical Applications. In *Nanostructured Materials for Biomedical Applications*; Tan, M. C., Chow, G. M., Ren, L., Ed.; Transworld Research Network: Kerala, 2009; pp 119–149.

Hohng, S.; Ha, T. Single-Molecule Quantum-Dot Fluorescence Resonance Energy Transfer. *Chemphyschem* **2005,** *6*(5), 956–960.

Holister, P.; Cristina, R. V.; Fullerenes, H. T. *Nanoparticles, Technology White papers nr. 3,* Cientifica: UK, **2003**.

Horák, D. Magnetic Microparticulate Carriers with Immobilized Selective Ligands in DNA Diagnostics. *Polymer* **2005,** *46*(4), 1245–1255.

Hu, F. Q.; Jiang, S. P.; Du, Y. Z.; Yuan, H.; Ye, Y. Q.; Zeng, S. Preparation and Characteristics of Monostearin Nanostructured Lipid Carriers. *Int. J. Pharm.* **2006,** *314*(1), 83–89.

Huang, Y. L.; Shen, C. K. F.; Luh, T. Y.; Yang, H. C.; Hwang, K. C.; Chou, C. K. Blockage of Apoptotic Signaling of Transforming Growth Factor-β in Human Hepatoma Cells by Carboxyfullerene. *Eur. J. Biochem.* **1998,** *254,* 38–43.

Irving, B. Nanoparticle Drug Delivery Systems. *Innovations Pharm. Biotechnol.* **2007,** *24,* 58–62.

Jain, D.; Banerjee, R. Comparison of Ciprofloxacin Hydrochloride-loaded Protein, Lipid, and Chitosan Nanoparticles for Drug Delivery. *J. Biomed. Mater. Res. B. Appl. Biomater.* **2008,** *86*(1), 105–112.

Jain, A.; Jain, S. K.; Ganesh, N.; Barve, J.; Beg, A. M. Design and Development of Ligand-Appended Polysaccharidic Nanoparticles for the Delivery of Oxaliplatin in Colorectal Cancer. *Nanomedicine* **2009a,** *6*(1), 179–190.

Jain, P.; Mishra, A.; Yadav, S. K.; Patil, U. K.; Baghel, U. S. Formulation Development and Characterization of Solid Lipid Nanoparticles Containing Nimesulide. *Int. J. Drug Deliver Technol.* **2009b,** *1*(1), 24–27.

Jain, N.; Jain, R.; Thakur, N.; Gupta, B. P.; Jain, D. K.; Banveeri, J.; Jain, S. Nanotechnology: a Safe and Effective Drug Delivery System. *Asian J. Pharm. Clin. Res.* **2010,** *3*(3), 159–164.

Jordan, A.; Scholz, R.; Wust, P.; Fähling, H.; Felix, R. Magnetic Fluid Hyperthermia (MFH): Cancer Treatment with AC Magnetic Field Induced Excitation of Biocompatible Superparamagnetic Nanoparticles. *J. Magn. Magn. Mater.* **1999,** *201*(1–3), 413–419.

Kaesermann, F.; Kempf, C. Photodynamic Inactivation of Enveloped Viruses by Buckminsterfullerene. *Antiviral Res.* **1997,** *34*(1), 65–70.

Kalele, S. A.; Ashtaputre, S. S.; Hebalkar, N. Y.; Gosavi, S. W.; Deobagkar, D. N.; Deobagkar, D. D.; Kulkarni, S. K. Optical Detection of Antibody Using Silica–Silver Core-Shell Particles. *Chem. Phys. Lett.* **2005,** *404*(1–3), 136–141.

Kalele, S. A.; Gosavi, S. W.; Urban, J.; Kulkarni, S. K. Nanoshell Particles: Synthesis, Properties and Applications. *Curr. Sci.* **2006a,** *91*(8), 1038–1052.

Kalele, S. A.; Kundu, A. A.; Gosavi, S. W.; Deobagkar, D. N.; Deobagkar, D. D.; Kulkarni, S. K. Rapid Detection of *Echerischia coli* Using Antibody Conjugated Silver Nanoshells. *Small* **2006b,** *2*(3), 335–338.

Kaur, C. D.; Nahar, M.; Jain, N. K. Lymphatic Targeting of Zidovudine Using Surface-Engineered Liposomes. *J. Drug Target.* **2008,** *16*(10), 798–805.

Kim, H. J.; Jones, M. N. The Delivery of Benzyl Penicillin to *Staphylococcus aureus* Biofilms by use of Liposomes. *J. Liposome. Res.* **2004,** *14*(3–4), 123–139.

Kobayashi, H.; Brechbiel, M. W. Dendrimer-Based Macromolecular MRI Contrast Agents: Characteristics and Application. *Mol. Imaging* **2003,** *2*(1), 1–10.

Kotelnikova, R. A.; Bogdanov, G. N.; Frog, E. C.; Kotelnikov, A. I.; Shtolko, V. N.; Romanova, V. S.; Andreev, S. M.; Kushch, A. A.; Fedorva, N. E.; Medzhidova, A. A.; Miller, G. G. Nanobionics of Pharmacologically Active Derivatives of Fullerene C60. *J. Nanopart. Res.* **2003,** *5,* 561–566.

Kreuter, J. *Encyclopaedia of Pharmaceutical Technology;* Marcel Dekker Inc.: New York, 1994.

Kreuter, J.; Ramge, P.; Petrov, V.; Hamm, S.; Gelperina, S. E.; Engelhardt, B.; Alyautdin, R.; von Briesen, H.; Begley, D. J. Direct Evidence that Polysorbate-80-Coated Poly(Butylcyanoacrylate) Nanoparticles Deliver Drugs to the CNS via Specific Mechanisms Requiring Prior Binding of Drug to the Nanoparticles. *Pharma. Res.* **2003,** *20,* 409–416.

Kroll, R. A.; Pagel, M. A.; Muldoon, L. L.; Roman-Goldstein, S.; Fiamengo, S. A.; Neuwelt, E. A. Improving Drug Delivery to Intracerebral Tumor and Surrounding Brain in a Rodent Model: a Comparison of Osmotic Versus Bradykinin Modification of the Blood-Brain and/ or Blood-Tumor Barriers. *Neurosurgery* **1998,** *43,* 879–886.

Krusic, P. J.; Wasserman, E.; Keizer, P. N.; Morton, J. R.; Preston, K. F. Radical Reactions of C60. *Science* **1991,** *254*(5035), 1183–1185.

Lamprecht, A.; Ubrich, N.; Yamamoto, H.; Schäfer, U.; Takeuchi, H.; Maincent, P.; Kawashima, Y.; Lehr, C. M. Biodegradable Nanoparticles for Targeted Drug Delivery in Treatment of Inflammatory Bowel Disease. *J. Pharmacol. Exp. Ther.* **2001**, *299*, 775–781.

Lockman, P. R.; Mumper, R. J.; Khan, M. A.; Allen, D. D. Nanoparticle Technology for Drug Delivery Across the Blood-Brain Barrier. *Drug. Dev. Ind. Pharm.* **2002**, *28*, 1–13.

Loo, C.; Lin, A.; Hirsch, L.; Lee, M.; Barton, J.; Halas, N.; West, J.; Drezek, R. Nanoshell-Enabled Photonics-Based Imaging and Therapy of Cancer. *Technol. Cancer Res. Treat.* **2004**, *3*(1), 33–40.

Magallanes, M.; Dijkstra, J.; Fierer, J. Liposome-Incorporated Ciprofloxacin in Treatment of Murine Salmonellosis. *Antimicrob. Agents Chemother.* **1993**, *37*(11), 2293–2297.

Magenheim, B. A New in Vitro Technique for the Evaluation of Drug Release Profile from Colloidal Carriers—Ultrafiltration Technique at Low Pressure. *Int. J. Pharm.* **1993**, *94*, 115–123.

Marchesan, S.; Da Ros, T.; Spalluto, G.; Balzarini, J.; Prato, M. Anti-HIV Properties of Cationic Fullerene Derivatives. *Bioorg. Med. Chem. Lett.* **2005**, *15*(15), 3615–3618.

Mashino, T.; Shimotohno, K.; Ikegami, N.; Nishikawa, D.; Okuda, K.; Takahashi, K.; Nakamura, S.; Mochizuki, M. Human Immunodeficiency Virus-Reverse Transcriptase Inhibition and Hepatitis C Virus RNA-Dependent RNA Polymerase Inhibition Activities of Fullerene Derivatives. *Bioorg. Med. Chem. Lett.* **2005**, *15*(4), 1107–1109.

Matsumura, Y.; Maeda, H. A New Concept for Macromolecular Therapeutics in Cancer Chemotherapy: Mechanism of Tumoritropic Accumulation of Proteins and the Antitumor Agent Smancs. *Cancer Res.* **1986**, *46*(12), 6387–6392.

Mazzola, L. Commercializing Nanotechnology. *Nat. Biotechnol.* **2003**, *21*, 1137–1143.

McIlwain, C. C. *XPharm: the Comprehensive Pharmacology;* Elsevier: Amsterdam, 2008.

Meng, J.; Fan, J.; Galiana, G.; Branca, R. T.; Clasen, P. L.; Ma, S.; Zhou, J.; Leuschner, C.; Kumar, C. S. S. R.; Hormes, J.; Otiti, T.; Beye, A. C.; Harmer, M. P.; Kiely, C. J.; Warren, W.; Haataja, M. P.; Soboyejo, W. O. LHRH-Functionalized Superparamagnetic Iron Oxide Nanoparticles for Breast Cancer Targeting and Contrast Enhancement in MRI. *Mat. Sci. Eng. C.* **2009**, *29*(4), 1467–1479.

Michalet, X.; Pinaud, F. F.; Bentolila, L. A.; Tsay, J. M.; Doose, S.; Li, J. J.; Sundaresan, G.; Wu, A. M.; Gambhir, S. S.; Weiss, S. Quantum Dots for Live Cells, in Vivo Imaging, and Diagnostics. *Science* **2005**, *307*(5709), 538–544.

Mimoso, I. M.; Francisco, A. P. G.; Cruz, M. E. M. Liposomal Formulation of Netilmicin. *Int. J. Pharm.* **1997**, *147*(1), 109–117.

Minard, K. R. *Magnetic Particle Imaging. Encyclopedia of Spectroscopy and Spectrometry,* 2nd ed.; Elsevier: Amsterdam, 2009.

Mirza, A. Z.; Siddiqui, F. A. Nanomedicine and Drug Delivery: a Mini Review. *Int. Nano Lett.* **2014**, *4*(1), 1–7.

Mitchell, D. R.; Brown, R. M.; Jr Spires, T. L.; Romanovicz, D. K.; Lagow, R. J. The Synthesis of Megatubes: New Dimensions in Carbon Materials. *Inorg. Chem.* **2001**, *40*(12), 2751–2755.

Moghimi, S. M.; Hunter, A. C.; Murray, J. C. Long-Circulating and Target-Specific Nanoparticles: Theory to Practice. *Pharmacol. Rev.* **2001**, *53*, 283–318.

Morishige, K.; Kacher, D. F.; Libby, P.; Josephson, L.; Ganz, P.; Weissleder, R.; Aikawa, M. High-Resolution Magnetic Resonance Imaging Enhanced with Superparamagnetic Nanoparticles Measures Macrophage Burden in Atherosclerosis. *Circulation* **2010**, *122*(17), 1707–1715.

Müller, R. H.; Maabenb, S.; Weyhersa, H.; Spechtb, F.; Lucksb, J. S. Cytotoxicity of Magnetite-Loaded Polylactide, Polylactide/Glycolide Particles and Solid Lipid Nanoparticles. *Int. J. Pharm.* **1996a,** *138*(1), 85–94.

Müller, R. H.; Maassen, S.; Weyhers, H.; Mehnert, W. Phagocytic Uptake and Cytotoxicity of Solid Lipid Nanoparticles (SLN) Sterically Stabilized with Poloxamine 908 and Poloxamer 407. *J. Drug Targeting* **1996b,** *4*, 161–170.

Müller, R. H.; Radtke, M.; Wissing, S. A. Solid Lipid Nanoparticles (SLN) and Nanostructured Lipid Carriers (NLC) in Cosmetic and Dermatological Preparations. *Adv. Drug Delivery Rev.* **2002,** *54*, S131–S155.

Nikalje, A. P. Nanotechnology and Its Applications in Medicine. *Med. Chem.* **2015,** *5*, 81–89.

Nikam, A. P.; Mukesh, P. R.; Haudhary, S. P. Nanoparticles—an Overview. *J. Drug. Delivery Ther.* **2014,** *3*, 1121–1127.

Oldenberg, S. J.; Averitt, R. D.; Westcott, S. L.; Halas, N. J. Nanoengineering of Optical Resonances. *Chem. Phys. Lett.* **1998,** *288*(2–4), 243–247.

Olivier, J. C. Drug Transport to Brain with Targeted Nanoparticles. *NeuroRx* **2005,** *2*, 108–119.

Omri, A.; Suntres, Z. E.; Shek, P. N. Enhanced Activity of Liposomal Polymyxin B Against *Pseudomonas aeruginosa* in a Rat Model of Lung Infection. *Biochem. Pharmacol.* **2002,** *64*(9), 1407–1413.

Onyeji, C. O.; Nightingale, C. H.; Marangos, M. N. Enhanced Killing of Methicillin-Resistant *Staphylococcus aureus* in Human Macrophages by Liposome-Entrapped Vancomycin and Teicoplanin. *Infection* **1994,** *22*(5), 338–342.

Pandey, R.; Khuller, G. K. Solid Lipid Particle-Based Inhalable Sustained Drug Delivery System Against Experimental Tuberculosis. *Tuberc. (Edinb)* **2005,** *85*(4), 227–234.

Panga, X.; Cui, F.; Tian, J.; Chen, J.; Zhou, J.; Zhou, W. Preparation and Characterization of Magnetic Solid Lipid Nanoparticles Loaded with Ibuprofen. *Asian. J. Pharm. Sci.* **2009,** *4*(2), 132–137.

Panyam, J.; Labhasetwar, V. Biodegradable Nanoparticles for Drug and Gene Delivery to Cells and Tissue. *Adv. Drug. Delivery Rev.* **2003,** *55*, 329–347.

Panyam, J.; Williams, D.; Dash, A.; Leslie-Pelecky, D.; Labhasetwar, V. Solid-State Solubility Influences Encapsulation and Release of Hydrophobic Drugs from PLGA/PLA Nanoparticles. *J. Pharm. Sci.* **2004,** *93*, 1804–1814.

Park, J. W. Liposome-Based Drug Delivery in Breast Cancer Treatment. *Breast Cancer Res.* **2002,** *4*(3), 95–99.

Pathak, S.; Choi, S. K.; Arnheim, N.; Thompson, M. E. Hydroxylated Quantum Dots as Luminescent Probes for in Situ Hybridization. *J. Am. Chem. Soc.* **2001,** *123*(17), 4103–4104.

Peracchia, M. T. PEG-Coated Nanospheres from Amphiphilic Diblock and Multiblock Copolymers: Investigation of their Drug Encapsulation and Release Characteristics. *J. Controlled Release* **1997,** *46*, 223–231.

Pragati, S.; Ashok, S.; Kuldeep, S. Recent Advances in Periodontal Drug Delivery Systems. *Int. J. Drug. Delivery* **2009,** *1*, 1–14.

Qi, L.; Gao, X. Emerging Application of Quantum Dots for Drug Delivery and Therapy. *Expert Opin. Drug Delivery* **2008,** *5*(3), 63–67.

Radtke, M.; Müller, R. H. Novel Concept of Topical Cyclosporine Delivery with Supersaturated SLN™ Creams. *Int. Symp. Con. Rel. Bioact. Mater.* **2001,** *28*, 470–471.

Redhead, H. M.; Davis, S. S.; Illum, L. Drug Delivery in Poly(Lactide-Co-Glycolide) Nanoparticles Surface Modified with Poloxamer 407 and Poloxamine 908: in Vitro Characterisation and in Vivo Evaluation. *J. Control Rel.* **2001,** *70*, 353–363.

Ruckmani, K.; Sivakumar, M.; Ganeshkumar, P. A. Methotrexate Loaded Solid Lipid Nanoparticles (SLN) for Effective Treatment of Carcinoma. *J. Nanosci. Nanotechnol.* **2006,** *6*(9–10), 2991–2995.

Rupp, R.; Rosenthal, S. L.; Stanberry, L. R. VivaGel (SPL7013 Gel): a Candidate Dendrimer–Microbicide for the Prevention of HIV and HSV Infection. *Int. J. Nanomed.* **2007,** *2*(4), 561–566.

Sahoo, S. K.; Sawa, T.; Fang, J.; Tanaka, S.; Miyamoto, Y.; Akaike, T.; Maeda, H. Pegylated Zinc Protoporphyrin: a Water-Soluble Heme Oxygenase Inhibitor with Tumor-Targeting Capacity. *Bioconjugate Chem.* **2002,** *13,* 1031–1038.

Salata, O. Applications of Nanoparticles in Biology and Medicine. *J. Nanobiotechnol.* **2004,** *2*, 3.

Sanna, V.; Gavini, E.; Cossu, M.; Rassu, G.; Giunchedi, P. Solid Lipid Nanoparticles (SLN) as Carriers for the Topical Delivery of Econazole Nitrate: In-Vitro Characterization, Ex-Vivo and in-Vivo Studies. *J. Pharm. Pharmacol.* **2007,** *59*(8), 1057–1064.

Sano, N.; Wang, H.; Chhowalla, M.; Alexandrou, I.; Amaratunga, G. A. J. Synthesis of Carbon 'Onions' in Water. *Nature* **2001,** *414*(6863), 506–507.

Sarmento, B.; Martins, S.; Ferreira, D.; Souto, E. B. Oral Insulin Delivery by Means of Solid Lipid Nanoparticles. *Int. J. Nanomed.* **2007,** *2*(4), 743–749.

Schumacher, I.; Margalit, R. Liposome-Encapsulated Ampicillin: Physicochemical and Antibacterial Properties. *J. Pharm. Sci.* **1997,** *86*(5), 635–641.

Schumann, H.; Wassermann, B. C.; Schutte, S.; Velder, J.; Aksu, Y.; Krause, W. Synthesis and Characterization of Water-Soluble Tin-Based Metallodendrimers. *Organometallics* **2003,** *22*(10), 2034–2041.

Schuster, D. I.; Wilson, S. R.; Kirschner, A. N.; Schinazi, R. F.; Schluter-Wirtz, S.; Tharnish, P.; Barnett, T.; Ermolieff, J.; Tang, J.; Brettreich, M.; Hirsch, A. Evaluation of the Anti-HIV Potency of a Water-Soluble Dendrimeric Fullerene. *Proc. Electrochem. Soc.* **2000,** *9*, 267–270.

Serpe, L.; Catalano, M. G.; Cavalli, R.; Ugazio, E.; Bosco, O.; Canaparo, R.; Muntoni, E.; Frairia, R.; Gasco, M. R.; Eandi, M.; Zara, G. P. Cytotoxicity of Anticancer Drugs Incorporated in Solid Lipid Nanoparticles on HT-29 Colorectal Cancer Cell Line. *Eur. J. Pharm. Biopharm.* **2004,** *58*(3), 673–680.

Silva, A. C.; Santos, D.; Ferreira, D. C.; Souto, E. B. Minoxidil-Loaded Nanostructured Lipid Carriers (NLC): Characterization and Rheological Behaviour of Topical Formulations. *Pharmazie* **2009,** *64*(3), 177–182.

Singh, R.; Lillard, J. W. Nanoparticle-Based Targeted Drug Delivery. *Exp. Mol. Pathol.* **2009,** *86*(3), 215–223.

Singh, S.; Kumar, V.; Pandey, Tewari, R. P.; Agarwal, V. Nanoparticle Based Drug Delivery System: Advantages and Applications. *Indian J. Sci. Technol.* **2011,** *4*(3), 177–180.

Slater, T. F.; Cheeseman, K. H.; Ingold, K. U. Carbon Tetrachloride Toxicity as a Model for Studying Free-Radical Mediated Liver Injury. *Philos. Trans. R. Soc. London B: Biol. Sci.* **1985,** *311*(1152), 633–645.

Smith, A. M.; Dave, S.; Nie, S.; True, L.; Gao, X. Multicolor Quantum Dots for Molecular Diagnostics of Cancer. *Expert Rev. Mol. Diagn.* **2006,** *6*(2), 231–244.

Smith, B. R.; Heverhagen, J.; Knopp, M.; Schmalbrock, P.; Shapiro, J.; ShioSmithmi, M.; Moldovan, N. I.; Ferrari, M.; Lee, S. C. Localization to Atherosclerotic Plaque and Biodistribution of Biochemically Derivatized Superparamagnetic Iron Oxide Nanoparticles (SPIONs) Contrast Particles for Magnetic Resonance Imaging (MRI). *Biomed. Microdevices* **2007,** *9*(5), 719–727.

Souto, E. B.; Müller, R. H. SLN and NLC for Topical Delivery of Ketoconazole. *J. Microencapsulation* **2005**, *22*(5), 501–510.

Souto, E. B.; Wissing, S. A.; Barbosa, C. M.; Müller, R. H. Development of a Controlled Release Formulation Based on SLN and NLC for Topical Clotrimazole Delivery. *Int. J. Pharm.* **2004**, *278*(1), 71–77.

Spangler, R. S. Insulin Administration via Liposomes. *Diabetes Care* **1990**, *13*(9), 911–922.

Sparnacci, K.; Laus, M.; Tondelli, L.; Magnani, L.; Bernardi, C. Core-Shell Microspheres by Dispersion Polymerization as Drug Delivery Systems. *Macromol. Chem. Phys.* **2002**, *203*(10–11), 1364–1369.

Svenson, S.; Tomalia, D. A. Dendrimers in Biomedical Applications—Reflections on the Field. *Adv. Drug Delivery Rev.* **2005**, *57*(15), 2106–2129.

Takemoto, K.; Yamamoto, Y.; Ueda, Y.; Sumita, Y.; Yoshida, K.; Niki, Y. Comparative Studies on the Efficacy of AmBisome and Fungizone in a Mouse Model of Disseminated Aspergillosis. *J. Antimicrob. Chemother.* **2004**, *53*(2), 311–317.

Taniguchi, N. *On the Basic Concept of Nanotechnology.* In Proceedings of the International Congress on Prod Eng.: Tokyo, 1974.

Taton, T. A. Nanostructures as Tailored Biological Probes. *Trends. Biotechnol.* **2002**, *20*, 277–279.

Thassu, D.; Deleers, M.; Pathak Y. V. *Nanoparticulate Drug Delivery Systems.* Informa Healthcare: New York, 2007.

Thaxton, C. S.; Rosi, N. L.; Mirkin, C. A. Optically and Chemically Encoded Nanoparticle Materials for DNA and Protein Detection. *MRS. Bull.* **2005**, *30*(5), 376–380.

The Royal Society. *Nanoscience and Nanotechnologies: Opportunities and Uncertainties;* Royal Society: London, 2004.

Thomas, T. P.; Patri, A. K.; Myc, A.; Myaing, M. T.; Ye, J. Y.; Norris, T. D.; Baker, J. R. In Vitro Targeting of Synthesized Antibody—Conjugated Dendrimer Nanoparticles. *Biomacromolecules* **2004**, *5*(6), 2269–2274.

Thrash, T. P.; Cagle, D. W.; Alford, J. M.; Ehrhardt, G. J.; Wright, K.; Mirzadeh, S.; Wilson, L. J. Toward Fullerene-Based Radiopharmaceuticals: High-Yield Neutron Activation of Endohedral 165HO Metallofullerenes. *Chem. Phys. Lett.* **1999**, 308(3–4), 329–336.

Tolia, G. T.; Choi, H. H.; Ahsan, F. The Role of Dendrimers in Topical Drug Delivery. *Pharm. Tech.* **2008**, *32*(11), 88–98.

Tomalia, D. A.; Reyna, L. A.; Svenson, S. Dendrimers as Multi-Purpose Nanodevices for Oncology Drug Delivery and Diagnostic Imaging. *Biochem. Soc. Trans.* **2007**, *35*(1), 61–67.

Ung, T.; Liz-Marzan, L. M.; Mulvaney, P. Controlled Method For Silica Coating of Silver Colloids. Influence of Coating on the Rate of Chemical Reactions. *Langmuir* **1998**, *14*, 3740–3748.

Vemuri, S.; Rhodes, C. T. Preparation and Characterization of Liposomes as Therapeutic Delivery Systems: a Review. *Pharm. Acta. Helv.* **1995**, *70*(2), 95–111.

Wadher, K.; Kalsait, R.; Umekar, M. Alternate Drug Delivery System: Recent Advancement and Future Challenges. *Arch. Pharm. Sci. Res.* **2009**,*1*(2), 97–105.

Weissig, V.; Pettinger, T. K.; Murdock, N. Nanopharmaceuticals (Part 1): Products on the Market. *Int. J. Nanomed.* **2014**, *9*, 4357–4373.

Wiener, E. C.; Brechbiel, M. W.; Brothers, H.; Magin, R. L.; Gansow, O. A.; Tomalia, D. A.; Lauterbur, P. C. Dendrimer-Based Metal Chelates: a New Class of MRI Contrast Agents. *Magn. Reson. Med.* **1994**, *31*(1), 1–8.

Wissing, S. A.; Kayser, O.; Müller, R. H. Solid Lipid Nanoparticles for Parenteral Drug Delivery. *Adv. Drug Delivery Rev.* **2004**, *56*(9), 1257–1272.

Wong, H. L.; Rauth, A. M.; Bendayan, R. A.; Manias, J. L.; Ramaswamy, M.; Liu, Z.; Erhan, S. Z.; Wu, X. Y. New Polymer-Lipid Hybrid Nanoparticle System Increases Cytotoxicity of Doxorubicin against Multidrug-Resistant Human Breast Cancer Cells. *Pharm. Res.* **2006,** *23*(7), 1574–1585.

Woodward, S. C.; Brewer, P. S.; Moatamed, F.; Schindler, A.; Pitt, C. G. The Intracellular Degradation of Poly(Epsilon-Caprolactone). *J. Biomed. Mater. Res.* **1985,** *19,* 437–444.

Xia, Y.; Gates, B.; Yin, Y.; Lu, Y. Monodispersed Colloidal Spheres: Old Materials with New Applications. *Adv. Mater.* **2000,** *12*(10), 693–713.

Yang, L. J.; Li, Y. B. Simultaneous Detection of Escherichia Coli O157:H7 and Salmonella Typhimurium Using Quantum Dots as Fluorescence Labels. *Analyst* **2006,** *131*(3), 394–401.

Yoo, M. K.; Kim, I. Y.; Kim, E. M.; Jeong, H. J.; Lee, C. M.; Jeong, Y. Y.; Akaike, T.; Cho, C. S. Superparamagnetic Iron Oxide Nanoparticles Coated with Galactose-Carrying Polymer for Hepatocyte Targeting. *J. Biomed. Biotechnol.* **2007,** *2007,* 9.

Zhang, L.; Granick, S. How to Stabilize Phospholipid Liposomes (Using Nanoparticles). *Nano. Lett.* **2006,** *6*(4), 694–698.

Zhang, L.; Gu, F. X.; Chan, J. M.; Wang, A. Z.; Langer, R. S.; Farokhzad, O. C. Nanoparticles in Medicine: Therapeutic Applications and Developments. *Clin. Pharmacol. Ther.* **2008,** *83*(5), 761–769.

Zhang, L.; Pornpattananangkul, D.; Hu, C. M. J.; Huang, C. M. Development of Nanoparticles for Antimicrobial Drug Delivery. *Curr. Med. Chem.* **2010,** *17*(6), 585–594.

Zhuo, R. X.; Du, B.; Lu, Z. R. In Vitro Release of 5-Fluorouracil with Cyclic Core Dendritic Polymer. *J. Control Release* **1999,** *57*(3), 249–257.

Zur Mühlen, A.; Mehnert, W. Drug Release and Release Mechanism of Prednisolone Loaded Solid Lipid Nanoparticles. *Pharmazie* **1998,** *53,* 552–555.

Zur Mühlen, A.; von Elverfeldt D.; Bassler N.; Neudorfer I.; Steitz B.; Petri-Fink A.; Hofmann H.; Bode C.; Peter K. Superparamagnetic Iron Oxide Binding and Uptake as Imaged by Magnetic Resonance is Mediated by the Integrin Receptor Mac-1 (CD11b/ CD18): Implications on Imaging of Atherosclerotic Plaques. *Atherosclerosis* **2007,** *193*(1), 102–111.

CHAPTER 4

NANOPARTICLES ADVANCED DRUG DELIVERY FOR CANCER CELLS

HANI NASSER ABDELHAMID[2,*] and HUI-FEN WU[1,3,4,5,*]

[1]*Department of Chemistry and Center for Nanoscience and Nanotechnology, National Sun Yat-Sen University, 80424 Kaohsiung, Taiwan, Tel.: 886752520003955, Fax: 88675253909*

[2]*Department of Chemistry, Assuit University, Assuit 71515, Egypt, Tel.: 00201279744643, Fax: 0022342708*

[3]*School of Pharmacy, College of Pharmacy, Kaohsiung Medical University, 807 Kaohsiung, Taiwan*

[4]*Institue of Medical Science and Technology, National Sun Yat-Sen University, 80424 Kaohsiung, Taiwan*

[5]*National Sun Yat-Sen University and Academia Sinica, 80424 Kaohsiung, Taiwan*

Corresponding author. E-mail: hany.abdelhameed@science.au.edu.eg; chemist.hani@yahoo.com; hwu@faculty.nsysu.edu.tw

ABSTRACT

Nanoparticles (NPs) have been widely applied for drug delivery. NPs circumvent many challenges of the traditional drug. They improve the drug biosafety, enhance the drug efficacy, provide a targeted delivery system for the investigated drugs, improve bioavailability, offer synergic effect using photothermal or photodynamic effect, and improve the stability of therapeutic agents against chemical and enzymatic degradation. However, they show disadvantages of intrinsic toxicities of the nanocarrier, pathogenic tendency of bio-vesicles, lack of understanding for the mechanistic pathways of the system, and accumulation in the body, tissue, or organs. Thus, the factors influencing the performance of the delivery system should be carefully investigated. This chapter is a valuable reference source for those scientists working in the field of pharmaceutical sciences, medicine, bionanotechnology, materials science, biomedical sciences, and related areas of life sciences. We have focused on the most relevant references and concisely summarized the findings with illustrated examples.

4.1   INTRODUCTION

Nanoparticles (NPs) deal with emerging new technologies for developing customized solutions for drug delivery systems (Baeza et al., 2015; Blanco et al., 2015; Pathak and Thassu, 2009; Yang et al., 2015). The drug delivery systems have positive impact on the rate of absorption, distribution, metabolism, and excretion of the drug. It should also allow the drug to bind to its target receptor and show no influence effect on the receptor's signaling or activity. Nanomedicines showed considerable good clinical and commercial success applications. Annual revenues showed several billion dollars for the investment of drug delivery. There are several approval nano-based medicines according to the Food and Drug Administration (FDA) in the USA as reviewed by Eifler and Thaxton (2011). Typically, the entire process for a new drug required a long time (about 10–15 years) to appear in the pharmaceutical market and that costs several billion dollars. Thus, the progress in the field is slow.

Cancer increases human mortality. The World Health Organization (WHO) announced that annual cancer cases are expected to rise from 14 million in 2012 to 22 million in the next two decades (WHO, 2016). WHO expected that the number of new cancer cases will be raised by about 70% over the next two decades. They found that the five most common sites of cancer for male in 2012 were lung, prostate, colorectum, stomach, and liver cancer. While the five most common sites for female were breast, colorectum, lung, cervix, and stomach cancer. They observed that tobacco is the most important risk factor and that caused around 20% of global cancer deaths and around 70% of global lung cancer deaths. They found that the regions of Africa, Asia, and Central and South America account for 70% of the world's cancer deaths. However, according to the American Cancer Society, cancer deaths in the United States have dropped for a second straight year, which is attributed to the decrease in smoking rates and to earlier detection and more effective treatment of tumors (Jemal et al., 2007).

This chapter summarizes the material-based technologies in the field of drug delivery. An overview of NPs, include inorganic NPs, solid lipid NPs

(SLN), liposomes, polysaccharides, polyesters, micelles, exosomes, and others, are addressed. It introduces most of the NPs that can be applied for maximum benefits and therapeutic application includes chemotherapy, gene, and vaccine delivery. It is not possible to cover the entire literature, thus this chapter throw the light on the important points of drug delivery using NPs.

4.2 LIMITATIONS OF CONVENTIONAL CHEMOTHERAPY

Nowadays, the treatment of cancers with the use of chemical drugs, referred as chemotherapy is the primary therapeutic approach. Although, chemo-therapeutic drugs possess many limitations that are discussed below:

4.2.1 TOXICITY OF CHEMOTHERAPY

Chemotherapeutic agents are among the most effective treatment for cancer cells. Among those agents, doxorubicin (DOX) has been widely used. However, the use of DOX is limited by its renal, hepatic, and most importantly cardiac toxicity (Takemura and Fujiwara, 2007).

4.2.2 LACK OF SPECIFICITY TOWARD THE TARGET CELLS

Chemotherapeutic agents lack specificity as they kill all cells, not just the cancerous ones, but also the normal cells. These agents cause significant damage to non-cancerous cells leading to severe surplus side effects such as mucositis, suppression of bone marrow activity (immuno and myelosuppression), nausea, secondary neoplasms, and infertility. In addition, the high distribution volume of chemotherapeutics makes the drug delivery nonspecific to tumors (Giordano and Jatoi, 2005). This leads to an abnormal concentration of the antineoplastic drugs in healthy tissues. The lack of selectivity toward cancer cells is a major challenge for chemotherapy (Rothenberg et al., 2003). These effects reveal dramatic limitations for chemotherapeutic agents (Mays et al., 2010).

4.2.3 DOSE-DEPENDENT SIDE EFFECTS

Chemotherapeutic agents are low-molecular-weight compounds, so that it can be excreted easily from the body. They are effective and induce effi-cient therapeutic treatment. However, they induce cytotoxicity due to high pharmacokinetic volume of distribution. For this reason, a higher concentration is required to achieve a therapeutic effect. The use of high concentration leads to toxicity in some cases. Chemotherapeutic drugs have low therapeutic index. These low indexes reveal that the needed concentration for the effective treatment is often high. The high concentration causes systemic dose-dependent side effects (Torchilin, 2000).

4.2.4   SOLUBILITY

Formulating of the chemotherapeutic drugs is challenging due to their poor aqueous solubility. The low solubility makes the drug's uptake a difficult task (Gao et al., 2015). The chemotherapeutic application of paclitaxel is limited due to high hydrophobicity and poor solubility in water (<0.5 mg/l). Chemotherapeutic agents with poor solubility have the tendency to the inclusion of lipophilic groups (Lipinski, 2000; Lipinski et al., 2012). This inclusion shows affinity toward the target receptor. The poorly soluble drugs may cause embolization of blood vessels upon intravenous injection. Thus, it causes aggregation of the insoluble drugs and often causes a local toxicity.

4.2.5   CHEMORESISTANCE

Resistance of the cancer cells to the chemotherapy drugs limit the efficacy of anticancer drugs (Szakács et al., 2006). The interstitium of a tumor tissue has high hydrostatic pressure. This pressure can push the drug away from the tumor unlike from normal tissues. Another challenge is the multidrug resistance (MDR) due to overexpression of the plasma membrane P-glycoprotein (P-gp) that keeps the drugs away from the cell (Brigger et al., 2002). Several strategies have been proposed to avoid P-gp-mediated MDR, including the encapsulation of anticancer drugs in NPs and the coadministration of P-gp inhibitors (Krishna and Mayer, 2000). The drug efflux transporters of lipophilic drugs decrease the therapeutic drug concentration at the site of action. It leads to suboptimal concentration and limit the success of chemotherapy (Serpe, 2007).

4.2.6   TRANSPORT OF THE DRUGS TO THE CELLS

Conventional chemotherapy suffers from low tumor cell uptake. Physico-chemical properties of the drug, including size, molecular weights, charge, and function groups which plays major role in the drug transport to the cells (Park et al., 2008). Further hurdle is the pathophysiological tumor heterogeneity, which inhibits a uniform drug delivery into the whole tumoral mass. In addition, acidic tumor microenvironment causes degradation of the acid-sensitive drugs (Kumar, 2007). Conventional formulations, that is, solutions, suspensions, and ointment, for ocular diseases showed that only <5% of the applied dose penetrates the cornea and reaches intraocular tissues. While a major fraction of the instilled dose lost due to the presence of many ocular barriers such as external barriers (Ali et al., 2014). It causes rapid loss of the instilled solution from the precorneal area and nasolacrimal drainage system

4.3 THERAPEUTIC APPLICATION OF NANOPARTICLES (NPS)

The last decade was witness of the great advances in cancer-targeted drug delivery using nanotechnology that enabled more effective drug design and development. The publication analysis of the NPs applications as drug delivery register in PubMed using the search term "nanoparticle drug delivery" is shown in Figure 4.1. Figure 4.1 shows that the growth of publication number is exponentially increased over the last two decades. The advantages of NPs for drug delivery justify the exponential growth in the number of publications dealing with NPs for drug delivery applications (Fig. 4.1). Nanotechnology shows a revolutionized chemotherapy. There are more than 5000 academic journal articles in this area. A total of 1381 nanomedicine formulations for cancer therapy were registered for clinical trials by December 2014 (Wicki et al., 2015).

There are several NPs that can sever as carrier for drug delivery (Badrzadeh et al., 2016). We would like to classify these wide types into: (1) small vesicles (extracellular matrix [ECM] and ECM-like materials), such as apoptotic bodies, microvesicles (MVs), and exosomes (Aryani and Denecke, 2016), hyaluronic acid (Han et al., 2015; Dosio et al., 2016); (2) biomolecules, such as liposomes (Agarwal et al., 2014; Zylberberg and Matosevic, 2016), peptide-based nanotubes and nanogels (NGs), chitosan (Elgadir et al., 2015), gelatin, sodium alginate, and other hydrophilic/biodegradable polymers (Soppimath et al., 2001), SLN (Müller et al., 2000; Wissing et al., 2004), polysaccharides (Liu et al., 2008b); (3) proteins-based NPs, such as cell penetrating peptides (Lehto et al., 2016); (4) polymers, such as amphiphilic copolymers (Gref et al., 1994), poly(D, L-lactide-co-glycolide) (PLGA) (Panyam and Labhasetwar, 2003), hydrogel (Hamidi et al., 2008), dendrimers (Tomalia, 2005), and polymer–drug conjugates

(Jhaveri et al., 2014; Kanapathipillai et al., 2014; Markman et al., 2013), (5) metallic NPs, such as gold NPs (Paciotti et al., 2016; Ruan et al., 2015); (6) metal oxides such as hollow mesoporous silica (Liu et al., 2016), mesoporous silica (Baeza et al., 2015), and magnetic NPs (Arruebo et al., 2007; Wu et al., 2015); (7) metal organic frameworks (Horcajada et al., 2010; Sun et al., 2013) is hybrid material of inorganic–organic moieties with porous frameworks (Chen et al., 2012); (8) surfactant-based NPs such as microemulsions (Sharma et al., 2015). More than 51 FDA-approved nanomedicines were reported and more than 77 products in clinical trials, with ~40% of trials listed in clinicaltrials.gov started in 2014 or 2015 (Bobo et al., 2016). NPs are applied for drug delivery of chemotherapy (Bartoş et al., 2016; Steichen et al., 2013; Tian et al., 2016; Tseng et al., 2016), gene transfer or delivery (Loh et al., 2016), and vaccine (Aburahma, 2014; Calderón-Gonzalez et al., 2016; López-Sagaseta et al., 2016; Saso and Kampmann, 2016; Skwarczynski and Toth, 2016; Torres-Sangiao et al., 2016).

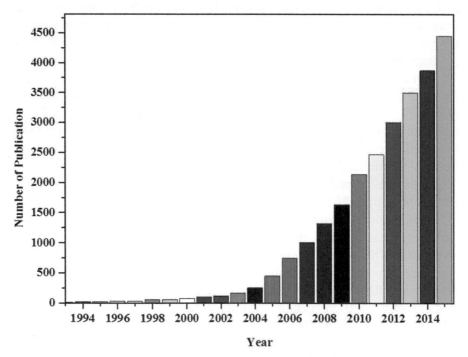

FIGURE 4.1   Temporal evolution in the number of scientific papers published involving drug delivery using nanoparticles (NPs).

Source: PubMed; using "nanoparticle drug delivery," Date of search: December 2016.

4.4   REQUIREMENTS OF NPS FOR DRUG DELIVERY APPLICATIONS

In order to serve as carrier for drug delivery, there are certain requirements that should be fulfilled for NPs. We summarize some of these points as below

4.4.1   BIOCOMPATIBILITY

Generally speaking, NPs should be biocompatible for biological applications (Abdelhamid and Wu, 2013, 2015; Abdelhamid et al., 2016a, 2016b; Khan et al. 2015). This requirement is very important and usually required careful investigation. Biocompatibility is varied based on the therapeutic agents that are applied. For instance, the toxicity of expression vectors limit the gene therapy (Clark and Hersh, 1999). The intrinsic toxicity of the therapeutic agent or nanocarrier limits the dose of agents that can be delivered.

4.4.2   SUITABLE INTERACTIONS BETWEEN THE DRUG AND NPS

The suitable interactions between the NPs and the therapeutic agents should be available. The main forces among the drug and the nanocarries could be chemical bonding or physical interactions. Modification methods as well as surface characterization were summarized from 1990 through mid-2000 (Soppimath et al., 2001). These forces depend on the properties of the drug and the nanocarrier. The parameters such as size, NPs morphology, surface charge, function groups, and drug function groups would have remarkable effects on the interaction between the NPs and the drug. Physical adsorption of DOX on carbon nanotube (CNT) and single-wall nanotube (SWNTs) was reported (Liu et al., 2007). Data showed that SWNTs offer high loading of drug (400% by weight) (Liu et al., 2007). However, paclitaxel formed unstable drug delivery system. This is due to the bulky structure of paclitaxel. Authors found that bulky drugs are usually conjugated to CNT-dispersing polymers at their distal ends for CNTs surface functionalization (Liu et al., 2008a). Physical interactions are simple for the modification, but they are weak and can be destroyed very fast with the changes in temperature, pH value, pressure, or other force.

Chemical interaction through the formation of covalent bond leads to strong interactions between the drug and NPs. This modification usually required extra steps during the synthesis and sometimes expensive chemical reagents. Surface modification of the NPs with the drug requires the presence of suitable function groups. Sometimes cross-linking between both the entities is required. The chemical modification offers stable drug delivery system. Chemical modification of the nanocarrier with the drug requires strong changes in the biological system in order to release the drug.

4.4.3   BIODEGRADABLE

The body should be able to remove or excrete the drug delivery system (NPs-drug). In recent years, biodegradable polymeric NPs have attracted considerable attention as potential degradable drug systems (Kamaly et al., 2016). Among the wide number of inorganic NPs, supermagnetic iron oxide NPs are considered to be biodegradable. Iron can be reused and recycled by cells using normal biochemical pathways for iron metabolism (Bulte and Kraitchman, 2004).

4.4.4   NO INTERACTIONS WITH OTHER BIOMOLECULES

The NPs for drug delivery should show no interactions with the biomolecules surrounded by the target cells. Photon correlation spectroscopy or dynamic light scattering and atomic force microscopy confirmed the interactions of SLN with the major circulatory protein and serum albumin (Gualbert et al., 2003). Data revealed an increase of the particle size and formation of layers of the protein on the surface of NPs. These interactions not only changed the structure but also the surface chemical of the NPs. These new entities required further characterization and clear investigations in order to understand their bioactivities.

4.5   FACTORS AFFECT THE EFFICIENCY OF NPS FOR DRUG DELIVERY

There are many key parameters that affect the performance of NPs to serve as an effective drug delivery system. The parameters such as the composition of the NPs, size, morphology, type of the interactions between NPs, and the drug and surface adsorption would have remarkable effects on the performance of the NPs for the application of drug delivery system (Moghimi et al., 2005; Pokropivny and Skorokhod, 2008). We briefly summarized some of these key parameters in the following paragraphs. These parameters can be classified as (1) NPs; (2) therapeutic agents; (3) the interactions among the NPs and the drug; and (4) the type of the investigated cells (Fig. 4.2).

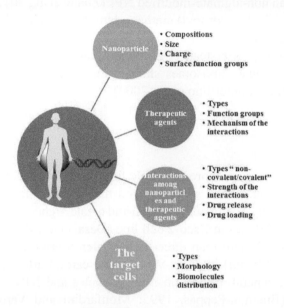

FIGURE 4.2   Key parameters affect the performance of the drug delivery system.

4.6   TYPES OF THERAPEUTIC

There are different types of therapeutic molecules such as compounds with small molecular weight (chemotherapeutic agents) (Chen et al., 2013; Ma et al., 2016; Mujokoro et al., 2016), proteins (Almeida et al., 1997), oligo-nucleotides (Cheng and Lee, 2016; Dass, 2002; Juliano, 2016; Moroz et al., 2016; Turner et al., 2016; Zatsepin et al., 2016), and vaccine (Fontana et al., 2017). These therapeutic agents have different chemistry and are targeted to different organs.

4.6.1   NPS

4.6.1.1   COMPOSITION

The compositions of NPs affect their performance. They dramatically show differences based on their composition. For example, inorganic NPs were more efficient than those made from organic materials (0.8 versus 0.6%) (Wilhelm et al., 2016). In vivo results showed that the concentration of DOX reaching the liver from glycyrrhetinic acid-modified alginate NPs was 4.7 times higher than non-alginate-modified NPs (Zhang et al., 2012). Polymers such as poly[N(2-hydroxypropyl) methacrylamide], poly(vinylpyrrolidone), poly(2-methyl-2-oxazoline), (poly(N, N-dimethylacrylamide)), poly(N-acryloylmorpholine) (Kierstead et al., 2015), and poloxamer 188 (Zhang et al., 2015), conjugate liposomes showed an increase of the circulation times of liposomes (Immordino et al., 2006).

4.6.1.2   SIZE

NPs with smaller size are more favorable for drug delivery compared to the NPs with the bigger size. A study showed that 100-nm sized NPs offered 2.5-fold greater uptake compared to 1 mm and 6-fold higher uptake compared to 10-mm microparticles in Caco-2 cell line (Desai et al., 1997). PLGA NPs (~100 nm) showed more than three-fold greater arterial uptake compared to larger NPs (~275 nm) in an ex vivo canine carotid artery model (Song et al., 1998). A general aspects on microparticles and NPs was reviewed in references (Brannon-Peppas, 1995; Monfardini and Veronese, 1998; Torchilin, 1998; Uhrich et al., 1999; Zimmer and Kreuter, 1995) for more details. Small NPs (<10 nm) are cleared through the kidneys, and larger particles can be captured by liver and the mononuclear phagocyte system (Fox et al., 2009; Rolfe et al., 2014; Sadauskas et al., 2007; Soo Choi et al., 2007). The nanocarrier in the range of 10–50 nm is recommended (Alexis et al., 2008). It is important to keep in mind that NPs interact with the protein and form protein corona "opsonization". They can be removed through the opsonization and phagocytosis by macrophages following the mechanism of receptor-mediated endocytosis (Maeda, 2001).

4.6.1.3   INTERACTIONS OF NPS AND BIOLOGICAL CELLS

Interactions between NPs drug loaded and biological cells are critical for the performance of the drug delivery. However, current limitations in the field of bionano science, focusing on material biological interactions and directions slow down the pace of scientific discovery (Björnmalm et al., 2016). These interactions determine the mechanistic uptake of the drug delivery system. The pH-responsive biomaterials undergo conformational changes through various mechanisms such as protonation, charge reversal, or cleavage of a chemical bond. These changes offer tumor-specific cell uptake or drug release (Kanamala et al., 2016). These nanocarriers can be designed based on the following three different mechanisms; through the introduction of (1) protonatable groups, (2) acid-labile bonds in the polymers of the carriers, and (3) pH-responsive materials.

Nanocarriers conjugate a drug that have the capability to achieve targeted delivery of their payload to solid tumors (Cui et al., 2010; Kaasgaard and Andresen, 2010) through the "enhanced permeability and retention (EPR) effect" (Fukumura et al., 2010; Matsumura and Maeda, 1986). EPR effect is arguably the most important strategy for improving the delivery of chemotherapeutic agents to tumors (Kanapathipillai et al., 2014, Torchilin, 2009). This effect required the nanocarriers to stay in systemic circulation for a prolonged period of time to be extravasated into tumor tissue (Liu et al., 2014; Torchilin, 2009). Therefore, long circulation of nanocarriers is essential for maximum tumor targeting (Shen et al., 2012). The most common approach in prolonging circulation is achieved by coating the surface of nanocarriers with a hydrophilic polymer, such as polyethylene glycol (PEG), a process known as PEGylation or PEG attachment (Jiang et al., 2013; Shen et al., 2012). However, it is important to mention that this hydrophilic coating sterically hinders opsonization (Moghimi et al., 2001). This approach is known as "stealth technology" (Immordino et al., 2006). "Stealth" nanocarriers, most notably Doxil® or Caelyx® (PEGylated liposomal DOX) have been approved for clinical use with notable success.

4.6.2 DRUG LOADING

NPs offer high loading of the target drug due to its large surface area. This feature offers a high loading capacity to reduce the quantity of the carrier required for administration. This process is achieved by two methods: (1) one-pot synthesis of NPs in the presence of the investigated drug; (2) adsorbing the drug after the formation of NPs by incubating them in the drug solution. The latter method showed lower amount of drug compared to the former method (Alonso et al., 1991; Ueda et al., 1998). Factors determining the loading capacity of the drug in the nanocarriers are, for example:

i.   Solubility of the drug and the dispersion of NPs
ii. Available function groups of nanocarriers and the drug

iii. Chemical and physical structure of the drug and nanocarriers
iv. Synthesis conditions such as temperature, precursors, reaction time, concentration, and so forth

4.6.3 DRUG RELEASE

Drugs are released from the nanocarriers due to the change of their environment (stimuli). These stimuli can be classified to:

i.   Physically induced release: physical action could be temperature, electric field, light, pressure, sound, and magnetic field.
ii. Chemically induced release systems: such as pH, dissolvation, fused in the cell membrane, solvent composition, ions, and specific molecular recognition.
iii. Other stimuli-induced release systems: such as degradation and enzymatic effect opsonization.

The release rates of drug from NPs into the cells depend upon: (i) desorption of the surface bound or adsorbed drug; (ii) diffusion of the drug through the matrix of the nanocarrier; (iii) diffusion (in case of nanocapsules) through the polymer wall; (iv) the type of stimuli, (v) the cell types; and (vi) incubation time. To delay degradation, the surface of the NPs has been decorated using a biocompatible and non-immunogenic hydrophilic polymer, PEG that reduces nanoparticle binding to opsonins, avoiding reticuloendothelial degradation (Owens and Peppas, 2006).

Nanocarrier-mediated drug delivery with insufficient and delayed drug release from the nanocarriers to cancer cells could develop drug resistance. Quick drug release in the cytoplasm as well as delivering combination therapy using multiple drugs or drug/nucleic acids are required to overcome MDR (Akita et al., 2009; Biswas et al., 2013).

4.6.4 DESIGN OF THE DRUG DELIVERY SYSTEM

The design of drug delivery system affects the drug release and could be controlled. Based on the current designs (Fig. 4.3); the drug incorporation models may be:

i.   Solid solution model

ii. Core-shell model; drug-enriched shell
iii. Core-shell model; drug-enriched core
iv. Drug attached or modified the surface of the nanocarrier

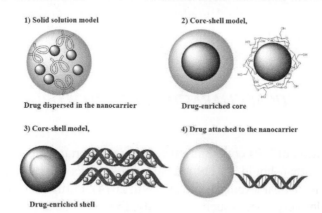

FIGURE 4.3 Design forms of the drug delivery system. Figure is not considered the object real size.

4.6.5 RESIDUALS IN THE DRUG SYSTEM

During the synthesis or drug's incorporation with the nanocarrier, there are residual with lack of characterization. This residual may play a critical role and required serious characterization to avoid wrong conclusion. Relatively high water content of the dispersions (70–99.9%) has been observed (Wissing et al., 2004). This challenge is annoying especially for the drug delivery system that comes from extraction or isolation. For instance, most of the published data are from impure preparations of exosomes (Baglio et al., 2012; Lai et al., 2012; Théry et al., 2006). The common protocol for exosome preparation is differential centrifugation. This method also includes other vesicles in the size of exosomes such as MVs and large protein aggregates. It is also arguably that astrocyte-derived exosomes might contain mitochondrial deoxyribonucleic acid (DNA) (Guescini et al., 2010a, 2010b). However, a study claimed that exosomes contain no DNA (Valadi et al., 2007). This discrepancy can be explained due to several facts: (i) the most often used technique for diagnostic and assessing the nucleic acid content of exosomes is microarray technology (Ioannidis et al., 2009). (ii) There are variations of isolation techniques that can greatly influence the relative amount of the specific exosome cargo. Another example is the production of

SLN through precipitation that arise solvent residues in the final dispersion. Sjöström et al. calculated the amount of toluene residues as 20–100 ppm in final dispersions (Sjöström et al., 1995). Methods that include ultrasonication may contain potential metal contamination or degradation. Storage of the drug delivery may cause degradation and leads to undesirable impurities that alter their performance. These residuals are very critical and required further characterization.

4.6.6 ANALYTICAL CHARACTERIZATION OF DRUG DELIVERY SYSTEM

Characterization of NPs containing the drug is important and still far away from the point that should be. The real structure of NPs–drug or NPs–protein interactions is unclear (Abdelhamid and Wu, 2014). Further characterization of the drug delivery system such as stability, forces inside the system, and its properties are required. The current analytical methods may require change of the status of drug delivery system. For instance, extraction of the drug is required in order to analyze the content of the drug system. Thus, the current analytical methods show no clear information of the real structure of the drug delivery system.

4.6.7 PHARMACEUTICAL FORMS OF THE DRUG DELIVERY SYSTEM

The different administrations of the drug delivery systems have influential effect on the performance of the drug. The route-based delivery systems such as oral, pulmonary, ocular, and dermal are discussed in reference (Gibson, 2005). These different routes required certain pharmaceutical forms. These forms affect the stability and performance of the drug delivery system.

4.6.8 TYPE OF THE TARGET CELLS

The cells types and its structure affect the efficiency of the drug delivery. Factor arises from the heterogeneity of the tumor tissues affect the delivery and efficiency of the drug. Central part of the tumor comprised of tumor stem cells show less accumulation of nanocarriers compared to other parts of the tumor.

4.7   ADVANTAGES OF NPS AS DRUG DELIVERY

NPs advance the applications for drug delivery. We summarized the most advantages points as below

4.7.1   INCREASE OF THE DRUG'S EFFICIENCY

NPs increase the efficiency of the conventional drug. The drug delivery system using NPs showed greater impact to maximize ocular drug absorption, and minimize systemic absorption and side effects (Ali et al., 2014). NPs offer reduction of the quantity of drug needed to attain a particular concentration in the vicinity of the target. NPs increase drug efficiency by achieving steady state therapeutic levels of drugs. NPs improve the therapeutic index of the loaded chemotherapeutic agents compared to the drugs delivered through conventional dosage forms. The drug delivery efficiency can be increased through the combination therapy by utilizing chemotherapeutic, photothermal, and photodynamic effects.

4.7.2   REDUCE SIDE EFFECT

NPs as carrier system offer the reduction of side effects caused at the injection site. NPs-based drug delivery reduces toxic side effects of the incorporated highly potent drugs and increase the efficacy of the treatment. The ability to target specific locations in the body using NPs decreases the side effect of the target drug. The reduction of the concentration of the drug at nontarget sites reduce severe side effects on the healthy cells (Ritter et al., 2004). Nanoparticle decreases the drug toxicity due to controlled drug release and improves drug's pharmacokinetics by increasing solubility and stability. These improvements lead to apply new formulations of DOX including liposome technology (Myocet® and Caelyx®) (Takemura and Fujiwara, 2007). These new formulations improve DOX's efficacy and reduce its toxicity.

4.7.3   SITE-SPECIFIC TARGETING

NPs provide site-specific targeting and controlled release of the incorporated drugs. The site-specific targeting can be achieved through different methodologies, that is, ligand-mediated or antibody (Ab)-mediated targeting of the nanocarriers to the cancer cells (Torchilin, 2010). Ligand-mediated targeting of nanocarriers improves the therapeutic index of the drug, increases the drug's efficacy, and reduces the nonspecific toxicity (Malam et al., 2009; Perumal et al., 2011). Alternative to these expensive ligands, a wide range of targeting ligands such as nanocarrier surface modifier, including low-molecular-weight ligands (vitamin [folic acid], thiamine, sugars), peptides (arginine–glycine–aspartic acid or RGD, LHRD), proteins (glycoprotein [e.g., transferrin], ABs, lectins, bovine serum albumin, cytokines, fibrinogen, thrombin), polysaccharides (hyaluronic acid, chitosan), polyunsaturated fatty acids, DNA, and aptamer are currently being exploited extensively for the development of cancer-targeted nanocarriers (Brody and Gold, 2000; Juliano and Stamp, 1976; Heath et al., 1983; Rui et al., 1998; Schiffelers et al., 2003). Another targeted method is the use of external magnetic field for targeting drug delivery system using magnetic NPs as nanocarrier (Ulbrich et al., 2016). Magnetic nanostructures make delivery of NPs easier with the application of an external magnetic field (Jain, 1987).

4.7.4   INCREASE THE DRUG BIOAVAILABILITY

Drug delivery using NPs can pass through the smallest capillary vessels because of their ultra-tiny volume. Small NPs showed slow clearance that prolong the duration in blood stream. NPs provide controlled release properties due to the biodegradability, pH, ion, and temperature sensibility of materials.

4.7.5   OVERCOME MULTIDRUG RESISTANCE

MDR of breast cancer limits the treatment and prognosis of breast cancer. Thus, MRD leads to a serious threat on women's health. However, using NPs for drug delivery of anticancer agents open a new technology to be exploited in the treatment of the patients. These resistances are due to multi-specific drug efflux transporters such as P-gp (ABCB1), MDR protein-1 (ABCC1) and breast cancer resistance protein (ABCG2). Drugs can be delivered to tumor tissue by passive and active tumor targeting strategies. These methods reduce or reverse drug resistance. For more details regarding to MDR-associated proteins, as well as various nanoparticle formulations developed to overcome MDR in breast cancer, we recommend reader to read this article (Yuan et al., 2016) for more details.

4.7.6   LOADED POORLY SOLUBLE HYDROPHOBIC DRUGS

NPs improve the pharmacokinetics and offer successful loading for the poorly soluble hydrophobic drugs (Kumari et al., 2016). The large surface area of NPs allows useful dispersion of insoluble therapeutic agents (Abdelhamid et al., 2014). Thus, the modification could be done in routes of administration of available chemotherapeutic agents by optimizing drug delivery systems (Langer, 1998). Alternatively, surfactants may be employed in the formulation to solubilize the drug. However, surfactants may cause the drug to precipitate in vivo, because their critical micelle concentrations in physiological fluids are too low. Another strategy based on the formation of stable polymeric micelles composed of a hydrophobic core surrounded by a hydrophilic shell is also reported for poorly soluble drugs (Kim et al., 2006; Torchilin, 2002).

4.7.7   MULTIFUNCTIONAL

NPs provide multifunctional role in the drug delivery. For instance, exosomes are involved in different procedures such as communication and regulation. Their role in communication is supported by the facts that (i) exosomes are intercellular vehicles for ribonucleic acid transfer within the central nervous system (Valadi et al., 2007) and (ii) exosomes also contain abundant proteins such as heat shock proteins and tetraspanins that are involved in cell–cell communications (Frühbeis et al., 2012; Smalheiser, 2007; Théry et al., 2009). NPs provide feasibility to deliver multiple small molecule chemotherapeutics using nanocarriers (Hu et al., 2016). Thus, they provide synergistic, additive, and potentiation effects to the drug. NPs provide the opportunity for the application of combination therapy using chemotherapeutic and photothermal effects. Furthermore, NPs, for example, multifunctional hybrid-CNTs, offer multifunctional tools for diagnosis (Mehra and Jain, 2016).

4.7.8   INCREASE THE DRUG STABILITY

NPs encapsulation of the drug provides a protective insulation of drug molecules. The drug encapsulation enhances their stability and minimizes the drug degradation.

4.8   DISADVANTAGES OF NPS AS DRUG DELIVERY

4.8.1   INTRINSIC TOXICITY OF NANOCARRIERS

The currently available formulation paclitaxel comprises of Cremophor EL (polyethoxylated castor oil) and dehydrated ethanol showed toxicity and causes serious side effects, including hypersensitivity reactions, nephrotoxicity, and cardiotoxicity (Brannon-Peppas and Blanchette, 2012). The selection of the biocompatible NPs is highly important.

4.8.2   LOW EFFICIENCY

The efficiency of the drug using NPs as nanocarrier for drug delivery is still an area for argumentation. A review article claimed that the median delivery efficiency using NPs is low (only 0.7% of an injected dose of NPs ends up in a tumor) (Wilhelm et al., 2016). This low efficiency according to the author's point of view is a hurdle for translating nanomedicines into the clinical trials. Interstitial fluid pressure due to the high protein content in the interstitial space prevents the penetration of nanocarriers inside the tissues. However, the drug efficiency using NPs may increase using stimuli (Felber et al., 2012; Fleige et al., 2012). Thus, we need further investigation to understand the efficiency of the drug delivery.

4.8.3   PATHOGENIC TENDENCY

Some drug nanocarriers such as exosomes which are derived from bacteria or virus-infected cells, may contain factors such as pathogen-derived antigens or cytokines that activate a pro-inflammatory pathway (van Niel et al., 2006).

4.8.4 ACCUMULATION OF NPS IN THE BODY

Nanomedicines are preferentially accumulated in the interstitial fluid of tumor compared to other normal tissues. The phenomena of eventual accumulation of nanomedicines to the tumor are referred to as EPR effect. However, the NPs tend to accumulated in lung, liver, and kidney based on their size. This drawback slows down the progress of NPs applications for in vivo studies.

4.8.5 DRUG-LOADING CAPACITY

Some of the NPs showed potential disadvantages such as insufficient loading capacity (Wissing et al., 2004). The drug-loading capacity of conventional SLN is limited (generally up to approximately 25% with regard to the lipid matrix, up to 50% for special actives such as ubidecarenone) by the solubility of drug in the lipid melt.

4.8.6 STABILITY

NPs could affect the stability of the conventional drug. NPs may decrease the critical physical stability of the drug-containing emulsions due to a reduction of the zeta potential (ZP). The decrease in ZP leads to agglomeration, drug expulsion, and eventually breaking of the emulsion. Drug expulsion after polymorphic transition during storage has also been observed (Wissing et al., 2004).

4.8.7 LACK OF A CLEAR MECHANISM

So far, there is no clear mechanism of the uptake of NPs conjugated the drug. It could be due to many hypotheses such as internalization by endocytosis, members of the integrin family, through receptor–ligand interaction (Aryani and Denecke, 2016), and endothelial uptake by phagocytosis. General view of what could take place is represented in Figure 4.4. The drug uptake for pH-sensitive drug delivery system is governed by the pH variations as shown in Figure 4.4. The drug releases from the delivery system after uptake. Then, it targets mitochondria, cytoplasm, or nucleus(Fig. 4.4). The lack of knowledge about the distribution and location of targeted NPs after cell uptake limit the development of effective drug delivery system.

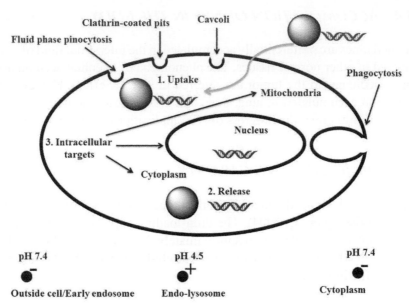

FIGURE 4.4   Schematic representation shows the different intracellular uptake pathways, intracellular targets for NPs and pH values.

4.9   OUTLOOK AND PROSPECTIVE

The applications of nanoparticle for the drug delivery system should focus on (1) the selection of effective and combination of carrier materials to obtain suitable drug release speed; (2) complete understanding of the drug delivery system and well-defined characterization; (3) the surface modification of NPs to improve their targeting ability; (4) key structure–property relationship examination on cytotoxicity (acute, subacute, and chronic toxicity, teratogenicity, and mutagenicity) and the efficiency of the drug delivery; (5) mechanistic study of the cell uptake and the fate of the drug and the nanocarrier; (6) the drug stability for storage; (7) the optimal pharmaceutical formulation and the effect of the additives on the efficiency of the drug delivery; the optimization of the preparation of NPs to increase their drug delivery capability; (8) the investigation of in vivo dynamic process to disclose the interaction of NPs with blood and targeting tissues/organs (hematocompatibility); (9) biodegradation; (10) immunogenicity; and (11) pharmacokinetics (body distribution, metabolism, bioavailability, elimination, organ-specific toxicity). There are many questions which are to be answered for new NPs as carrier before the clinical trials. NPs impact the drug delivery and that can be proved from the several approved nanocarriers as shown in Figure 4.5. The wide number of different NPs helps to select the best nanocarriers that are easy for modification of the drug.

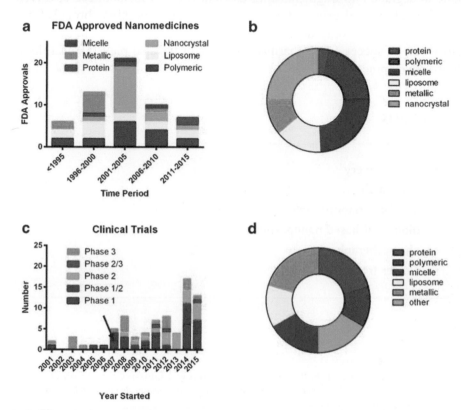

FIGURE 4.5 The development of nanomedicines for approved or clinical trials, (a) FDA-approved nanomedicines stratified by category; (b) FDA-approved nanomedicines stratified by category overall; (c) clinical trials identified in clinicaltrials.gov from 2001 to 2015 with arrow indicating approximate start date of US law (FDAAA 801) requiring reporting to FDA database; (d) nanomedicines under clinical trial investigation stratified by category overall.

Source: Reprinted from Bobo, D., Robinson, K.J., Islam, J. et al. Pharm Res (2016) 33: 2373. https://doi.org/10.1007/s11095-016-1958-5. With permission of Springer.

4.10 CONCLUSIONS

NPs as drug delivery or nanocarriers offer many advantages. Drug delivery using NPs improve safety and efficacy of the drugs, provide targeted delivery of drugs, enhance the drug bioavailability, increase the drug efficiency using photothermal and photodynamical effects, and improve the stability of therapeutic agents against chemical/enzymatic degradation. Drug delivery materials should be compatible, easy to bind with a particular drug, and able to degrade into fragments after use that is either metabolized or driven out through normal excretory routes. There will be no breakthrough for a delivery system if only academic research groups are developing it. The story can be successful if pharmaceutical industry also plays a key role in these developments.

KEYWORDS

- **drug delivery**
- **nanocarriers**
- **inorganic nanoparticles**
- **biological-based nanoparticles**
- **chemotheraphy**
- **cancer treatment**

REFERENCES

Abdelhamid, H. N.; Wu, H.-F. Probing the Interactions of Chitosan Capped CdS Quantum Dots with Pathogenic Bacteria and Their Biosensing Application. *J. Mater. Chem. B* **2013**, *1*(44), 6094–6106.

Abdelhamid, H. N.; Wu, H.-F. Proteomics Analysis of the Mode of Antibacterial Action of Nanoparticles and their Interactions with Proteins. *TrAC Trends Anal. Chem.* **2014**, *65*, 30–46.

Abdelhamid, H. N.; Wu, H.-F. Synthesis and Characterization of Quantum Dots for Application in Laser Soft Desorption/ionization Mass Spectrometry to Detect Labile Metal–drug Interactions and their Antibacterial Activity. *RSC Adv.* **2015**, *5*(93), 76107–76115.

Abdelhamid, H. N.; Khan, M. S.; Wu, H.-F. Graphene Oxide as a Nanocarrier for Gramicidin (GOGD) for High Antibacterial Performance. *RSC Adv.* **2014**, *4*(91), 50035–50046.

Abdelhamid, H. N.; Kumaran, S.; Wu, H.-F. One-Pot Synthesis of CuFeO$_2$ Nanoparticles Capped with Glycerol and Proteomic Analysis of Their Nanocytotoxicity Against Fungi. *RSC Adv.* **2016a**, *6*(100), 97629–97635.

Abdelhamid, H. N.; Talib, A.; Wu, H.-F. One Pot Synthesis of Gold – Carbon Dots Nanocomposite and Its Application for Cytosensing of Metals for Cancer Cells. *Talanta* **2016b**, *166*, 357–363.

Aburahma, M. H. Bile Salts-Containing Vesicles: Promising Pharmaceutical Carriers for Oral Delivery of Poorly Water-Soluble Drugs and Peptide/Protein-Based Therapeutics or Vaccines. *Drug Delivery* **2014**, *23*, 1–21.

Agarwal, R.; Iezhitsa, I.; Agarwal, P.; Abdul Nasir, N. A.; Razali, N.; Alyautdin, R.; Ismail, N. M. Liposomes in Topical Ophthalmic Drug Delivery: An Update. *Drug Delivery* **2014,** *23*(4), 1–17.

Akita, H.; Kudo, A.; Minoura, A.; Yamaguti, M.; Khalil, I. A.; Moriguchi, R.; Masuda, T.; Danev, R.; Nagayama, K.; Kogure, K. Multi-Layered Nanoparticles for Penetrating the Endosome and Nuclear Membrane via a Step-Wise Membrane Fusion Process. *Biomaterials* **2009,** *30*(15), 2940–2949.

Alexis, F.; Pridgen, E.; Molnar, L. K.; Farokhzad, O. C. Factors Affecting the Clearance and Biodistribution of Polymeric Nanoparticles. *Mol. Pharm.* **2008,** *5*(4), 505–515.

Ali, J.; Fazil, M.; Qumbar, M.; Khan, N.; Ali, A.; Ali, A. Colloidal Drug Delivery System: Amplify the Ocular Delivery. *Drug Delivery.* **2014,** *23*(3)1–17.

Almeida, A. J.; Runge, S.; Müller, R. H. Peptide-Loaded Solid Lipid Nanoparticles (SLN): Influence of Production Parameters. *Int. J. Pharm.* **1997,** *149*(2), 255–265.

Alonso, M. J.; Losa, C.; Calvo, P.; Vila-Jato, J. Approaches to Improve the Association of Amikacin Sulphate to Poly (alkylcyanoacrylate) Nanoparticles. *Int. J. Pharm.* **1991,** *68*(1), 69–76.

Arruebo, M.; Fernández-Pacheco, R.; Ibarra, M. R.; Santamaría, J. Magnetic Nanoparticles for Drug Delivery. *Nano Today* **2007,** *2*(3), 22–32.

Aryani, A.; Denecke, B. Exosomes as a Nanodelivery System: A Key to the Future of Neuromedicine? *Mol. Neurobiol.* **2016,** *53*(2), 818–834.

Badrzadeh, F.; Rahmati-Yamchi, M.; Badrzadeh, K.; Valizadeh, A.; Zarghami, N.; Farkhani, S. M.; Akbarzadeh, A. Drug Delivery and Nanodetection in Lung Cancer. *Artif. Cells Nanomed. Biotechnol.* **2016,** *44*(2), 618–634.

Baeza, A.; Colilla, M.; Vallet-Regí, M. Advances in Mesoporous Silica Nanoparticles for Targeted Stimuli-Responsive Drug Delivery. *Expert Opin. Drug Delivery* **2015,** *12*(2), 319–337.

Baglio, S. R.; Pegtel, D. M.; Baldini, N. Mesenchymal Stem Cell Secreted Vesicles Provide Novel Opportunities in (Stem) Cell-Free Therapy. *Front. Physiol.* **2012,** *3,* 359.

Bartoş, A.; Bartoş, D.; Szabo, B.; Breazu, C.; Opincariu, I.; Mironiuc, A.; Iancu, C. Recent Achievements in Colorectal Cancer Diagnostic and Therapy by the Use of Nanoparticles. *Drug Metab. Rev.* **2016,** *48*(1), 27–46.

Biswas, S.; Deshpande, P. P.; Navarro, G.; Dodwadkar, N. S.; Torchilin, V. P. Lipid Modified Triblock PAMAM-Based Nanocarriers for siRNA Drug Co-Delivery. *Biomaterials* **2013,** *34*(4), 1289–1301.

Björnmalm, M.; Faria, M.; Caruso, F. Increasing the Impact of Materials in and beyond Bio-Nano Science. *J. Am. Chem. Soc.* **2016,** *138*(41), 13449–13456.

Blanco, E.; Shen, H.; Ferrari, M. Principles of Nanoparticle Design for Overcoming Biological Barriers to Drug Delivery. *Nat. Biotechnol.* **2015,** *33*(9), 941–951.

Bobo, D.; Robinson, K. J.; Islam, J.; Thurecht, K. J.; Corrie, S. R. Nanoparticle-Based Medicines: A Review of FDA-Approved Materials and Clinical Trials to Date. *Pharm. Res.* **2016,** *33,* 2373–2387.

Brannon-Peppas, L. Recent Advances on the Use of Biodegradable Microparticles and Nanoparticles in Controlled Drug Delivery. *Int. J. Pharm.* **1995,** *116*(1), 1–9.

Brannon-Peppas, L.; Blanchette, J. O. Nanoparticle and Targeted Systems for Cancer Therapy. *Adv. Drug Delivery Rev.* **2012,** *64*, 206–212.

Brigger, I.; Dubernet, C.; Couvreur, P. Nanoparticles in Cancer Therapy and Diagnosis. *Adv. Drug Delivery Rev.* **2002,** *54*(5), 631–651.

Brody, E. N.; Gold, L. Aptamers as Therapeutic and Diagnostic Agents. *J. Biotechnol.* **2000,** *74*(1), 5–13.

Bulte, J. W. M.; Kraitchman, D. L. Iron Oxide MR Contrast Agents for Molecular and Cellular Imaging. *NMR Biomed.* **2004,** *17*(7), 484–499.

Calderón-Gonzalez, R.; Terán-Navarro, H.; Frande-Cabanes, E.; Ferrández-Fernández, E.; Freire, J.; Penadés, S.; Marradi, M.; García, I.; Gomez-Román, J.; Yañez-Díaz, S.; et al. Pregnancy Vaccination with Gold Glyco-Nanoparticles Carrying Listeria Monocytogenes Peptides Protects Against Listeriosis and Brain- and Cutaneous-Associated Morbidities. *Nanomaterials* **2016,** *6*(8), 151.

Chen, S.; Chen, L.; Wang, J.; Hou, J.; He, Q.; Liu, J.; Wang, J.; Xiong, S.; Yang, G.; Nie, Z. 2,3,4,5-Tetrakis (3′,4′-Dihydroxylphenyl) Thiophene: A New Matrix for the Selective Analysis of Low Molecular Weight Amines and Direct Determination of Creatinine in Urine by MALDI-TOF MS. *Anal. Chem.* **2012,** *84*(23), 10291–10297.

Chen, Y.-C.; Huang, X.-C.; Luo, Y.-L.; Chang, Y.-C.; Hsieh, Y.-Z.; Hsu, H.-Y. Non-Metallic Nanomaterials in Cancer Theranostics: A Review of Silica- and Carbon-Based Drug Delivery Systems. *Sci. Technol. Adv. Mater.* **2013,** *14*(4), 44407.

Cheng, X.; Lee, R. J. The Role of Helper Lipids in Lipid Nanoparticles (LNPs) Designed for Oligonucleotide Delivery. *Adv. Drug Delivery Rev.* **2016,** *99,* 129–137.

Clark, P. R.; Hersh, E. M. Cationic Lipid-Mediated Gene Transfer: Current Concepts. *Curr. Opin. Mol. Ther.* **1999,** *1*(2), 158–176.

Cui, Y.; Wu, Z.; Liu, X.; Ni, R.; Zhu, X.; Ma, L.; Liu, J. Preparation, Safety, Pharmacokinetics, and Pharmacodynamics of Liposomes Containing *Brucea Javanica* Oil. *AAPS Pharm. Sci. Tech.* **2010,** *11*(2), 878–884.

Dass, C. R. Vehicles for Oligonucleotide Delivery to Tumours. *J. Pharm. Pharmacol.* **2002,** *54*(1), 3–27.

Desai, M. P.; Labhasetwar, V.; Walter, E.; Levy, R. J.; Amidon, G. L. The Mechanism of Uptake of Biodegradable Microparticles in Caco-2 Cells Is Size Dependent. *Pharm. Res.* **1997,** *14*(11), 1568–1573.

Dosio, F.; Arpicco, S.; Stella, B.; Fattal, E. Hyaluronic Acid for Anticancer Drug and Nucleic Acid Delivery. *Adv. Drug Delivery Rev.* **2016,** *97,* 204–236.

Eifler, A. C.; Thaxton, C. S. Nanoparticle Therapeutics: FDA Approval, Clinical Trials, Regulatory Pathways, and Case Study. *Methods Mole. Biol.* **2011,** *726,* 325–338.

Elgadir, M. A.; Uddin, M. S.; Ferdosh, S.; Adam, A.; Chowdhury, A. J. K.; Sarker, M. Z. I. Impact of Chitosan Composites and Chitosan Nanoparticle Composites on Various Drug Delivery Systems: A Review. *J. Food Drug Anal.* **2015,** *23*(4), 619–629.

Felber, A. E.; Dufresne, M.-H.; Leroux, J.-C. pH-Sensitive Vesicles, Polymeric Micelles, and Nanospheres Prepared with Polycarboxylates. *Adv. Drug Delivery Rev.* **2012,** *64*(11), 979–992.

Fleige, E.; Quadir, M. A.; Haag, R. Stimuli-Responsive Polymeric Nanocarriers for the Controlled Transport of Active Compounds: Concepts and Applications. *Adv. Drug Delivery Rev.* **2012,** *64*(9), 866–884.

Fontana, F.; Liu, D.; Hirvonen, J.; Santos, H. A. Delivery of Therapeutics with Nanoparticles: What's New in Cancer Immunotherapy? *Wiley Interdiscip. Rev. Nanomed. Nanobiotechnol.* **2017,** *9*(1), e1421.

Fox, M. E.; Szoka, F. C.; Fréchet, J. M. J. Soluble Polymer Carriers for the Treatment of Cancer: The Importance of Molecular Architecture. *Acc. Chem. Res.* **2009,** *42*(8), 1141–1151.

Frühbeis, C.; Fröhlich, D.; Krämer-Albers, E.-M. Emerging Roles of Exosomes in Neuron–Glia Communication. *Front. Physiol.* **2012**, *3,* 119.

Fukumura, D.; Duda, D. G.; Munn, L. L.; Jain, R. K. Tumor Microvasculature and Microenvironment: Novel Insights Through Intravital Imaging in Pre-Clinical Models. *Microcirculation* **2010**, *17*(3), 206–225.

Gao, Y.; Li, Z.; Xie, X.; Wang, C.; You, J.; Mo, F.; Jin, B.; Chen, J.; Shao, J.; Chen, H.; et al. Dendrimeric Anticancer Prodrugs for Targeted Delivery of Ursolic Acid to Folate Receptor-Expressing Cancer Cells: Synthesis and Biological Evaluation. *Eur. J. Pharm. Sci.* **2015**, *70,* 55–63.

Gibson, M. *Pharmaceutical Preformulation and Formulation: A Practical Guide from Candidate Drug Selection to Commercial Dosage Form;* Informa Healthcare: New York, 2005; Vol. 9.

Giordano, K. F.; Jatoi, A. The Cancer Anorexia/Weight Loss Syndrome: Therapeutic Challenges. *Curr. Oncol. Rep.* **2005**, *7*(4), 271–276.

Gref, R.; Minamitake, Y.; Peracchia, M.; Trubetskoy, V.; Torchilin, V.; Langer, R. Biodegradable Long-Circulating Polymeric Nanospheres. *Science* **1994**, *263*(5153) 1600–1603.

Gualbert, J.; Shahgaldian, P.; Coleman, A. W. Interactions of Amphiphilic calix [4] Arene-Based Solid Lipid Nanoparticles with Bovine Serum Albumin. *Int. J. Pharm.* **2003**, *257*(1), 69–73.

Guescini, M.; Genedani, S.; Stocchi, V.; Agnati, L. F. Astrocytes and Glioblastoma Cells Release Exosomes Carrying mtDNA. *J. Neural Transm.* **2010a**, *117*(1), 1–4.

Guescini, M.; Guidolin, D.; Vallorani, L.; Casadei, L.; Gioacchini, A. M.; Tibollo, P.; Battistelli, M.; Falcieri, E.; Battistin, L.; Agnati, L. F.; et al. C2C12 Myoblasts Release Micro-Vesicles Containing mtDNA and Proteins Involved in Signal Transduction. *Exp. Cell Res.* **2010b**, *316*(12), 1977–1984.

Hamidi, M.; Azadi, A.; Rafiei, P. Hydrogel Nanoparticles in Drug Delivery. *Adv. Drug Delivery Rev.* **2008**, *60*(15), 1638–1649.

Han, H. S.; Thambi, T.; Choi, K. Y.; Son, S.; Ko, H.; Lee, M. C.; Jo, D.-G.; Chae, Y. S.; Kang, Y. M.; Lee, J. Y.; et al. Bioreducible Shell-Cross-Linked Hyaluronic Acid Nanoparticles for Tumor-Targeted Drug Delivery. *Biomacromolecules* **2015**, *16*(2), 447–456.

Heath, T. D.; Montgomery, J. A.; Piper, J. R.; Papahadjopoulos, D. Antibody-Targeted Liposomes: Increase in Specific Toxicity of Methotrexate-Gamma-Aspartate. *Proc. Natl. Acad. Sci. U. S. A.* **1983**, *80*(5), 1377–1381.

Horcajada, P.; Chalati, T.; Serre, C.; Gillet, B.; Sebrie, C.; Baati, T.; Eubank, J. F.; Heurtaux, D.; Clayette, P.; Kreuz, C.; et al. Porous Metal–organic-Framework Nanoscale Carriers as a Potential Platform for Drug Delivery and Imaging. *Nat. Mater.* **2010**, *9*(2), 172–178.

Hu, Q.; Sun, W.; Wang, C.; Gu, Z. Recent Advances of Cocktail Chemotherapy by Combination Drug Delivery Systems. *Adv. Drug Delivery Rev.* **2016**, *98,* 19–34.

Immordino, M. L.; Dosio, F.; Cattel, L. Stealth Liposomes: Review of the Basic Science, Rationale, and Clinical Applications, Existing and Potential. *Int. J. Nanomed.* **2006**, *1*(3), 297–315.

Ioannidis, J. P. A.; Allison, D. B.; Ball, C. A.; Coulibaly, I.; Cui, X.; Culhane, A. C.; Falchi, M.; Furlanello, C.; Game, L.; Jurman, G.; et al. Repeatability of Published Microarray Gene Expression Analyses. *Nat. Genet.* **2009**, *41*(2), 149–155.

Jain, R. K. Transport of Molecules in the Tumor Interstitium: A Review. *Cancer Res.* **1987**, *47*(12), 3039–3051.

Jemal, A.; Siegel, R.; Ward, E.; Murray, T.; Xu, J.; Thun, M. J. Cancer Statistics. *CA. Cancer J. Clin.* **2007**, *57*(1), 43–66.

Jhaveri, A.; Deshpande, P.; Torchilin, V. Stimuli-Sensitive Nanopreparations for Combination Cancer Therapy. *J. Control. Release* **2014**, *190,* 352–370.

Jiang, T.; Li, Y.-M.; Lv, Y.; Cheng, Y.-J.; He, F.; Zhuo, R.-X. Amphiphilic Polycarbonate Conjugates of Doxorubicin with pH-Sensitive Hydrazone Linker for Controlled Release. *Colloids Surf. B* **2013**, *111,* 542–548.

Juliano, R. L. The Delivery of Therapeutic Oligonucleotides. *Nucleic Acids Res.* **2016**, *44*(14), 6518–6548.

Juliano, R. L.; Stamp, D. Lectin-Mediated Attachment of Glycoprotein-Bearing Liposomes to Cells. *Nature* **1976**, *261*(5557), 235–238.

Kaasgaard, T.; Andresen, T. L. Liposomal Cancer Therapy: Exploiting Tumor Characteristics. *Expert Opin. Drug Delivery* **2010**, *7*(2), 225–243.

Kamaly, N.; Yameen, B.; Wu, J.; Farokhzad, O. C. Degradable Controlled-Release Polymers and Polymeric Nanoparticles: Mechanisms of Controlling Drug Release. *Chem. Rev.* **2016**, *116*(4), 2602–2663.

Kanamala, M.; Wilson, W. R.; Yang, M.; Palmer, B. D.; Wu, Z. Mechanisms and Biomaterials in pH-Responsive Tumour Targeted Drug Delivery: A Review. *Biomaterials* **2016**, *85,* 152–167.

Kanapathipillai, M.; Brock, A.; Ingber, D. E. Nanoparticle Targeting of Anti-Cancer Drugs That Alter Intracellular Signaling or Influence the Tumor Microenvironment. *Adv. Drug Delivery Rev.* **2014**, *79,* 107–118.

Khan, S. M.; Abdelhamid, H. N.; Wu, H.-F. Near Infrared (NIR) Laser Mediated Surface Activation of Graphene Oxide Nanoflakes for Efficient Antibacterial, Antifungal and Wound Healing Treatment. *Colloids Surf. B* **2015**, *127C,* 281–291.

Kierstead, P. H.; Okochi, H.; Venditto, V. J.; Chuong, T. C.; Kivimae, S.; Fréchet, J. M. J.; Szoka, F. C. The Effect of Polymer Backbone Chemistry on the Induction of the Accelerated Blood Clearance in Polymer Modified Liposomes. *J. Control. Release* **2015**, *213,* 1–9.

Kim, J.-H.; Kim, Y.-S.; Kim, S.; Park, J. H.; Kim, K.; Choi, K.; Chung, H.; Jeong, S. Y.; Park, R.-W.; Kim, I.-S.; et al. Hydrophobically Modified Glycol Chitosan Nanoparticles as Carriers for Paclitaxel. *J. Control. Release* **2006**, *111*(1–2), 228–234.

Krishna, R.; Mayer, L. D. Multidrug Resistance (MDR) in Cancer. *Eur. J. Pharm. Sci.* **2000**, *11*(4), 265–283.

Kumar, C. S. S. R. *Nanomaterials for Medical Diagnosis and Therapy;* Wiley-VCH: Weinheim, Germany, 2007.

Kumari, P.; Ghosh, B.; Biswas, S. Nanocarriers for Cancer-Targeted Drug Delivery. *J. Drug Target.* **2016**, *24*(3), 179–191.

Lai, R. C.; Tan, S. S.; Teh, B. J.; Sze, S. K.; Arslan, F.; de Kleijn, D. P.; Choo, A.; Lim, S. K. Proteolytic Potential of the MSC Exosome Proteome: Implications for an Exosome-Mediated Delivery of Therapeutic Proteasome. *Int. J. Proteomics* **2012**, *2012,* 1–14.

Langer, R. Drug Delivery and Targeting. *Nature* **1998**, *392*(Suppl. 6679), 5–10.

Lehto, T.; Ezzat, K.; Wood, M. J. A.; EL Andaloussi, S. Peptides for Nucleic Acid Delivery. *Adv. Drug Delivery Rev.* **2016**, *106,* 172–182.

Lipinski, C. A. Drug-like Properties and the Causes of Poor Solubility and Poor Permeability. *J. Pharmacol. Toxicol. Methods* **2000**, *44*(1), 235–249.

Lipinski, C. A.; Lombardo, F.; Dominy, B. W.; Feeney, P. J. Experimental and Computational Approaches to Estimate Solubility and Permeability in Drug Discovery and Development Settings. *Adv. Drug Delivery Rev.* **2012**, *64,* 4–17.

Liu, Z.; Sun, X.; Nakayama-Ratchford, N.; Dai, H. Supramolecular Chemistry on Water-Soluble Carbon Nanotubes for Drug Loading and Delivery. *ACS Nano* **2007**, *1*(1), 50–56.

Liu, Z.; Chen, K.; Davis, C.; Sherlock, S.; Cao, Q.; Chen, X.; Dai, H. Drug Delivery with Carbon Nanotubes for in Vivo Cancer Treatment. *Cancer Res.* **2008a,** *68*(16), 6652–6660.

Liu, Z.; Jiao, Y.; Wang, Y.; Zhou, C.; Zhang, Z. Polysaccharides-Based Nanoparticles as Drug Delivery Systems. *Adv. Drug Delivery Rev.* **2008b,** *60*(15), 1650–1662.

Liu, J.; Huang, Y.; Kumar, A.; Tan, A.; Jin, S.; Mozhi, A.; Liang, X.-J. pH-Sensitive Nano-Systems for Drug Delivery in Cancer Therapy. *Biotechnol. Adv.* **2014,** *32*(4), 693–710.

Liu, J.; Luo, Z.; Zhang, J.; Luo, T.; Zhou, J.; Zhao, X.; Cai, K. Hollow Mesoporous Silica Nanoparticles Facilitated Drug Delivery via Cascade pH Stimuli in Tumor Microenvironment for Tumor Therapy. *Biomaterials* **2016,** *83,* 51–65.

Loh, X. J.; Lee, T.-C.; Dou, Q.; Deen, G. R. Utilising Inorganic Nanocarriers for Gene Delivery. *Biomater. Sci.* **2016,** *4*(1), 70–86.

López-Sagaseta, J.; Malito, E.; Rappuoli, R.; Bottomley, M. J. Self-Assembling Protein Nanoparticles in the Design of Vaccines. *Comput. Struct. Biotechnol. J.* **2016,** *14,* 58–68.

Ma, Y.; Fan, X.; Li, L. pH-Sensitive Polymeric Micelles Formed by Doxorubicin Conjugated Prodrugs for Co-Delivery of Doxorubicin and Paclitaxel. *Carbohydr. Polym.* **2016,** *137,* 19–29.

Maeda, H. The Enhanced Permeability and Retention (EPR) Effect in Tumor Vasculature: The Key Role of Tumor-Selective Macromolecular Drug Targeting. *Adv. Enzyme Regul.* **2001,** *41,* 189–207.

Malam, Y.; Loizidou, M.; Seifalian, A. M. Liposomes and Nanoparticles: Nanosized Vehicles for Drug Delivery in Cancer. *Trends Pharmacol. Sci.* **2009,** *30*(11), 592–599.

Markman, J. L.; Rekechenetskiy, A.; Holler, E.; Ljubimova, J. Y. Nanomedicine Therapeutic Approaches to Overcome Cancer Drug Resistance. *Adv. Drug Delivery Rev.* **2013,** *65*(13–14), 1866–1879.

Matsumura, Y.; Maeda, H. A New Concept for Macromolecular Therapeutics in Cancer Chemotherapy: Mechanism of Tumoritropic Accumulation of Proteins and the Antitumor Agent Smancs. *Cancer Res.* **1986,** *46*(12 Pt 1), 6387–6392.

Mays, A. N.; Osheroff, N.; Xiao, Y.; Wiemels, J. L.; Felix, C. A.; Byl, J. A. W.; Saravanamuttu, K.; Peniket, A.; Corser, R.; Chang, C.; et al. Evidence for Direct Involvement of Epirubicin in the Formation of Chromosomal Translocations in t (15;17) Therapy-Related Acute Promyelocytic Leukemia. *Blood* **2010,** *115*(2), 326–330.

Mehra, N. K.; Jain, N. K. Multifunctional Hybrid-Carbon Nanotubes: New Horizon in Drug Delivery and Targeting. *J. Drug Targeting* **2016,** *24*(4), 294–308.

Moghimi, S. M.; Hunter, A. C.; Murray, J. C. Long-Circulating and Target-Specific Nanoparticles: Theory to Practice. *Pharmacol. Rev.* **2001,** *53*(2), 283–318.

Moghimi, S. M.; Hunter, A. C.; Murray, J. C. Nanomedicine: Current Status and Future Prospects. *FASEB J.* **2005,** *19*(3), 311–330.

Monfardini, C.; Veronese, F. M. Stabilization of Substances in Circulation. *Bioconjugate Chem.* **1998,** *9*(4), 418–450.

Moroz, E.; Matoori, S.; Leroux, J.-C. Oral Delivery of Macromolecular Drugs: Where We Are after Almost 100 years of Attempts. *Adv. Drug Delivery Rev.* **2016,** *101,* 108–121.

Mujokoro, B.; Adabi, M.; Sadroddiny, E.; Adabi, M.; Khosravani, M. Nano-Structures Mediated Co-Delivery of Therapeutic Agents for Glioblastoma Treatment: A Review. *Mater. Sci. Eng. C* **2016,** *69,* 1092–1102.

Müller, R. H.; Mäder, K.; Gohla, S. Solid Lipid Nanoparticles (SLN) for Controlled Drug Delivery – a Review of the State of the Art. *Eur. J. Pharm. Biopharm.* **2000,** *50*(1), 161–177.

van Niel, G.; Porto-Carreiro, I.; Simoes, S.; Raposo, G. Exosomes: A Common Pathway for a Specialized Function. *J. Biochem.* **2006,** *140*(1), 13–21.

Owens, D.; Peppas, N. Opsonization, Biodistribution, and Pharmacokinetics of Polymeric Nanoparticles. *Int. J. Pharm.* **2006**, *307*(1), 93–102.

Paciotti, G. F.; Zhao, J.; Cao, S.; Brodie, P. J.; Tamarkin, L.; Huhta, M.; Myer, L. D.; Friedman, J.; Kingston, D. G. I. Synthesis and Evaluation of Paclitaxel-Loaded Gold Nanoparticles for Tumor-Targeted Drug Delivery. *Bioconjugate Chem.* **2016**, *27*(11), 2646–2657.

Panyam, J.; Labhasetwar, V. Biodegradable Nanoparticles for Drug and Gene Delivery to Cells and Tissue. *Adv. Drug Delivery Rev.* **2003**, *55*(3), 329–347.

Park, J. H.; Lee, S.; Kim, J.-H.; Park, K.; Kim, K.; Kwon, I. C. Polymeric Nanomedicine for Cancer Therapy. *Prog. Polym. Sci.* **2008**, *33*(1), 113–137.

Pathak, Y.; Thassu, D. *Drug Delivery Nanoparticles Formulation and Characterization;* Informa Healthcare: New York, 2009.

Perumal, V.; Banerjee, S.; Das, S.; Sen, R. K.; Mandal, M. Effect of Liposomal Celecoxib on Proliferation of Colon Cancer Cell and Inhibition of DMBA-Induced Tumor in Rat Model. *Cancer Nanotechnol.* **2011**, *2*(1–6), 67–79.

Pokropivny, V. V.; Skorokhod, V. V. New Dimensionality Classifications of Nanostructures. *Phys. E Low-Dimensional Syst. Nanostruct.* **2008**, *40*(7), 2521–2525.

Ritter, J. A.; Ebner, A. D.; Daniel, K. D.; Stewart, K. L. Application of High Gradient Magnetic Separation Principles to Magnetic Drug Targeting. *J. Magn. Magn. Mater.* **2004**, *280*(2), 184–201.

Rolfe, B. E.; Blakey, I.; Squires, O.; Peng, H.; Boase, N. R. B.; Alexander, C.; Parsons, P. G.; Boyle, G. M.; Whittaker, A. K.; Thurecht, K. J. Multimodal Polymer Nanoparticles with Combined 19 F Magnetic Resonance and Optical Detection for Tunable, Targeted, Multimodal Imaging In Vivo. *J. Am. Chem. Soc.* **2014**, *136*(6), 2413–2419.

Rothenberg, M. L.; Carbone, D. P.; Johnson, D. H. Opinion: Improving the Evaluation of New Cancer Treatments: Challenges and Opportunities. *Nat. Rev. Cancer* **2003**, *3*(4), 303–309.

Ruan, S.; Yuan, M.; Zhang, L.; Hu, G.; Chen, J.; Cun, X.; Zhang, Q.; Yang, Y.; He, Q.; Gao, H. Tumor Microenvironment Sensitive Doxorubicin Delivery and Release to Glioma Using Angiopep-2 Decorated Gold Nanoparticles. *Biomaterials* **2015**, *37,* 425–435.

Rui, Y.; Wang, S.; Low, P. S.; Thompson, D. H. Diplasmenylcholine–Folate Liposomes: An Efficient Vehicle for Intracellular Drug Delivery†. *J. Am. Chem. Soc.* **1998**, *120*(44), 11213–11218.

Sadauskas, E.; Wallin, H.; Stoltenberg, M.; Vogel, U.; Doering, P.; Larsen, A.; Danscher, G. Kupffer Cells Are Central in the Removal of Nanoparticles from the Organism. *Part. Fibre Toxicol.* **2007**, *4,* 10.

Saso, A.; Kampmann, B. Vaccination against Respiratory Syncytial Virus in Pregnancy: A Suitable Tool to Combat Global Infant Morbidity and Mortality? *Lancet Infect. Dis.* **2016**, *16*(8), e153–163.

Schiffelers, R. M.; Koning, G. A.; Ten Hagen, T. L. M.; Fens, M. H. A. M.; Schraa, A. J.; Janssen, A. P. C. A.; Kok, R. J.; Molema, G.; Storm, G. Anti-Tumor Efficacy of Tumor Vasculature-Targeted Liposomal Doxorubicin. *J. Controlled Release* **2003**, *91*(1–2), 115–122.

Serpe, L. Conventional Chemotherapeutic Drug Nanoparticles for Cancer Treatment. In *Nanotechnologies for the Life Sciences;* Challa S. S. R., Ed. Wiley-VCH Verlag GmbH & Co. KGaA: Weinheim, Germany, 2007; Vol 6, p –39.

Sharma, A. K.; Garg, T.; Goyal, A. K.; Rath, G. Role of Microemuslsions in Advanced Drug Delivery. *Artif. Cells Nanomed. Biotechnol.* **2015**, *44,* 1–9.

Shen, M.; Huang, Y.; Han, L.; Qin, J.; Fang, X.; Wang, J.; Yang, V. C. Multifunctional Drug Delivery System for Targeting Tumor and Its Acidic Microenvironment. *J. Controlled Release* **2012**, *161* (3), 884–892.

Sjöström, B.; Kaplun, A.; Talmon, Y.; Cabane, B. Structures of Nanoparticles Prepared from Oil-in-Water Emulsions. *Pharm. Res.* **1995,** *12*(1), 39–48.

Skwarczynski, M.; Toth, I. Peptide-Based Synthetic Vaccines. *Chem. Sci.* **2016,** *7*(2), 842–854.

Smalheiser, N. R. Exosomal Transfer of Proteins and RNAs at Synapses in the Nervous System. *Biol. Direct* **2007,** *2*(1), 35.

Song, C.; Labhasetwar, V.; Cui, X.; Underwood, T.; Levy, R. J. Arterial Uptake of Biodegradable Nanoparticles for Intravascular Local Drug Delivery: Results with an Acute Dog Model. *J. Controlled Release* **1998,** *54*(2), 201–211.

Soo Choi, H.; Liu, W.; Misra, P.; Tanaka, E.; Zimmer, J. P.; Itty Ipe, B.; Bawendi, M. G.; Frangioni, J. V. Renal Clearance of Quantum Dots. *Nat. Biotechnol.* **2007,** *25*(10), 1165–1170.

Soppimath, K. S.; Aminabhavi, T. M.; Kulkarni, A. R.; Rudzinski, W. E. Biodegradable Polymeric Nanoparticles as Drug Delivery Devices. *J. Controlled Release* **2001,** *70*(1), 1–20.

Steichen, S. D.; Caldorera-Moore, M.; Peppas, N. A. A Review of Current Nanoparticle and Targeting Moieties for the Delivery of Cancer Therapeutics. *Eur. J. Pharm. Sci.* **2013,** *48*(3), 416–427.

Sun, C.-Y.; Qin, C.; Wang, X.-L.; Su, Z.-M. Metal-Organic Frameworks as Potential Drug Delivery Systems. *Expert Opin. Drug Delivery* **2013,** *10*(1), 89–101.

Szakács, G.; Paterson, J. K.; Ludwig, J. A.; Booth-Genthe, C.; Gottesman, M. M. Targeting Multidrug Resistance in Cancer. *Nat. Rev. Drug Discovery* **2006,** *5*(3), 219–234.

Takemura, G.; Fujiwara, H. Doxorubicin-Induced Cardiomyopathy. *Prog. Cardiovasc. Dis.* **2007,** *49*(5), 330–352.

Théry, C.; Amigorena, S.; Raposo, G.; Clayton, A. Isolation and Characterization of Exosomes from Cell Culture Supernatants and Biological Fluids. *Curr. Protoc. Cell Biol.* **2006,** *3*(1), 3.22.1–3.22.

Théry, C.; Ostrowski, M.; Segura, E. Membrane Vesicles as Conveyors of Immune Responses. *Nat. Rev. Immunol.* **2009,** *9*(8), 581–593.

Tian, J.; Rodgers, Z.; Min, Y.; Wan, X.; Qiu, H.; Mi, Y.; Tian, X.; Wagner, K. T.; Caster, J. M.; Qi, Y.; et al. Nanoparticle Delivery of Chemotherapy Combination Regimen Improves the Therapeutic Efficacy in Mouse Models of Lung Cancer. *Nanomed. Nanotechnol. Biol. Med.* **2016,** *13*(3), 1301–1307.

Tomalia, D. A. Birth of a New Macromolecular Architecture: Dendrimers as Quantized Building Blocks for Nanoscale Synthetic Polymer Chemistry. *Prog. Polym. Sci.* **2005,** *30*(3), 294–324.

Torchilin, V. P. Polymer-Coated Long-Circulating Microparticulate Pharmaceuticals. *J. Microencapsul.* **1998,** *15*(1), 1–19.

Torchilin, V. P. Drug Targeting. *Eur. J. Pharm. Sci.* **2000,** *11,* S81–S91.

Torchilin, V. P. PEG-Based Micelles as Carriers of Contrast Agents for Different Imaging Modalities. *Adv. Drug Delivery Rev.* **2002,** *54*(2), 235–252.

Torchilin, V. Multifunctional and Stimuli-Sensitive Pharmaceutical Nanocarriers. *Eur. J. Pharm. Biopharm.* **2009,** *71*(3), 431–444.

Torchilin, V. P. Passive and Active Drug Targeting: Drug Delivery to Tumors as an Example. In *Handbook of experimental pharmacology;* Schäfer-Korting, M., Ed.; Springer: Berlin, Heidelberg, 2010; Vol. 197, pp 3–53.

Torres-Sangiao, E.; Holban, A.; Gestal, M. Advanced Nanobiomaterials: Vaccines, Diagnosis and Treatment of Infectious Diseases. *Molecules* **2016,** *21*(7), 867.

Tseng, Y.-Y.; Kau, Y.-C.; Liu, S.-J. Advanced Interstitial Chemotherapy for Treating Malignant Glioma. *Expert Opin. Drug Delivery* **2016,** *13*(11), 1533–1544.

Turner, C. T.; Hasanzadeh Kafshgari, M.; Melville, E.; Delalat, B.; Harding, F.; Mäkilä, E.; Salonen, J. J.; Cowin, A. J.; Voelcker, N. H. Delivery of Flightless I siRNA from Porous Silicon Nanoparticles Improves Wound Healing in Mice. *ACS Biomater. Sci. Eng.* **2016,** *2*(12), 2339–2346.

Ueda, M.; Iwara, A.; Kreuter, J. Influence of the Preparation Methods on the Drug Release Behaviour of Loperamide-Loaded Nanoparticles. *J. Microencapsul.* **1998,** *15*(3), 361–372.

Uhrich, K. E.; Cannizzaro, S. M.; Langer, R. S.; Shakesheff, K. M. Polymeric Systems for Controlled Drug Release. *Chem. Rev.* **1999,** *99*(11), 3181–3198.

Ulbrich, K.; Holá, K.; Šubr, V.; Bakandritsos, A.; Tuček, J.; Zbořil, R. Targeted Drug Delivery with Polymers and Magnetic Nanoparticles: Covalent and Noncovalent Approaches, Release Control, and Clinical Studies. *Chem. Rev.* **2016,** *116*(9), 5338–5431.

Valadi, H.; Ekström, K.; Bossios, A.; Sjöstrand, M.; Lee, J. J.; Lötvall, J. O. Exosome-Mediated Transfer of mRNAs and microRNAs Is a Novel Mechanism of Genetic Exchange between Cells. *Nat. Cell Biol.* **2007,** *9*(6), 654–659.

WHO. WHO – Cancer. *WHO.* World Health Organization 2016.

Wicki, A.; Witzigmann, D.; Balasubramanian, V.; Huwyler, J. Nanomedicine in Cancer Therapy: Challenges, Opportunities, and Clinical Applications. *J. Controlled Release* **2015,** *200,* 138–157.

Wilhelm, S.; Tavares, A. J.; Dai, Q.; Ohta, S.; Audet, J.; Dvorak, H. F.; Chan, W. C. W. Analysis of Nanoparticle Delivery to Tumours. *Nat. Rev. Mater.* **2016,** *1*(5), 16014.

Wissing, S.; Kayser, O.; Müller, R. Solid Lipid Nanoparticles for Parenteral Drug Delivery. *Adv. Drug Delivery Rev.* **2004,** *56*(9), 1257–1272.

Wu, W.; Wu, Z.; Yu, T.; Jiang, C.; Kim, W.-S. Recent Progress on Magnetic Iron Oxide Nanoparticles: Synthesis, Surface Functional Strategies and Biomedical Applications. *Sci. Technol. Adv. Mater.* **2015,** *16*(2), 23501.

Yang, D.; Ma, P.; Hou, Z.; Cheng, Z.; Li, C.; Lin, J. Current Advances in Lanthanide Ion (Ln 3+)-Based Upconversion Nanomaterials for Drug Delivery. *Chem. Soc. Rev.* **2015,** *44*(6), 1416–1448.

Yuan, Y.; Cai, T.; Xia, X.; Zhang, R.; Chiba, P.; Cai, Y. Nanoparticle Delivery of Anticancer Drugs Overcomes Multidrug Resistance in Breast Cancer. *Drug Delivery* **2016,** *23*(9), 3350–3357.

Zatsepin, T. S.; Kotelevtsev, Y. V; Koteliansky, V. Lipid Nanoparticles for Targeted siRNA Delivery – Going from Bench to Bedside. *Int. J. Nanomed.* **2016,** *11,* 3077–3086.

Zhang, C.; Wang, W.; Liu, T.; Wu, Y.; Guo, H.; Wang, P.; Tian, Q.; Wang, Y.; Yuan, Z. Doxorubicin-Loaded Glycyrrhetinic Acid-Modified Alginate Nanoparticles for Liver Tumor Chemotherapy. *Biomaterials* **2012,** *33*(7), 2187–2196.

Zhang, W.; Wang, G.; See, E.; Shaw, J. P.; Baguley, B. C.; Liu, J.; Amirapu, S.; Wu, Z. Post-Insertion of Poloxamer 188 Strengthened Liposomal Membrane and Reduced Drug Irritancy and in Vivo Precipitation, Superior to PEGylation. *J. Controlled Release* **2015,** *203,* 161–169.

Zimmer, A.; Kreuter, J. Microspheres and Nanoparticles Used in Ocular Delivery Systems. *Adv. Drug Delivery Rev.* **1995,** *16*(1), 61–73.

Zylberberg, C.; Matosevic, S. Pharmaceutical Liposomal Drug Delivery: A Review of New Delivery Systems and a Look at the Regulatory Landscape. *Drug Delivery* **2016,** *23*(9), 3319–3329.

CHAPTER 5

NANOTECHNOLOGY-BASED FORMULATIONS FOR DRUG TARGETING TO THE CENTRAL NERVOUS SYSTEM

JOSEF JAMPÍLEK[1,*] and KATARÍNA KRÁĽOVÁ[2]

[1]Department of Pharmaceutical Chemistry, Faculty of Pharmacy, Comenius University, Odbojárov 10, 83232 Bratislava, Slovakia

[2]Institute of Chemistry, Faculty of Natural Sciences, Comenius University, Ilkovičova 6, 842 15 Bratislava, Slovakia

*Corresponding author. E-mail: josef.jampilek@gmail.com

ABSTRACT

Central nervous system (CNS) disorders have been incorporated into our daily lifestyle now and in general, are faced by people of all age-groups and affect both men and women across the world. The drugs for the treatment of these disorders must overcome the blood–brain barrier, serious obstacle for the permeation of drugs that require CNS action. Rapid development of various nanosystems employed for nano-based drug delivery systems have great potential to facilitate the movement of drugs across all barriers and thus these drug carriers have been extensively studied as a strategy of direct drug delivery to the CNS. This contribution is focused on the effects and CNS targeting of nanoscale formulations containing drugs such as antiepileptics, antipsychotics, anxiolytics, antidepressants as well as drugs applied for the treatment of schizophrenia and Parkinson's and Alzheimer's diseases. The benefits connected with their application (e.g., reduction of required drug dose at bioavailability increase, reduced side effects due to decreased toxicity against nontarget cells, prolongation of time in circulation) are highlighted as well.

5.1 INTRODUCTION

Central nervous system (CNS) disorders, especially neurological (epilepsy, Parkinson's disease [PD], Alzheimer's disease [AD]) or psychiatric (depressions, anxiety, schizophrenia) diseases, have been incorporated in our daily lifestyle now and in general, are faced by people of all age-groups and affect both men and women across the world (McGovern, 2014). They are more frequently caused by various stress factors, traumas, low-quality/ unfavorable social surroundings, and so forth, as well as by excessive load/ strain in job and family. Psychiatric drugs have become one of the best-selling and most profitable classes of drugs, generating $22 billion in global sales in 2008 (Healthcare Finance News, 2009). For example, Abilify® (aripiprazole [APZ]), used to treat schizophrenia and bipolar disorder (Aripiprazole, 2017), and Lyrica® (pregabalin), used to treat epilepsy, neuropathic pain, fibromyalgia, and generalized anxiety disorder and prevent migraines and social anxiety disorder (Pregabalin, 2017), belong among top 20 prescription drugs based on the U.S. sales in 2015 (Statista, 2017). It can be stated that brain disorders, including developmental, psychiatric, and neurodegenerative diseases, represent an enormous disease burden, in terms of human suffering and economic costs (McGovern, 2014).

Drugs for the treatment of these disorders require delivery to the CNS—to the brain, that is, they must overcome all barriers to achieve the brain tissue, and the blood–brain barrier (BBB) is the last, critical, and serious obstacle for the permeation of drugs that require CNS action. The BBB represents a structure with complex cellular organization that separates the brain from the systemic circulation. The BBB acts not only as a mechanic barrier but also as a metabolic barrier due to the presence of numerous enzymes. These enzymes can either metabolize potentially harmful drugs to CNS-inactive compounds or convert inactive drugs to their CNS-active metabolites or degrade them into metabolites or substrates of specific efflux transporters, such as P-glycoprotein (P-gp)/multidrug resistance proteins (Abbott et al., 2010; Alavijeh et al., 2005; Jampílek et al., 2015; Khanbabaie and Jahan-shahi, 2012; Passeleu-Le Bourdonnec et al., 2013).

To circumvent the BBB and allow active CNS drugs to reach their target, many strategies exist. They can be classified with respect to the BBB as either invasive (direct injection into the cerebrospinal fluid or therapeutic or osmotic opening of the BBB) or noninvasive such as the use of alternative routes of administration (e.g., intraventricular/intrathecal route, nose-to-brain route, olfactory, and trigeminal pathways to brain), inhibition of efflux transporters, chemical modification of drugs (lipophilic analogues, prodrugs or bioprecursors), and encapsulation of drugs into nanocarriers (Černíková and Jampílek, 2014; Jampílek et al., 2015; Lu et al., 2014; Passeleu-Le Bourdonnec et al., 2013). Nanoparticles (NPs) as drug carriers have been extensively studied for last few decades as a strategy of direct drug delivery to the CNS. It has been shown that nanodelivery systems have great potential to facilitate the movement of drugs across all barriers. Their uptake into the brain is hypothesized to occur through adsorptive transcytosis and receptor-mediated endocytosis (Wong et al., 2012; Yang et al., 2010a). Particle size, surface affinity, and stability in circulation are important factors influencing the brain distribution of colloidal particles (Bhaskar et al., 2010; Simko and Mattson, 2014).

Rapid development of nanomedicine and various nanosystems employed for nano-based drug delivery systems has offered new opportunities in designing and evaluating novel treatment strategies for neurological and psychiatric diseases. The application of nanomaterials as drug carriers has potentially a number of advantages, primarily in terms of the increased bioavailability, and thus, the therapeutic efficiency of the medicaments, and reduction of toxicity (Dimitrijevic and Pantic, 2014; Khanbabaie and Jahanshahi, 2012). Nano-based drug delivery systems for drug formulations include polymeric NPs, nanospheres, nanosuspensions, nanoemulsions (NEs), nanogels, nano-micelles, nanoliposomes, carbon nanotubes, nanofibers, solid lipid nanoparticles (SLNPs), nanostructured lipid carriers (NLCs), and lipid drug conjugates. The unique properties of NPs such as nanoscale size, large surface, higher solubility, and multifunctionality represent the capability to interact with composite cellular functions in new ways (Alyautdin et al., 2014; Cacciatore et al., 2016; Jampílek and Kráľová, 2017a; 2017b; 2017c; Masserini, 2013; Saraiva et al., 2016). In fact, nanotechnology has now emerged as an area of research for the invention of newer approaches for the CNS drug delivery and a revolutionary method to improve diagnosis and therapy of neurodegenerative disorders (Dimitrijevic and Pantic, 2014; Khanbabaie and Jahanshahi, 2012).

This chapter is focused on the effects and CNS targeting of nanoscale formulations containing a wide range of drugs such as antiepileptics, antipsychotics, anxiolytics, antidepressants as well as drugs applied for the treatment of schizophrenia and PD and AD. The benefits connected with their application (e.g., reduction of required drug dose at bioavailability increase, reduced side effects due to decreased toxicity against non-target cells, prolongation of time in circulation) are highlighted as well.

5.2   ANTIEPILEPTICS

Epilepsy is a disorder of the CNS characterized by periodic loss of consciousness with or without convulsions associated with abnormal electrical activity in the brain. It could be caused due to brain damage, but in most cases, the cause of this disease is unknown. It is estimated that there are more than 50 million people with epilepsy worldwide (Sharma and Dixit, 2013). Apart from benzodiazepines, specifically designed drugs used for treatment of epilepsy can be classified according to their structure as follows: (i) barbiturate-like monocyclic antiepileptics, (ii) dibenzazepines, (iii) γ-aminobutyric acid (GABA)-based drugs, and (iv) various structures (Steinhilber et al., 2010). Some drugs from these groups have been reformulated as NPs for higher bioavailability and better therapeutic index.

5.2.1   MONOCYCLIC ANTIEPILEPTICS

Phenytoin (PHT) is an anti-seizure drug used for the prevention of tonic-clonic and partial seizures. Pluronic P85-coated PHT poly(butylcyanoacrylate) NPs were found to increase PHT concentrations in the brain of rats with drug-resistant epilepsy (Fang et al., 2016). Under salivary conditions, the application of the stable crystalline nanosuspension of PHT (330 nm) prepared through wet media milling using Tween 80 as a stabilizer resulted in a significant increase of the solubility characteristics of the drug as well as an increase of permeability across the buccal mucosa (Baumgartner et al., 2016). The coating with fully degradable synthetic polymer, poly(vinyl alcohol) (PVA), effectively stabilized liposomes, and the effect increased with an increasing number of deposited layers until the polymeric film reached the optimal thickness; bilayer-coated liposomes released PHT less rapidly than uncoated ones indicating that liposomes coated with ultrathin film prepared from PVA derivatives are suitable as drug nanocarriers (Zasada et al., 2015).

Comparison of PHT-loaded angiopep-2 (ANG)-electroresponsive hydrogel NPs (ANG-PHT-ERHNPs), PHT-loaded nonelectroresponsive hydrogel NPs, and PHT solution showed that ANG-PHT-ERHNPs lowered the effective therapeutic doses of drug and demonstrated improved antiseizure effects as compared to the two other formulations in both electrical- (maximal electrical shock) and chemical-induced (pentylenetetrazole and pilocarpine) seizure models (Wang et al., 2016).

PHT loaded in ordered mesoporous silica and TiO_2 nanotubes showed a good stability inside the used materials, whereby the adsorption and desorption of PHT were faster in nanostructured TiO_2 tubes than in mesoporous silica matrix (Lopez et al., 2011). The controlled release of PHT from nanostructured TiO_2 reservoirs prepared by the solgel method combined with hydrothermal treatment, using NPs of five different sizes ranging from 6.8 to 16.4 nm in diameter and loaded with 5 wt.% PHT, which could be used as implanted reservoirs for the treatment of chronic diseases, was studied by Heredia-Cervera et al. (2009). Previously, the potential of encapsulated valproic acid (VPA) and PHT-Na in ordered mesoporous SiO_2 solids to deliver the drug without causing secondary effects for the treatment of temporal lobe epilepsy was also investigated. These ordered SiO_2 nanostructures used as drug-containing reservoirs implanted in rat brain did not cause necrosis and inflammation and showed a good biocompatibility of the reservoir with brain tissue (Lopez et al., 2006). The application of PHT loaded in the silica core of iron oxide NPs to P-gp-overexpressing rats resulted in reduced prevalence of clonus (40%; $p<0.05$) and tonic-clonic seizures (20%; $p<0.02$), while these effects did not occur without loading PHT into NPs, which indicates that NPs can be used as drugs carriers to the brain with pharmacoresistant seizures (Rosillo-de la Torre et al., 2015).

Ethosuximide (ETX) is a succinimide anticonvulsant, used mainly in the absence of seizures. Huang et al. (2012) designed self-assembling PVA-F127 thermosensitive nanocarriers by the incorporation of iron oxide NPs and hydrophobic ETX molecules into a thermosensitive matrix composed of poly(ethylene oxide)–poly(propylene oxide)–poly(ethylene oxide) (PEO–PPO–PEO) triblock-copolymer and PVA using a mini-emulsion process for highly sensitive magnetically triggered drug release for epilepsy therapy in vivo. A preliminary in vivo study using the Long-Evans rat model has demonstrated a significant reduction in the spike–wave discharge after ETX was burst released from the thermosensitive nanocarriers, and it was concluded that these nanocarriers, which can be considered as a remotely triggered drug delivery platform with a tunable burst drug release profile due to the structure deformation after an application of an external magnetic field, may provide significant advantages as highly temperature-responsive nanocarriers for the treatment of acute diseases. Using electrophoretic deposition of drug-carrying magnetic core–shell Fe_3O_4 at SiO_2 NPs onto an electrically conductive flexible poly(ethyleneterephtalate) substrate, a flexible chip-like drug delivery device was designed for controlled delivery of ETX, in which the release of the drug can be controlled by directly modulating the magnetic field. The flexible and membrane-like drug delivery chip utilized drug-carrying magnetic NPs as building blocks that ensure a rapid and precise response to magnetic stimulus. The potential of the application of the drug delivery chip supported the results of a preliminary in vivo study using Long-Evans rat model showing a considerable reduction in spike–wave discharge after ETX was burst released from the chip under the same induction of the magnetic field as in vitro (Huang et al., 2009).

Encapsulation of ETX as well as 5,5-diphenylhydantoin and carbamazepine (CBZ) in the polyethyl cyanoacrylate nanocapsules (100–400 nm) prepared by adding the monomer to an organic phase, consisting of Miglyol® 812 and an organic solvent (ethanol, acetone, or acetonitrile), and subsequent mixing the organic phase with the aqueous phase containing Pluronic® F68, resulted in controlled drug release, where the mechanism of drug release from nanocapsules was mainly diffusion from the oil core through the intact polymer barrier (Fresta et al., 1996).

Amphiphilic chitosan (CS) nanocapsule-based thermo-gelling biogel with sustained in vivo release of ETX and suppressing spike–wave discharges in Long-Evans rat model was designed by Hsiao et al. (2012), and it was considered to show properties, which are highly desirable for injectable depot gels for drug delivery.

5.2.2 DIBENZAZEPINES

CBZ is a drug used both in the treatment of epilepsy and neuropathic pain. CBZ is a lipophilic drug that shows its antiepileptic activity by inactivating sodium channels; however, due to its low solubility in water, if administered orally, its gastrointestinal absorption is slow and irregular, leading to delayed brain uptake with consequent peripheral side actions (Samia et al., 2012). The evaporation-assisted solvent–antisolvent interaction method was used to prepare spherical NPs of CBZ with particle size below 50 nm and CBZ NPs stabilized by polyvinylpyrrolidone (PVP) showing solubility ca. 12- and 22-fold higher than raw CBZ, and almost 100% of the drug was released from these NPs in in vitro dissolution in less than 60 min, while only 34% of the drug was released from raw CBZ even after 180 min (Kumar and Siril, 2014). Tan et al. (2016) reported that CBZ encapsulated in a parenteral oil-in-water NE improved its bioavailability for brain targeting. Samia et al. (2012) studied the brain targeting of CBZ through the olfactory mucosa in the form of an intranasal (i.n.) mucoadhesive oil/water nanoemulgel (MNEG) containing oleic acid/labrasol in a ratio of 1:5 as oil/surfactant and 0.1% xanthan gum as anionic mucoadhesive polymer and observed very low in vitro release of CBZ from MNEG, while CBZ uptake through liposomal membrane reached 65% within 1 h, and application of

MNEG to animals considerably prolonged the onset times for convulsion of chemically convulsive mice and protected the animals from two electric shocks. These facts indicate that in this manner formulated CBZ decreases the peripheral side actions.

CS SLNPs of CBZ with mean diameter 168 nm and zeta potential −28.9 mV were found to exhibit high encapsulation efficiency and high physical stability and demonstrated the controlled-release patterns of the drug for prolonged period, indicating that such formulation could increase CBZ therapeutic efficacy in the treatment of epilepsy (Nair et al., 2012).

Oxcarbazepine (OBZ) is an anticonvulsant drug that is especially used in the treatment of epilepsy. Emulsomes in which Compritol®, tripalmitin, tristearin, and triolein were used as nano-triglyceride cores and phospholipid sheath consisted of soy phosphatidylcholine showing up to 96% encapsulation of OBZ, which were found to be stable nanoformulations decreasing the drug toxicity and were suitable nanocarriers for nasal delivery (El-Zaafarany et al., 2016).

OBZ-loaded NPs with size 140–170 nm prepared by a modified solvent displacement method from biocompatible polymers poly(lactic-co-glycolic acid) (PLGA) with or without surfactant and polyethylene glycol (PEG) decorated PLGA exhibited encapsulation efficiency above 69% and showed rapid drug release kinetics and the permeability across in vitro models of the BBB (the human brain endothelial capillary hCMEC/D3 cells) and human placental trophoblast cells (BeWo b30 cells) comparable with that of the free drug; an increased permeability of surfactant-coated NPs was observed (Lopalco et al., 2015). Stable OBZ NLC ensuring drug encapsulation efficiency ≥90% was patented by Cheng et al. (2014).

5.2.3 GABA-BASED DRUGS

Gabapentin (GBP) is a drug used in the treatment of epilepsy, neuropathic pain, hot flashes, and restless leg syndrome. Microspheres of cyclodextrin (CD)-based nanosponge–GBP complexes coated with ethyl cellulose and Eudragit RS-100 were found to be suitable for controlled drug release and delivery. These nanosponges effectively masked the taste of drug and the coating polymers provided controlled release of the drug and enhanced taste masking, causing an increase in bioavailability by 24% as compared to the pure drug. Dry powder suspension loaded with microspheres of such nanosponge complexes could be used as a suitable controlled-release formulation for GBP delivery (Rao and Bhingole, 2015). Tween 80-coated GBP-loaded albumin NPs significantly reduced the duration of all phases of convulsion in both maximal electroshock-induced and pentylenetetrazole-induced convulsion models in comparison with the free drug and formulations of the drug bound with NPs, and increased the drug concentration in the brain about threefold compared to free GBP (Wilson et al., 2014).

Pregabalin is used in the treatment of epilepsy, neuropathic pain, and generalized anxiety disorder. The controlled release of an anticonvulsant model drug, pregabalin, under neutral pH condition from a pH-responsive composite hydrogel prepared by visible-light-induced synthesis and consisting of the pH-responsive layer based on poly(methacrylic acid-g-ethylene glycol) as a macromer, eosin Y as a photoinitiator, and triethanol-amine as a co-initiator that was functionalized with hydrophobic domains through incorporation of cross-linked styrene-butadiene-styrene copolymer into the pH-responsive prepolymer was reported by Cevik et al. (2015).

5.2.4 VARIOUS STRUCTURES

VPA is a drug used especially in the treatment of epilepsy and bipolar disorder. It is important to note that Barzago et al. (1996) found significantly reduced transplacental transfer and placental uptake of VPA in vitro after VPA was encapsulated in liposomes. Darius et al. (2000) studied the suitability of NPs as a drug carrier system for antiepileptic drug VPA in mice and found that the NPs inhibited the metabolic degradation of VPA through mitochondrial β-oxidation but not influenced any other metabolic pathway and also did not alter the drug brain tissue levels.

Varshosaz et al. (2013) studied factors affecting the production of nano-structure lipid carriers of VPA for nasal delivery using statistical methods and found that the optimum formulation was obtained using 1% of Poloxamer 188 as a surfactant, organic/aqueous phase volume ratio of 1:5 and acetone/ethanol volume ratio of 3:1 and showed a mean particle size of 154 nm, 47% payload, and 75% of drug content released within 21 days. Such carriers might be used as a delivery system in the treatment of seizures through the nasal route of administration. Hamidi et al. (2011) tested encapsulation of valproate-loaded nanogels inside human erythrocytes as a novel drug delivery system with an intravenous (i.v.) sustained drug delivery characteristic. Erythrocytes treated by valproate-loaded NPs showed entrapment efficiency (EE) of 42% and demonstrated a prolonged drug release behavior over 3 weeks. Eskandari et al. (2011) investigated brain delivery of VPA through i.n. administration of NLCs. They prepared NLCs with particle size ca. 154 nm, drug loading of 47%, and drug release of 75% after 21 days. It was found that plasma concentration ratio in rats was much higher after i.n. administration of NLCs of VPA than in the positive control group (intraperitoneal [i.p.] route), and i.n. administration of NLCs of VPA provided a better protection against maximal electroshock seizure.

Lamotrigine (LTG) is an anticonvulsant drug used to treat epilepsy (treatment of primary and secondary tonic-clonic seizure, focal seizures, and seizures associated with Lennox–Gastaut syndrome) and bipolar disorder. LTG, a sodium and calcium channel blocker, has also demonstrated efficacy for the treatment of neuropathic pain (Lalani et al., 2015).

The NLCs with a mean particle size of 151 nm, polydispersity index of 0.249, zeta potential of 11.75 mV, and EE of ca. 96% carrying LTG exhibited sustained drug concentration following i.n. administration after 24 h and provided high accumulation of the drug in the brain in Wistar rat model (Alam et al., 2015). The LTG NPs functionalized by the lactoferrin and transferrin showed preferential targeting to brain and reduced accumulation in nontarget organs over a prolonged duration of time in an in vivo biodistribution in mice, and also increased the pharmacodynamic response as antinociceptive effect (Lalani et al., 2015).

LTG nanosuspension prepared using an emulsification-solvent diffusion method with spherical NPs showing little surface-adsorbed drug provided a burst drug release profile during the first hour, followed by a controlled release extending up to 24 h, and the formulation remained stable up to 3 months (Mishra et al., 2010). LTG-loaded Pluronic® P123 (P123) polymeric micelles showing the mean encapsulating efficiency for the optimized formulation 98%, drug-loading ca. 6%, and particle size approx. 18 nm exhibited sustained release property in vitro, and after i.v. administration they were accumulated in the brain of rats at 0.5, 1, and 4 h to a greater extent than the free drug, indicating that P123 micelles have the potential to overcome the activity of P-gp expressed on the BBB and are promising for the targeted delivery of antiepileptic drugs (Liu et al., 2014). LTG loaded in copolymeric NPs of the partially water-soluble monomer ethyl methacrylate and the water-soluble monomer 2-hydroxyethyl methacrylate prepared from emulsions containing sodium dodecyl sulfate through free-radical polymerization with a particle size <50 nm exhibited controlled drug release (Shah et al., 2008).

Zonisamide (ZNS) is an anticonvulsant drug used as a supportive therapy in adults with partial-onset seizures, myoclonic, generalized tonic-clonic, and mixed seizure types of Lennox–Gastaut syndrome seizure. Oil-in-water microemulsion (ME) based on peceol with labrasol as a surfactant and transcutol as a cosurfactant (2:1) significantly increased ZNS diffusion across the nasal mucosa, and the addition of 1% citric acid as a permeation enhancer and 0.5% Carbopol® 934P as a mucoadhesive agent further enhanced drug delivery across the mucosa, indicating that this ME formulation can be used in effective i.n. delivery of ZNS resulting in its enhanced brain delivery for prosperous treatment of epilepsy (Shahiwala and Dash, 2010).

5.3   ANTIPARKINSON DRUGS

Antiparkinsonics are central nervous-effective drugs that are able to remove symptoms of the degenerative disorder of the extrapyramidal system of the CNS known as Parkinson syndrome that is characterized by shaking, rigidity, slowness of movement, and difficulty with walking. The exact cause of PD is not known. At PD, a progressive degeneration of pigment cells in substantia nigra in mesencephalon leads to dopamine (DA) deficiency resulting in acetylcholine excess. Current pharmacotherapy only helps to correct these disbalances and is not able to treat the disease (Steinhilber et al., 2010). Most widely used modern antiparkinsonics for nanoformulations from dopaminergic compounds divided into five classes are mentioned below.

Although bulk DA due to its physicochemical properties and enzyme instability cannot be used for the treatment of PD, DA bearing positively charged small liposomes prepared by sonicating multilamellar vesicles could effectively deliver DA to the brain and protect it against degradation in circulation. Moreover, they were found to be superior compared to plain levodopa (LD) as well as the marketed formulation of LD containing carbidopa (Syndopa®) (Jain et al., 1998). DA-loaded PLGA NPs were found to be able to cross the BBB and capillary endothelium in the striatum and substantia nigra in a 6-hydroxydopamine (6-OHDA)-induced rat model of PD, which considerably increased levels of DA and its metabolites, and reduced dopamine-D2 receptor supersensitivity in the striatum of parkinsonian rats, and markedly recovered neurobehavioral abnormalities in 6-OHDA-induced parkinsonian rats without negative effects on the heart rate, blood pressure, brain, and other peripheral organs (Pahuja et al., 2015).

5.3.1   LEVODOPA AND DOPA-DECARBOXYLASE INHIBITORS

LD is a precursor to neurotransmitters DA, noradrenaline, and adrenaline. It mediates neurotrophic factor released by the brain and the CNS and is used in the clinical treatment of PD. LD prodrug-loaded lipid nanocarriers (LNCs) provided a controlled prodrug release and attenuated parkinsonian disabilities in two behavioral tests specific to akinesia (bar test) or akinesia/bradykinesia (drag test) performed in 6-OHDA hemilesioned mice and showed slightly reduced maximal efficacy but a longer lasting action (up to 24 h) compared to the equal dose of LD (Ravani et al., 2015). Ngwuluka et al. (2015) reported about interpolymeric blend nano-enabled gastroretentive LD-loaded drug delivery systems exhibiting zero-order drug release over a prolonged period of time, which could be used as a sustained delivery system for the management of PD.

Following i.n. administration of CSNPs loaded with LD, which were incorporated in a thermo-reversible gel prepared using Pluronic PF 127 and suspended in saline, maximum recovery of the drug in brain was estimated, slightly exceeding that of the drug dispersed in plain pluronic gel (Sharma et al., 2014). The CS-coated LD nanoliposomes considerably decreased scores of abnormal involuntary movement in rats compared to LD treatment and levels of phosphorylated extracellular signal-regulated kinase (ERK1/2), DA- and cAMP-regulated phosphoprotein of 32 kDa (DARPP-32), and FosB/ΔFosB in striatum was found to decrease considerably in liposome group lesion side compared with LD group (Cao et al., 2016).

Zhou et al. (2013) reported that LD-loaded PLGA NPs with the particle size of 256 nm and the EE 62% showing spherical shape with porous outer skin, in which LD and PLGA polymer maintained its backbone structure, might represent a favorable formulation for brain delivery of the drug. PLGA-loaded LD methyl ester/benserazide-loaded microspheres, which released LD and benserazide in a sustained manner, were found to be suitable to reduce established LD-induced dyskinesia in a rat's model of PD (Yang et al., 2012a). The results of further experiments of Yang et al. (2012b) showed that dyskinetic rat treated with such microspheres showed lower abnormal involuntary movements scores than LD plus benserazide-treated dyskinetic rats, and reduced LD-induced dyskinesia was connected with downregulating phosphorylated GluR1 expression in 6-OHDA-lesioned rats.

Chlorotoxin (ClTox)-modified stealth liposomes encapsulating LD markedly facilitated the uptake of liposomes by brain microvascular endothelial cells in vitro, and after i.p. injection to mice, considerable increase in the distribution of LD metabolites (DA and dihydroxyphenyl acetic acid) was estimated in the substantia nigra and striatum. Moreover, this liposomal formulation markedly attenuated the serious behavioral disorders in the methyl-phenyl-tetrahydropyridine-induced PD mice model, and

ClTox-modified liposomes loaded with LD considerably depressed serious behavioral disorders and diminished the methyl-phenyl-tetrahydropyridine-induced loss of tyrosine hydroxylase-positive dopaminergic neurons (Xiang et al., 2012).

Neither LD-loaded nanocomposite of Zn/Al-layered hydroxide nor the un-intercalated nanocomposite disturbed the cytoskeletal structure of the neurogenic cells at their IC50 concentration, and the drug release from the nanocomposite showed slow sustained and controlled character in contrast to the pure drug treatment, exhibiting the burst uptake and release system (Kura et al., 2014a). The 3T3 cells (fibroblasts) exposed to the nanocomposite, in which LD was intercalated into the inorganic interlayers of a Zn/Al-layered double hydroxide, showed increased cell viability after 72 h of exposure compared with those exposed to pure LD (Kura et al., 2013). Coating of the external surfaces of nanocomposites of LD incorporated in Zn/Al-layered double hydroxides by Tween 80 resulted in slower drug release and improved viability of PC12 cells compared to nanocomposite without surfactant coating (Kura et al., 2014b).

Supramolecular complexes of carboxylated, single-walled carbon nanotubes with LD exhibited excellent slow sustained-release characteristics as a drug carrier with a release period exceeding 20 h, and such nanohybrid did not compromise the cell viability; PC12 cells that are widely used in in vitro Parkinson's model for neurotoxicity studies remained stable during the experiments up to 72 h after treatment (Tan et al., 2015).

Carbidopa is able to inhibit peripheral metabolisms of LD and thus it is administered to people with PD together with LD. The formulation of carbidopa bio-NPs for brain targeting through ear using a novel biopolymer isolated from *Sesamum indicum* (white variety) was patented by Madhav (2012).

5.3.2   CATECHOL-O-METHYLTRANSFERASE(COMT)INHIBITORS

Entacapone (ETC) is a selective and reversible inhibitor of the catechol-O-methyltransferase enzyme that is used in combination with other drugs for the treatment of PD. The solid dispersion of ETC prepared by spray drying with the mean volume diameter of the particles collected from drying chamber and cyclone ca. 12 nm and ca. 69 nm, respectively, increased the saturation solubility for ETC 4.6- and 1.8-fold compared to the free drug and the spray-dried drug, respectively. The release of the drug from the formulation was immediate and completely independent of pH and demonstrated a significant improvement of bioavailability (area under the curve [AUC] = 54,048 ng/h/ml) compared to the drug suspension (AUC = 9438 ng/h/mL) (Prasad et al., 2010).

5.3.3   MONOAMINE OXIDASE B (MAO-B) INHIBITORS

Selegiline (SGL) as an MAO-B inhibitor is used not only for the treatment of PD, but also for the treatment of AD and depression. In vitro drug release profile suggesting two-phase drug release, an initial burst release followed by a slow release over an extended period of time (about 10 h), was observed with nanospheres of SGL prepared using gelatin which would be suitable for oral delivery of the drug (Al-Dhubiab, 2013). An optimized formulation of NEs loaded with SGL for direct nose-to-brain delivery for the treatment of PD demonstrated a spherical shape of NE with a droplet size of 61 nm and showed approximately fourfold enhancement in the drug permeation as compared to the drug suspension. Intranasally administered SGL NE also significantly improved the behavioral activities in rats with haloperidol (HPL)-induced PD compared to the orally administered drug (Hassanzadeh et al., 2015).

SGL-hydrochloride-loaded CSNPs with mean size 303 nm, zeta potential + 32.50 mV, and EE 86% released about 82% of the drug in phosphate buffer saline (pH 5.5) using goat nasal mucosa up to 28 h; however, it could be assumed that the drug release was controlled by more than one process, that is, the superposition of two phenomena, the diffusion controlled as well as swelling controlled release (Gulati et al., 2014). SGL-loaded thiolated CSNPs with particle size of 215 nm, zeta potential + 17.06 mV, and EE 70% were found to reduce prolonged immobility time in the forced swim and the tail suspension tests used to evaluate the antidepressant activity, and such nanoformulation could be considered as suitable for nose-to-brain delivery of this drug (Singh et al., 2016a).

Rasagiline (RGL), an irreversible inhibitor of MAO-B, is used as a monotherapy to treat symptoms in the early stage of PD or as an adjunct therapy in more advanced cases. Intranasal administration of RGL-loaded CS glutamate NPs with a mean particles size of 151 nm and encapsulation efficiency of ca. 96% resulted in significantly higher drug concentrations in the brain compared to their i.v. administration or i.n. application of the pure drug, indicating that this nanoformulation could be used in direct nose-to-brain targeting in PD therapy (Mittal et al., 2016). Optimized preparation of RGL-mesylate-loaded SLNPs composed of stearic acid as a lipid matrix with mean particle size of 169 nm and quasispherical shape with smooth surface fabricated by ME technique was described by Kunasekaran and Krishnamoorthy (2015).

5.3.4   DOPAMINE RECEPTORS AGONISTS

An ergoline derivative, bromocriptine (BRC), is a DA agonist. BRC encapsulated in NLCs prepared using tristearin/tricaprin mixture was released from this formulation in a prolonged fashion for 48 h and, similarly to free BRC, reduced the time spent on the blocks (i.e., attenuated akinesia) in the bar test, although the action of encapsulated BRC was more rapid in onset and prolonged (Esposito et al., 2008). Although BRC could be encapsulated with high EE both in monoolein aqueous dispersions and in NLCs, only NLCs provided long-lasting therapeutic effects, possibly extending BRC half-life in vivo. By studying the effects of applied formulations on motor disabilities in 6-OHDA hemilesioned rats in vivo, the BRC NLCs, similarly to free BRC reduced the immobility time in the behavioral test specific for akinesia (bar test) and enhanced the number of steps in the test specific for akinesia/bradykinesia (drag test), the effect of BRC NLCs being longer lasting (5 h) (Esposito et al., 2012).

Thongrangsalit et al. (2015) reported that BRC tablets of self-microemulsifying system adsorbed onto porous carrier stimulated lipoproteins secretion for brain cellular uptake. Although significantly lower amount of drug permeated from such tablets, increased drug uptake was estimated.

BRC-loaded CSNPs with mean size 161 nm, zeta potential +40.3 mV, loading capacity ca. 37% and EE ca. 84% showed the brain/blood ratio of 0.69 at 0.5 h when administered i.n. in mice, while by the administration of BRC solution (i.n.), this ratio was only 0.47, and i.v. administration of these NPs was significantly less effective. That means that the direct nose-to-brain transport bypasses the BBB. Moreover, animals receiving BRC-loaded NPs showed reversal in catalepsy and akinesia behavior when compared to HPL-treated mice, which was more pronounced than for treatment with BRC solution (Md et al., 2013). The permeability coefficient of BRC-loaded CSNPs ($0.9997 \times 10^{-2}/cm^2/h$) through the nasal mucosa was higher than that of drug solution ($0.409 \times 10^{-2}/cm/h$) (Md et al., 2012). BRC-loaded CSNPs administered i.n. showed a significantly higher DA concentration (20.65 ng/ml) as compared to HPL-treated mice (10.94 ng/ml) and could markedly revert selective degeneration of dopaminergic neurons in HPL-treated mice (Md et al., 2014). BRC alginate nanocomposite was found to be effective in reducing the PD symptoms in transgenic flies when mixed in diet, which was reflected in a considerable dose-dependent delay in the loss of climbing activity and activity pattern following treatment with 0.5, 1.0, and 1.5 μM of BRC nanocomposite during 24 days (Siddique et al., 2016).

Apomorphine (APM) is a nonselective DA agonist that activates primarily D2-receptors but at the same time is able to antagonize 5-HT2 and α-adrenergic receptors. Hsu et al. (2010) compared APM-loaded NLCs of the size 370–430 nm, SLNPs, and lipid emulsions (LEs) from the aspect of brain targeting and accumulation of the drug in the brain using i.v. administration and found that the lowest drug release was observed from LEs; however, NLCs could be targeted, through certain vessels to selected brain regions. APM-loaded PEGylated liposomes incorporated with nonionic surfactant exhibited slower release behavior compared to the drug in an aqueous solution as well as greater stability in plasma than free APM, and the in vivo brain uptake of these liposomes after an i.v. bolus injection into rats was rapid and prolonged (Hsu et al., 2011). Sesame oil/cetyl palmitate as lipid matrices was used to prepare NLCs for loading APM diester prodrugs, diacetyl apomorphine (DAA) and diisobutyryl apomorphine (DIA), into the brain. By the addition of PEG to NLCs, particles with diameter 250 nm were fabricated, which largely accumulated in the brain, and the synergistic effect of integrating strategies of prodrugs and NLCs resulted in sustained release. Slower release was observed with the longer carbon chain (DIA<DAA) (Liu et al., 2012). Diester prodrugs DAA and DIA were also found to exhibit superior skin permeation compared to the parent drug when formulated into nanosized LEs suggesting the feasibility of these prodrugs for the transdermal delivery of APM (Liu et al., 2011). Using incorporation of glyceryl monostearate (GMS) or PEG monostearate (PMS) into SLNPs as emulsifiers, NPs with mean diameters 155 and 63 nm, respectively, were fabricated, which showed 12- to 13-fold higher bioavailability after loading with APM than the reference solution in rats after administration per os in dose 26 mg/kg, and also increased drug distribution in the striatum following application of SLNPs was estimated. Moreover, an increase of the total number of rotations was observed when the drug was administered from SLNPs containing GMS and PMS (from 20 to 94 and from 20 to 115, respectively) (Tsai et al., 2011a).

Acoustically active perfluorocarbon nanobubbles (PNBs) with particle size ranging from 150 to 380 nm were tested for encapsulation of both APM HCl and base forms to circumvent delivery problems connected with drug instability and the need for frequent injections. Both drug forms incorporated in PNBs were protected from degradation and showed retarded and sustained release profiles. However, in contrast to APM base showing a decreased release profile with ultrasound application, following the application of 1 MHz, a two- to fourfold increase of the APM HCl release was observed compared to the non-ultrasound group (Hwang et al., 2009).

To eliminate uptake by the liver and enhance brain targeting, Wen et al. (2012) incorporated quantum dots (QDs) and APM into liposomes and estimated a 2.4-fold increase of drug accumulation in the brain due to application of liposomal formulation, and theranostic liposomes with a QD–drug hybrid were found to be suitable for in vivo bioimaging.

Nanostructured solgel silica–DA reservoirs tested for controlled drug release in the CNS exhibited two regimes of release, a fast and sustained DA delivery up to 24 h, and then the rate of delivery became constant. The in vivo evaluation of such reservoirs showed that intrastriatal silica–DA implants reversed the rotational asymmetry induced by APM, a DA agonist, in hemiparkinsonian rats, and no dyskinesias or other motor abnormalities were observed in animals implanted with silica or silica–DA (Lopez et al., 2010).

Pramipexole is a non-ergoline DA agonist suitable for treating PD and restless legs syndrome. It was shown that in CSNPs loaded with pramipexole hydrochloride, potent interactions between the drug and CS matrix were developed, and the NPs exhibited mucoadhesive properties which were reduced with increasing drug content and showed sustained in vitro drug release in simulated intestinal fluid, indicating that they could be further evaluated for the controlled oral delivery of pramipexole hydrochloride (Papadimitriou et al., 2008).

Ropinirole (RPN) is, similarly to pramipexole, a non-ergoline DA agonist. While, in aqueous solution with or without penetration enhancer hydrophilic drug RPN HCl could not be transported across rat skin after 12 h of application, NE formulations used as carrier vehicle significantly increased the permeation rate of the drug across rat skin compared to the drug solution from 0 to 63.23 $\mu g/cm^2/h$, and the lag time was shortened from more than 12 h to about ca. 3 h, and also physicochemical stability of drug-loaded NE formulation after 3-month storage at 25°C was confirmed (Tsai et al., 2014). Azeem et al. (2012) tested the transdermal RPN-loaded NE gel in rats with Parkinson-lesioned brain induced by 6-OHDA and observed more extended drug release from this gel compared to that from a conventional gel and oral marketed tablet Ropitor®, and the relative bioavailability of RPN has been enhanced more than twofold. ME of RPN with globule size 160 nm and zeta potential −4.24 mV that was investigated for transdermal application enhanced the drug permeation across the rat skin and the porcine ear skin

3.5- and 2-fold, respectively, compared to the hydrogel; they antagonized the catalepsy in the HPL-induced catalepsy rat model 10-fold as compared to the marketed tablets and improved motor function in rotenone-induced Parkinsonism rat model by 76%, while the application of oral tablets resulted in only 5% restoration of the normal function (Patel et al., 2014).

The CS-coated RPN oil in water NEs delivered i.n. in HPL-induced PD rat models showed a significantly high mucoadhesive potential, deep localization in the brain, and significantly high AUC0→24 and amplified C_{max} in Wistar rat brain and plasma over i.v. treatment group; they were also the most effective in DA recovery in HPL-induced Wistar rats (Mustafa et al., 2015). NE gel showing a 7.5-fold increase in RPN skin permeation rate when compared to the conventional hydrogel was reported by Azeem et al. (2009). Investigating the effect of homogenization on the fate of RPN-loaded true NE in brain translocation, Mustafa et al. (2012a) found that formulation must be directly transported from the nasal cavity into the cerebrospinal fluid, and the homogenization effect drastically improved brain uptake of the drug. The CS-coated i.n. RPN NE for better management option of PD showing considerably high drug translocation in different parts of Wister rat brain in ex vivo evaluation was described by Mustafa et al. (2012b). RPN-loaded CSNPs prepared by an ionic gelation method showed sustained release profiles for up to 18 h, and their i.n. administration resulted in significantly higher RPN concentration in the brain of rats and higher brain/blood ratio at 0.5 h (0.386) compared with the i.n. application of RH solution reaching the brain/blood ratio of 0.251, indicating that direct nose-to-brain transport bypasses the BBB (Jafarieh et al., 2015).

Gabal et al. (2014) investigated the effect of surface charge on the brain delivery of optimized NLCs showing size <200 nm and absolute zeta potential value approximately 34 mV incorporated in poloxamer in situ gels which were loaded with RPN and administered through the nasal route and found that the absolute bioavailability of the drug-loaded anionic and cationic NLCs in situ gels was enhanced compared to that of the i.n. applied drug solution, and the anionic NLCs in situ gel gave nearly 1.2-fold higher drug targeting efficiency in the brain (158%) than the cationic NLCs in situ gel.

The formulation of ethosomal gel prepared by the incorporation of optimized ethosomal suspension into gel base and loaded with RPN was capable to deliver the drug into the systemic circulation by the transdermal route in amounts equal to those delivered orally, however delivered at a rate slow enough to achieve longer blood levels (Mishra et al., 2013).

Amphiphilic triblock copolymers of poly(propylene succinate) (PPSu) and PEG were used to prepare core–shell NPs with hydrophobic PPSu and hydrophilic PEG forming the core and shell and loaded with hydrophilic (RPN) or hydrophobic (tibolone) drugs. These NPs of the mean particle size of 150–300 nm released hydrophilic RPN at a much higher rate than hydrophobic tibolone, indicating that these copolymers can be useful especially in controlled drug delivery applications involving relatively hydrophobic drugs (Vassiliou et al., 2010).

Rotigotine (RTG) is a DA (D1, D2, and D3 receptors) agonist of the non-ergoline class used for treating PD and restless legs syndrome. By addition of 1% Carbomer 1342 to the ME containing water (68%), Labrafil® (6.8%), Cremophor® RH40 (13.44%), Labrasol® (6.72%), and Transcutol® HP (5.04%) ME-based hydrogel for transdermal RTG delivery with lower application site reactions was designed by Wang et al. (2015), which showed 105% bioavailability with respect to the marketed RTG patch Neupro® and also less skin irritation. The RTG flexible liposome formulations promoting drug absorption and having skin penetration rate higher by 34% than that of pure RTG were patented by Liu et al. (2007).

Piribedil is a D2 and D3 receptor agonist as well as an α-2-adrenergic antagonist. Peroral administration of suspensions of piribedil-loaded SLNPs exhibiting controlled-release rate to rabbits resulted in higher drug bioavailability than the application of the bulk drug (Demirel et al., 2001).

Amantadine (AMT), besides other effects, increases DA release and blocks DA reuptake. AMT-based ion pair amphiphiles prepared using oleic acid surfactant, which can self-assemble into vesicles with the size of 200–300 nm in aqueous solution loaded into PLGA–PEG–PLGA copolymer hydrogel, represent a suitable drug delivery system with long-term controlled drug release behavior (Yang et al., 2014). Wu et al. (2014) reported that AMT can be associated with zwitterionic PC bilayers but has a negligible influence on the flip-flop behavior of PC molecules unless at high concentrations. On the other hand, in negatively charged dipalmitoylphosphatidylglycerol (DPPG), the low concentration of AMT (e.g., 0.20 mM) in the subphase could immediately disturb the outer lipid leaflet; however, subsequently, the outer leaflet returns to the original orderly packed state, while by the application of a higher concentration (e.g., 5.0 mM), the packing state of the outer lipid leaflet was immediately disordered. Using electron paramagnetic resonance (EPR) spectroscopy, it was found that spin-labeled AMT can penetrate into the gel-phase membrane of multilamellar liposomes made of l-α-dimyristoyl-, l-α-dipalmitoyl-, and l-α-distearoylphosphatidylcholine with the same partitioning as when penetrating into the fluid-phase membrane, and at least part of the spin-labeled AMT molecule is deeply buried in the hydrocarbon chain region of the membrane (Subczynski et al., 1998).

5.3.5 MISCELLANEOUS

Citicoline (CTC) is a psychostimulant/nootropic that could be considered as a valuable coadjuvant for the treatment of cognitive impairment in chronic degenerative CNS diseases such as Alzheimer's, PD-associated dementia, and ischemic stroke (e.g., Milani, 2013; Eberhardt et al., 1990). CTC sodium liposome solid preparation useful for protecting brain and neurons and preventing and/or treating diseases was patented by Liao (2011); sustained release liposomal injection comprising CTC entrapped in ammonium sulfate liposomes used for enhancing brain uptake efficiency was patented by Misra et al. (2010). Encapsulation of CTC in transferrin-coupled liposomes significantly improved the radioprotective effect approximately eightfold in epithelial ovarian cancer cell line OVCAR-3 and twofold in human umbilical vein endothelial cells (HUVEC) as compared to the free drug, which could be connected with the entry of CTC into cells through transferrin-receptor-mediated endocytosis (Reddy et al., 2006).

Domperidone (DMP) is a peripherally selective DA D2 receptor antagonist that can be used to relieve gastrointestinal symptoms in PD. Clemens-Hemmelmann et al. (2016) estimated that 10 min after i.p. injection of DMP noncovalently absorbed into micellar aggregates of poly(N-(2-hydroxypropyl)-methacrylamide) (poly(HPMA)) that was copolymerized with hydrophobic lauryl methacrylate, the drug was detected in blood and in the brain of mice, and 40 min after the injection, the highest serum and brain DMP levels were estimated representing a 48-fold increase in serum in contrast to mice injected with the bare drug. DMP encapsulated into these amphiphilic copolymers was also found to be able to cross the BBB and affected motor behavior in animals (Hemmelmann et al., 2011). Spherical DMP-loaded SLNPs and NLCs prepared by hot homogenization followed by ultrasonication technique using trimyristin as a solid lipid, cetyl ricinoleate as a liquid lipid, and a mixture of soy phosphatidylcholine (99%) and Tween 80 as a surfactant showing the particle size of about 30 nm and EE 87.84 and 90.49%, respectively, exhibited controlled release over a period of 24 h (Thatipamula et al., 2011).

Transdermal application of the NE formulation containing oleic acid (4% w/w), Tween 20 (10% w/w), diethylene glycol monoethyl ether (20% w/w), and water (64% w/w) with small droplet size (<90 nm), uniform size distribution, and low viscosity (<160 mP) loaded with DMP exhibited a 3.5-fold increase in relative drug bioavailability compared to the oral drug suspension, and the effective drug plasma concentration was maintained for 16 h after the transdermal application, indicating that such formulation is suitable for transdermal DMP delivery for a prolonged period (Akhter et al., 2008).

It is important to note that experiments with a model drug Rhodamine 123 exhibiting, like DMP, low brain permeability showed that amphiphilic copolymers based on hydrophilic poly(HPMA) possessing randomly distributed hydrophobic poly(lauryl methacrylate) could be considered as a promising delivery system for neurological therapeutics (Hemmelmann et al., 2012).

5.4 DRUGS AGAINST ALZHEIMER'S DISEASE (AD)

AD is a complex and progressive neurodegenerative disorder manifested by cognitive memory deterioration and variety of neuropsychiatric symptoms, and represents the most common form of dementia. The number of people affected by AD is estimated to be doubled by the year of 2050, and more than 100 million people worldwide will be affected by this disease. Drugs approved by the Food and Drug Administration belong to (i) cholinesterase inhibitors competitive such as galantamine, rivastigmine (RVG), and noncompetitive such as tacrine (TCR), donepezil (DNP) and (ii) N-methyl-d-aspartate (NMDA) receptor antagonists, such as memantine (MMTN) (Bajic et al., 2016).

The progression of AD is accompanied by disturbances of the endosome/lysosome (EL) system, and there is accumulation of peptides of the AD-associated amyloid beta (Aβ) type in EL vesicles of affected neurons (Kanazirska et al., 2012). The function and action of arginine metabolizing enzymes with respect to the formation of senile plaques and amyloid peptide aggregation are discussed in the review paper of Whiteley (2014). The possibility to use hydrophobic plant antioxidants for the prevention of amyloid transformation of proteins and other neurodegenerative processes was outlined by Muronetz et al. (2014) who analyzed the role of misfolded proteins in the regulation of the chaperone system involved in the genesis of amyloid neurodegenerative diseases and also the role of modification of the glycolytic enzyme glyceraldehyde-3-phosphate dehydrogenase in the inhibition of glycolysis and the induction of nerve cell apoptosis.

Nanotechnologies may prove to be a promising contribution in drug delivery strategies in future, particularly drug carrier nano- or microsystems, which can limit the side effects of drugs, including anti-AD drugs (Ahmad et al., 2014; Burkhart et al., 2014; Di Stefano et al., 2011). Current trends and advances in targeted drug delivery to the CNS for the treatment of neurodegenerative disorders were summarized by Goyal et al. (2014), while Ruozi et al.

(2014) focused their attention on the most outstanding nanomedicine-driven approaches in AD imaging/detection and treatment. Current progresses on nanodelivery systems for the treatment of neuropsychiatric diseases, AD, and schizophrenia, were presented by Silva et al. (2013).

5.4.1 DRUGS APPROVED FOR AD

5.4.1.1 COMPETITIVE CHOLINESTERASE INHIBITORS

Galantamine (GLT) is used to treat mild to moderate AD and other memory damages. GLT-loaded polymeric NPs prepared from GLT-loaded NEs followed by solvent evaporation having hydrodynamic radius approximately 20 nm, negative surface charge, stability more than 3 months, and encapsulation efficiencies >90 wt.% showed a sustained drug release profile as compared to that from aqueous and micellar solutions, and the enzymatic activity of GLT was maintained at 80% after its encapsulation into NPs (Fornaguera et al., 2015).

Mufamadi et al. (2013) designed spherical ligand-functionalized nanoliposomes for effective intracellular delivery of GLT into PC12 neuronal cells, in which GLT and the peptide ligand were incorporated into the inner core and surface of the nanoliposomes, respectively, and such optimized formulation showed sustained drug release, releasing 30% of the drug within 48 h. GLT hydrobromide-loaded spherical SLNPs with size <100 nm and maximum drug EE 83% showed >90% in vitro drug release for a period of 24 h in a controlled manner, and this nanoformulation demonstrated in vivo significant memory restoration capability in cognitive deficit rats and offered approximately twofold bioavailability compared to that of the plain drug (Misra et al., 2016).

Complexation of GLT with CS resulting in GLT/CS complex NPs was found to be a promising approach to enhance the entrapment of a cationic drug into the CSNPs because it had insignificant effect on the physicochemical properties of CSNPs, but GLT/CS complex NPs with diameter 190 nm and zeta potential +31.6 mV showing prolonged drug release were successfully delivered to different brain regions of male Wistar rats shortly after i.n. administration, suggesting their potential as a delivery system for AD management (Hanafy et al., 2015).

The efficiency of acetylcholinesterase inhibition of GLT was found to be enhanced by i.n. administration compared to oral administration; especially GLT loaded in flexible liposomes could readily transport GLT into brain tissues, suggesting some promise for this approach in successful brain drug targeting in AD treatment (Li et al., 2012).

A nanocomposite prepared from GLT ceria nanodots-containing hydroxyapatite (GLT–Ce-HAp), which was i.p. injected into ovariectomized AD albino rats, successfully upregulated oxidative stress markers and secured total recovery of degenerated neurons in hippocampal and cerebral tissues and disappearance of Aβ plaques. These nanocomposites also exhibited optimizable in vitro release of GLT and nanoceria, suggesting that rod-like hydroxyapatite particles could be applied for selective delivery of GLT and nanoceria to AD-affected brain areas (Wahba et al., 2016).

Rivastigmine (RVG) is used for the treatment of mild to moderate dementia of the Alzheimer's type and dementia due to PD. Wilson et al. (2008a) reported that the brain concentration of i.v. injected RVG can be enhanced over 3.82-fold by binding to poly(butyl cyanoacrylate) (PBCA) NPs coated with 1% nonionic surfactant Tween 80. RVG-loaded NPs of biodegradable polymers, PLGA and PBCA with particle size and drug EE 135 nm and 74% for PLGA NPs and 146 nm and 57% for PBCA NPs showing ca. 30% and ca. 43% release from PLGA and PBCA NPs in 72 h, respectively, were found to cause faster regain of memory loss in amnesic mice compared to RVG solution, suggesting higher extent of drug transport into the mice brain (Joshi et al., 2010a). Scialabba et al. (2012) described amphiphilic polyaspartamide copolymer-based micelles showing a nanometric hydrodynamic diameter with narrow size distribution and negative surface charge for RVG delivery to neuronal cells, which incorporated a large amount of drug, and the system maintained the stability of RVG after incubation in human plasma. The micelles were internalized by neuroblastoma cell lines with drug uptake depending on the micelles concentration, and no cytotoxic effects of either empty or loaded micelles on the mouse neuroblastoma cells (Neuro2a) were observed in an in vitro biological assay.

Ismail et al. (2013) reported about the excellent therapeutic effect of RVG liposomes over RVG solution that was evidenced by nearly preventing amyloid plaque formation in the brain of AlCl$_3$-treated rats suggesting that such liposomes could be a potential drug delivery system for ameliorating AD. RVG liposomes and especially cell-penetrating peptide-modified liposomes showing very mild nasal toxicity were found to enhance the permeability across the BBB in murine brain microvascular endothelial cells model in vitro, and the nasal olfactory pathway into the brain after i.n. administration simultaneously decreased the hepatic first-pass metabolism and gastrointestinal adverse effects (Yang et al., 2013).

RVG-loaded human serum albumin (HSA) NPs prepared using glutaraldehyde as a cross-linking agent and coated with Tween 80 to facilitate brain targeting through endocytosis, in which the drug was completely entrapped in HSA NPs, showed 55.59±3.80% release of drug from HSA NPs in 12 h in vitro, indicating that HSA NPs used as a carrier could provide sustained delivery of RVG (Avachat et al., 2014). The i.v. administration of l-lactide-depsipeptide polymeric NPs loaded with RVG for targeting the drug to brain with mean particle size and drug EE 142 nm and 60%, respectively, resulted in a fivefold and twofold increase in the brain concentration of the drug as compared to plain RVG solution administered by oral and i.v. routes, respectively (Pagar et al., 2014).

NLCs prepared using glyceryl monostearate, Capmul MCM C8, lecithin, and Tween 80 with average particle size 123 nm and RVG EE 68% incorporated into an in situ gelling system using 0.8% gellan gum and 15% Lutrol F 127 showed a twofold increase in the nasal permeation of the drug over plain RVG solution and a threefold increase in enzyme inhibition efficacy (Wavikar and Vavia, 2015).

The brain concentration after the i.n. administration of RVG-loaded CSNPs was significantly higher (ca. 966 ng/ml; t_{max} 60 min) compared with i.v. administration of RVG solution (ca. 3871 ng/ml; t_{max} 30 min) or i.n. administration of RVG solution (ca. 508 ng/ml; t_{max} 60 min), whereby the drug transport efficiency represented 355% and direct transport percentage was 71.80% (Fazil et al., 2012).

5.4.1.2 NONCOMPETITIVE CHOLINESTERASE INHIBITORS

TCR is an indirect cholinergic agonist and a centrally acting anticholinesterase used to treat AD. Eslami et al. (2016) investigated the compatibility of PBCA and CS with different degrees of polymerization versus a TCR unit using molecular dynamics simulation and found that the TCR molecule exhibited higher compatibility with PBCA than CS, and in contrast to CS/TCR systems, the interaction between TCR molecules and PBCA NPs became stronger with the increasing length of the polymer chain. On the other hand, TCR-loaded CSNPs coated with Tween 80 slightly reduced the drug release from the NPs; the release was found to be diffusion-controlled and showed optimal pharmacokinetic characteristics in a rat model (Wilson et al., 2010). The encapsulation of TCR did not change the mean size and the polydispersity index of unloaded CSNPs prepared by ionic gelation using sodium tripolyphosphate as a cross-linking agent, whereas the zeta potential was increased to $+38$ mV due to the positive charge of TCR; TCR encapsulation efficiency into the NPs was about 66%, and the NPs were stable during 25 days in an acidic medium at 4 or 25°C (Hassani et al., 2015). Elmizadeh et al. (2013) described preparation and optimization of CSNPs and magnetic CSNPs with an average particle size from 33.64 to 74.87 nm as delivery systems for the anti-Alzheimer's drug TCR using Box–Behnken statistical design. The magnetic CSNPs significantly increased the concentration of TCR in the brain of animals after i.v. administration compared to the free drug (Wilson et al., 2009).

The investigation of TCR hydrochloride nasal delivery by means of albumin NPs carrying β-CD and two different β-CD derivatives (hydroxypropyl β-CD and sulfobutylether β-CD) showed that the presence of different β-CDs in the polymeric network affected drug loading and could differently modulate NPs mucoadhesiveness and drug permeation behavior (Luppi et al., 2011).

Compared to the uncoated NPs and the free drug, the brain concentration of i.v. injected TCR can be enhanced by binding to PBCA NPs coated with 1% nonionic surfactant Tween 80, while the accumulation of the drug in the liver and spleen is reduced (Wilson et al., 2008b).

DNP is used in the palliative treatment of AD for the improvement of cognition and behavior but without any effect on the progression of the disease. DNP-loaded CS nanosuspension with average size 150–200 nm and polydispersity index 0.341 for direct olfactory administration was developed by Bhavna et al. (2014a). DNP suspension i.n. instilled into the nostrils of rats using a dose of 0.5 mg/ml resulted in 7.2 and 82.8 ng/ml DNP concentration in brain and plasma, while for DNP-loaded CS nanosuspension when administered i.n. in the same dose, these concentrations dramatically increased to 147.54 and 183.45 ng/ml, respectively, indicating that such nanosuspensions transported through the olfactory nasal pathway to the brain could be used as a delivery system for the treatment of AD. In another study, Bhavna et al. (2014b) showed that DNP-loaded spherical CSNPs with particle size within 100–200 nm and smooth morphology showed higher drug transport efficiency (191%) and direct transport percentage (1834%) after i.n. administration in rats as compared to DNP solution.

The release behavior of DNP from PLGA NPs with Tween 80 coating on the NPs surface exhibited a biphasic pattern characterized by an initial burst release followed by a slower and continuous sustained release, and in vivo studies using gamma scintigraphy techniques showed that the i.v. application of these NPs resulted in a higher percentage of radioactivity per gram in the brain compared with the drug solution. Consequently, it could be assumed that high concentrations of DNP uptaken in the brain were due to the coated NPs (Bhavna et al., 2014c). Electrostatic spinning was applied to prepare DNP HCl-loaded nanofibers with a diameter of 100 and 300 nm with narrow distribution as a potential orally dissolving dosage form using different polymers in order to fabricate a consistent and removable web on the collector, and PVA of low molecular weight was found to be the most appropriate. In vitro drug release of the webs occurred immediately (<30 s) after immersion independently of their drug content owing to the huge surface area formed, while for complete dissolution of cast films with the same compositions and commercial tablets ≥30 min was required (Nagy et al., 2010).

5.4.1.3   NMDA RECEPTOR ANTAGONISTS

MMTN is a drug used to treat AD that acts on the glutamatergic system by blocking NMDA receptors. Kanazirska et al. (2012) reported the beneficial effects of lysosome-modulating and other pharmacological and nanocar- rier agents on Aβ-treated cells, some of which facilitated the anti-amyloid actions of MMTN. Moreover, tests focused on the improvement of the solu- bility and absorption through the gastrointestinal tract of lipoyl-MMTN, a potential anti-Alzheimer codrug, by SLNPs showed that SLNPs are not cytotoxic and are able to release the free codrug, indicating their potential at application as drug delivery systems for the brain (Laserra et al., 2015). Using   (3-(4,5-dimethylthiazol-2-yl)-2,5-diphenyltetrazolium   bromide) assay in murine fibroblast 3T3 cell line, Mittapelly et al. (2016) reported that nanocrystals of MMTN-palmoic acid salt are less cytotoxic and more tolerable than plain MEM HCl, and when the nanocrystals were adminis- tered as i.m. injection at three different doses in female Sprague Dawley rats, the plasma levels lasted till the 24th day of the study, suggesting that injectable nanocrystals could represent a therapeutic alternative for the treatment of AD.

5.4.2   NANOFORMULATIONS OF EXPERIMENTAL ANTI-AD COMPOUNDS

It was reported that some of nonsteroidal anti-inflammatory drugs, such as indomethacin-loaded lipid core nanocapsules (Bernardi et al., 2012), meloxicam-loaded polymeric nanocapsules (Ianiski et al., 2012; Ianiski et al.,

2016), or tarenflurbil-loaded PLGA NPs and SLNPs (Muntimadugu et al., 2016) have positive effects on progress of disorder in AD models. Tiwari et al. (2014) reported that curcumin (CCM)-loaded PLGA NPs induced adult neurogenesis and reversed cognitive deficits in AD model through canonical Wnt/β-catenin pathway. PLGA NPs coupled with Tet-1 peptide, which has the affinity to neurons and possesses retrograde transportation properties, were able to destroy amyloid aggregates, exhibited antioxidative property and were not cytotoxic, indicating that they could represent a potential therapeutic tool against AD (Mathew et al., 2012). In self-assembled nanogels of CCM–hyaluronic acid conjugates which were found to inhibit β-fibrillogenesis and mitigate the amyloid cytotoxicity more efficiently than the free drug, CCM encapsulation into nanogels protected cells from the toxicity of free CCM; the hydrogel network hindered the interactions between Aβ molecules, and the counteraction of the hydrophobic binding between Aβ and the conjugated CCM against the electrostatic repulsion between the like-charged Aβ and hyaluronic acid was proved (Jiang et al., 2016). Stable CCM-conjugated nanoliposomes with CCM exposed at the surface downregulated the secretion of amyloid peptide, partially prevented Aβ-induced toxicity, and were able to specifically stain the Aβ deposits in vivo suggesting that they could find application in the diagnosis and targeted drug delivery in AD (Lazar et al., 2013). A novel low-density lipoprotein-mimic NLC modified with high level of lactoferrin and loaded with CCM showed the effective uptake in the brain capillary endothelial cells using an AD model of rats; it effectively permeated the BBB and preferentially accumulated in the brain in approximately threefold higher extent than with unmodified NLC (Meng et al., 2015).

Single-wall carbon nanotubes (SWCNTs) successfully delivered acetyl-choline into the brain for the treatment of experimentally induced AD with a moderate safety range by precisely controlling the doses, indicating that SWCNTs preferentially enter lysosomes, the target organelles, and not mito-chondria (Yang et al., 2010b). Baysal et al. (2013) reported that SGL-loaded PLGA-b-PEG NPs seem to be a promising drug carrier for destabilizing the β-amyloid fibrils in AD patients.

B6 peptide-modified PEG–polylactic acid (PLA) NPs exhibited signifi-cantly higher accumulation in brain capillary endothelial cells through lipid raft-mediated and clathrin-mediated endocytosis than PEG–PLA NPs, and administration of neuroprotective peptide NAPVSIPQ encapsulated in B6 peptide-modified NPs to AD mouse models resulted in excellent amelioration in learning impairments, cholinergic disruption, and loss of hippocampal neurons even at a lower dose (Liu et al., 2013a). Moreover, lactoferrin-modified PEG-co-poly-ε-caprolactone (PCL) NPs were found to be suitable for enhancing brain delivery of the neuroprotective peptide NAPVSIPQ following i.n. administration, where memory improvement effect was estimated even at a lower dose (Liu et al., 2013b). The conjugation of lactoferrin to the generation 3 polypropylenimine dendrimer resulted in enhanced twofold DNA uptake in bEnd3 murine brain capillary endothelial cells compared to the unmodified dendriplex in vitro. Following the i.v. administration in vivo increased gene expression in the brain was estimated (more than sixfold) compared to that of unmodified dendriplex, while in the lung and the kidneys, the gene expression decreased (Somani et al., 2015). The ability of liposomes bifunctionalized with phosphatidic acid and with an ApoE-derived peptide to withdraw amyloid peptides from the brain was presented in the paper of Mancini et al. (2016).

Zhang et al. (2014a) developed a dual-functional NP drug delivery system based on a PEGylated PLA polymer to the surface of which two targeting peptides, TGN (specifically targeting ligands at the BBB) and QSH (showing good affinity with Aβ(1–42), which is the main component of amyloid plaque) were conjugated; this might be a valuable targeting system for AD diagnosis and therapy. Basic fibroblast growth factor entrapped in PEG–PLGA NPs modified with *Solanum tuberosum* lecithin that selectively binds to N-acetylglucosamine on the nasal epithelial membrane significantly improved the spatial learning and memory of AD rats compared with AD model group, and their i.n. administration could be utilized for brain delivery of basic fibroblast growth factor to treat AD (Zhang et al., 2014b). Loureiro et al. (2016) developed PLGA NPs surface functionalized with anti-transferrin receptor monoclonal antibody (OX26) and anti-Aβ (DE2B4) to deliver encapsulated peptide iAβ5 into the brain. Using porcine brain capillary endothelial cells as a BBB, it was estimated that the uptake of these immune NPs showing controlled delivery of the peptide iAβ5 substantially increased compared to the NPs without monoclonal antibody functionalization.

A nanostructure of monosialotetrahexosylganglioside-modified reconstituted high-density lipoprotein (GM1-rHDL), possessing antibody-like high binding affinity to Aβ, loaded with neuroprotective peptide (NAP) protected neurons from Aβ(1–42) oligomer/glutamic acid-induced cell toxicity in vitro and reduced Aβ deposition, ameliorated neurological changes, and rescued memory loss more efficiently than both αNAP solution and GM1–rHDL in AD model mice following i.n. administration (Huang et al., 2015).

Gao et al. (2016) designed a polyoxometalate-based nanozyme with both protease-like activity for depleting Aβ aggregates and superoxide dismutase-like activity for scavenging Aβ-mediated reactive oxygen species (ROS), which can also remove Cu from Cu-induced Aβ oligomers by chelation.

Intranasally administered liposomes loaded with H102, a novel β-sheet breaker peptide, showing 2.92-fold larger AUC in the hippocampus than that of solution group, excellently ameliorated spatial memory impairment of AD model rats and inhibited plaque deposition, even in a lower dosage compared with H102 i.n. solution (Zheng et al., 2015).

As phospholipids play an essential role in memory and learning abilities and also act as a source of choline in acetylcholine synthesis, formulations of CS/phospholipid/β-CD microspheres were designed and administered to rats. From this, it was found that they significantly improved the learning and memory abilities of rats, attenuated the expression of protein kinase C-δ, and inhibited the activation of microglia, indicating their potential to be used in the treatment of AD (Shan et al., 2016). Nanoliposomes containing phosphatidylcholine, cholesterol, and phosphatidic acid were found to prevent Aβ 42 amyloid formation, reversed Aβ 42-induced human microvascular endothelial dysfunction, and may be useful in AD therapy (Truran et al., 2016). Pulmonary administration of liposomes functionalized with phosphatidic acid and an ApoE-derived peptide significantly reduced β-amyloid in the brain of AD mice, suggesting that lung administration could be exploited as an alternative for non-invasive brain delivery of NPs designed for AD therapy (Sancini et al., 2016).

Functionalization of gadolinium-based NPs with peptides highly specific for Aβ amyloid fibrils, (Pro18, Asp21)-β-amyloid (17–21) (binds to the central hydrophobic region of Aβ) and β-amyloid (16–20) (crucial for the formation of β-sheet structures and binds to the full-length Aβ peptide through atypical antiparallel β-sheet motif) were reported as a useful multimodal imaging tool to selectively discriminate and diagnose amyloidoses (Plissonneau et al., 2016).

5.5   ANTIPSYCHOTICS

Schizophrenia is a mental disease that is characterized by abnormal social behavior and failure to understand what is real. Hearing voices, unclear or confused thinking, false beliefs, reduced social engagement and emotional expression, and a lack of motivation are its common symptoms. Schizophrenia is a debilitating mental illness that affects 1% of the population in all cultures, about equal numbers of men and women, but the onset is often later in women than in men (Millier et al., 2014; Schultz et al., 2007). Bipolar disorder is a mental disorder with periods of depression and periods of elevated mood known as mania (Anderson et al., 2012). Long-acting injectable formulations of antipsychotic drugs for the treatment of schizophrenia useful for improving medication compliance with a better therapeutic option to treat patients who lack insight or adhere poorly to oral medication were reviewed by Park et al. (2013). Antipsychotics have been classified into two separate classes: typical (the first-generation antipsychotics) and atypical (the second-generation antipsychotics). Typical drugs (former named as neuroleptics) can cause extrapyramide syndrome after long application. They are weak 5HT2 blockers and rather weaker postsynaptic D2 blockers with strong bond on muscarinic, histaminic, and adrenergic receptors. Atypical drugs are selective D2 (sometimes D3) antagonists and 5HT2 antagonists. A specific group of drugs from the atypical class, Multi-acting Receptor-Targeted Antipsychotics, are able to affect especially DA and serotonin receptors. From the viewpoint of side effects, typical drugs cause dyskinesias, including tardive dyskinesia, while atypical drugs have metabolic side effects (Nandra et al., 2012; Steinhilber et al., 2010). As antipsychotics possess similar structural features, although different effects to individual receptors, below-selected drugs used in nanoformulations are divided according to their chemical structure.

5.5.1   TRICYCLIC NEUROLEPTICS

Chlorpromazine (CPZ) is a phenothiazine antipsychotic, primarily used to treat schizophrenia. CPZ-loaded self-assembled PLGA NPs, the surface of which was modified with CS, showed improved mucoadhesive potential in vitro as well as satisfactory ex vivo permeation (Chalikwar et al., 2013). Using a novel melt-dispersion technique, Govender et al. (2015) designed CPZ-hydrochloride-loaded PCL-based nanocapsules intended for site-specific delivery to the frontal lobe showing particle size from 132 to 566 nm, zeta potential in the range of 15.1–28.8 mV, and biphasic in vitro drug release with an initial burst release followed by pseudo-steady controlled release over 30 days.

Perphenazine (PHP) is a phenothiazine typical antipsychoic used for the treatment of schizophrenia, the manic phases of bipolar disorder and other psychotic disorders. PLGA biodegradable NPs containing PHP or CPZ hydrochloride were formulated using an emulsion solvent evaporation technique by Halayqa and Domanska (2014). Fibers of two hydrophilic polymers, PVP and polyvinyl caprolactam-polyvinyl acetate-PEG graft copolymer, prepared by an electrospinning method and loaded with insoluble drug PHP effectively promoted the drug dissolution rate in water due to the fine dispersion of the drug into polymeric matrices (Bruni et al., 2016).

Zuclopenthixol is a thioxanthene typical antipsychotic used to treat schizophrenia. Polyvinyl phosphonic acid hydrogel NPs (nanogels) that were rendered magnetic field responsive by the inclusion of silica-coated Fe_3O_4 NPs loaded with zuclopenthixol and phenazopyridine hydrochloride were reported as drug delivery devices in biomedical applications with targetable ability (Sengel and Sahiner, 2016).

Clozapine (CZP) is an atypical tricyclic antipsychotic with central 1,4-diazepine ring mainly used for the treatment of schizophrenia that does not improve following the use of other antipsychotic drugs. Polymeric poly-l-lysine-poly-l-glutamic acid CZP NPs surface modified by PEG grafting were found to be very promising nanovehicles for improving CZP delivery (Lukasiewicz et al., 2016). PLGA NPs entrapping both CZP and risperidone (RPD) formulated by spray drying showing EE of drugs in PLGA NPs 94% for CZP and 93% for RPD released ca. 80% of the entrapped drugs over 10 days of time if low-molecular-weight PLGA was used, and this polymeric nanoformulation was found to be suitable for delivery of dual drugs for the treatment of schizophrenia (Panda et al., 2016). Nanosystems with isopropyl myristate as the oil core of self-assembly nanovesicles constituted of CS and lecithin, in which CS formed the hydrophilic shell layer protecting the core comprising lecithin and the hydrophobic groups of oil with encapsulated CZP, showed almost threefold higher AUC (0→infinity) than the free drug (Haas et al., 2014).

Olanzapine (OZP) is the most frequently used atypical tricyclic antipsychotic based on the central 1,4-diazepine ring for the treatment of schizophrenia and bipolar disorder. Asadi et al. (2016) designed magnetic fluorescent multi-core shell structure molecularly imprinted polymer (MIP) through coprecipitation polymerization in the presence of OZP as a template and fructose as a monomer and a cross-linker, suitable for application in brain drug delivery, where the magnetic structure of prepared MIP facilitated the aggregation of carrier near target tissue under magnetic field, and fructose produced during degradation of MIP could be used by brain cells as a fuel. By rapid expansion of supercritical OZP solutions (RESS) into air or into an aqueous solution, NPs with a mean particle size of 191 nm, representing a 200-fold size reduction compared to the original anhydrous material, and nanosuspensions with similar particle size were prepared which after freeze-drying, similar to NPs prepared by RESS procedure, were characterized by improved drug dissolution rate (Paisana et al., 2016). Sucrose-based microfibers containing OZP produced by a temperature-controlled solvent-free centrifugal spinning process were found to increase the drug bioavailability caused by its low aqueous solubility, and these microfiber-based dispersions possessed characteristics that are favorable for the enhanced dissolution and oral absorption of drugs (Marano et al., 2016).

Mucoadhesive amphiphilic methacrylic copolymer-functionalized PCL nanocapsules for nose-to-brain delivery of OZP designed by Fonseca et al. (2015) tested in vitro were able to interact with mucin (up to 17% increment in particle size and 30% reduction in particle concentration) and nasal mucosa (2-fold higher force for detaching) and increase the retention of OZP (ca. 40%) on the nasal mucosa after continuous wash, while in vivo they enhanced the amount of drug in the brain of rats 1.5-fold compared to the drug solution and improved the prepulse inhibition impairment induced by APM, which is considered as an operational measure of pre-attentive sensorimotor gating impairment present in schizophrenia.

Salama et al. (2012) investigated the brain delivery of OZP by the i.n. administration of transfersomal vesicles having mainly spherical shape and diameters ranging from 310 to 885 nm and they found that the deformability index could be considered as a parameter having a direct relation to the amount of the drug delivered to the brain by the nasal route.

Oral administration of SLNPs of OZP prepared using lipids (stearic acid and GMS), soy lecithin, poloxamer 188, and charge-modifier stearylamine to Wistar rats resulted in up to fourfold increase of the area under curve and a decrease of clearance compared to application of OZP suspension, suggesting an improvement of drug oral bioavailability (Sood et al., 2013).

OZP loaded into lipid-core nanocapsules (LCNs) with particle size 142 nm and zeta potential -19.6 mV was administered with dose 10 mg/kg by i.p. route to male Wistar rats showed 226% of bioavailability in the plasma compared with the free drug, which resulted in pronounced and long-lasting effects on the CNS, while the i.p. application of these nanocapsules significantly diminished the stereotyped behavior induced by d, l-amphetamine up to 12 h, treatment with a lower dose, 1.0 mg/kg, i.p., has shown a marked sedative effect and prevented the prepulse inhibition disruption induced by APM at a lower dose than OZP in the free form (2.5 mg/kg, i.p.) (Dimer et al., 2015). The application of OZP-loaded LCNs with mean diameter 156 nm, polydispersity index <0.1, pH value 6.12, zeta potential -17.0 mV, and encapsulation efficiency close to 100% to male Wistar rats resulted in lower weight gain and total cholesterol levels as compared to free OZP administration and exhibited more prolonged antipsychotic action in the stereotyped behavior animal model induced by d, l-amphetamine (Dimer et al., 2014).

CSNPs used as a delivery system enhanced the systemic bioavailability of OZP following i.n. administration in rabbits compared to the free drug showing the systemic absorption with ca. 51% absolute bioavailability as compared to 28% after i.n. administration of drug solution (Baltzley et al., 2014).

The surface modification of mesoporous hydroxyapatite by hydrophobic compounds enabling loading of OZP by physical absorption was found to be a useful treatment strategy to achieve long-term drug release with a single i.m. injection. When injected into muscle, this carrier maintained constant medication release over 2 weeks indicating that such nanoformulation could help to solve the problem of nonadherent medication intake that often occurs in antidepressant therapy (Shyong et al., 2015).

Quetiapine (QTP) is an atypical 1,4-thiazepine antipsychotic used to treat schizophrenia, bipolar disorder, and major depressive disorder. An increased half-life was observed after i.v. administration of 5 mg/kg of QTP encapsulated into a polymeric nanocarrier to male Wistar rats as compared to the free drug due to a significant decrease in total clearance, and an increase in QTP liver exposure was observed after nanoencapsulation, probably due to a reduction in drug metabolization process (Carreno et al., 2016a). Carreno et al. (2016b) also investigated the role of LCNs on brain penetration of QTP in male Wistar rats using microdialysis in the presence of drug transporters inhibitor probenecid using QTP-loaded LCNs with particle size 143 nm and high encapsulation efficiency (95%). It was found that LNCs produced a short in vivo sustained release of the drug, and in contrast to the free drug, at the application of nanoencapsulated QTP, the inhibition of influx transporters by probenecid had no effect on the brain penetration factor, probably because QTP loaded into LNC was not able to interact with transporters.

Li et al. (2015) investigated the brain delivery in rat by nasal QTP fumarate loaded with SLNPs in situ gel and compared the brain delivery of QTP fumarate administered through an oral, nasal, or tail vein approach to rats, in which schizophrenia was established by i.p. injection of dizocilpine. If administered i.n., the transparent gels with the size of 117 nm and 97% EE caused significantly higher QTP fumarate concentration in the blood and brain of rats as compared to the rats, to which the drug was administered per os, and also more effectively ameliorated the hippocampal morphology changes induced by dizocilpine compared to tail vein QTP fumarate. Narala and Veerabrahma (2013) reported that the relative bioavailability of QTP fumarate from optimized SLNPs preparation was increased approximately fourfold when compared with the bulk drug suspension, indicating the improvement of drug bioavailability by minimizing first-pass metabolism.

The liposomal dispersion of QTP fumarate has been proved superior for diffusion through nasal route compared to simple dispersion, showing greater percentage diffusion of ca. 32% and very high permeability with a coefficient value 4.13×10^{-5} cm/s (Upadhyay et al., 2016).

Optimized QTP-fumarate-loaded CSNPs prepared by an ionic gelation method with mean particle size 131 nm and 65% EE exhibited significantly higher brain/blood ratio and twofold higher nasal bioavailability in the brain compared to QTP fumarate solution following i.n. administration, indicating preferential nose-to-brain transport bypassing the BBB and prolonged retention of the drug at the site of action connected with superior permeability enhancer properties of CS (Shah et al., 2016a). Moreover, the noninvasive i.n. delivery of QTP-fumarate-loaded ME was found to be advantageous for brain targeting (Shah et al., 2016b).

Asenapine (ANP) is a modern atypical antipsychotic used to treat schizophrenia and acute mania associated with bipolar disorder, in which the central seven-membered oxepine ring was extended by ortho-condensated pyrrole. The treatment with ANP-loaded NLCs having spherical shape with smooth surface and particle size 167 nm, EE 83%, and zeta potential −4.33 mV through i.n. route resulted in a higher drug concentration and residence time in the brain of rats compared to the free drug and showed a significant decrease in extrapyramidal side effects with increasing antipsychotic effect after 1–2 week(s) of treatment (Singh et al., 2016b). Shreya et al. (2016) reported a beneficial synergistic effect of the combined strategy of chemical and nanocarrier (transfersomal)-based approaches of permeation enhancement in increasing the transdermal permeation and bioavailability of ANP, whereby optimized spherical transfersomes with drug:soy phosphatidylcholine:sodium deoxycholate weight ratio of 5:75:10, average size 126 nm, and EE ca. 55%, which were combined with ethanol (20% v/v), were evaluated as the best skin permeation enhancer.

5.5.2 ARYL-4-(4-ARYLPIPERIDINYL)BUTANONE-LIKE DRUGS

HPL is a typical antipsychotic used for the treatment of schizophrenia and mania in bipolar disorder. Dendrimer-based formulation of HPL developed by Katare et al. (2015) increased its aqueous solubility more than 100-fold and showed significantly higher distribution of the drug in the brain and plasma compared to the control formulation of HPL administered through i.p. injection. Moreover, produced behavioral (cataleptic and locomotor) responses following the i.n. administration of this dendrimer-based formulation was comparable to those induced by HPL formulations administered through i.p. injection. HPL-loaded lipid-core polymeric nanocapsules with the size about 250 nm administered to adult male Wistar rats for a period of 28 days at dose 0.5 mg/kg/day-i.p. reduced DNA damage in blood and oxidative stress in liver and kidneys of rats (Roversi et al., 2015). El-Setouhy et al. (2016) developed HPL-loaded NE for brain targeting with globule size 209 nm which showed approximately twofold higher and sixfold faster peak brain levels after i.n. administration to mice than after i.v. administration, and the HPL concentration of 275.6 ng/g brain 8 h post i.n. instillation was found to be higher than the therapeutic concentration range of HPL (0.8–5.15 ng/ ml). The i.n. administration of lecithin-functionalized PEG-PLGA NPs loaded with HPL with sizes <135 nm was found to increase brain tissue HPL concentrations 1.5–3-fold as compared to nonfunctionalized NPs and other routes of administration (Piazza et al., 2014). Intranasally administered HPL-loaded SLNPs that were tested on albino Wistar rats showed a brain/ blood ratio 1.61 at 0.5 h compared to 0.17 and 0.031 estimated for the i.n. and i.v. treatment with HPL solution. Similarly, the maximum concentration (C_{max}) in the brain achieved from i.n. administration of these SLNPs (329.17 ng/mL, t_{max} 2 h) was also significantly higher than that estimated for treatments with HPL solution, and the HPL-SLNPs showed the highest drug-targeting efficiency (2362%) and direct transport percentage (95%) (Yasir and Sara, 2014).

RPD is an antipsychotic used for the treatment of schizophrenia, bipolar disorder, and irritability in people with autism. Korzhikov et al. (2016) investigated the patterns of RPD entrapment and release using polyester-based microparticles of different hydrophobicity and found that the drug entrapment is greater for the hydrophobic polymers. Drug release was more rapid from crystalline particles (PLA, PCL, poly(ω-pentadecalactone)) than from amorphous poly (ω-pentadecalactone) and PLGA. RPD-loaded NEs with mean size about 160 nm, containing sodium oleate in the aqueous phase and Tween 80, poloxamer 188, or Solutol® HS15 as a co-emulsifier, demonstrated erratic brain profiles of the drug following i.p. administration to rats, which could be connected with the different composition of the surface stabilizing layer. Tween 80 costabilized NE showed 1.4–7.4-fold higher RPD brain availability compared to other NEs and the pure drug (Dordevic et al., 2015). High-selective imprinted NP polymers prepared by a miniemulsion polymerization technique, using RPD as a template, methacrylic acid as a functional monomer, and trimethylolpropane trimethacrylate as a cross-linker loaded with RPD exhibited controlled drug release from NPs; however; the release rate was slower than from nonimprinted NPs (Asadi et al., 2014).

Thermal properties and morphology changes in the degradation process of PLGA matrices with RPD described by Turek et al. (2015) indicated that matrices exhibited stable process of degradation, which may be advantageous for the development of prolonged drug release systems. Microspheres prepared using biodegradable 50504A PLGA or blends of 5050-type PLGAs exerted spherical and smooth morphology, high RPD encapsulation efficiency, and nearly zero-order release kinetics and were considered to have a great potential for better depot preparation than marketed Risperdal Consta™, which could further improve the patient compliance (Su et al., 2011). PLGA NPs of RPD with particle size ranging from 85 to 219 nm and thermal-responsive in situ gel containing RPD NPs (loaded at 1.7–8.3% by weight of the polymer) for parenteral delivery were designed by Muthu et al. (2009) and administered subcutaneously to mice. In vivo investigation showed significantly prolonged antipsychotic effect for up to 72 h with fewer extrapyramidal side effects after administration of RPD formulations as compared to that of drug solution, indicating that application of such formulations could result in dose reduction at the treatment of psychotic disorders. Suspensions of polymeric NPs containing RPD made of poly(d,l-lactide) with particle size ranging between 78 and 184 nm exhibited sustained drug release for more than 24 h with 75% drug, in contrast to 1.5 h estimated for the release from RPD solution, indicating the feasibility of such nanoformulations for the treatment of psychotic disorders (Muthu and Singh, 2009). A hybrid solid lipid/PLGA implant formed in situ following i.m. injection for long-acting RPD resulting in improved antipsychotic treatment was designed by Dong et al. (2011). This hybrid system was characterized by improved in vitro sustained drug release profiles, significantly reducing the burst effect, showing in vivo relatively high bioavailability (88%) and prolonged mean residence time (86.8 h) compared to that estimated for the RPD solution and PLGA implant (5.8 and 32.6 h, respectively); the system underwent almost complete biodegradation, with less than 2% remaining after 28 days.

A combined pattern of diffusion and erosion release mechanism (anomalous non-Fickian transport) was found for RPD-loaded SLNPs, which shows the ability of the system for controlled drug release (Silva et al., 2012a). The RPD-loaded SLNPs were stable and had high encapsulation efficiency and similar shape to placebo formulations before and after storage (Silva et al., 2012b). SLNPs-based hydrogel formulations developed as mucoadhesive systems for RPD oral transmucosal delivery revealed more pronounced drug release after SLNP hydrogel entrapment when compared to the dispersions alone, and pH-dependent release was observed as well (Silva et al., 2012c). Varshosaz et al. (2012) designed stealth SLNPs of RPD suitable for controlled delivery through the i.v. route enabling reducing the frequency of administration, dose, and adverse effects during the short-term management of manifestation of psychotic disorders.

Narayan et al. (2016) tested functionalized RPD liposomes with the mean vesicle size ranging from 90 to 100 nm, in which their surface was modified by coating with stearylamine and PEG monomethyl ether-distearoyl phosphatidylethanolamine for brain targeting through nasal route for effective therapeutic management of schizophrenia. All formulations showed prolonged diffusion-controlled drug release and provided enhanced brain exposure in vivo, whereby PEGylated liposomes exhibited greater uptake of RPD into the brain than plasma. Testing of RPD-loaded proniosomal formulations as potential transdermal delivery systems showed that they exhibited significantly (4.4-fold) higher skin permeation through rat skin than conventional liposomes, whereby the drug bioavailability was found to be 1.3-fold higher than that of oral dosage form (Imam et al., 2015).

Paliperidone (PPD) as a DA antagonist and 5-TH2A antagonist is the atypical antipsychotic used for the treatment of schizophrenia and bipolar disorder. PPD palmitate-loaded d-α-tocopheryl PEG 1000 succinate micelles with particle size 26.5 nm and encapsulation efficiency ca. 92% showing sustained drug release for more than 24 h with 40% of drug release provided an improved and prolonged antipsychotic effect in comparison to control PPD palmitate formulation (Muthu et al., 2016).

Patel et al. (2016) evaluated PPD ME and mucoadhesive PPD ME for a nose-to-brain targeted drug delivery system by performing pharmacodynamic assessments (APM-induced compulsive behavior and spontaneous motor activity) using mice, pharmacokinetic evaluation of drug in the brain using Swiss albino rats, and brain scintigraphy imaging in rabbits and found that the brain-to-blood ratio 8 h following i.n. administration of PPD mucoadhesive ME was six- to eightfold higher than after i.v. PPD ME administration, indicating a greater extent of drug distribution in brain, which was also supported by using rabbit brain scintigraphy. The mucoadhesive ME formulation of PPD containing 0.5% (w/w) of polycarbophil was found to display a higher in vitro mucoadhesive potential (ca. 18 min) and diffusion coefficient (ca. 3.83×10^{-6}) than ME; it was found to be free from nasal ciliotoxicity, showed stability for 6 months, and could be considered as a suitable formulation for i.n. PPD delivery (Patel et al., 2013). Darville et al. (2014) reported that the i.m. administration of PPD palmitate extended-release injectable microsuspension induced a subclinical inflammatory reaction modulating the pharmacokinetics in rats. The effect of macrophage and angiogenesis inhibition on the drug release and absorption from an i.m. sustained-release PPD palmitate suspension in rats was analyzed by Darville et al. (2016a), and another paper of Darville et al. (2016b) is focused on modeling the time course of tissue responses to i.m. long-acting PPD palmitate nano-/microcrystals and polystyrene microspheres in the rat. Leng et al. (2014) designed PPD palmitate nanosuspensions by a wet media milling method and found that following single-dose i.m. administration to beagle dogs, the release of drug lasted for nearly 1 month indicating long-acting effect of these suspensions, better results being obtained with nanosuspension containing larger particle size (1041 nm) compared to nanosuspension in which the particle size was 505 nm.

Spherical PPD-loaded stearic acid NPs with the average particle size of SLNPs of 230 nm and drug EE in the lipid 42% (w/w) exhibited controlled-release pattern in vitro; however, in vitro cell culture studies against RAW 264 murine macrophages revealed that they have some cytotoxicity (Kumar et al., 2015). SLNPs stabilized by sodium deoxycholate which can also act as a permeability enhancer with average size ca. 200 nm and EE 55% with 4% of PPD loading in lipid matrix were prepared by Kumar and Randhawa (2013).

Iloperidone (IPD) is an atypical antipsychotic used to treat schizophrenia. Mandpe and Pokharkar (2015) reported that IPD NLCs with drug EE between 63 and 96% demonstrated sustained release profile and an eightfold increase in oral bioavailability compared to pure drug suspension over the period of 24 h in the pharmacokinetic study in Wistar rats, and this nanoformulation remarkably improved the oral bioavailability of IPD and demonstrated a promising perspective for oral delivery of poorly water-soluble drugs.

5.5.3 2-METHOXYBENZAMIDES

Sulpiride (SPD) is an antipsychotic used to treat psychosis associated with schizophrenia and major depressive disorder. To reach enhanced intestinal permeability of SPD resulting in an increase of the overall oral absorption of the drug, Ibrahim et al. (2014) loaded SPD into uniform spherical SLNPs with the size ranging from 147 to 298 nm and found that SLNPs prepared by combination of triglycerides with stearic acid secured a significant increase in zeta potential, EE, and drug loading even though the particle size was increased. SPD-loaded PEGylated NPs prepared using methoxy PEG-PLA and maleimide-PEG-PLA, to which thiolated canonized bovine serum albumin was conjugated, with mean particle size 329 nm were administered to Sprague Dawley rats (10 mg/kg) through tail vein and showed a higher accumulation in brain as compared to bovine serum albumin NPs and uncoated NPs (Parikh et al., 2010).

5.5.4 INDOLE-, QUINOLINONE-FRAGMENT-BASED DRUGS

Ziprasidone (ZPD) is an atypical antipsychotic used to treat schizophrenia and bipolar disorder. Dening et al. (2016) investigated three lipid-based drug delivery systems (LBDDS), namely a self-nanoemulsifying drug delivery system (SNEDDS), a solid SNEDDS formulation, and silica lipid hybrid microparticles, for reducing the fed/fasted variations of ZPD in vitro and found that the pure drug exhibited the lowest rate and extent of drug solubilization under fasting conditions and a significant 2.4-fold increase in drug solubilization under fed conditions, while all three LBDDS significantly enhanced the extent of drug solubilization under fasting conditions between 18- and 43-fold in comparison to pure ZPD and could be considered as an appropriate formulation strategy to explore further the improved oral drug delivery.

Thombre et al. (2012a) focused their attention on solubilized formulations of ZPD, an amorphous inclusion complex of ZPD mesylate and CD, a nanosuspension of crystalline ZPD-free base, and jet-milled ZPD HCl-coated crystals made by spray drying the drug with hypromellose acetate succinate, in an effort to improve absorption in the fasted state, thereby resulting in a reduced food effect. When these formulations were administered to fasted beagle dogs, the amorphous complex and the nanosuspension showed increased absorption in the fasted state compared to Geodon® capsules, indicating that solubilized formulations of ZPD have the potential to reduce the food effect in humans. Complete fasted-state absorption of ZPD, achieving the desired improvement in the fed/fasted ratio, was obtained with solid nanocrystalline dispersion consisting of a high-energy crystalline form of ZPD in domains approximately 100 nm in diameter but with crystal grain sizes in the order of 20 nm that was administered orally in capsules to beagle dogs (Thombre et al., 2012b).

Lurasidone (LRD) is an atypical antipsychotic used to treat schizophrenia and depressive episodes associated with bipolar disorder. In optimized LRD hydrochloride nanosuspensions containing 0.21% (w/v) LRD, 0.06% (w/v) sodium dodecyl sulfate, and 0.16% (w/v) F68 with particle size 124 nm and polydispersity index 0.097, the drug solubility and in vitro dissolution rate were found to be significantly improved with reducing particle size, and oral administration of nanosuspensions resulted in 1.5-fold higher C_{max} and AUC (024) compared to the raw drug (Lu et al., 2016). LRD-SNEDDSs prepared using Capmul MCM, Tween 80, and glycerol as an oil phase, a surfactant and a cosurfactant, respectively, showed improved drug release profiles. The release behavior was not affected by the medium pH with total drug release over 90% within 5 min, and after administration to beagle dogs in fed and fasted state, the AUC and C_{max} were similar, indicating that no food effect on the drug absorption was proved (Miao et al., 2016).

APZ is an atypical antipsychotic especially used to treat schizophrenia and bipolar disorder. APZ-loaded PCL NPs with particle size 199 nm, zeta potential -21.4 mV and EE 69% released in vitro 90% drug after 8 h, and i.n. administration of these NPs resulted in higher drug concentration in the brain compared to i.v. administration (Sawant et al., 2016). Abdelbary et al. (2013) reported that with an increase in fixed aqueous layer thickness (FALT) of polymers used as polymeric stabilizers for preparation of APZ nanosuspensions, the stability of nanosuspensions was improved and a linear correlation between the FALT and the length of hydrophilic chains in Pluronics was estimated. The enhancement of the encapsulation efficiency of NE-containing APZ for the treatment of schizophrenia was reported by Masoumi et al. (2015), and Samiun et al. (2016) developed artificial neural network models for the prediction of the particle size of NEs loaded with APZ and a stable NE system suitable for effective i.v. administration.

Investigation of the effects of nanomilling and coprecipitation of APZ with hydroxypropylcellulose, PVP K17, and Pluronic F127 on the enhancement of the in vitro dissolution rate showed that the crystallinity of APZ decreased from physical mixtures to coprecipitates and further to NPs, and the increased drug dissolution rate was due to decreased crystallinity in coprecipitate compositions and disruption of crystallinity in nanomilled compositions. Nanomilling more markedly affected the increase in the dissolution rate than coprecipitation, and the dissolution rate was preserved after compression of NPs into tablets (Abdelbary et al., 2014).

A nanohybrid system based on a bentonite clay material, montmorillonite (MMT), which could both mask the taste and enhance the solubility of APZ, which was coated with a cationic polymer polyvinylacetal diethylamino acetate exhibited about 20% higher systemic exposure of APZ compared with Abilify® in in vivo experiments by using Sprague Dawley rats (Oh et al., 2013).

5.6 ANXIOLYTICS

Drugs denoted as anxiolytics (or antineurotics or antiphobics) are compounds of various structure that are able to reduce feelings of anxiety and stress. Most of these drugs are used in higher doses as hypnotics/sedatives or anti-depressants as well as myorelaxants, anticonvulsants, or H1-antihistamines (Steinhilber et al., 2010).

Alprazolam (ALPZ) is a short-acting benzodiazepine used to treat panic disorder, generalized anxiety, or social anxiety disorder. Jana et al. (2013) prepared NPs for oral delivery of ALPZ using interpolymeric complexation of cationic CS and anionic egg albumin stabilized with PEG 400 with an average diameter of 259 nm and EE ranging from 68 to 99% showing sustained drug release over a period of 24 h. The effectiveness of nose-to-brain delivery of ALPZ using ALPZ-loaded SLNPs investigated by estimating the distribution of the drug to different organs in male Wistar rats and using gamma scintigraphy imaging in New Zealand rabbits by tagging the formulation with radioactive substance Tc-99m showed that the drug was rapidly transferred to the brain through i.n. route, bypassing the BBB and a direct nose-to-brain transfer could contribute to the reduction of the dose and dosing frequency (Singh et al., 2012).

Hydroxyzine (HXZ) is an antihistamine showing antagonistic effects on several receptor systems in the brain, and due to its strong anxiolytic and mild antiobsessive as well as antipsychotic properties, it is used to treat anxiety and tension associated with psychoneurosis. Elzainy et al. (2005) studied the effect of the duration of film hydration, freeze-thawing, and buffer pH change on the extent of the entrapment of HXZ and cetirizine into liposomes and the stability of these liposomes. The entrapment of HXZ into small unilamellar vesicles showed a significant increase from 53 to 84% when the pH of the buffer was increased from 5.0 to 5.5, and a further pH increase to 7.0 resulted in 94% drug entrapment, but in multilamellar vesicles, it increased slightly from 82 to 94% when the buffer pH values increased from 5.0 to 7.0. While the effect on the stability of liposomes stored at temperatures > 37°C caused by the freeze-thawing processes was estimated after 24 months, no effects of pH on the percentage entrapment of HXZ for formulation stored at 10°C were observed. Small unilamellar vesicles and multilamellar vesicles containing HXZ prepared using l-α-phosphatidylcholine significantly suppressed histamine-induced wheal formation by 75–95% for up to 24 h, whereby mean maximum suppression (85% to 94%) occurred from 2 to 6 h. Only 0.02–0.06% of the initial HXZ dose remained on the skin after 24 h, whereby cetirizine formed in vivo contributed to some H1-antihistaminic activity (Elzainy et al., 2003).

Buspirone (BPR) is an azapirone anxiolytic psychotropic drug used for the treatment of generalized anxiety disorder. The optimized SLNPs of BPR prepared by Varshosaz et al. (2010) with a particle size 345 nm, loading efficiency 33%, and zeta potential -6.8 mV released about 90% of drug during 4.5 h in vitro and notably increased the relative bioavailability of BPR compared to that of the drug solution. Permeation parameters of BPR from MEs were greatly affected by the composition of MEs. MEs containing surfactant with HLB value of 11.16 possessed higher flux. The increasing amount of surfactant in MEs resulted in the ME viscosity increase, flux decrease, and lag time prolongation, and application of ethanol as cosurfactant resulted in a higher permeation rate. As these BPR MEs with higher flux can provide the therapeutic minimum effective concentration at workable administrated area about 3.3–5.8 cm^2, they could be considered as a suitable drug carrier for transdermal delivery systems (Tsai et al., 2011b).

Lyophilized transfersomal gel containing transfersomes prepared using Tween 80 as a flexibility imparting agent to the vesicular walls and oleic acid as a permeation enhancer was considered as a promising transdermal delivery system for BPR hydrochloride (Shamma and Elsayed, 2013). The i.n. administration of thiolated CSNPs of BPR with particle size 226 nm, drug EE 81%, and loading capacity 49% to albino Wistar rats resulted in the brain concentration of 797.46 ng/ml with t_{max} 120 min, while significantly lower concentration was estimated after i.v. and i.n. administration of BPR solution (384.15 ng/ml, t_{max} = 120 min and 417.77 ng/ml, t_{max} 60 min, respectively) (Bari et al., 2015). Increased access of BPR to the blood and brain from i.n. mucoadhesive formulation containing CS and hydroxypropyl-β-CD administered to rats was reported by Khan et al. (2009).

The release behavior of BPR from MMT–BPR composite as well as from composite which was further compounded with Eudragit® L 100 55 (methacrylic acid copolymer) studied in in vitro conditions using buffer media of pH 6.8 showed controlled release of the intercalated drug from MMT–BPR composite, whereby the application of Eudragit was found to decrease the rate and amount of drug released within 12 h by about a half (23–29% compared to 54% from MMT–BPR) (Joshi et al., 2010b).

5.7 ANTIDEPRESSANTS

Depression is a serious, long-lasting mental disorder that is manifested by reduction or even disappearance of the ability to experience pleasure, depressed moods, and pathological sadness. Patients often feel frustration and hopelessness, lack of motivation, inability to feel pleasure (anhedonie), anxiety and loneliness, feelings of worthlessness or guilt, low self-esteem, fatigue, and impaired attention and concentration. This is a serious, sometimes even life-threatening condition. The reaction of the affected individual to this state of mind is different. Patient may be irritable, aggressive, and spiteful, or vice versa tired, quiet, and peaceful. Depression has been linked to disruption in the cerebral levels of specific neurotransmitters, and l-tyrosine is a precursor of more than one of the neurotransmitters affected by depression (Alabsi et al., 2016). This disorder can be treated with antidepressant drugs that can be classified, excluding obsolete and rarely used monoamine oxidase inhibitors, as follows: (i) first- to fourth-generation thymoleptics and (ii) melatonin agonists and selective serotonin antagonists (Steinhilber et al., 2010).

5.7.1 THYMOLEPTICS

The dibenzoazepine derivative imipramine (IPM) is a thymoleptic of the first generation, structurally close to tricyclic antipsychotics. It is mainly used in the treatment of major depression. He et al. (2010) designed four-arm poly(ethylene oxide)-b-poly(methacrylic acid) block copolymer, which self-assembled into core–shell micelles and extended unimers at low and high pH, respectively. The negatively charged carboxylate groups on the polymer chains interacted with the cationic drug through electrostatic interaction forming polymer/IPM complexes, and the hydrodynamic radius of the polymer/drug complexes ranged from 46 to 84 nm and from 32 to 55 nm at pH of 4.6 and 8.0, respectively. The release behavior was connected mainly with chain relaxation induced by ion exchange that was dependent on pH. Drug release from IPM hydrochloride-loaded pH-responsive nanogels consisting of methacrylic acid–ethyl acrylate cross-linked with diallyl phthalate (DAP) decreased with pH and DAP content, and IPM was found to be electrostatically bounded onto the nanogels, which was further enhanced by hydrogen bonding (Tan et al., 2007). The NE formulations composed of propylene glycol, Transcutol®, water, Labrasol®, Plurol® Oleique, isostearyl isostearate, oleic acid, and d-limonene containing 3% (w/w) concentration of IPM or doxepin were found to be stable for a period of 3 months, and the evaluation of their permeation behavior through human skin performed ex vivo indicated that topically applied IPM and doxepin-loaded NEs are safe for a local effect. Stronger in vivo analgesic and anti-allodynic activity in rats estimated for the doxepin-loaded NE suggested its potential as an alternative analgesic therapy with a potential clinical application (Sandig et al., 2013).

Clomipramine (CPM), a thymoleptic of the first generation, an IPM chloro derivative, is a drug used to treat major depressive disorder, panic disorder, obsessive compulsive disorder, and chronic pain. Cave et al. (2013) reported that LE and tailored liposomes administered i.v. in New Zealand white rabbits improved hemodynamic recovery compared with bicarbonate application in CPM-induced cardiotoxicity in rabbits, whereby greater 30-minute mean arterial pressure was observed for the treatment with 20% i.v. LE. EPR investigation of CPM interaction with saturated dimyristoyl phosphatidylcholine and dipalmitoyl phosphatidylcholine membranes in the presence and absence of cholesterol showed that the presence of 30 mol.% cholesterol increased the fluidizing effect of CPM, affected the lateral diffusion of nitroxide in the inner part of the membrane, and the changes in the central part of the membrane were even stronger than in the upper part of the membrane, indicating that CPM was incorporated into the membrane with its hydrophobic ring parallel to phospholipid chains (Yonar et al., 2011). The interaction of CPM with pig ear stratum corneum (SC) and model membranes was studied using EPR spin labeling and fluidizing effect of CPM on pig ear SC throughout the whole membrane layers indicated that the drug penetrated into the SC, which is essential for its transdermal delivery (Yonar et al., 2013).

Moreover, amitriptyline (ATL), a dibenzocycloheptadiene derivative based on tricyclic antipsychotics, is a thymoleptic of the first generation. It is a drug used for the treatment of major depressive disorder and anxiety disorder and, less commonly, also bipolar disorder. By encapsulation of ATL, doxepin, and IPM PLGA NPs with mean particle size of 420, 480, and 373 nm and drug loadings of 40, 31, and 32% were prepared, which exhibited long-lasting and higher antinociceptive and anti-allodynic activity following local infiltration in rats than application of drug solution, encapsulated doxepin being the most effective (Garcia et al., 2011). On the other hand, Dhanikula et al. (2007) tested phospholipid-based nanosized vesicles (spherulites) to reverse the cardiotoxicity induced by ATL in an isolated rat heart perfusion model and found that they have a protective effect against acute cardiovascular failure following intoxication with ATL and possibly other cardiotoxic drugs because they caused restoration of both hemodynamic and metabolic oxygen demands on the heart. Howell and Chauhan (2008) reported that predominantly anionic liposomes incorporated with PEG are excellent candidates for ATL overdose treatment. The study of interactions between tricyclic antidepressants and phospholipid bilayer membranes supported the hypothesis that these drugs could have the effect on affective disorders partially through binding to the lipid part of the membrane resulting in changes of lipid–protein interactions (Fisar, 2005). Damitz and Chauhan (2015) summarized the current state and progress in the use of emulsions and liposomes for the treatment of drug overdose, and an emerging role of nanomaterials in drug intoxication treatment was analyzed by Graham et al. (2011).

Citalopram (CLP), one of the most popular antidepressant drugs, was approved for the treatment of depressive episodes and panic disorder with or without agoraphobia. It belongs to the thymoleptics of the third generation, also known as selective serotonin reuptake inhibitors or serotonin-specific reuptake inhibitors (SSRIs). Nowadays, this generation of drugs has become a class of drugs typically used as antidepressants in the treatment of major depressive disorder and anxiety disorders. Casolaro and Casolaro (2015) loaded antidepressant drugs (CLP and trazodone [TRZ]) into multiple stimuli-sensitive hydrogels based on α-aminoacid (l-phenylalanine or l-valine) residues and found that the release of CLP in phosphate-buffered saline, pH 7.4 (4 days), was slower than that of TRZ (24 h). At acidic pH (4.6), the lower capacity of ionization and swelling of the hydrogel resulted in much slower and durable drug release. Moreover, if magnetic NPs ($CoFe_2O_4$) were embedded into hydrogels bearing l-phenyl-alanine, additional remote control of drug release was possible through the stimulation of an appropriate alternating magnetic field (20 kHz and 50 W), resulting in substantially increased kinetics of the drug released. MEs were found to be a suitable drug vehicle for the transdermal delivery of CLP, causing a significant increase of the drug permeation rate and shortening of the lag time compared to the aqueous control of 40% isopropyl alcohol solution containing 3% CLP, and in the animal study, an optimized formulation containing 3% CLP with an application area of 3.46 cm^2 was able to reach a minimum effective therapeutic concentration without erythematous reaction (Huang et al., 2013). PEGylated and glycosylated CLP hydrobromide liposomes showed acceptable cell viability with preserved monolayer integrity and an enhanced flux and permeability in brain endothelial cell models, indicating that such formulations could be promising for brain targeting (Kamal et al., 2015).

Paroxetine (PXT) is an SSRI antidepressant used for the treatment of major depressive disorder, panic disorder, obsessive-compulsive disorder, general anxiety disorder, social anxiety disorder, and posttraumatic stress disorder. PXT-loaded NE (oil/water type) with spherical droplets with mean diameter 58 nm and zeta potential -33 mV administered i.n. to depressed rats considerably improved the behavioral activities in comparison with orally administered PXT suspension, effectively enhanced the depressed levels of glutathione, and reduced the elevated levels of thiobarbituric acid reactive substances (TBARS) (Pandey et al., 2016).

Fluoxetine (FXT) is an SSRI antidepressant used for similar indications as PXT. In addition, FXT is used for the treatment of bulimia nervosa or premenstrual dystrophic disorder. Using behavioral tests in acute and chronic mild stress models of depression in rats, it was found that l-tyrosine-loaded NPs and FXT applied at concentration 10 mg/kg markedly decreased the immobility time in the forced swim tests, concomitant with restoration of the basal levels of locomotor activity, distance travelled, and rearing counts, and after application of 10 mg/kg of l-tyrosine-loaded NPs or FXT, an increase of the sucrose consumption in the sucrose preference test was estimated (Alabsi et al., 2016). FXT-hydrochloride-bearing NPs prepared by an emulsion solvent (internal phase) evaporation method, which were compacted into small diskettes for facilitating oral mucosal application, showed a great improvement in pharmacokinetic parameters like C_{max}, t_{max}, and AUC in an in vivo experiment with white male albino rats. This indicates that such mucoadhesive system for buccal route could effectively secure the drug release at modulated rate compared to oral dosage forms resulting in better antidepressant response (Sapre et al., 2009). The interactions of FXT with multilamellar liposomes of pure phosphatidylcholine containing cholesterol (10% M) led to a general damage of the membrane even at low FXT concentrations, both in terms of packing and cooperativity, and the formation of any new phase was no longer observable. Using EPR spectroscopy, it was found that FXT lowered the order of the lipid chains (Momo et al., 2005). It could be mentioned that FXT in liposomal formulations was investigated also as adjuvant agent for anticancer therapy. Haeri et al. (2014) tested a nanoliposomal formulation of FXT as potential adjuvant therapy for drug-resistant tumors. PEGylated liposomes composed of 1,2-distearoyl-sn-glycero-3-phosphorylethanolamine (DSPE)-PEG, 1,2-distearoyl-sn-glycero-3-phosphorylcholine (DSPC), and cholesterol at respective molar ratio of 5:70:25 with particle size 101 ± 12 nm and zeta potential -9.0 mV exhibited 83% drug encapsulation efficiency, released approximately 20% FXT 48 h in vitro, and the optimum formulation was stable for 9 days when incubated at 37°C. As a promising anticancer formulation stealth liposomes coencapsulating doxorubicin (DOX) and FXT were presented. The formulation demonstrated drug release in synergistic ratios in cell culture media, better cytotoxicity than liposomal DOX in both MCF-7 and MCF-7/ADR cells, efficaciously prolonged drug circulation time, and reduced tissue biodistribution (Ong et al., 2011). Spherical NPs of starch-g-lactic acid/ MMT nanocomposite with mean size around 355 nm showed good FXT encapsulation efficiency and a sustained drug release pattern (Namazi and Belali, 2016).

Sertraline (STR) is another SSRI drug that is used for the treatment of major depressive disorder, obsessive-compulsive disorder, panic disorder, and social anxiety disorder. To eliminate problems associated with poor aqueous solubility and vulnerability to enzymatic degradation in liver and problems associated with salt formation, STR-loaded SNEDDS s were prepared by Rahman et al. (2012), which enabled considerably higher drug release than from drug suspensions. Zidan and Aldawsari (2015) investigated ultrasound effects on mannosylated liposomes with STR and used an in vitro BBB transport model to assess the transendothelial capacity of the optimized mannosylated vesicles. Optimized formulation was characterized with a mean particle size 46 nm and 77% STR entrapment, and the transendothelial ability increased 2.5-fold by mannosylation through binding with glucose transporters. STR could also act as a clinically relevant cancer multiple-drug resistance inhibitor, and the application of its combination with PEGylated liposomal DOX in an aggressive and highly resistant human ovarian xenograft mouse model generated strong reduction in tumor progression and extended the median survival of tumor-bearing mice (Drinberg et al., 2014).

TRZ is an antidepressant of the serotonin antagonist and reuptake inhibitorbelonging to the third generation of thymoleptics and also showing anxiolytic and hypnotic effects. Coadministration of 70-nm silica particles with TRZ did not increase a biochemical marker of liver injury (Li et al., 2011). Using 5- and 16-doxyl stearic acid spin labels, it was found that TRZ incorporation caused changes in the physical properties of phosphatidylcholine liposomes, whereby the changes in lipid structure and dynamics caused by TRZ may modulate the biophysical activity of membrane-associated receptors, which is reflected in the pharmacological action of the drug (Yonar et al., 2014).

Venlafaxine (VLF) is an antidepressant belonging to the fourth generation of thymoleptics, the serotonin-noradrenaline reuptake inhibitors (SNRI), and is used for the treatment of major depressive disorder, generalized anxiety disorder, panic disorder, and social phobia. Shah et al. (2016c) developed VLF-loaded NLC using quality by design and risk assessment approach. They applied Compritol® 888 ATO and Capmul® MCM EP as solid and liquid lipid or their mixture, and the optimized formulation of VLF NLC having particle size 77 nm and EE 81% did not show toxicity to nasal mucosa and exhibited a higher flux value across goat nasal mucosa compared to VLF solution.

After vein injection of VLF SLNPs with 186 nm and EE 75%, the brain uptake of the drug was significantly higher than that of VLF solution, VLF solution with empty SLNPs and VLF solution with Verapamil, and it was shown that SLNPs could overcome P-gp and achieve brain target by i.v. administration (Zhou et al., 2015).

Intranasally administered VLF-loaded alginate NPs markedly improved the behavioral analysis parameters, that is, swimming, climbing, and immobility, as well as locomotor activity in albino Wistar rats in comparison to VLF solution (i.n.) and VLF tablets (oral), and it was confirmed that these NPs delivered a greater drug amount to the brain than the two other formulations, indicating that such NPs formulation could be suitable for the treatment of depression (Haque et al., 2014).

Ibrahim and Salah (2016) reported a formulation of VLF for once daily administration using drug–resin complexation followed by polymer encapsulation. The formulation prepared using 50:50 PLA/Eudragit at 1:1 encapsulation ratio exhibited sustained drug release up to 24 h with low burst release and secured higher VLF absorption in rabbits compared to the commercial capsules.

After i.n. administration of VLF-loaded CSNPs, the brain/blood ratio of VLF at 0.5 h was 0.1612, compared to the corresponding values of 0.0293 and 0.0700 obtained after i.v. and i.n. application of the pure drug, indicating a better brain uptake of VLF (Haque et al., 2012).

Jain and Datta (2016) designed MMT alginate biopolymeric composites as microbeads with 97% VLF encapsulation efficiency, which showed substantially less burst release with cumulative release of 20% (over a period of 26 h) and 22% (over a period of 29 h) in the gastric and intestinal fluid, respectively, compared to administration of the free drug for which burst release was followed by 100% cumulative release within 5.5 and 3.5 h in the gastric and intestinal fluid, respectively. Thus, application of clay not only reduced the burst effect but also resulted in the extended release of the drug, and the above formulation could be utilized for oral extended release dosage forms of VLF, for which the repeated intake (every 3–4 h) of the drug would not be necessary. Moreover, MMT-based VLF-PLGA nanocomposites could be recommended for designing the oral controlled-release formulations, which could minimize the drug administration frequency and the occurrence of side effects, thereby increasing the effectiveness of the drug and hence improving patient compliance (Jain and Datta, 2014). In vitro drug release study showed that also microcomposite spheres obtained by intercalation of nortriptyline and VLF in an interlayer gallery of Na^+–MMT, which was further compounded with poly(l-lactide), exhibited a controlled-release pattern (Rajkumar et al., 2015).

Upto 90.87%(w/w) of VLF could be loaded into mesocellular foams-based mesoporous silica nanospheres, which released 36% of drug after 1 h of incubation in simulated gastric fluid, and 53% of drug was released after 12 h in simulated intestinal fluid, and appropriate amounts of Pluronic 123 template preserved in mesopores showed an efficient synergetic effect on increasing VLF loading capacity and controlled-release property (Tang et al., 2012). Mesoporous silica nanospheres coated with cholic acid-cross-linked PLA significantly delayed the release of VLF in intestinal condition compared with gastric acid surrounding due to the fast decomposition rate of PLA in gastric acid, and their application also eliminated to a certain extent the initial burst in the simulated gastric fluid (Tang et al., 2011).

Duloxetine (DXT) is another SNRI used to treat major depressive disorder, generalized anxiety disorder, and neuropathic pain. DXT-loaded NLCs administered i.n. in albino Wistar rats exhibited improved behavioral analysis results (swimming, climbing, and immobility) compared to DXT solution after 24 h and demonstrated improved locomotor activity. This nanoformulation not only significantly increased the total swimming and climbing time when compared with the control but also considerably reduced the immobility period and secured higher drug concentration in the brain compared to DXT solution (Alam et al., 2012).

DXT-hydrochloride-loaded SLNPs of 91 nm with high drug EE (87%) and excellent stability in acidic medium showed great enhancement in antidepressant activity at 24 h when administered orally to mice in comparison to drug solution and the SLNPs were found to enhance chemical stability and improve the drug efficacy through oral route (Patel et al., 2012). The dispersion of NLC containing duloxetine (DXT–NLC) with particle size 125 nm and EE 80% and lyophilized DXT-NLC with particle size 137 nm and EE 79% showed sustained drug release, and permeation studies through the porcine nasal mucosal membrane confirmed the DXT-NLC could be used for delivery through nasal route (Alam et al., 2011).

Mesoporous silica NPs investigated for their potential use as a carrier for DXT hydrochloride that is prone to acid degradation were found to release 90% DXT in phosphate buffer of pH 7.4 but exhibited only 40% release in acidic pH, and the optimized formulation showed sustained release for 140 h (Ganesh et al., 2015). Mani et al. (2014) investigated the DXT-loaded mesoporous silica NPs prepared using a combination of pharmaceutical surfactants, Triton X-100 and Tween 40 as a template, and found that such mesoporous silica NPs could be considered as a better choice of the reservoir for the controlled delivery of drugs that require sustained release. Calcined and DXT-loaded mesoporous silica NPs characterized by large surface area and pore volume due to modification with Tween 20 displayed in vitro an initial burst release followed by sustained release for up to 140 h and could be utilized as a carrier for sustained release of active pharmaceutical ingredients (Ganesh et al., 2012a). Moreover, mesoporous TiO_2 was established as a viable, biocompatible DXT reservoir for its controlled release (Ganesh et al., 2012b).

5.7.2 MELATONIN AGONIST AND SELECTIVE SEROTONIN ANTAGONIST

Agomelatine is a melatonergic antidepressant used for the treatment of major depressive disorder. It is the first approved agent of new class of antidepressants—melatonin agonist and selective serotonin antagonist (MASSA). It is used for the treatment of serious depression and treatment of circadian rhythm. Liposome solid preparation used for antidepressant medicine, comprising agomelatine, distearoyl phosphatidylethanolamine, and soy sterol was patented by Wang (2012).

5.8 CONCLUSION

Based on the abovementioned statistics, neurological and psychiatric diseases have been more frequent in the population and, especially, disorders such as depressions, anxieties, and dementia occur more frequently in younger age. Thus, it can be stated that brain disorders, including developmental, psychiatric, and neurodegenerative diseases, represent an enormous disease burden, in terms of human suffering and economic costs. Based on these facts, new/innovative neurological and psychiatric drugs without various side effects are needed. These drugs have to circumvent the BBB and allow active CNS drugs to reach their target. An application of NPs as drug carriers can be considered as one of the possible strategies of direct drug delivery to the CNS. The applications of nanomaterials as drug carriers have a number of advantages, especially the increased bioavailability and thus better therapeutic efficiency of medicaments as well as decreased toxicity, and thus not only modern drugs but also relatively old drugs have been reformulated into nano-based drug delivery systems, which is supposed to serve revolution in the treatment of CNS diseases.

KEYWORDS

- nanoparticles
- nanoformulations
- central nervous system
- targeted delivery
- antiepileptics
- antiparkinsonics
- anti-Alzheimer's disease drugs
- antipsychotics
- anxiolytics
- antidepressants

REFERENCES

Abbott, N. J.; Patabendige, A. A. K.; Dolman, D. E. M.; Yusof, S. R.; Begley, D. J. Structure and Function of the Blood–Brain Barrier. *Neurobiol. Dis.* **2010,** *37*(1), 13–25.

Abdelbary, A. A.; Li, X. L.; El-Nabarawi, M.; Elassasy, A.; Jasti, B. Effect of Fixed Aqueous Layer Thickness of Polymeric Stabilizers on Zeta Potential and Stability of Aripiprazole Nanosuspensions. *Pharm. Dev. Technol.* **2013,** *18*(3), 730–735.

Abdelbary, A. A.; Li, X. L.; El-Nabarawi, M.; Elassasy, A.; Jasti, B. Comparison of Nanomilling and Coprecipitation on the Enhancement of in Vitro Dissolution Rate of Poorly Water-Soluble Model Drug Aripiprazole. *Pharm. Dev. Technol.* **2014,** *19*(4), 491–500.

Ahmad, M. Z.; Ahmad, J.; Amin, S.; Rahman, M.; Anwar, M.; Mallick, N.; Ahmad, F. J.; Rahman, Z.; Kamal, M. A.; Akhter, S. Role of Nanomedicines in Delivery of Anti-Acetylcholinesterase Compounds to the Brain in Alzheimer's Disease. *CNS Neurol. Disord. Drug Targets* **2014,** *13*(8), 1315–1324.

Akhter, S.; Jain, G. K.; Ahmad, F. J.; Khar, R. K.; Jain, N.; Khan, Z. I.; Talegaonkar, S. Investigation of Nanoemulsion System for Transdermal Delivery of Domperidone: Ex-Vivo and in Vivo Studies. *Curr. Nanosci.* **2008,** *4*(4), 381–390.

Alabsi, A.; Khoudary, A. C.; Abdelwahed, W. The Antidepressant Effect of L-Tyrosine-Loaded Nanoparticles: Behavioral Aspects. *Ann. Neurosci.* **2016,** *23*(2), 89–99.

Alam, M. I.; Baboota, S.; Ahuja, A.; Ali, M.; Ali, J.; Sahni, J. K. Nanostructured Lipid Carrier Containing CNS Acting Drug: Formulation, Optimization and Evaluation. *Curr. Nanosci.* **2011,** *7*(6), 1014–1027.

Alam, M. I.; Baboota, S.; Ahuja, A.; Ali, M.; Ali, J.; Sahni, J. K. Intranasal Administration of Nanostructured Lipid Carriers Containing Cns Acting Drug: Pharmacodynamic Studies and Estimation in Blood and Brain. *J. Psychiatr. Res.* **2012,** *46*(9), 1133–1138.

Alam, T.; Pandit, J.; Vohora, D.; Aqil, M.; Ali, A.; Sultana, Y. Optimization of Nanostructured Lipid Carriers of Lamotrigine for Brain Delivery: in Vitro Characterization and in Vivo Efficacy in Epilepsy. *Exp. Opin. Drug Delivery* **2015**, *12*(2), 181–194.

Alavijeh, M. S.; Chishty, M.; Qaiser, M. Z.; Palmer, A. M. Drug Metabolism and Pharmacokinetics, the Blood–Brain Barrier, and Central Nervous System Drug Discovery. *Neurotherapeutics* **2005**, *2*(4), 554–571.

Al-Dhubiab, B. E. Formulation and in Vitro Evaluation of Gelatin Nanospheres for the Oral Delivery of Selegiline. *Curr. Nanosci.* **2013**, *9*(1), 2125.

Alyautdin, R.; Khalin, I.; Nafeeza, M. I.; Haron, M. H.; Kuznetsov, D. Nanoscale Drug Delivery Systems and the Blood-Brain Barrier. *Int. J. Nanomed.* **2014**, *9*, 795–811.

Anderson, I. M.; Haddad, P. M.; Scott, J. Bipolar Disorder. *BMJ* **2012**, *345*, e8508.

Aripiprazole – DrugBank. https://www.drugbank.ca/drugs/DB01238 (accessed Jan 7, 2017).

Asadi, E.; Azodi-Deilami, S.; Abdouss, M.; Kordestani, D.; Rahimi, A.; Asadi, S. S. Recognition and Evaluation of Molecularly Imprinted Polymer Nanoparticle Using Miniemulsion Polymerization for Controlled Release and Analysis of Risperidone in Human Plasma Samples. *Korean J. Chem Eng.* **2014**, *31*(6), 1028–1035.

Asadi, E.; Abdouss, M.; Leblanc, R. M.; Ezzati, N.; Wilson, J. N.; Kordestani, D. Synthesis, Characterization and in Vivo Drug Delivery Study of a Biodegradable Nano-Structured Molecularly Imprinted Polymer Based on Cross-Linker of Fructose. *Polymer* **2016**, *97*, 226–237.

Avachat, A. M.; Oswal, Y. M.; Gujar, K. N.; Shah, R. D. Preparation and Characterization of Rivastigmine Loaded Human Serum Albumin (HSA) Nanoparticles. *Curr. Drug Delivery* **2014**, *11*(3), 359–370.

Azeem, A.; Ahmad, F. J.; Khar, R. K.; Talegaonkar, S. Nanocarrier for the Transdermal Delivery of an Antiparkinsonian Drug. *AAPS PharmSciTech* **2009**, *10*(4), 1093–1103.

Azeem, A.; Talegaonkar, S.; Negi, L. M.; Ahmad, F. J.; Khar, R. K.; Iqbal, Z. Oil Based Nanocarrier System for Transdermal Delivery of Ropinirole: a Mechanistic, Pharmacokinetic and Biochemical Investigation. *Int. J. Pharm.* **2012**, *422*(12), 436–444.

Bajic V.; Milovanovic, E. S.; Spremo-Potparevic, B.; Zivkovic, L.; Miliccivc, Z.; Stanimirovic, J.; Bogdanovic, N.; Isenovic, E. R. Treatment of Alzheimer's Disease: Classical Therapeutic Approach. *Curr. Pharm. Anal.* **2016**, *12*(2), 8290.

Baltzley, S.; Mohammad, A.; Malkawi, A. H.; Al-Ghananeem, A. M. Intranasal Drug Delivery of Olanzapine-Loaded Chitosan Nanoparticles. *AAPS PharmSciTech.* **2014**, *15*(6), 1598–1602.

Bari, N. K.; Fazil, M.; Hassan, M. Q.; Haider, M. R.; Gaba, B.; Narang, J. K.; Baboota, S.; Ali, J. Brain Delivery of Buspirone Hydrochloride Chitosan Nanoparticles for the Treatment of General Anxiety Disorder. *Int. J. Biol. Macromol.* **2015**, *81*, 49–59.

Barzago, M. M.; Bortolotti, A.; Stellari, F. F.; Diomede, L.; Algeri, M.; Efrati, S.; Salmona, M.; Bonati, M. Placental Transfer of Valproic Acid After Liposome Encapsulation During in Vitro Human Placenta Perfusion. *J. Pharmacol. Exp. Ther.* **1996**, *277*(1), 79–86.

Baumgartner, R.; Teubl, B. J.; Tetyczka, C.; Roblegg, E. Rational Design and Characterization of a Nanosuspension for Intraoral Administration Considering Physiological Conditions. *J. Pharm. Sci.* **2016**, *105*(1), 257–267.

Baysal, I.; Yabanoglu-Ciftci, S.; Tunc-Sarisozen, Y.; Ulubayram, K.; Ucar, G. Interaction of Selegiline-Loaded PLGA-B-PEG Nanoparticles with Beta-Amyloid Fibrils. *J. Neural. Transm.* **2013**, *120*(6), 903–910.

Bernardi, A.; Frozza, R. L.; Meneghetti, A.; Hoppe, J. B.; Battastini, A. M. O.; Pohlmann, A. R.; Guterres, S. S.; Salbego, C. G. Indomethacin-Loaded Lipid-Core Nanocapsules Reduce the Damage Triggered by Aβ 1–42 in Alzheimer's Disease Models. *Int. J. Nanomed.* **2012,** *7,* 4927–4942.

Bhaskar, S.; Tian, F.; Stoeger, T.; Kreyling, W.; de la Fuente, J. M.; Grazú, V.; Borm, P.; Estrada, G.; Ntziachristos, V.; Razansky, D. Multifunctional Nanocarriers for Diagnostics, Drug Delivery and Targeted Treatment Across Blood–Brain Barrier: Perspectives on Tracking and Neuroimaging. *Part. Fibre Toxicol.* **2010,** *7,* 3.

Bhavna Md, S.; Ali, M.; Ali, R.; Bhatnagar, A.; Baboota, S.; Ali, J. Donepezil Nanosuspension Intended for Nose to Brain Targeting: In Vitro and in vivo Safety Evaluation. *Int. J. Biol. Macromol.* **2014a,** *67,* 418–425.

Bhavna Md, S.; Ali, M.; Bhatnagar, A.; Baboota, S.; Sahni, J.K.; Ali, J. Design Development, Optimization and Characterization of Donepezil Loaded Chitosan Nanoparticles for Brain Targeting to Treat Alzheimer's Disease. *Sci. Adv. Mater.* **2014b,** *6*(4), 720–735.

Bhavna Md, S.; Ali, M.; Baboota, S.; Sahni, J.K.; Bhatnagar, A.; Ali, J. Preparation, Characterization, in Vivo Biodistribution and Pharmacokinetic Studies of Donepezil-Loaded PLGA Nanoparticles for Brain Targeting. *Drug. Dev. Ind. Pharm.* **2014c,** *40*(2), 278–287.

Bruni, G.; Maggi, L.; Tammaro, L.; Di Lorenzo, R.; Friuli, V.; D'Aniello, S.; Maietta, M.; Berbenni, V.; Milanese, C.; Girella, A.; Marini, A. Electrospun Fibers as Potential Carrier Systems for Enhanced Drug Release of Perphenazine. *Int. J. Pharm.* **2016,** *511*(1), 190–197.

Burkhart, A.; Azizi, M.; Thomsen, M. S.; Thomsen, L. B.; Moos, T. Accessing Targeted Nanoparticles to the Brain: the Vascular Route. *Curr. Med. Chem.* **2014,** *21*(36), 4092–4099.

Cacciatore, I.; Ciulla, M.; Fornasari, E.; Marinelli, L.; Di Stefano, A. Solid Lipid Nanoparticles as a Drug Delivery System for the Treatment of Neurodegenerative Diseases. *Exp. Opin. Drug Delivery.* **2016,** *13*(8), 1121–1131.

Cao, X. B.; Hou, D. Z.; Wang, L.; Li, S.; Sun, S. G.; Ping, Q. N.; Xu, Y. Effects and Molecular Mechanism of Chitosan-Coated Levodopa Nanoliposomes on Behavior of Dyskinesia Rats. *Biol. Res.* **2016,** *49,* 32.

Carreno, F.; Paese, K.; Silva, C. M.; Guterres, S. S.; Dalla Costa, T. Pre-Clinical Investigation of the Modulation of Quetiapine Plasma Pharmacokinetics and Tissues Biodistribution by Lipid-Core Nanocapsules. *J. Pharm. Biomed. Anal.* **2016a,** *119,* 152–158

Carreno, F.; Paese, K.; Silva, C. M.; Guterres, S. S.; Dalla Costa, T. Pharmacokinetic Investigation of Quetiapine Transport Across Blood-Brain Barrier Mediated by Lipid Core Nanocapsules Using Brain Microdialysis in Rats. *Mol. Pharm.* **2016b,** *13*(4), 1289–1297.

Casolaro, M.; Casolaro, I. Controlled Release of Antidepressant Drugs by Multiple Stimuli-Sensitive Hydrogels Based on Alpha-Aminoacid Residues. *J. Drug. Delivery. Sci. Technol.* **2015,** *30,* 82–89.

Cave, G.; Harvey, M.; Shaw, T.; Damitz, R.; Chauhan, A. Comparison of Intravenous Lipid Emulsion, Bicarbonate, and Tailored Liposomes in Rabbit Clomipramine Toxicity. *Acad. Emerg. Med.* **2013,** *20*(10), 1076–1079.

Černíková, A.; Jampílek, J. Structure Modification of Drugs Influencing Their Bioavailability and Therapeutic Effect. *Chem. Listy.* **2014,** *108*(1), 716.

Cevik, O.; Gidon, D.; Kizilel, S. Visible-Light-Induced Synthesis of Ph-Responsive Composite Hydrogels for Controlled Delivery of the Anticonvulsant Drug Pregabalin. *Acta Biomater.* **2015,** *11,* 151–161.

Chalikwar, S. S.; Mene, B. S.; Pardeshi, C. V.; Belgamwar, V. S.; Surana, S. J. Self-assembled, Chitosan Grafted PLGA Nanoparticles for Intranasal Delivery: Design, Development and ex Vivo Characterization. *Polym-Plast Technol.* **2013,** *52*(4), 368–380.

Cheng, L.; Jin, Y.; Wang, Y.; Wuc, C.; Zhang, J.; Zhao, Y. Oxcarbazepine Nanostructured Lipid Carrier Comprises Oxcarbazepine, Solid Lipid Material, Liquid Lipid Material, Emulsifying Agent and Balanced Amount of Water, and Solid Lipid Material is Selected from Lecithin or Soybean Phospholipid. CN Patent 104,523,696 A, December 9, 2014.

Clemens-Hemmelmann, M.; Kuffner, C.; Metz, V.; Kircher, L.; Schmitt, U.; Hiemke, C.; Postina, R.; Zentel, R. Amphiphilic Copolymers Shuttle Drugs Across the Blood-Brain Barrier. *Macromol. Biosci.* **2016,** *16*(5), 655–665.

Damitz, R.; Chauhan, A. Parenteral Emulsions and Liposomes to Treat Drug Overdose. *Adv. Drug Delivery Rev.* **2015,** *90*, 12–23.

Darius, J.; Meyer, F. P.; Sabel, B. A.; Schroeder, U. Influence of Nanoparticles on the Brain-To-Serum Distribution and the Metabolism of Valproic Acid in Mice. *J. Pharm. Pharmacol.* **2000,** *52*(9), 1043–1047.

Darville, N.; van Heerden, M.; Vynckier, A.; De Meulder, M.; Sterkens, P.; Annaert, P.; Van den Mooter, G. Intramuscular Administration of Paliperidone Palmitate Extended-Release Injectable Microsuspension Induces a Subclinical Inflammatory Reaction Modulating the Pharmacokinetics in Rats. *J. Pharm. Sci.* **2014,** *103*(7), 2072–2087.

Darville, N.; van Heerden, M.; Marien, D.; De Meulder, M.; Rossenu, S.; Vermeulen, A.; Vynckier, A.; De Jonghe, S.; Sterkens, P.; Annaert, P.; Van der Mooter, G. The Effect of Macrophage and Angiogenesis Inhibition on the Drug Release and Absorption from an Intramuscular Sustained-Release Paliperidone Palmitate Suspension. *J. Controlled Release* **2016a,** *230*, 95–108

Darville, N.; van Heerden, M.; Erkens, T.; De Jonghe, S.; Vynckier, A.; De Meulder, M.; Vermeulen, A.; Sterkens, P.; Annaert, P.; Van den Mooter, G. Modeling the Time Course of the Tissue Responses to Intramuscular Long-Acting Paliperidone Palmitate Nano-/Microcrystals and Polystyrene Microspheres in the Rat. *Toxicol. Pathol.* **2016b,** *44*(2), 189–210.

Demirel, M.; Yazan, Y.; Muller, R. H.; Kilic, F.; Bozan, B. Formulation and in Vitro-In Vivo Evaluation of Piribedil Solid Lipid Micro- and Nanoparticles. *J. Microencapsulation* **2001,** *18*(3), 359–371.

Dening, T. J.; Rao, S. S.; Thomas, N.; Prestidge, C. A. Silica Encapsulated Lipid-Based Drug Delivery Systems for Reducing the Fed/Fasted Variations of Ziprasidone in Vitro. *Eur. J. Pharm. Biopharm.* **2016,** *101*, 33–42.

Dhanikula, A. B.; Lamontagne, D.; Leroux, J. C. Rescue of Amitriptyline-Intoxicated Hearts with Nanosized Vesicles. *Cardiovasc. Res.* **2007,** *74*(3), 480–486.

Dimer, F. A.; Ortiz, M.; Pase, C. S.; Roversi, K.; Friedrich, R. B.; Pohlmann, A. R.; Burger, M. E.; Guterres, S. S. Nanoencapsulation of Olanzapine Increases Its Efficacy in Antipsychotic Treatment and Reduces Adverse Effects. *J. Biomed. Nanotechnol.* **2014,** *10*(6), 1137–1145.

Dimer, F. A.; Pigatto, M. C.; Boque, C. A.; Pase, C. S.; Roversi, K.; Pohlmann, A. R.; Burger, M. E.; Rates, S. M. K.; Dalla Costa, T.; Guterres, S. S. Nanoencapsulation Improves Relative Bioavailability and Antipsychotic Effect of Olanzapine in Rats. *J. Biomed. Nanotechnol.* **2015,** *11*(8), 1482–1493.

Dimitrijevic, I.; Pantic, I. Application of Nanoparticles in Psychophysiology and Psychiatry Research. *Rev. Adv. Mater. Sci.* **2014,** *37*, 83–88.

Di Stefano, A.; Iannitelli, A.; Laserra, S.; Sozio, P. Drug Delivery Strategies for Alzheimer's Disease Treatment. *Exp. Opin Drug Delivery* **2011,** *8*(5), 581–603.

Dong, S. Y.; Wang, S.; Zheng, C. H.; Liang, W. Q.; Huang, Y. Z. An in Situ-Forming, Solid Lipid/PLGA Hybrid Implant for Long-Acting Antipsychotics. *Soft. Matter.* **2011,** *7*(12), 5873–5878.

Dordevic, S. M.; Cekic, N. D.; Savic, M. M.; Isailovic, T. M.; Randelovic, D. V.; Markovic, B. D.; Savic, S. R.; Stamenic, T. T.; Daniels, R.; Savic, S. D. Parenteral Nanoemulsions as Promising Carriers for Brain Delivery of Risperidone: Design, Characterization and in Vivo Pharmacokinetic Evaluation. *Int. J. Pharm.* **2015,** *493*(12), 40–54.

Drinberg, V.; Bitcover, R.; Rajchenbach, W.; Peer, D. Modulating Cancer Multidrug Resistance by Sertraline in Combination with Ananomedicine. *Cancer Lett.* **2014,** *354*(2), 290–298.

Eberhardt, R.; Birbamer, G.; Gerstenbrand, F.; Rainer, E.; Traegner, H. Citicoline in the Treatment of Parkinson's Disease. *Clin. Ther.* **1990,** *12*(6), 489–495.

Elmizadeh, H.; Khanmohammadi, M.; Ghasemi, K.; Hassanzadeh, G.; Nassiri-Asl, M.; Garmarudi, A. B. Preparation and Optimization of Chitosan Nanoparticles and Magnetic Chitosan Nanoparticles as Delivery Systems Using Box-Behnken Statistical Design. *J. Pharm. Biomed. Anal.* **2013,** *80*, 141–146.

El-Setouhy, D. A.; Ibrahim, A. B.; Amin, M. M.; Khowessah, O. M.; Elzanfaly, E. S. Intranasal Haloperidol-Loaded Miniemulsions for Brain Targeting: Evaluation of Locomotor Suppression and In-Vivo Biodistribution. *Eur. J. Pharm. Sci.* **2016,** *92*, 244–254.

El-Zaafarany, G. M.; Soliman, M. E.; Mansour, S.; Awad, G. A. S. Identifying Lipidic Emulsomes for Improved Oxcarbazepine Brain Targeting: in Vitro and Rat in Vivo Studies. *Int. J. Pharm.* **2016,** *503*(12), 127–140.

Elzainy, A. W.; Gu, X. C.; Simons, F. E. R.; Simons, K. J. Hydroxyzine from Topical Phospholipid Liposomal Formulations: Evaluation of Peripheral Antihistaminic Activity and Systemic Absorption in a Rabbit Model. *AAPS PharmSci.* **2003,** *5*(4), E28.

Elzainy, A. A. W.; Gu, X. C.; Simons, F. E. R.; Simons, K. J. Hydroxyzine- and Cetirizine-Loaded Liposomes: Effect of Duration of Thin Film Hydration, Freeze-Thawing, and Changing Buffer Ph on Encapsulation and Stability. *Drug Dev. Ind. Pharm.* **2005,** *31*(3), 281–291.

Eskandari, S.; Varshosaz, J.; Minaiyan, M.; Tabbakhian, M. Brain Delivery of Valproic Acid Via Intranasal Administration of Nanostructured Lipid Carriers: in Vivo Pharmacodynamic Studies Using Rat Electroshock Model. *Int. J. Nanomed.* **2011,** *6*, 363–371.

Eslami, M.; Nikkhah, S. J.; Hashemianzadeh, S. M.; Sajadi, S. A. S. The Compatibility of Tacrine Molecule with Poly(N-Butylcyanoacrylate) and Chitosan as Efficient Carriers for Drug Delivery: a Molecular Dynamics Study. Eur. *J. Pharm. Sci.* **2016,** *82*, 79–85.

Esposito, E.; Fantin, M.; Marti, M.; Drechsler, M.; Paccamiccio, L.; Mariani, P.; Sivieri, E.; Lain, F.; Menegatti, E.; Morari, M.; Cortesi, R. Solid Lipid Nanoparticles as Delivery Systems for Bromocriptine. *Pharm. Res.* **2008,** *25*(7), 1521–1530.

Esposito, E.; Mariani, P.; Ravani, L.; Contado, C.; Volta, M.; Bido, S.; Drechsler, M.; Mazzoni, S.; Menegatti, E.; Morari, M.; Cortesi, R. Nanoparticulate Lipid Dispersions for Bromocriptine Delivery: Characterization and in Vivo Study. Eur. J. Pharm. Biopharm. **2012,** *80*(2), 306–314.

Fang, Z. Y.; Chen, S. D.; Qin, J. M.; Chen, B.; Ni, G. Z.; Chen, Z. Y.; Zhou, J. Q.; Li, Z.; Ning, Y. P.; Wu, C. B.; Zhou, L. Pluronic P85-Coated Poly(Butylcyanoacrylate) Nanoparticles Overcome Phenytoin Resistance in P-Glycoprotein Overexpressing Rats with Lithium-Pilocarpine-Induced Chronic Temporal Lobe Epilepsy. *Biomaterials* **2016,** *97*, 110–121.

Fazil, M.; Md, S.; Hague, S.; Kumar, M.; Baboota, S.; Sahni, J. K.; Ali, J. Development and Evaluation of Rivastigmine Loaded Chitosan Nanoparticles for Brain Targeting. *Eur. J. Pharm. Sci.* **2012,** *47*(1), 6–15.

Fisar, Z. Interactions Between Tricyclic Antidepressants and Phospholipid Bilayer Membranes. *Gen. Physiol. Biophys.* **2005,** *24*(2), 161–180.

Fonseca, F. N.; Betti, A. H.; Carvalho, F. C.; Gremiao, M. P. D.; Dimer, F. A.; Guterres, S. S.; Tebaldi, M. L.; Rates, S. M. K.; Pohlmann, A. R. Mucoadhesive Amphiphilic Methacrylic Copolymer-Functionalized Poly(Epsilon-Caprolactone) Nanocapsules for Nose-To-Brain Delivery of Olanzapine. *J. Biomed. Nanotechnol.* **2015,** *11*(8), 1472–1481.

Fornaguera, C.; Feiner-Gracia, N.; Caldero, G.; Garcia-Celma, M. J.; Solans, C. Galantamine-loaded PLGA Nanoparticles, From Nano-Emulsion Templating, as Novel Advanced Drug Delivery Systems to Treat Neurodegenerative Diseases. *Nanoscale* **2015,** *7*(28), 12076–12084.

Fresta, M.; Cavallaro, G.; Giammona, G.; Wehrli, E.; Puglisi, G. Preparation and Characterization of Polyethyl-2-Cyanoacrylate Nanocapsules Containing Antiepileptic Drugs. *Biomaterials* **1996,** *17*(8), 751–758.

Gabal, Y. M.; Kamel, A. O.; Sammour, O. A.; Elshafeey, A. H. Effect of Surface Charge on the Brain Delivery of Nanostructured Lipid Carriers in Situ Gels Via the Nasal Route. *Int. J. Pharm.* **2014,** *473*(12), 442–457.

Ganesh, M.; Hemalatha, P.; Mei, P. M.; Rajasekar, K.; Jang, H. T. A New Fluoride Mediated Synthesis of Mesoporous Silica and Their Usefulness in Controlled Delivery of Duloxetine Hydrochloride a Serotonin Re-Uptake Inhibitor. *J. Ind. Eng. Chem.* **2012a,** *18*(2), 684–689.

Ganesh, M.; Hemalatha, P.; Peng, M. M.; Cha, W. S.; Palanichamy, M.; Jang, H. T. Drug Release Evaluation of Mesoporous TiO2: A Nano Carrier for Duloxetine. In *Computer Applications for Modeling, Simulation, and Automobile. Communications in Computer and Information Science*; Kim, T.H., Ramos, C., Abawajy, J., Kang, B. H., Slezak, D., Adeli, H. Eds.; Springer: Berlin-Heidelberg, 2012b; Vol. 341, pp 237–243.

Ganesh, M.; Ubaidulla, U.; Hemalatha, P.; Peng, M. M.; Jang, H. T. Development of Duloxetine Hydrochloride Loaded Mesoporous Silica Nanoparticles: Characterizations and in Vitro Evaluation. *AAPS PharmSciTech.* **2015,** *16*(4), 944–951.

Gao, N.; Dong, K.; Zhao, A. D.; Sun, H. J.; Wang, Y.; Ren, J. S.; Qu, X. G. Polyoxometalate-Based Nanozyme: Design of a Multifunctional Enzyme for Multi-Faceted Treatment of Alzheimer's Disease. *Nano Res.* **2016,** *9*(4), 1079–1090.

Garcia, X.; Escribano, E.; Colom, H.; Domenech, J.; Queralt, J. Tricyclic Antidepressants-Loaded Biodegradable PLGA Nanoparticles: in Vitro Characterization and in Vivo Analgesic and Anti-Allodynic Effect. *Curr. Nanosci.* **2011,** *7*(3), 345–353.

Govender, T.; Choonara, Y. E.; Kumar, P.; du Toit, L. C.; Modi, G.; Naidoo, D.; Pillay, V. A Novel Melt-Dispersion Technique for Simplistic Preparation of Chlorpromazine-Loaded Polycaprolactone Nanocapsules. *Polymers.* **2015,** *7*(6), 1145–1176.

Goyal, K.; Koul, V.; Singh, Y.; Anand, A. Targeted Drug Delivery to Central Nervous System (Cns) for the Treatment of Neurodegenerative Disorders: Trends and Advances. *Cent. Nerv. Syst. Agents Med. Chem.* **2014,** *14*(1), 43–59.

Graham, L. M.; Nguyen, T. M.; Lee, S. B. Nanodetoxification: Emerging Role of Nanomaterials in Drug Intoxication Treatment. *Nanomedicine* **2011,** *6*(5), 921–928.

Gulati, N.; Nagaich, U.; Saraf, S. Fabrication and in Vitro Characterization of Polymeric Nanoparticles for Parkinson's Therapy: a Novel Approach. *Braz. J. Pharm. Sci.* **2014,** *50*(4), 869–876.

Haas, S. E.; de Andrade, C.; Sansone, P. E. D.; Guterres, S.; Dalla Costa, T. Development of Innovative Oil-Core Self-Organized Nanovesicles Prepared with Chitosan and Lecithin Using a 2(3) Full-Factorial Design. *Pharm. Dev. Technol.* **2014,** *19*(7), 769–778.

Haeri, A.; Alinaghian, B.; Daeihamed, M.; Dadashzadeh, S. Preparation and Characterization of Stable Nanoliposomal Formulation of Fluoxetine as a Potential Adjuvant Therapy for Drug-Resistant Tumors. *IJPR* **2014,** *13*, 3–14.

Halayqa, M.; Domanska, U. PLGA Biodegradable Nanoparticles Containing Perphenazine Or Chlorpromazine Hydrochloride: Effect of Formulation and Release. *Int. J. Mol. Sci.* **2014,** *15*(12), 23909–23923.

Hamidi, M.; Rafiei, P.; Azadi, A.; Mohammadi-Samani, S. Encapsulation of Valproate-Loaded Hydrogel Nanoparticles in Intact Human Erythrocytes: a Novel Nano-Cell Composite for Drug Delivery. *J. Pharm. Sci.* **2011,** *100*(5), 1702–1711.

Hanafy, A. S.; Farid, R. M.; ElGamal, S. S. Complexation as an Approach to Entrap Cationic Drugs Into Cationic Nanoparticles Administered Intranasally for Alzheimer's Disease Management: Preparation and Detection in Rat Brain. *Drug Dev. Ind. Pharm.* **2015,** *41*(12), 2055–2068.

Haque, S.; Md, S.; Fazil, M.; Kumar, M.; Sahni, J. K.; Ali, J.; Baboota, S. Venlafaxine Loaded Chitosan NPs for Brain Targeting: Pharmacokinetic and Pharmacodynamic Evaluation. *Carbohydr. Polym.* **2012,** *89*(1), 72–79.

Haque, S.; Md, S.; Sahni, J. K.; Ali, J.; Baboota, S. Development and Evaluation of Brain Targeted Intranasal Alginate Nanoparticles for Treatment of Depression. *J. Psychiatr. Res.* **2014,** *48*(1), 1–12.

Hassani, S.; Laouini, A.; Fessi, H.; Charcosset, C. Preparation of Chitosan-TPP Nanoparticles Using Microengineered Membranes–Effect of Parameters and Encapsulation of Tacrine. *Colloids Surf. A Physicochem. Eng. Asp.* **2015,** *482*, 34–43.

Hassanzadeh, K.; Nikzaban, M.; Moloudi, M. R.; Izadpanah, E. Effect of Selegiline on Neural Stem Cells Differentiation: a Possible Role for Neurotrophic Factors. *Iran. J. Basic Med. Sci.* **2015,** *18*(6), 548–554.

He, E.; Yue, C. Y.; Tam, K. C. Binding and Release Studies of a Cationic Drug from a Star-Shaped Four-Arm Poly(Ethylene Oxide)-B-Poly(Methacrylic Acid). *J. Pharm. Sci.* **2010,** *99*(2), 782–793.

Healthcare Finance News: Pipeline antipsychotic drugs to drive next market evolution. https://web.archive.org/web/20120218005726/http://www.healthcarefinancenews.com/press-release/pipeline-antipsychotic-drugs-drive-next-market-evolution (accessed Dec 27, 2016).

Hemmelmann, M.; Knoth, C.; Schmitt, U.; Allmeroth, M.; Moderegger, D.; Barz, M.; Koynov, K.; Hiemke, C.; Roesch, F.; Zentel, R. HPMA Based Amphiphilic Copolymers Mediate Central Nervous Effects of Domperidone. *Macromol. Rapid. Commun.* **2011,** *32*(910), 712–717.

Hemmelmann, M.; Metz, V. V.; Koynov, K.; Blank, K.; Postina, R.; Zentel, R. Amphiphilic HPMA-LMA Copolymers Increase the Transport of Rhodamine 123 Across a BBB Model Without Harming Its Barrier Integrity. *J. Controlled Release* **2012,** *163*(2), 170–177.

Heredia-Cervera, B. E.; Gonzalez-Azcorra, S. A.; Rodriguez-Gattorno, G.; Lopez, T.; Ortiz-Islas, E.; Oskam, G. Controlled Release of Phenytoin from Nanostructured Tio2 Reservoirs. *Sci. Adv. Mater.* **2009,** *1*(1), 63–68.

Howell, B.; Chauhan, A. Amitriptyline Overdose Treatment by PEGylated Anionic Liposomes. *J. Colloid Interface Sci.* **2008,** *324*(12), 61–70.

Hsiao, M. H.; Larsson, M.; Larsson, A.; Evenbratt, H.; Chen, Y. Y.; Chen, Y. Y.; Liu, D. M. Design and Characterization of a Novel Amphiphilic Chitosan Nanocapsule-Based Thermo-Gelling Biogel with Sustained in Vivo Release of the Hydrophilic Anti-Epilepsy Drug Ethosuximide. *J. Controlled Release* **2012,** *161*(3), 942–948.

Hsu, S. H.; Wen, C. J.; Al-Suwayeh, S. A.; Chang, H. W.; Yen, T. C.; Fang, J. Y. Physicochemical Characterization and in Vivo Bioluminescence Imaging of Nanostructured Lipid Carriers for Targeting the Brain: Apomorphine as a Model Drug. *Nanotechnology* **2010,** *21*(40), 405–101.

Hsu, S. H.; Al-Suwayeh, S. A.; Chen, C. C.; Chi, C. H.; Fang, J. Y. PEGylated Liposomes Incorporated with Nonionic Surfactants as an Apomorphine Delivery System Targeting the Brain: in Vitro Release and in Vivo Real-Time Imaging. *Curr. Nanosci.* **2011,** *7*(2), 191–199.

Huang, W. C.; Hu, S. H.; Liu, K. H.; Chen, S. Y.; Liu, D. M. A Flexible Drug Delivery Chip for the Magnetically-Control Led Release of Anti-Epileptic Drugs. J. Controlled Release **2009,** *139*(3), 221–228.

Huang, H. Y.; Hu, S. H.; Chian, C. S.; Chen, S. Y.; Lai, H. Y.; Chen, Y. Y. Self-Assembling Pva-F127 Thermosensitive Nanocarriers with Highly Sensitive Magnetically-Triggered Drug Release for Epilepsy Therapy in Vivo. *J. Mater Chem.* **2012,** *22*(17), 8566–8573.

Huang, C. T.; Tsai, M. J.; Lin, Y. H.; Fu, Y. S.; Huang, Y. B.; Tsai, Y. H.; Wu, P. C. Effect of Microemulsions on Transdermal Delivery of Citalopram: Optimization Studies Using Mixture Design and Response Surface Methodology. *Int. J. Nanomed.* **2013,** *8*, 2295–2304.

Huang, M.; Hu, M.; Song, Q. X.; Song, H. H.; Huang, J. L.; Gu, X.; Wang, X. L.; Chen, J.; Kang, T.; Feng, X. Y.; Jiang, D.; Zheng, G.; Chen, H.; Gao, X. Gm1-Modified Lipoprotein-Like Nanoparticle: Multifunctional Nanoplatform for the Combination Therapy of Alzheimer's Disease. *ACS Nano.* **2015,** *9*(11), 10801–10816.

Hwang, T. L.; Lin, Y. K.; Chi, C. H.; Huang, T. H.; Fang, J. Y. Development and Evaluation of Perfluorocarbon Nanobubbles for Apomorphine Delivery. *J. Pharm. Sci.* **2009,** *98*(10), 3735–3747.

Ianiski, F. R.; Alves, C. B.; Souza, A. C. G.; Pinton, S.; Roman, S. S.; Rhoden, C. R. B.; Alves, M. P.; Luchese, C. Protective Effect of Meloxicam-Loaded Nanocapsules Against Amyloid-Beta Peptide-Induced Damage in Mice. *Behav. Brain Res.* **2012,** *230*(1), 100–107.

Ianiski, F. R.; Alves, C. B.; Ferreira, C. F.; Rech, V. C.; Savegnago, L.; Wilhelm, E. A.; Luchese, C. Meloxicam-Loaded Nanocapsules as an Alternative to Improve Memory Decline in an Alzheimer's Disease Model in Mice: Involvement of Na+, K+-ATPase. *Metab. Brain Dis.* **2016,** *31*(4), 793–802.

Ibrahim, W. M.; Al Omrani, A. H.; Yassin, A. E. B. Novel Sulpiride-Loaded Solid Lipid Nanoparticles with Enhanced Intestinal Permeability. *Int. J. Nanomed.* **2014,** *9*, 129–144.

Ibrahim, H. K.; Salah, S. Formulation of Venlafaxine for Once Daily Administration Using Polymeric Material Hybrids. *J. Microencapsulation* **2016,** *33*(4), 299–306.

Imam, S. S.; Aqil, M.; Akhtar, M.; Sultana, Y.; Ali, A. Formulation by Design-Based Proniosome for Accentuated Transdermal Delivery of Risperidone: in Vitro Characterization and in Vivo Pharmacokinetic Study. *Drug Delivery* **2015,** *22*(8), 1059–1070.

Ismail, M. F.; ElMeshad, A. N.; Salem, N. A. H. Potential Therapeutic Effect of Nanobased Formulation of Rivastigmine on Rat Model of Alzheimer's Disease. *Int. J. Nanomed.* **2013,** *8*, 393–406.

Jafarieh O.; Md, S.; Ali, M.; Baboota, S.; Sahni, J. K.; Kumari, B.; Bhatnagar, A.; Ali, J. D. Characterization, and Evaluation of Intranasal Delivery of Ropinirole-Loaded Mucoadhesive Nanoparticles for Brain Targeting. *Drug Dev. Ind. Pharm.* **2015,** *41*(10), 1674–1681.

Jain, N. K.; Rana, A. C.; Jain, S. K. Brain Drug Delivery System Bearing Dopamine Hydrochloride for Effective Management of Parkinsonism. *Drug Dev. Ind. Pharm.* **1998,** *24*(7), 671–675.

Jain, S.; Datta, M. Montmorillonite-PLGA Nanocomposites as an Oral Extended Drug Delivery Vehicle for Venlafaxine Hydrochloride. *Appl. Clay Sci.* **2014,** *99*, 42–47.

Jain, S.; Datta, M. Montmorillonite-Alginate Microspheres as a Delivery Vehicle for Oral Extended Release of Venlafaxine Hydrochloride. *J. Drug Delivery Sci. Technol.* **2016,** *33,* 149–156.

Jampílek, J.; Záruba, K.; Oravec, M.; Kuneš, M.; Babula, P.; Ulbrich, P.; Brezaniová, I.; Opatřilová, R.; Tříska, J.; Suchý, P. Preparation of Silica Nanoparticles Loaded with Nootropics and Their in Vivo Permeation Through Blood–Brain Barrier. *Biomed. Res. Int.* **2015,** *2015,* 812–673.

Jampílek, J.; Kráľová, K. Nano-antimicrobials: Activity, Benefits and Weaknesses. Nanostructures in Therapeutic Medicine. In *Nanostructures for antimicrobial therapy; Grumezescu,* A. M., Ed.; Elsevier: London, 2017a; Chapter 2, Vol. 2, in press.

Jampílek, J.; Kráľová, K. Application of Nanobioformulations for Controlled Release and Targeted Biodistribution of Drugs. In *Recent advances and applications of nanobiomaterials*; Keservani, R. K. Ed.; CRC Press: Boca Raton, 2017b, in press.

Jampílek, J.; Kráľová, K. Impact of nanoparticles on living organisms and human health. In Encyclopedia of Nanoscience and Nanotechnology; Nalwa, H. S., Ed.; American Scientific Publishers: Valencia, 2017c, in press.

Jana, S.; Maji, N.; Nayak, A. K.; Sen, K. K.; Basu, S. K. Development of Chitosan-Based Nanoparticles Through Inter-Polymeric Complexation for Oral Drug Delivery *Carbohydr. Polym.* **2013,** *98*(1), 870–876.

Jiang, Z. Q.; Dong, X. Y.; Liu, H.; Wang, Y. J.; Zhang, L.; Sun, Y. Multifunctionality of Self-Assembled Nanogels of Curcumin-Hyaluronic Acid Conjugates on Inhibiting Amyloid Beta-Protein Fibrillation and Cytotoxicity. *React. Funct. Polym.* **2016,** *104,* 22–29.

Joshi, S. A.; Chavhan, S. S.; Sawant, K. K. Rivastigmine-loaded PLGA and PBCA Nanoparticles: Preparation, Optimization, Characterization, in Vitro and Pharmacodynamic Studies. *Eur. J. Pharm. Biopharm.* **2010a,** *76*(2), 189–199.

Joshi, G. V.; Kevadiya, B. D.; Bajaj, H. C. Design and Evaluation of Controlled Drug Delivery System of Buspirone Using Inorganic Layered Clay Mineral. *Microporous Mesoporous Mater* **2010b,** *132*(3), 526–530.

Kamal, N.; Cutie, A. J.; Habib, M. J.; Zidan, A.S. QbD Approach to Investigate Product and Process Variabilities for Brain Targeting Liposomes. *J. Liposome Res.* **2015,** *25*(3), 175–190.

Kanazirska, M. V.; Fuchs, P. M.; Chen, L. P.; Lal, S.; Verma, J.; Vassilev, P. M. Beneficial Effects of Lysosome-Modulating and Other Pharmacological and Nanocarrier Agents on Amyloid-Beta-Treated Cells. *Curr. Pharm. Biotechnol.* **2012,** *13*(15), 2761–2767.

Katare, Y. K.; Daya, R. P.; Gray, C. S.; Luckham, R. E.; Bhandari, J.; Chauhan, A. S.; Mishre, R. K. Brain Targeting of a Water Insoluble Antipsychotic Drug Haloperidol Via the Intranasal Route Using PAMAM Dendrimer. *Mol. Pharm.* **2015,** *12*(9), 3380–3388.

Khan, S.; Patil, K.; Yeole, P.; Gaikwad, R. Brain Targeting Studies on Buspirone Hydrochloride After Intranasal Administration of Mucoadhesive Formulation in Rats. *J. Pharm. Pharmacol.* **2009,** *61*(5), 669–675.

Khanbabaie, R.; Jahanshahi, M. Revolutionary Impact of Nanodrug Delivery on Neuroscience. *Curr. Neuropharmacol.* **2012,** *10*(4), 370–392.

Korzhikov, V.; Averianov, I.; Litvinchuk, E.; Tennikova, T. B. Polyester-Based Microparticles of Different Hydrophobicity: the Patterns of Lipophilic Drug Entrapment and Release. *J. Microencapsul.* **2016,** *33*(3), 199–208.

Kumar, S.; Randhawa, J. K. Preparation and Characterization of Paliperidone Loaded Solid Lipid Nanoparticles. *Colloids Surf., B Biointerfaces* **2013,** *102,* 562–568.

Kumar, R.; Siril, P. F. Ultrafine Carbamazepine Nanoparticles with Enhanced Water Solubility and Rate of Dissolution. *RSC Adv.* **2014**, *4*(89), 48101–48108.

Kumar, S.; Randhawa, J. K. Solid Lipid Nanoparticles of Stearic Acid for the Drug Delivery of Paliperidone. *RSC Adv.* **2015**, *5*(84), 68743–68750.

Kunasekaran, V.; Krishnamoorthy, K. Experimental Design for the Optimization of Nanoscale Solid Lipid Particles Containing Rasagiline Mesylate. *J. Young Pharm.* **2015**, *7*(4), 285–295.

Kura, A. U.; Al Ali, S. H. H.; Hussein, M. Z.; Fakurazi, S.; Arulselvan, P. Development of a Controlled-Release Anti-Parkinsonian Nanodelivery System Using Levodopa as the Active Agent. *Int. J. Nanomed.* **2013**, *8*, 1103–1110.

Kura, A. U.; Ain, N. M.; Hussein, M. Z.; Fakurazi, S.; Hussein-Al-Ali, S. H. Toxicity and Metabolism of Layered Double Hydroxide Intercalated with Levodopa in a Parkinson's Disease Model. *Int. J. Mol. Sci.* **2014a**, *15*(4), 5916–5927.

Kura, A. U.; Hussein-Al-Ali, S. H.; Hussein, M. Z.; Fakurazi, S. Preparation of Tween 80-Zn/Al-Levodopa-Layered Double Hydroxides Nanocomposite for Drug Delivery System. *Sci. World J.* **2014b**, 104–246.

Lalani, J.; Patil, S.; Kolate, A.; Lalani, R.; Misra, A. Protein-Functionalized Plga Nanoparticles of Lamotrigine for Neuropathic Pain Management. *AAPS PharmSciTech* **2015**, *16*(2), 413–427.

Laserra, S.; Basit, A.; Sozio, P.; Marinelli, L.; Fornasari, E.; Cacciatore, I.; Ciulla, M.; Turkez, H.; Geyikoglu, F.; Di Stefano, A. Solid Lipid Nanoparticles Loaded with Lipoyl-Memantine Codrug: Preparation and Characterization. *Int. J. Pharm.* **2015**, *485*(12), 183–191.

Lazar, A. N.; Mourtas, S.; Youssef, I.; Parizot, C.; Dauphin, A.; Delatour, B.; Antimisiaris, S. G.; Duyckaerts, C. Curcumin-Conjugated Nanoliposomes with High Affinity for a Beta Deposits: Possible Applications to Alzheimer Disease. *Nanomed Nanotechnol. Biol. Med.* **2013**, *9*(5), 712–721.

Leng, D. L.; Chen, H. M.; Li, G. J.; Guo, M. R.; Zhu, Z. L.; Xu, L.; Wang, Y. J. Development and Comparison of Intramuscularly Long-Acting Paliperidone Palmitate Nanosuspensions with Different Particle Size. *Int. J. Pharm.* **2014**, *472*(12), 380–385.

Li, X.; Kondoh, M.; Watari, A.; Hasezaki, T.; Isoda, K.; Tsutsumi, Y.; Yagi, K. Effect of 70-Nm Silica Particles on the Toxicity of Acetaminophen, Tetracycline, Trazodone, and 5-Aminosalicylic Acid in Mice. *Pharmazie* **2011**, *66*(4), 282–286.

Li, W. Z.; Zhou, Y. Q.; Zhao, N.; Hao, B. H.; Wang, X. N.; Kong, P. Pharmacokinetic Behavior and Efficiency of Acetylcholinesterase Inhibition in Rat Brain After Intranasal Administration of Galanthamine Hydrobromide Loaded Flexible Liposomes. *Environ. Toxicol. Pharmacol.* **2012**, *34*(2), 272–279.

Li, J. C.; Zhang, W. J.; Zhu, J. X.; Zhu, N.; Zhang, H. M.; Wang, X.; Zhang, J.; Wang, Q. Q. Preparation and Brain Delivery of Nasal Solid Lipid Nanoparticles of Quetiapine Fumarate in Situ Gel in Rat Model of Schizophrenia. *Int. J. Clin. Exp. Med.* **2015**, *8*(10), 17590–17600.

Liao, A Citicoline Sodium Liposome Solid Preparation Useful for Protecting Brain and Neurons, and Preventing and/or Treating Diseases, Comprises Citicoline Sodium, Phospholipid, Additive and Auxiliary Materials. CN Patent 102,078,299 A, January 4, 2011.

Liu, W.; Ma, S.; Zhang, X.; Hua, J.; Yan, L.; Zhou, F. Rotigotine Flexible Liposome as Pharmaceuticals, Contains Solid Components Including Rotigotine, Phospholipids and Surface Activating Agent, and Solvents Including Ether, Chloroform, Ethanol, Aqueous Ethanol Solution or Water. CN Patent 101,317,819 A, June 5, 2007.

Liu, K. S.; Sung, K. C.; Al-Suwayeh, S. A.; Ku, M. C.; Chu, C. C.; Wang, J. J.; Fang, J. Y. Enhancement of Transdermal Apomorphine Delivery with a Diester Prodrug Strategy. *Eur. J. Pharm. Biopharm.* **2011,** *78*(3), 422–431.

Liu, K. S.; Wen, C. J.; Yen, T. C.; Sung, K. C.; Ku, M. C.; Wang, J. J.; Fang, J. Y. Combined Strategies of Apomorphine Diester Prodrugs and Nanostructured Lipid Carriers for Efficient Brain Targeting. Nanotechnology. **2012,** *23*(9), 095–103.

Liu, Z. Y.; Gao, X. L.; Kang, T.; Jiang, M. Y.; Miao, D. Y.; Gu, G. Z.; Hu, Q. Y.; Song, Q. X.; Yao, L.; Tu, Y. F.; Chen, H. Z.; Jiang, X. G.; Chen, J. B6 Peptide-Modified PEG-PLA Nanoparticles for Enhanced Brain Delivery of Neuroprotective Peptide. *Bioconjugate Chem.* **2013a,** *24*(6), 997–1007.

Liu, Z. Y.; Jiang, M. Y.; Kang, T.; Miao, D. Y.; Gu, G. Z.; Song, Q. X.; Yao, L.; Hu, Q. Y.; Tu, Y. F.; Pang, Z. Q.; Chen, H.; Jiang, X.; Gao, X.; Chen, J. Lactoferrin-Modified PEG-co-PCL Nanoparticles for Enhanced Brain Delivery of NAP Peptide Following Intranasal Administration. Biomaterials **2013b,** *34*(15), 3870–3881.

Liu, J. S.; Wang, J. H.; Zhou, J.; Tang, X. H.; Xu, L.; Shen, T.; Wu, X. Y.; Hong, Z. Enhanced Brain Delivery of Lamotrigine with Pluronic (R) Pi23-Based Nanocarrier. *Int. J. Nanomed.* **2014,** *9*, 3923–3935.

Lopalco, A.; Ali, H.; Denora, N.; Rytting, E. Oxcarbazepine-Loaded Polymeric Nanoparticles: Development and Permeability Studies Across in Vitro Models of the Blood-Brain Barrier and Human Placental Trophoblast. *Int. J. Nanomed.* **2015,** *10*, 1985–1996.

Lopez, T.; Basaldella, E. I.; Ojeda, M. L.; Manjarrez, J.; Alexander-Katz, R. Encapsulation of Valproic Acid and Sodic Phenytoin in Ordered Mesoporous Sio2 Solids for the Treatment of Temporal Lobe Epilepsy. *Opt. Mater.* **2006,** *29*(1), 75–81.

Lopez, T.; Bata-Garcia, J. L.; Esquivel, D.; Ortiz-Islas, E.; Gonzalez, R.; Ascencio, J.; Quintana, P.; Oskam, G.; Alvarez-Cervera, F. J.; Heredia-Lopez, F. J.; Gongora-Alfaro, J. L. Treatment of Parkinson's Disease: Nanostructured Sol-Gel Silica-Dopamine Reservoirs for Controlled Drug Release in the Central Nervous System. *Int. J. Nanomed.* **2010,** *6*, 19–31.

Lopez, T.; Ortiz, E.; Meza, D.; Basaldella, E.; Bokhimi, X.; Magana, C.; Sepulveda, A.; Rodriguez, F.; Ruiz, J. Controlled Release of Phenytoin for Epilepsy Treatment from Titania and Silica Based Materials. *Mater. Chem. Phys.* **2011,** *126*(3), 922–929.

Loureiro, J. A.; Gomes, B.; Fricker, G.; Coelho, M. A. N.; Rocha, S.; Pereira, M. C. Cellular Uptake of Plga Nanoparticles Targeted with Anti-Amyloid and Anti-Transferrin Receptor Antibodies for Alzheimer's Disease Treatment. *Colloids Surf., B Biointerfaces* **2016,** *145*, 8–13.

Lu, C. T.; Zhao, Y. Z.; Wong, H. L.; Cai, J.; Peng, L.; Tian, X. Q. Current Approaches to Enhance CNS Delivery of Drugs Across the Brain Barriers. *Int. J. Nanomed.* **2014,** *9*, 2241–2257.

Lu, S.; Yu, P. P.; He, J. H.; Zhang, S. S.; Xia, Y. L.; Zhang, W. L.; Liu, J. P. Enhanced Dissolution and Oral Bioavailability of Lurasidone Hydrochloride Nanosuspensions Prepared by Antisolvent Precipitation-Ultrasonication Method. *RSC Adv.* **2016,** *6*(54), 49052–49059.

Lukasiewicz, S.; Szczepanowicz, K.; Podgorna, K.; Blasiak, E.; Majeed, N.; Ogren, S. O. O.; Nowak, W.; Warszynki, P.; Dziedzicka-Wasylewska, M. Encapsulation of Clozapine in Polymeric Nanocapsules and Its Biological Effects. *Colloids Surf., B Biointerfaces* **2016,** *140*, 342–352.

Luppi, B.; Bigucci, F.; Corace, G.; Delucca, A.; Cerchiara, T.; Sorrenti, M.; Catenacci, L.; Di Pietra, A. M.; Zecchi, V. Albumin Nanoparticles Carrying Cyclodextrins for Nasal Delivery of the Anti-Alzheimer Drug Tacrine. *Eur. J. Pharm. Sci.* **2011,** *44*(4), 559–565.

Madhav, S. N. V. Carbidopa Bio-Nanoparticles for Brain Targeting Via Ear. IN Patent 201,203,451 A1, November 7, 2012.

Mancini, S.; Minniti, S.; Gregori, M.; Sancini, G.; Cagnotto, A.; Couraud, P. O.; Ordonez-Gutierrez, L.; Wandosell, F.; Salmona, M.; Re, F. The Hunt for Brain a Beta Oligomers by Peripherally Circulating Multi-Functional Nanoparticles: Potential Therapeutic Approach for Alzheimer Disease. *Nanomed. Nanotechnol. Biol. Med.* **2016,** *12*(1), 43–52.

Mandpe, L.; Pokharkar, V. Quality by Design Approach to Understand the Process of Optimization of Iloperidone Nanostructured Lipid Carriers for Oral Bioavailability Enhancement. *Pharm. Dev. Technol.* **2015,** *20*(3), 320–329.

Mani, G.; Pushparaj, H.; Peng, M. M.; Muthiahpillai, P.; Udhumansha, U.; Jang, H. T. Synthesis and Characterization of Pharmaceutical Surfactant Templated Mesoporous Silica: Its Application to Controlled Delivery of Duloxetine. *Mater. Res. Bull.* **2014,** *51*, 228–235.

Marano, S.; Barker, S. A.; Raimi-Abraham, B. T.; Missaghi, S.; Rajabi-Siahboomi, A.; Craig, D. Q. M. Development of Micro-Fibrous Solid Dispersions of Poorly Water-Soluble Drugs in Sucrose Using Temperature-Controlled Centrifugal Spinning. *Eur. J. Pharm. Biopharm.* **2016,** *103*, 84–94.

Masoumi, H. R. F.; Basri, M.; Samiun, W. S.; Izadiyan, Z.; Lim, C. H. J. Enhancement of Encapsulation Efficiency of Nanoemulsion-Containing Aripiprazole for the Treatment of Schizophrenia Using Mixture Experimental Design. *Int. J. Namomed.* **2015,** *10*, 6469–6476.

Masserini, M. Nanoparticles for Brain Drug Delivery. *ISRN Biochem.* **2013,** *2013*, 238–428.

Mathew, A.; Fukuda, T.; Nagaoka, Y.; Hasumura, T.; Morimoto, H.; Yoshida, Y.; Maekawa, T.; Venugopal, K.; Kumar, D. S. Curcumin Loaded-Plga Nanoparticles Conjugated with Tet-1 Peptide for Potential Use in Alzheimer's Disease. *PLoS One* **2012,** *7*(3), e32616.

McGovern Institute for Brain Research at MIT: Brain Disorders: By the Numbers. https://mcgovern.mit.edu/brain-disorders/by-the-numbers (accessed Jan 23, 2014).

Md, S.; Kumar, M.; Baboota, S.; Sahni, J. K.; Ali, J. Preparation, Characterization and Evaluation of Bromocriptine Loaded Chitosan Nanoparticles for Intranasal Delivery. *Sci. Adv. Mater.* **2012,** *4*(9), 949–960.

Md, S.; Khan, R. A.; Mustafa, G.; Chuttani, K.; Baboota, S.; Sahni, J. K.; Ali, J. Bromocriptine Loaded Chitosan Nanoparticles Intended for Direct Nose to Brain Delivery: Pharmacodynamic, Pharmacokinetic and Scintigraphy Study in Mice Model. *Eur. J. Pharm. Sci.* **2013,** *48*(3), 393–405.

Md, S.; Haque, S.; Fazil, M.; Kumar, M.; Baboota, S.; Sahni, J. K.; Ali, J. Optimised Nanoformulation of Bromocriptine for Direct Nose-To-Brain Delivery: Biodistribution, Pharmacokinetic and Dopamine Estimation by Ultra-Hplc/Mass Spectrometry Method. *Exp. Opin. Drug Delivery* **2014,** *11*(6), 827–842.

Meng, F. F.; Asghar, S.; Gao, S. Y.; Su, Z. G.; Song, J.; Huo, M. R.; Meng, W. D.; Ping, Q. N.; Xiao, Y. Y. A Novel LDL-Mimic Nanocarrier for the Targeted Delivery of Curcumin Into the Brain to Treat Alzheimer's Disease. *Colloids Surf. B Biointerfaces* **2015,** *134*, 88–97.

Miao, Y. F.; Sun, J. Q.; Chen, G. G.; Lili, R.; Ouyang, P. Enhanced Oral Bioavailability of Lurasidone by Self-Nanoemulsifying Drug Delivery System in Fasted State. *Drug Dev. Ind. Pharm.* **2016,** *42*(8), 1234–1240.

Milani, M. D. M. Citicoline as Coadiuvant Treatment of Cognitive Impairment in Chronic Degenerative Central Nervous System Diseases and in Ischemic Stroke: a Review of Available Data. *Online J. Med. Med. Sci. Res.* **2013,** *2*(2), 13–18.

Millier, A.; Schmidt, U.; Angermeyer, M. C.; Chauhan, D.; Murthy, V.; Toumi, M.; Cadi-Soussi, N. Humanistic Burden in Schizophrenia: A Literature Review. 2014. *J. Psychiatr. Res. 54*, 85–93.

Mishra, B.; Arya, N.; Tiwari, S. Investigation of Formulation Variables Affecting the Properties of Lamotrigine Nanosuspension Using Fractional Factorial Design. *Daru J. Fac. Pharm.* **2010**, *18*(1), 1–8.

Mishra, A. D.; Patel, C. N.; Shah, D. R. Formulation and Optimization of Ethosomes for Transdermal Delivery of Ropinirole Hydrochloride. *Curr. Drug Delivery* **2013**, *10*(5), 500–516.

Misra, A. R.; Gandhi, N. I.; Bajaj, M. R.; Shah, B. B.; Samant, R. S.; Jamil, A. S. P. Liposomal Citicoline Injection. WO Patent 2,010,092,597 A2, Feb 10, 2010.

Misra, S.; Chopra, K.; Sinha, V. R.; Medhi, B. Galantamine-Loaded Solid-Lipid Nanoparticles for Enhanced Brain Delivery: Preparation, Characterization, in Vitro and in Vivo Evaluations. *Drug Delivery* **2016**, *23*(4), 1434–1443.

Mittal, D.; Md, S.; Hasan, Q.; Fazil, M.; Ali, A.; Baboota, S.; Ali, J. Brain Targeted Nanoparticulate Drug Delivery System of Rasagiline Via Intranasal Route. *Drug Delivery* **2016**, *23*(1), 130–139.

Mittapelly, N.; Rachumallu, R.; Pandey, G.; Sharma, S.; Arya, A.; Bhatta, R. S.; Mishra, P. R. Investigation of Salt Formation Between Memantine and Pamoic Acid: Its Exploitation in Nanocrystalline Form as Long Acting Injection. *Eur. J. Pharm. Biopharm.* **2016**, *101*, 62–71.

Momo, F.; Fabris, S.; Stevanato, R. Interaction of Fluoxetine with Phosphatidylcholine Liposomes. *Biophys. Chem.* **2005**, *118*(1), 15–21.

Mufamadi, M. S.; Choonara, Y. E.; Kumar, P.; Modi, G.; Naidoo, D.; van Vuuren, S.; Ndesendo, V. M. K.; du Toit, L. C.; Iyuke, S. E.; Pillay, V. Ligand-Functionalized Nanoliposomes for Targeted Delivery of Galantamine. *Int. J. Pharm.* **2013**, *448*(1), 267–281.

Muntimadugu, E.; Dhommati, R.; Jain, A.; Challa, V. G. S.; Shaheen, M.; Khan, W. Intranasal Delivery of Nanoparticle Encapsulated Tarenflurbil: a Potential Brain Targeting Strategy for Alzheimer's Disease. *Eur. J. Pharm. Sci.* **2016**, *92*, 224–234.

Muronetz, V.; Asryants, R.; Semenyuk, P.; Schmalhausen, E.; Saso, L. Hydrophobic Plant Antioxidants. Preparation of Nanoparticles and Their Application for Prevention of Neurodegenerative Diseases. Review and Experimental Data. *Curr. Top. Med. Chem.* **2014**, *14*(22), 2520–2528.

Mustafa, G.; Baboota, S.; Ali, J.; Kumar, N.; Singh, T.; Bhatnagar, A.; Ahuja, A. Effect of Homogenization on the Fate of True Nanoemulsion in Brain Translocation: A Gamma Scintigraphic Evaluation. *Sci. Adv. Mater.* **2012a**, *4*(7), 739–748.

Mustafa, G.; Baboota, S.; Ahuja, A.; Ali, J. Formulation Development of Chitosan Coated Intra Nasal Ropinirole Nanoemulsion for Better Management Option of Parkinson: An in Vitro ex Vivo Evaluation. *Curr. Nanosci.* **2012b**, *8*(3), 348–360.

Mustafa, G.; Ahuja, A.; Al Rohaimi, A. H.; Muslim, S.; Hassan, A. A.; Baboota, S.; Ali, J. Nano-Ropinirole for the Management of Parkinsonism: Blood-Brain Pharmacokinetics and Carrier Localization. *Expert Rev. Neurother.* **2015**, *15*(6), 695–710.

Muthu, M. S.; Singh, S. Poly (D, L-Lactide) Nanosuspensions of Risperidone for Parenteral Delivery: Formulation and In-Vitro Evaluation. *Curr. Drug Delivery* **2009**, *6*(1), 62–68.

Muthu, M. S.; Rawat, M. K.; Mishra, A.; Singh, S. Plga Nanoparticle Formulations of Risperidone: Preparation and Neuropharmacological Evaluation. Nanomed. *Nanotechnol.* **2009**, *5*(3), 323–333.

Muthu, M. S.; Sahu, A. K.; Sonali; Abdulla, A.; Kaklotar, D.; Rajesh, C. V.; Singh, S.; Pandey, B. L. Solubilized Delivery of Paliperidone Palmitate by D-Alpha-Tocopheryl Polyethylene Glycol 1000 Succinate Micelles for Improved Short-Term Psychotic Management. *Drug Delivery* **2016**, *23*(1), 230–237.

Nagy, Z. K.; Nyul, K.; Wagner, I.; Molnar, K.; Marosi, G. Electrospun Water Soluble Polymer Mat for Ultrafast Release of Donepezil HCL. Express *Polym. Lett.* **2010**, *4*(12), 763–772.

Nair, R.; Kumar, A. C. K.; Priya, V. K.; Yadav, C. M.; Raju, P. Y. Formulation and Evaluation of Chitosan Solid Lipid Nanoparticles of Carbamazepine. *Lipids Health Dis.* **2012**, *11*, 72.

Namazi, H.; Belali, S. Starch-G-Lactic Acid/Montmorillonite Nanocomposite: Synthesis, Characterization and Controlled Drug Release Study. *Starch-Starke.* **2016**, *68*(34), 177–187.

Nandra, K. S.; Agius, M. The Differences Between Typical and Atypical Antipsychotics: the Effects on Neurogenesis. *Psychiat. Danub* **2012**, *24*, 95–99.

Narala, A.; Veerabrahma, K. Preparation, Characterization and Evaluation of Quetiapine Fumarate Solid Lipid Nanoparticles to Improve the Oral Bioavailability. *J. Pharm.* **2013**, *2013*, Article ID: 265741, 7 pages http://dx.doi.org/10.1155/2013/265741.

Narayan, R.; Singh, M.; Ranjan, O. P.; Nayak, Y.; Garg, S.; Shavi, G. V.; Nayak, U. Y. Development of Risperidone Liposomes for Brain Targeting Through Intranasal Route. *Life Sci.* **2016**, *163*, 38–45.

Ngwuluka, N. C.; Choonara, Y. E.; Kumar, P.; du Toit, L. C.; Modi, G.; Pillay, V. An Optimized Gastroretentive Nanosystem for the Delivery of Levodopa. *Int. J. Pharm.* **2015**, *494*(1), 49–65.

Oh, Y. J.; Choi, G.; Choy, Y. B.; Park, J. W.; Park, J. H.; Lee, H. J.; Yoon, Y. J.; Chang, H. C.; Choy, J. H. Aripiprazole-Montmorillonite: a New Organic-Inorganic Nanohybrid Material for Biomedical Applications. *Chem. Eur. J.* **2013**, *19*(15), 4869–4875.

Ong, J. C. L.; Sun, F.; Chan, E. Development of Stealth Liposome Coencapsulating Doxorubicin and Fluoxetine. *J. Liposome Res.* **2011**, *21*(4), 261–271.

Pagar, K. P.; Sardar, S. M.; Vavia, P. R. Novel L-Lactide-Depsipeptide Polymeric Carrier for Enhanced Brain Uptake of Rivastigmine in Treatment of Alzheimer's Disease. *J. Biomed. Nanotechnol.* **2014**, *10*(3), 415–426.

Pahuja, R.; Seth, K.; Shukla, A.; Shukla, R. K.; Bhatnagar, P.; Chauhan, L. K. S.; Saxena, P. N.; Arun, J.; Chaudhari, B. P.; Patel, D. K.; Singh, S. P.; Shukla, R.; Khanna, V. K.; Kumar, P.; Chaturvedi, R. K.; Gupta, K. C. Trans-Blood Brain Barrier Delivery of Dopamine-Loaded Nanoparticles Reverses Functional Deficits in Parkinsonian Rats. *ACS Nano* **2015**, *9*(5), 4850–4871.

Paisana, M. C.; Muellers, K. C.; Wahl, M. A.; Pinto, J. F. Production and Stabilization of Olanzapine Nanoparticles by Rapid Expansion of Supercritical Solutions (RESS). *J. Supercrit. Fluids* **2016**, *109*, 124–133.

Panda, A.; Meena, J.; Katara, .R.; Majumdar, D. K. Formulation and Characterization of Clozapine and Risperidone Co-Entrapped Spray-Dried Plga Nanoparticles. *Pharm. Dev. Technol.* **2016**, *21*(1), 43–53.

Pandey, Y. R.; Kumar, S.; Gupta, B. K.; Ali, J.; Baboota, S. Intranasal Delivery of Paroxetine Nanoemulsion via the Olfactory Region for the Management of Depression: Formulation, Behavioural and Biochemical Estimation. *Nanotechnology* **2016**, *27*(2), 025102.

Papadimitriou, S.; Bikiaris, D.; Avgoustakis, K.; Karavas, E.; Georgarakis, M. Chitosan Nanoparticles Loaded with Dorzolamide and Pramipexole. *Carbohydr. Polym.* **2008**, *73*(1), 44–54.

Parikh, T.; Bommana, M. M.; Squillante, E. Efficacy of Surface Charge in Targeting Pegylated Nanoparticles of Sulpiride to the Brain. *Eur. J. Pharm. Biopharm.* **2010**, *74*(3), 442–450.

Park, E. J.; Amatya, S.; Kim, M. S.; Park, J. H.; Seol, E.; Lee, H.; Shin, Y. H.; Na, D. H. Long-Acting Injectable Formulations of Antipsychotic Drugs for the Treatment of Schizophrenia. *Arch. Pharm. Res.* **2013,** *36*(6), 651–659.

Passeleu-Le Bourdonnec, C.; Carrupt, P. A.; Scherrmann, J. M.; Martel, S. Methodologies to Assess Drug Permeation Through the Blood–Brain Barrier for Pharmaceutical Research. *Pharm. Res.* **2013,** *30*(1), 2729–2756.

Patel, K.; Padhye, S.; Nagarsenker, M. Duloxetine HCL Lipid Nanoparticles: Preparation, Characterization, and Dosage Form Design. *AAPS PharmSciTech.* **2012,** *13*(1), 125–133.

Patel, R. B.; Patel, M. R.; Bhatt, K. K.; Patel, B. G. Paliperidone-Loaded Mucoadhesive Microemulsion in Treatment of Schizophrenia: Formulation Consideration. *J. Pharm. Innov.* **2013,** *8*(3), 195–204.

Patel, P.; Pol, A.; More, S.; Kalaria, D. R.; Kalia, Y. N.; Patravale, V. B. Colloidal Soft Nanocarrier for Transdermal Delivery of Dopamine Agonist: Ex Vivo and in Vivo Evaluation. *J. Biomed. Nanotechnol.* **2014,** *10*(11), 3291–3303.

Patel, M. R.; Patel, R. B.; Bhatt, K. K.; Patel, B. G.; Gaikwad, R. V. Paliperidone Microemulsion for Nose-To-Brain Targeted Drug Delivery System: Pharmacodynamic and Pharmacokinetic Evaluation. *Drug Delivery* **2016,** *23*(1), 346–354.

Piazza, J.; Hoare, T.; Molinaro, L.; Terpstra, K.; Bhandari, J.; Selvaganapathy, P. R.; Gupta, B.; Mishra, R. K. Haloperidol-Loaded Intranasally Administered Lectin Functionalized Poly (Ethylene Glycol)-Block-Poly(D, L)-Lactic-Co-Glycolic Acid (Peg-Plga) Nanoparticles for the Treatment of Schizophrenia. *Eur. J. Pharm. Biopharm.* **2014,** *87*(1), 30–39.

Plissonneau, M.; Pansieri, J.; Heinrich-Nalard, L.; Morfin, J. F.; Stransky-Heilkron, N.; Rivory, P.; Mowat, P.; Dumoulin, M.; Cohen, R.; Allemann, E.; Tóth, E.; Saraiva, M. J.; Louis, C.; Tillement, O.; Forge, V.; Lux, F.; Marquette, C. Gd-Nanoparticles Functionalization with Specific Peptides for β-Amyloid Plaques Targeting. *J. Nanobiotechnol.* **2016,** *14*(1), 60.

Prasad, R. S.; Yandrapu, S. K.; Manavalan, R. Lipid Solid Dispersions for the Aqueous Solubility and Bioavailability Enhancement of Entacapone. *Asian J. Chem.* **2010,** *22*, 4549–4558.

Pregabalin – DrugBank. https://www.drugbank.ca/drugs/DB00230 (accessed Jan 14, 2017).

Rahman, M. A.; Iqbal, Z.; Hussain, A. Formulation Optimization and in Vitro Characterization of Sertraline Loaded Self-Nanoemulsifying Drug Delivery System (SNEDDS) for Oral Administration. *J. Pharm. Investig.* **2012,** *42*(4), 191–202.

Rajkumar, S.; Kevadiya, B. D.; Bajaj, H.C. Montmorillonite/Poly (L-Lactide) Microcomposite Spheres as Reservoirs of Antidepressant Drugs and Their Controlled Release Property. *Asian J. Pharm. Sci.* **2015,** *10*(5), 452–458.

Rao, M. R. P.; Bhingole, R.C. Nanosponge-Based Pediatric-Controlled Release Dry Suspension of Gabapentin for Reconstitution. *Drug Dev. Ind. Pharm.* **2015,** *41*(12), 2029–2036.

Ravani, L.; Sarpietro, M. G.; Esposito, E.; Di Stefano, A.; Sozio, P.; Calcagno, M.; Drechsler, M.; Contado, C.; Longo, F.; Giuffrida, M. C.; Castelli, F.; Morari, M.; Cortesi, R. Lipid Nanocarriers Containing a Levodopa Prodrug with Potential Antiparkinsonian Activity. *Mater. Sci. Eng. C Mater. Biol. Appl.* **2015,** *48*, 294–300.

Reddy, J. S.; Venkateswarlu, V.; Koning, G. A. Radioprotective Effect of Transferrin Targeted Citicoline Liposomes. *J. Drug Target.* **2006,** *14*(1), 13–19.

Rosillo-de la Torre, A.; Zurita-Olvera, L.; Orozco-Suarez, S.; Casillas, P. E. G.; Salgado-Ceballos, H.; Luna-Barcenas, G.; Rocha, L. Phenytoin Carried by Silica Core Iron Oxide Nanoparticles Reduces the Expression of Pharmacoresistant Seizures in Rats. *Nanomedicine* **2015,** *10*(24), 3563–3577.

Roversi, K.; Benvegnu, D. M.; Roversi, K.; Trevizol, F.; Vey, L. T.; Elias, F.; Fracasso, R.; Motta, M. H.; Ribeiro, R. F.; Hausen, B. D. S.; Moresco, R. N.; Garcia, S. C.; da Silva, C. B.; Burger, M. E. Haloperidol-Loaded Lipid-Core Polymeric Nanocapsules Reduce DNA Damage in Blood and Oxidative Stress in Liver and Kidneys of Rats. *J. Nanopart. Res.* **2015**, *17*(4), 199.

Ruozi, B.; Belletti, D.; Pederzoli, F.; Veratti, P.; Forni, F.; Vandelli, M. A.; Tosi, G. Nanotechnology and Alzheimer's Disease: What Has Been Done and What to Do. *Curr. Med. Chem.* **2014**, *21*(36), 4169–4185.

Salama, H. A.; Mahmoud, A. A.; Kamel, A. O.; Hady, M. A.; Awad, G. A. S. Brain Delivery of Olanzapine by Intranasal Administration of Transfersomal Vesicles. *J. Liposome Res.* **2012**, *22*(4), 336–345.

Samia, O.; Hanan, R.; Kamal, E. Carbamazepine Mucoadhesive Nanoemulgel (MNEG) as Brain Targeting Delivery System Via the Olfactory Mucosa. *Drug Delivery* **2012**, *19*(1), 58–67.

Samiun, W. S.; Basri, M.; Masoumi, H. R. F.; Khairudin, N. The Prediction of the Optimum Compositions of a Parenteral Nanoemulsion System Loaded with a Low Water Solubility Drug for the Treatment of Schizophrenia by Artificial Neural Networks. *RSC Adv.* **2016**, *6*(17), 14068–14076.

Sancini, G.; Dal Magro, R.; Ornaghi, F.; Balducci, C.; Forloni, G.; Gobbi, M.; Salmona, M.; Re, F. Pulmonary Administration of Functionalized Nanoparticles Significantly Reduces Beta-Amyloid in the Brain of an Alzheimer's Disease Murine Model. *Nano Res.* **2016**, *9*(7), 2190–2201.

Sandig, A. G.; Campmany, A. C. C.; Campos, F. F.; Villena, M. J. M.; Naveros, B. C. Transdermal Delivery of Imipramine and Doxepin from Newly Oil-In-Water Nanoemulsions for an Analgesic and Anti-Allodynic Activity: Development, Characterization and in Vivo Evaluation. *Colloids Surf. B Biointerfaces* **2013**, *103*, 558–565.

Sapre, A.; Parikh, R.; Gohel, M. Delivery of Fluoxetine Hydrochloride: Formulation and Characterization. In *Nanotech Conference and EXPO 2009 – Technical Proceedings*; Laudon, M., Romanowicz, B., Eds.; 2009; Vol. 2, pp 72–75.

Saraiva, C.; Praça, C.; Ferreira, R.; Santos, T.; Ferreira, L.; Bernardino, L. Nanoparticle-Mediated Brain Drug Delivery: Overcoming Blood-Brain Barrier to Treat Neurodegenerative Diseases. *J. Control. Release* **2016**, *235*, 34–47.

Sawant, K.; Pandey, A.; Patel, S. Aripiprazole Loaded Poly(Caprolactone) Nanoparticles: Optimization and in Vivo Pharmacokinetics. *Mater. Sci. Eng. C Mater. Biol. Appl.* **2016**, *66*, 230–243.

Schultz, S. H.; North, S. W.; Shields, C. G. Shizophrenia – a Review. *Am. Fam. Physician* **2007**, *75*(12), 1821–1829.

Scialabba, C.; Rocco, F.; Licciardi, M.; Pitarresi, G.; Ceruti, M.; Giammona, G. Amphiphilic Polyaspartamide Copolymer-Based Micelles for Rivastigmine Delivery to Neuronal Cells. *Drug Delivery* **2012**, *19*(6), 307–316.

Sengel, S. B.; Sahiner, N. Poly(Vinyl Phosphonic Acid) Nanogels with Tailored Properties and Their Use for Biomedical and Environmental Applications. *Eur. Polym. J.* **2016**, *75*, 264–275.

Shah, S.; Pal, A.; Rajyaguru, T.; Murthy, R. S. R.; Devil, S. Lamotrigine-Loaded Polyacrylate Nanoparticles Synthesized Through Emulsion Polymerization. *J. Appl. Polym. Sci.* **2008**, *107*(5), 3221–3229.

Shah, B.; Khunt, D.; Misra, M.; Padh, H. Application of Box-Behnken Design for Optimization and Development of Quetiapine Fumarate Loaded Chitosan Nanoparticles for Brain Delivery Via Intranasal Route. *Int. J. Biol. Macromol.* **2016a**, *89*, 206–218.

Shah, B.; Khunt, D.; Misra, M.; Padh, H. Non-Invasive Intranasal Delivery of Quetiapine Fumarate Loaded Microemulsion for Brain Targeting: Formulation, Physicochemical and Pharmacokinetic Consideration. *Eur. J. Pharm. Sci.* **2016b**, *91*, 196–207.

Shah, B.; Khunt, D.; Bhatt, H.; Misra, M.; Padh, H. Intranasal Delivery of Venlafaxine Loaded Nanostructured Lipid Carrier: Risk Assessment and QbD Based Optimization. *J. Drug Delivery Sci. Technol.* **2016c**, *33*, 37–50.

Shahiwala, A.; Dash, D. Preparation and Evaluation of Microemulsion Based Formulations for Rapid-Onset Intranasal Delivery of Zonisamide. *Adv. Sci Lett.* **2010**, *3*(4), 442–446.

Shamma, R. N.; Elsayed, I. Transfersomal Lyophilized Gel of Buspirone Hcl: Formulation, Evaluation and Statistical Optimization. *J. Liposome Res.* **2013**, *23*(3), 244–254.

Shan, L.; Tao, E. X.; Meng, Q. H.; Hou, W. X.; Liu, K.; Shang, H. C.; Tang, J. B.; Zhang, W. F. Formulation, Optimization, and Pharmacodynamic Evaluation of Chitosan/Phospholipid/beta-Cyclodextrin Microspheres. *Drug Des. Dev. Ther.* **2016**, *10*, 417–429.

Sharma, S.; Dixit, V. Epilepsy – a Comprehensive Review. *Int. J. Pharma Res. Rev.* **2013**, *2*(12), 61–80.

Sharma, S.; Lohan, S.; Murthy, R.S. Formulation and Characterization of Intranasal Mucoadhesive Nanoparticulates and Thermo-Reversible Gel of Levodopa for Brain Delivery. *Drug Dev. Ind. Pharm.* **2014**, *40*(7), 869–878.

Shreya, A. B.; Managuli, R. S.; Menon, J.; Kondapalli, L.; Hegde, A. R.; Avadhani, K.; Shetty, P. K.; Amirthalingam, M.; Kalthur, G.; Mutalik, S. Nano-Transfersomal Formulations for Transdermal Delivery of Asenapine Maleate: In Vitro and in Vivo Performance Evaluations. *J. Liposome Res.* **2016**, *26*(3), 221–232.

Shyong, Y. J.; Wang, M. H.; Tseng, H. C.; Cheng, C.; Chang, K. C.; Lin, F. H. Mesoporous Hydroxyapatite as Olanzapine Carrier Provides a Long-Acting Effect in Antidepression Treatment. *J. Med. Chem.* **2015**, *58*(21), 8463–8474.

Siddique, Y. H.; Khan, W.; Fatima, A.; Jyoti, S.; Khanam, S.; Naz, F.; Rahul; Ali, F.; Singh, B. R.; Naqvi, A. H. Effect of Bromocriptine Alginate Nanocomposite (Banc) on a Transgenic Drosophila Model of Parkinson's Disease. *Dis. Model Mech.* **2016**, *9*(1), 63–68.

Silva, A. C.; Lopes, C. M.; Fonseca, J.; Soares, M. E.; Santos, D.; Souto, E. B.; Ferreira, D. Risperidone Release from Solid Lipid Nanoparticles (SLN): Validated HPLC Method and Modelling Kinetic Profile. *Curr. Pharm. Anal.* **2012a**, *8*(4), 307–316.

Silva, A. C.; Kumar, A.; Wild, W.; Ferreira, D.; Santos, D.; Forbes, B. Long-Term Stability, Biocompatibility and Oral Delivery Potential of Risperidone-Loaded Solid Lipid Nanoparticles. *Int. J. Pharm.* **2012b**, *436*(12), 798–805.

Silva, A. C.; Amaral, M. H.; Gonzalez-Mira, E.; Santos, D.; Ferreira, D. Solid Lipid Nanoparticles (SLN)–Based Hydrogels as Potential Carriers for Oral Transmucosal Delivery of Risperidone: Preparation and Characterization Studies. *Colloids Surf. B Biointerfaces* **2012c**, *93*, 241–248.

Silva, A. C.; Gonzalez-Mira, E.; Lobo, J. M. S.; Amaral, M. H. Current Progresses on Nanodelivery Systems for the Treatment of Neuropsychiatric Diseases: Alzheimer's and Schizophrenia. *Curr. Pharm. Design.* **2013**, *19*(41), 7185–7195.

Simko, M.; Mattson, M. O. Interactions Between Nanonized Materials and the Brain. *Curr. Med. Chem.* **2014**, *21*(37), 4200–4214.

Singh, A. P.; Saraf, S. K.; Saraf, S. A. SLN Approach for Nose-To-Brain Delivery of Alprazolam. *Drug Delivery Transl. Res.* **2012**, *2*(6), 498–507.

Singh, D.; Rashid, M.; Hallan, S. S.; Mehra, N. K.; Prakash, A.; Mishra, N. Pharmacological Evaluation of Nasal Delivery of Selegiline Hydrochloride-Loaded Thiolated Chitosan

Nanoparticles for the Treatment of Depression. *Artif. Cells, Nanomed., Biotechnol.* **2016a,** *44*(3), 865–877.

Singh, S. K.; Dadhania, P.; Vuddanda, P. R.; Jain, A.; Velaga, S.; Singh, S. Intranasal Delivery of Asenapine Loaded Nanostructured Lipid Carriers: Formulation, Characterization, Pharmacokinetic and Behavioural Assessment. *RSC Adv.* **2016b,** *6*(3), 2032–2045.

Somani, S.; Robb, G.; Pickard, B. S.; Dufes, C. Enhanced Gene Expression in the Brain Following Intravenous Administration of Lactoferrin-Bearing Polypropylenimine Dendriplex. *J. Control. Release* **2015,** *217,* 235–242.

Sood, S.; Jawahar, N.; Jain, K.; Gowthamarajan, K.; Meyyanathan, S. N. Olanzapine Loaded Cationic Solid Lipid Nanoparticles for Improved Oral Bioavailability. *Curr. Nanosci.* **2013,** *9*(1), 26–34.

Statista – The Statistics Portal: U.S. top 20 Prescribed Drugs Based on Sales 2015. https://www.statista.com/statistics/258010/top-branded-drugs-based-on-retail-sales-in-the-us/ (accessed Jan 14, 2017).

Steinhilber, D.; Schubert-Zsilavecz, M.; Roth, H. J. *Medizinische Chemie: Targets, Arzneistoffe, Chemische Biologie.* Deutscher Apotheker Verlag: Stutgart, 2010.

Su, Z. X.; Shi, Y. N.; Teng, L. S.; Li, X.; Wang, L. X.; Meng, Q. F.; Teng, L. R.; Li, Y. X. Biodegradable Poly(D, L-Lactide-Co-Glycolide) (PLGA) Microspheres for Sustained Release of Risperidone: Zero-Order Release Formulation. *Pharm. Dev. Technol.* **2011,** *16*(4), 377–384.

Subczynski, W. K.; Wojas, J.; Pezeshk, V.; Pezeshk, A. Partitioning and Localization of Spin-Labeled Amantadine in Lipid Bilayers: an EPR Study. *J. Pharm. Sci.* **1998,** *87*(10), 1249–1254.

Tan, J. P. K.; Goh, C. H.; Tam, K. C. Comparative Drug Release Studies of Two Cationic Drugs from pH-Responsive Nanogels. *Eur. J. Pharm. Sci.* **2007,** *32*(45), 340–348.

Tan, J. M.; Foo, J. B.; Fakurazi, S.; Hussein, M. Z. Release Behaviour and Toxicity Evaluation of Levodopa from Carboxylated Single-Walled Carbon Nanotubes. *Beilstein J. Nanotechnol.* **2015,** *6,* 243–253.

Tan, S. L.; Stanslas, J.; Basri, M.; Karjiban, R. A. A.; Kirby, B. P.; Sani, D.; Bin Basri, H. Nanoemulsion-Based Parenteral Drug Delivery System of Carbamazepine: Preparation, Characterization, Stability Evaluation and Blood-Brain Pharmacokinetics. Curr. *Drug Delivery* **2016,** *12*(6), 795–804.

Tang, J.; Slowing, I.I.; Huang, Y. L.; Trewyn, B. G.; Hu, J.; Liu, H. L.; Lin, V. S. Y. Poly(Lactic Acid)-Coated Mesoporous Silica Nanosphere for Controlled Release of Venlafaxine. *J. Colloid Interface Sci.* **2011,** *360*(2), 488–496.

Tang, J.; Bian, Z. J.; Hu, J.; Xu, S. H.; Liu, H. L. The Effect of a P123 Template in Mesopores of Mesocellular Foam on the Controlled-Release of Venlafaxine. *Int. J. Pharm.* **2012,** *424*(12), 89–97.

Thatipamula, R. P.; Palem, C. R.; Gannu, R.; Mudragada, S.; Yamsani, M. R. Formulation and *in vitro* Characterization of Domperidone Loaded Solid Lipid Nanoparticles and Nanostructured Lipid Carriers. *DARU-J. Pharm. Sci.* **2011,** *19*(1), 23–32.

Thombre, A. G.; Shah, J. C.; Sagawa, K.; Caldwell, W. B. In vitro and in Vivo Characterization of Amorphous, Nanocrystalline, and Crystalline Ziprasidone Formulations. *Int. J. Pharm.* **2012a,** *428*(12), 8–17.

Thombre, A. G.; Caldwell, W. B.; Friesen, D. T.; McCray, S. B.; Sutton, S. C. Solid Nanocrystalline Dispersions of Ziprasidone with Enhanced Bioavailability in the Fasted State. *Mol. Pharm.* **2012b,** *9*(12), 3526–3534.

Thongrangsalit, S.; Phaechamud, T.; Lipipun, V.; Ritthidej, G. C. Bromocriptine Tablet of Self-Microemulsifying System Adsorbed onto Porous Carrier to Stimulate Lipoproteins Secretion for Brain Cellular Uptake. *Colloids Surf. B Biointerfaces* **2015**, *131*, 162–169.

Tiwari, S. K.; Agarwal, S.; Seth, B.; Yadav, A.; Nair, S.; Bhatnagar, P.; Karmakar, M.; Kumari, M.; Chauhan, L. K. S.; Patel, D. K.; Srivastava, V.; Singh, D.; Gupta, S. K.; Tripathu, A.; Chaturvedi, R. K.; Gupta, K. C. Curcumin-Loaded Nanoparticles Potently Induce Adult Neurogenesis and Reverse Cognitive Deficits in Alzheimer's Disease Model Via Canonical WNT/B-Catenin Pathway. *ACS Nano* **2014**, *8*(1), 76–103.

Truran, S.; Weissig, V.; Madine, J.; Davies, H. A.; Guzman-Villanueva, D.; Franco, D. A.; Karamanova, N.; Burciu, C.; Serrano, G.; Beach, T. G.; Migrino, R. Q. Nanoliposomes Protect Against Human Arteriole Endothelial Dysfunction Induced by Beta-Amyloid Peptide. *J. Cerebr. Blood F. Met.* **2016**, *36*(2), 405–412.

Tsai, M. J.; Huang, Y. B.; Wu, P. C.; Fu, Y. S.; Kao, Y. R.; Fang, J. Y.; Tsai, Y. H. Oral Apomorphine Delivery from Solid Lipid Nanoparticles with Different Monostearate Emulsifiers: Pharmacokinetic and Behavioral Evaluations. *J. Pharm. Sci.* **2011a,** *100*(2), 547–557.

Tsai, Y. H.; Chang, J. T.; Chang, J. S.; Huang, C. T.; Huang, Y. B.; Wu, P. C. The Effect of Component of Microemulsions on Transdermal Delivery of Buspirone Hydrochloride. *J. Pharm. Sci.* **2011b,** *100*(6), 2358–2365.

Tsai, M. J.; Fu, Y. S.; Lin, Y. H.; Huang, Y. B.; Wu, P. C. The Effect of Nanoemulsion as a Carrier of Hydrophilic Compound for Transdermal Delivery. *PLoS One.* **2014,** *9*(7), e102850.

Turek, A.; Kasperczyk, J.; Jelonek, K.; Borecká, A.; Janeczek, H.; Libera, M.; Gruchlik, A.; Dobrzynski, P. Thermal Properties and Morphology Changes in Degradation Process of Poly(L-Lactide-Co-Glycolide) Matrices with Risperidone. *Acta Bioeng. Biomech.* **2015,** *17*(1), 11–20.

Upadhyay, P.; Trivedi, J.; Pundarikakshudu, K.; Sheth, N. Comparative Study Between Simple and Optimized Liposomal Dispersion of Quetiapine Fumarate for Diffusion Through Nasal Route. *Drug Delivery* **2016,** *23*(4), 1214–1221.

Varshosaz, J.; Tabbakhian, M.; Mohammadi, M. Y. Formulation and Optimization of Solid Lipid Nanoparticles of Buspirone Hcl for Enhancement of Its Oral Bioavailability. *J. Liposome Res.* **2010,** *20*(4), 286–296.

Varshosaz, J.; Sadeghi, H.; Shafipour, F. Characteristics on Their Hydrophobicity, Haemolysis and Macrophage Phagocytosis. *Farmacia* **2012,** *60*(1), 64–79.

Varshosaz, J.; Eskandari, S.; Kennedy, R.; Tabbakhian, M.; Minaiyan, M. Factors Affecting the Production of Nanostructure Lipid Carriers of Valproic Acid. *J. Biomed. Nanotechnol.* **2013,** *9*(2), 202–212.

Vassiliou, A. A.; Papadimitriou, S. A.; Bikiaris, D. N.; Mattheolabakis, G.; Avgoustakis, K. Facile Synthesis of Polyester-Peg Triblock Copolymers and Preparation of Amphiphilic Nanoparticles as Drug Carriers. *J. Control. Release* **2010,** *148*(3), 388–395.

Wahba, S. M. R.; Darwish, A. S.; Kamal, S. M. Ceria-Containing Uncoated and Coated Hydroxyapatite-Based Galantamine Nanocomposites for Formidable Treatment of Alzheimer's Disease in Ovariectomized Albino-Rat Model. *Mater. Sci. Eng. C Mater. Biol. Appl.* **2016,** *65*, 151–163.

Wang, P. Agomelatine Liposome Solid Preparation Used for Antidepressant Medicine, Comprises Agomelatine, Distearoyl Phosphatidylethanolamine, and soyasterol. CN Patent 103,040,750 A, December 18, 2012.

Wang, Z.; Mu, H. J.; Zhang, X. M.; Ma, P. K.; Lian, S. N.; Zhang, F. P.; Chu, S. Y.; Zhang, W. W.; Wang, A. P.; Wang, W. Y.; Sun, K. X. Lower Irritation Microemulsion-Based Rotigotine

Gel: Formulation Optimization and in Vitro and in Vivo Studies. *Int. J. Nanomed.* **2015,** *10*, 633–644.

Wang, Y.; Ying, X. Y.; Chen, L. Y.; Liu, Y.; Wang, Y.; Liang, J.; Xu, C. L.; Guo, Y.; Wang, S.; Hu, W. W.; Du, Y.; Chen, Z. Electroresponsive Nanoparticles Improve Antiseizure Effect of Phenytoin in Generalized Tonic-Clonic Seizures. *Neurotherapeutics* **2016,** *13*(3), 603–613.

Wavikar, P. R.; Vavia, P. R. Rivastigmine-Loaded in Situ Gelling Nanostructured Lipid Carriers for Nose to Brain Delivery. *J. Liposome Res.* **2015,** *25*(2), 141–149.

Wen, C. J.; Zhang, L. W.; Al-Suwayeh, S. A.; Yen, T. C.; Fang, J. Y. Theranostic Liposomes Loaded with Quantum Dots and Apomorphine for Brain Targeting and Bioimaging. *Int. J. Nanomed.* **2012,** *7*, 1599–1611.

Whiteley, C. G. Arginine Metabolising Enzymes as Targets Against Alzheimers' Disease. *Neurochem. Int.* **2014,** *67*, 23–31.

Wilson, B.; Samanta, M. K.; Santhi, K.; Kumar, K. P. S.; Paramakrishnan, N.; Suresh, B. Poly(n-butylcyanoacrylate) Nanoparticles Coated with Polysorbate 80 for the Targeted Delivery of Rivastigmine into the Brain to Treat Alzheimer's Disease. *Brain Res.* **2008a,** *1200*, 159–168.

Wilson, B.; Samanta, M. K.; Santhi, K.; Kumar, K. P. S.; Paramakrishnan, N.; Suresh, B. Targeted delivery of Tacrine into the Brain with Polysorbate 80-Coated Poly(n-butylcyanoacrylate) Nanoparticles. *Eur. J. Pharm. Biopharm.* **2008b,** *70*(1), 75–84.

Wilson, B.; Samanta, M. K.; Santhi, K.; Kumar, K. P. S.; Ramasamy, M.; Suresh, B. Significant Delivery of Tacrine Into the Brain Using Magnetic Chitosan Microparticles for Treating Alzheimer's Disease. *J. Neurosci. Methods.* **2009,** *177*(2), 427–433.

Wilson, B.; Samanta, M. K.; Santhi, K.; Kumar, K. P. S.; Ramasamy, M.; Suresh, B. Chitosan Nanoparticles as a New Delivery System for the Anti-Alzheimer Drug Tacrine. *Nanomed. Nanotechnol. Biol. Med.* **2010,** *6*(1), 144–152.

Wilson, B.; Lavanya, Y.; Priyadarshini, S. R. B.; Ramasamy, M.; Jenita, J. L. Albumin Nanoparticles for the Delivery of Gabapentin: Preparation, Characterization and Pharmacodynamic Studies. *Int. J. Pharm.* **2014,** *473*(12), 73–79.

Wong, H. L.; Wu, X. Y.; Bendayan, R. Nanotechnological Advances for the Delivery of CNS Therapeutics. *Adv. Drug Delivery Rev.* **2012,** *64*(7), 686–700.

Wu, F. G.; Yang, P.; Zhang, C.; Li, B. L.; Han, X. F.; Song, M. H.; Chen, Z. Molecular Interactions Between Amantadine and Model Cell Membranes. *Langmuir.* **2014,** *30*(28), 8491–8499.

Xiang, Y.; Wu, Q.; Liang, L.; Wang, X. Q.; Wang, J. C.; Zhang, X.; Pu, X. P.; Zhang, Q. Chlorotoxin-Modified Stealth Liposomes Encapsulating Levodopa for the Targeting Delivery Against the Parkinson's Disease in the MPTP-Induced Mice Model. *J. Drug Target.* **2012,** *20*(1), 67–75.

Yang, Z.; Liu, Z. W.; Allaker, R. P.; Reip, P.; Oxford, J.; Ahmad, Z.; Ren, G. A Review of Nanoparticle Functionality and Toxicity on the Central Nervous System. *J. R. Soc. Interface* **2010a,** *7*(4), 411–422.

Yang, Z.; Zhang, Y. G.; Yang, Y. L. A.; Sun, L.; Han, D.; Li, H.; Wang, C. Pharmacological and Toxicological Target Organelles and Safe Use of Single-Walled Carbon Nanotubes as Drug Carriers in Treating Alzheimer Disease. *Nanomed. Nanotech. Biol. Med.* **2010b,** *6*(3), 427–441.

Yang, X. X.; Zheng, R. Y.; Cai, Y. P.; Liao, M. L.; Yuan, W. E.; Liu, Z. G. Controlled-Release Levodopa Methyl Ester/Benserazide-Loaded Nanoparticles Ameliorate Levodopa-Induced Dyskinesia in Rats. *Int. J. Nanomed.* **2012a,** *7*, 2077–2086.

Yang, X. X.; Chen, Y. H.; Hong, X. Y.; Wu, N.; Song, L.; Yuan, W. E.; Liu, Z. G. Levodopa/ Benserazide Microspheres Reduced Levodopa-Induced Dyskinesia by Downregulating Phosphorylated GluR1 Expression in 6-OHDA-Lesioned Rats. *Drug Des. Devel. Ther* **2012b**, *6*, 341–347.

Yang, Z. Z.; Zhang, Y. Q.; Wang, Z. Z.; Wu, K.; Lou, J. N.; Qi, X. R. Enhanced Brain Distribution and Pharmacodynamics of Rivastigmine by Liposomes Following Intranasal Administration. *Int. J. Pharm.* **2013**, *452*(12), 344–354.

Yang, X. X.; Ji, X. Q.; Shi, C. H.; Liu, J.; Wang, H. Y.; Luan, Y. X. Investigation on the Ion Pair Amphiphiles and Their in Vitro Release of Amantadine Drug Based on PLGA-PEG-PLGA Gel. *J. Nanopart Res.* **2014**, *16*(12), 27–80.

Yasir, M.; Sara, U. V. S. Solid Lipid Nanoparticles for Nose to Brain Delivery of Haloperidol: in Vitro Drug Release and Pharmacokinetics Evaluation. *Acta Pharm. Sin. B.* **2014**, *4*(6), 454–463.

Yonar, D.; Paktas, D. D.; Horasan, N.; Strancar, J.; Sentjurc, M.; Sunnetcioglu, M. M. Epr Investigation of Clomipramine Interaction with Phosphatidylcholine Membranes in Presence and Absence of Cholesterol. *J. Liposome Res.* **2011**, *21*(3), 194–202.

Yonar, D.; Horasan, N.; Paktas, D. D.; Abramovic, Z.; Strancar, J.; Sunnetcioglu, M. M.; Sentjurc, M. Interaction of Antidepressant Drug, Clomipramine, with Model and Biological Stratum Corneum Membrane as Studied by Electron Paramagnetic Resonance. *J. Pharm. Sci.* **2013**, *102*(10), 3762–3772.

Yonar, D.; Sunnetcioglu, M. M. Spectroscopic and Calorimetric Studies on Trazodone Hydrochloride-Phosphatidylcholine Liposome Interactions in the Presence and Absence of Cholesterol. *Biochim. Biophys. Acta Biomembranes* **2014**, *1838*(10), 2369–2379.

Zasada, K.; Lukasiewicz-Atanasov, M.; Klysik, K.; Lewandowska-Lancucka, J.; Gzyl-Malcher, B.; Puciul-Malinowska, A.; Karewicz, A.; Nowakowska, M. "One-Component" Ultrathin Multilayer Films Based on Poly(Vinyl Alcohol) as Stabilizing Coating for Phenytoin-Loaded Liposomes. *Colloids Surf. B Biointerfaces* **2015**, *135*, 133–142.

Zhang, C.; Wan, X.; Zheng, X. Y.; Shao, X. Y.; Liu, Q. F.; Zhang, Q. Z.; Qian, Y. Dual-Functional Nanoparticles Targeting Amyloid Plaques in the Brains of Alzheimer's Disease Mice. *Biomaterials* **2014a**, *35*(1), 456–465.

Zhang, C.; Chen, J.; Feng, C. C.; Shao, X. Y.; Liu, Q. F.; Zhang, Q. Z.; Pang, Z. Q.; Jiang, X. G. Intranasal Nanoparticles of Basic Fibroblast Growth Factor for Brain Delivery to Treat Alzheimer's Disease. *Int. J. Pharm.* **2014b**, *461*(12), 192–202.

Zheng, X. Y.; Shao, X. Y.; Zhang, C.; Tan, Y. Z.; Liu, Q. F.; Wan, X.; Zhang, Q. Z.; Xu, S. M.; Jiang, X. G. Intranasal H102 Peptide-Loaded Liposomes for Brain Delivery to Treat Alzheimer's Disease. *Pharm. Res.* **2015**, *32*(12), 3837–3849.

Zhou, Y. Z.; Alany, R. G.; Chuang, V.; Wen, J.Y. Optimization of Plga Nanoparticles Formulation Containing L-Dopa by Applying the Central Composite Design. *Drug Dev. Ind. Pharm.* **2013**, *39*(2), 321–330.

Zhou, Y.; Zhang, G. Q.; Rao, Z.; Yang, Y.; Zhou, Q.; Qin, H. Y.; Wei, Y. H.; Wu, X. A. Increased Brain Uptake of Venlafaxine Loaded Solid Lipid Nanoparticles by Overcoming the Efflux Function and Expression of P-gp. *Arch. Pharm. Res.* **2015**, *38*(7), 1325–1335.

Zidan, A. S.; Aldawsari, H. Ultrasound Effects on Brain-Targeting Mannosylated Liposomes: in Vitro and Blood-Brain Barrier Transport Investigations. *Drug Des. Devel. Ther.* **2015**, *9*, 3885–3898.

PART II
Metallic Nanoparticles

CHAPTER 6

MAGNETIC NANOPARTICLES FOR DRUG DELIVERY

MEENAKSHI PONNANA and LAKSHMI KIRAN CHELLURI*

Department of Transplant Biology, Immunology and Stem Cell Unit, Global Hospitals, Lakdi-ka-Pool, Hyderabad, Andhra Pradesh 500004, India, Tel.: 91-40-30244501; Fax: 91-40-2324 4455

*Corresponding author. E-mail: lkiran@globalhospitalsindia.com

ABSTRACT

Magnetic nanoparticles (MNPs) are submicron moieties made of inorganic or organic (e.g., polymeric) materials, which may or may not be biodegradable. Their importance relates to the fact that the characteristics of these nanoparticles are different from those of bulk materials of the same composition, which is principally because of size effects, the magnetic and electronic properties, and the role played by surface phenomena when the size is reduced. MNPs are used foremost in the targeted drug delivery with reduced side effects and restricted drug release. The toxicity of nanoparticles is lessened due to the outer coating elevating their applicability as drug targeting agents. The advantages of these nanoparticles for their exemplifying application in the drug delivery at the target site are highlighted. The stability of the nanoparticles made inroads in the medicine and their wide flexibility increases their chance of removing the contaminants. MNPs are classified as ferromagnetic and paramagnetic based on their response toward the magnetic field. One targeted delivery technique that has gained prominence in recent years is the use of MNPs for noninvasive imaging. In these systems, therapeutic compounds are attached to biocompatible MNPs and magnetic fields generated outside the body are focused on specific in vivo targets. Future prospective and challenges faced by MNPs in the drug delivery and their scope has also been addressed. The unique magnetic properties and their ability to function at the cellular and molecular level of biological interactions created an attractive platform for the development of advanced technologies and enabling the diagnosis and treatment of the disease with greater effectiveness than ever before.

6.1   INTRODUCTION

Nanoparticles are under investigation since last three decades. They are microscopic material, less than 1 μm in diameter, used for the targeted drug delivery. Since then, tremendous progress in their design and synthesis techniques followed by few clinical trials has been accomplished (Stuart et al., 2008). The inadequacies in the conventional diagnostic and therapeutic agents for the disease are tackled, by the development of nanoparticles with specific functional properties. This is made possible with the rapid advances in the fields of molecular biology and nanotechnology (Ferrari et al., 2005; Sanvicens et al., 2008). With the nanoparticle technology, imaging probes, which are intense, and tissue-specific, are being developed to envisage and facilitate disease diagnosis at its initial stages or even prior to disease manifestation. Magnetic nanoparticles (MNPs) are a kind of noninvasive material that displays certain responses, for example, their alignment toward the magnetic field and are used for magnetic resonance imaging (MRI).

The MNPs may be metallic, bimetallic, or superparamagnetic, though they possess other elements that are mostly iron oxides. Most widespread iron oxides are magnetite (Fe_3O_4), maghemite (γ-Fe_2O_3), hematite (α-Fe_2O_3), and geotite (Seyda Bucek et al., 2012). One or more of the iron oxide phases may form central to the type of experimental conditions. It is imperative to carefully monitor the experimental conditions to ensure the presence of a single phase. The superparamagnetic iron oxide nanoparticles (SPIONS) are mostly preferred due to its mild toxicity profile and reactive surface that can be easily customized upon targeting, imaging, biocompatible coatings, and therapeutic molecules (Laurent et al., 2008; Lewinski et al., 2008). This has made possible, the wide application of SPION as a biosensor, for the magnetic separation, for the medical imaging, and the drug delivery (Duran et al., 2008; Solanki et al., 2008).

Thorough engineering of the size, shape, coating, and surface adaptation is required for the making of next-generation SPIONs that can explicitly target and thereby eliminate or illuminate damaged tissue. A nanoparticle is produced by designing every parameter in such a way that it overcomes biological barriers and carries out its function. In doing so, the targeting molecules must be selected not only based on their binding characteristics but also based on their physical properties. These targets are then incorporated into the nanoparticle design system in such a way that they remain functionally active (Omid et al., 2010).

6.1.1 CURRENT TECHNOLOGIES AND APPLICATION OF MAGNETIC NANOPARTICLES (MNPS)

Nanoscience, one of the significant research areas in modern science led to the development of nanotechnology allowing professionals of broad arena such as scientists, chemists, physicians, and engineers to execute their work at the molecular and cellular aspects thereby producing significant advances in the healthcare and life sciences. Nanoparticles offer major advantage due to its uniqueness in size and physicochemical properties. Based on the widespread application of these nanoparticles in various fields such as biotechnology, material science, and biomedicine, we have classified under different sections.

6.1.1.1 BIOMEDICAL APPLICATIONS

For the nanoparticle to be applied in biology, diagnosis, and therapy, it has to be stable at neutral pH and physiological salinity, should demonstrate superparamagnetic performance at room temperature (Schütt et al., 2005; Morcos et al., 2007), even depends to be or not to be used for the in vivo and in vitro applications. These nanoparticles can be used in vitro for the selection and diagnostic partition, for the in vivo application of the nanoparticle; it may perhaps be further divided based on the hyperthermia and drug-targeting purpose or drug repurposing related to therapeutic use and in the magnetic resonance for the diagnostic application (Liu et al., 2005; Piao et al., 2008).

The in vivo applications of nanoparticles include therapeutics, drug delivery, and diagnostics based on their size and functionality. Magnetic energy is transferred in the form of heat known as hyperthermia when the SPIONs are placed in altering orientations, that is, in parallel and antiparallel directions. The pathological cells in the tumor cells are destroyed by the application of this hyperthermia technique in vivo (Kim et al., 2006). In the future, this technique might be versatile for the successful cancer therapy as hyperthermia is only restricted to the tumor area (Green et al., 2005). The basic technique of drug delivery involves the system wherein the MNP loaded with the required drug is transferred to the cancer site under the influence of the magnetic field depending on the interplay between the blood vessel and the magnetic forces. In the diagnostics, magnetopharmaceuticals came into light based on the nuclear MRI technology wherein drugs are administered in the patients to understand the status of their vital organs and to increase the image contrast to differentiate between the normal and the affected tissue (Abolfazl et al., 2012).

MNPs are applied in vitro in the diagnostics for its separation and selection mode, magnetorelaxometry, MRI, and bioseparation. They have a noteworthy impact on the sample extraction wherein it extracts the target from a large volume of solution based on the magnetic adsorbent properties (Gao et al., 2004; Pellegrino et al., 2004). Magnetorelaxometry measures the magnetic viscosity in the absence of magnetic field and is used for the assessment of immunoassays based on its ability to distinguish between the free and bound conjugates (Sun et al., 2006). The higher photostability and stronger fluorescence make the applicability of the SPIONS as a contrast agent in the MRI (Lu et al., 2002). For the analysis of certain biological entities such as cells, nucleic acids, and proteins, their separation is required that is performed by the superparamagnetic colloids. MPNs bearing catalytic sites for various reactions such as hydrogenation, polymerization, and cross-coupling reactions have emerged recently and overcome the barrier of heterogeneous catalysis (Lee et al., 2008). MNPs have displayed promising performance in the mitigation of toxicity through pollutant removal and have wide environmental applications (Tratnyek and Johnson 2006). In situ application and huge flexibility enhances its ability in remediation of various groundwater, air, and soil contaminants (Wang et al., 2005).

6.1.1.2 ANALYTICAL APPLICATIONS

Magnetic luminescent nanoparticles (MLNPs) such as polystyrene magnetic beads, iron oxide nanoparticles, quantum dots due to their small size show a larger surface area-to-volume ratio leading to perfect reaction homogeneity and swift reaction kinetics. The preparations are simple and are applied in quantitative bioanalysis (Savva et al., 1999). The optical characteristics of the MLNPs remove the need of centrifugation facilitating the execution of internal calibration in the detection system. This technique can be applied for the detection and diagnosis of the disease using the conventional organic dyes based on the protein analysis. By precipitation and surface reaction, shells/hybrid coatings are formed on the colloids. Iron compounds are produced at elevated temperatures and polymer colloid dispersion. Polymer capsules of sub-micrometer size are acquired by the removal of the template by calcinations (Quellec et al., 1998).

MNPs are encapsulated in polymeric matrixes to reduce susceptibility and enhance compatibility with organic ingredients. They are prepared in the presence of albumin, chitosan, and polymers by applying mechanical energy for the dispersion of magnetite (Jeong et al., 2000). Silica, an inorganic material is used for the MNP encapsulation to produce stable and ideal anchorage dispersions that favor covalent binding of the ligands (Velge-Roussel et al., 1996).

6.2 CHARACTERISTICS/FEATURES OF MNPS

Nanoparticles are of enormous scientific interest, since longtime to date as they act as a link between massive and molecular structures. Nanoparticles thus exhibit many physical and chemical properties, among which morphology, charge, surface properties, hydrodynamic size, high magnetic susceptibility, nontoxicity, biocompatibility, and coercivity are the crucial properties of MNPs.

6.2.1 PHYSICAL CHARACTERISTICS

Movements in the particles such as electrons, holes, protons, and positive and negative ions with both mass and electric charges are caused by the magnetic effects. A spinning electric-charged particle called magneton creates a magnetic dipole; these magnetons are associated, and aligned in ferromagnetic materials that distinguish them from the paramagnetic particles. The reliance of size of the ferromagnetic particles on the magnetic behavior is determined by the domain structures. The MNPs are small in the range of 10–50 nm in size which helps to avoid precipitation, and have large surface area-to-volume ratio which enhances coating efficiency and targeting process. Nanoparticles display unexpected properties by confining electrons within their minute size and produce quantum effects (Hewakuruppu et al., 2013). Owing to these quantum effect nanoparticles, for example, gold nanoparticles appear deep red to black in color. Magnetization is uniform for particles below critical size known as single domain particles and nonuniform for larger particles referred to as multi-domain particles. Various factors such as magnetic saturation, potency of crystal anisotropy, and exchange forces, surface energy and particle shape affect the critical size (Hines et al., 1996; Qu et al., 2001). The large surface area of these particles made them to dominate the contributions made by other material. Superparamagnetic properties are achieved by reducing the size of the nanoparticle to improve colloidal strength and avoid aggregation due to the decline in their dipole interactions (Yadollahpour 2015).

MNPs are classified as ferromagnetic and paramagnetic particles based on their behavior in the presence of an external magnetic field. Superparamagnetism caused by thermal effect is demagnetized due to thermal fluctuations; hence, possess zero coercivity (Peng et al., 1997). The presence of external magnet turns the nanoparticle magnetic while its absence revert it to a nonmagnetic state, avoiding an "active" behavior of the particles when there is no applied field. Crystalline materials such as Fe, Co, or Ni exhibit magnetic property; however, ferrite oxide-magnetite (Fe_3O_4) is widely used as superpara-MNPs for all types of biological applications due to its high magnetic property (Dabbousi et al., 1997; Rockenberger et al., 1999).

6.2.2   CHEMICAL CHARACTERISTICS

Different applications need specific characteristics of MNPs, such as in data storage purpose the particles must be stable, should possess flexible magnetic property, and have to sustain fluctuations in the temperature. They have to be stable at particular pH of the physiological environment and in water to be applied in diagnosis and therapy (Ersoy et al., 2007). To apply these nanoparticles in the biomedical field they must be encapsulated with a polymer to obstruct the biodegradation (Sosnovik et al., 2007).

The saturation of magnetization is high for the MNPs enabling the external magnetic field to control the mobility of the nanoparticles in the blood. This even enables the nanoparticles to move in close proximity to the pathological tissue (Issa et al., 2013). The biocompatibility, stability, and nontoxicity are dependent on the modification in the size and coating properties of the nanoparticles (Nitin et al., 2004). MNPs made from iron, nickel, and cobalt, are oxidized leading to acid erosion, and hence coating is required (Shenoy and Amiji 2005). MNPs on an applied field depend on the remanence and coercivity (related to the thickness of the curve). Coercivity is the only property, which is size dependent, as it increases to maximum and decreases to zero as the particle size decreases by turning the magnetic property.

MNPs should be capable of escaping from reticuloendothelial system to reach their target to be applied in the biomedical field. Opsonization occurs in the bloodstream on the introduction of the nanoparticles coated with the plasma proteins and are shortly removed by phagocytic cells preventing their exposure to the target cells (Briley-Saebo et al., 2004). In order to avert this event, inorganic substances such as silica and carbon and organic substances such as polymers and surfactants are used as a coating on these nanoparticles to increase its circulation time and colloidal stability (Owens and Peppas 2006).

6.3 PREPARATION/MANUFACTURE OF MNPS

Nanoparticles with nanoscale dimensions, magnetic property, and its ability in carrying out specific tasks can drive them to the target organ for the drug delivery. There are various routes to attain highly stable, shape-controlled, and narrow size distributed MNPs. Certain techniques and design parameters assist in synthesizing functional MNPs consisting of a magnetic core, protective coating, and surface functionality (Lazaro et al., 2005). There are single step and multistep procedures in fabricating the MNPs described below.

MNPs commonly iron nanoparticles in use, are prepared by wet precipitation wherein the pH of the iron salt solution is maintained to generate fine suspension of particles and coprecipitation is carried out in preparing magnetite and ferrites whose coating without aggregation is difficult (Liu et al., 2005). High-quality magnetite and maghemite nanoparticles are organized by oxidation of iron nanoparticles by chemical vapor condensation wherein certain metal compounds that are volatile are heated in a volatile atmosphere for their decomposition (Sun and Zeng 2002). Thermal decomposition and reduction is another technique where metal oxy-salts such as carbonates, acetates, and nitrates are heated to certain temperatures leading to their decomposition to metal oxide nanoparticles, which are further reduced to metal in the presence of hydrogen or carbon monoxides (Nurmi et al., 2005). Liquid phase reduction is another method where the magnetic alloy nanoparticles are formed using reducing agents such as sodium borohydride and lithium borohydride (Srivastava et al., 2006).

6.3.1   MAGNETIC CORE MATERIAL

Magnetic materials have a wide array of magnetic properties. They ought to be nontoxic for them to be applied in biomedicine. Protective coating needs to be applied on iron oxide nanoparticles such as magnetite and maghemite with enormous mechanical strength, to protect from toxicity. Magnetite, maghemite, iron-based metal oxides, iron alloys, and other materials such as rare earth metal alloys and transition metal clusters can be used as magnetic cores for manufacturing MNPs. Magnetite and maghemite are ferromagnetic materials with strong magnetization due to lattice vacancies that lead to the uncompensated electron spins within the structure. Iron-based metal oxides are almost similar to magnetite in structure and due to their vulnerability toward corrosion; they are coated with nonporous substance (Stuart et al., 2008).

6.3.2   COATING MATERIAL

Natural polymers such as carbohydrates and proteins are biocompatible and therefore suitable for the coating of nanoparticles. Dextran, a carbohydrate, is used in cancer treatment and MRI while its combination with chitosan, poly-L-lactic acid, and silica, forms blended coating (Subramani et al., 2006). The synthetic polymers such as polyethylene glycol (PEG) and polyvinyl alcohol (PVA) fulfill the lack of mechanical strength in natural polymers. PEG, a polyether, lacks an apparent site for organic functionalization and PVA has a hydroxyl group ensuring the hydrophobic property of the coating (Schulze et al., 2006).

Gold nanoparticles are the most commonly used as they can be easily functionalized with thiol linkers which possess high affinity toward the gold surface. This ability is employed in biotechnology for the antigen binding in the immunoassays (Ameur et al., 2000). Organic linkers with a wide range of surface properties create electrostatic interactions. Quite a large number of negatively charged deoxyribonucleic acid (DNA) molecules are bound to the nanoparticles preferably by electrostatic interactions and released into the cell after internalization of nanoparticles. This aspect makes the MNPs suitable for the gene delivery (McBain et al., 2007). The synthesis of nanoparticles is depicted in the form of flowchart (Fig. 6.1).

Synthesis of magnetic nanoparticles

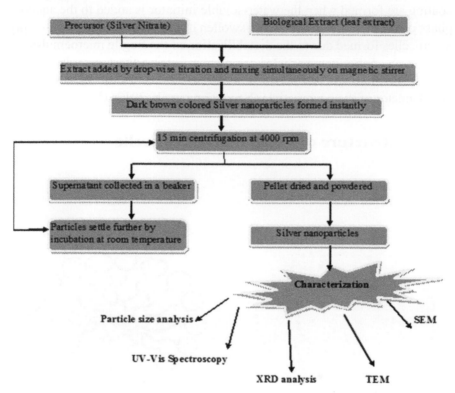

FIGURE 6.1   Synthesis of magnetic nanoparticles (MNPs).

6.3.1   IN-VITRO PROTOCOL-MICELLE FORMATION

Drug delivery agents, that is, micelles or liposomes are phospholipid bilay-ered membrane vesicles and are considered as the most primitive forms of nanomedicine. Micelle formation is an earlier trend of surfactant chemistry where surfactants are structures with a hydrophilic head and an elongated hydrophobic tail (Meyers 2005). When the surfactant molecules reach a critical concentration known as the critical micelle concentration, micelles are formed which are of two types. Normal micelles (which are formed in the aqueous phase, that is, in detergents for cleaning purpose) and reverse micelles (which are formed in the oily medium like in hexane), the size of which determines the nanoparticles size which is in the range of 20–500 nm. Beyond this range, nanoparticles cannot be synthesized using the reverse micelle method. A novel and effective approach for the synthesis of nanosized colloidal particles is the microemulsion technique. MNPs with the surfactant coating are formed when the water-soluble initiator is added to the aqueous phase of microemulsion containing swollen micelles. In certain cases, empty are micelles formed due to the destabilization of the fragile microemulsions. Hollow magnetite nanoparticles are prepared using the core substance as oleylamine micelles. The micelle concentration dictates the shell thickness and diameter of the nanoparticle (Nagavarma et al., 2012) (Fig. 6.2).

Structure of Liposome and Micelle

Liposome   **Micelle**

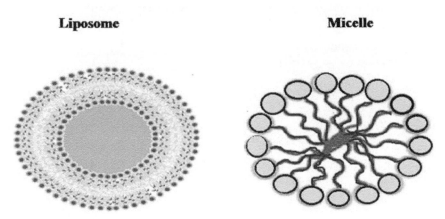

FIGURE 6.2   Structure of liposome and micelle.

They are exploited for the delivery of miniature molecules, DNA, proteins, and peptides and imaging contrast agents as their size is in the range from 100 nm to 5 μm (Torchilin et al., 2005). They are capable of encapsulating quite a large number of cores of MNPs and carry them together by avoiding their dilution at the target site. PEGylation markedly establishes the in vivo performance of the liposome encapsulation leading to their long circulation times. Another favorable feature of these delivery vehicles is that their multifunctionality is supplemented in combination with a therapeutic agent. In the same way, micelles with numerous functions shaped with amphiphilic blocks are used for the MNP entrapment for various applications (Lecommandoux et al., 2006; Nasongkla et al., 2006).

Maghemite nanocrystals when encapsulated within egg phosphatidylcholine, unilamellar vesicles undergo film hydration and attached by sequential extrusion, producing magnetic fluid-loaded liposomes with hydrodynamic size of 193 ± 33 nm. The magnetic fluid-loaded liposomes that are capable of encapsulating up to 1.67 molecules of iron per molecule of lipid, upon in vivo assessment in mice using magnetic resonance angiography established their presence in the blood 24 h after intravenous injection confirming their long circulating behavior (Martina et al., 2005). SPIONs illustrated certain advantages during the drug delivery when coated with the micelles such as effortless surface modification, easy encapsulation of drugs within the substructures and their sequestration and protection of the drug until they are degraded at the target site (Mulder et al., 2006). PEG, one of the well-known hydrophilic synthetic polymers, is used as a drug carrier. However, it is also used as a coating for the other delivery vehicles, for example, the nanoparticles wherein it acts as a copolymer in the manufacture of polymeric micelles. It even works as a spacer between the biologically active molecule and the drug carrier (Thakur et al., 2013).

Reverse micelles are hydrophilic and store the reaction mixtures of inorganic components. Iron oxide-based MNPs are synthesized by the dissolution of inorganic precursors such as ferric chloride in an aqueous medium and its further addition to the oily reaction mixture in the presence of surfactants. Later pH regulators such as ammonia and sodium hydroxide and inorganic coating materials such as silica and gold are added to the above mixture. Micelles control the size of the nanoparticles and accordingly they have a propensity to be uniform in size. During this process, inorganic coating materials can be added so that they act as a protective layer on the nanoparticles, for example, in magnetite (Santra et al., 2001). However, this technique has a drawback, as synthetic organic coatings are not possible since monomers stay behind in the organic state of micelle mixture, that is, outside the micelles.

Entrapped hydrophilic drugs + Lipid bilayer = Liposomes

Lipophilic drugs + Lipid layer (Single) = Micelles

6.3.2 MECHANISMS FOR TARGETED DELIVERY OF NANOPARTICLES

Drug delivery targeting to the affected tissue or organs is a better and inexpensive approach, which is further classified into active targeting, passive targeting, and physical targeting. Various disorders especially in cancer this technique is employed for the targeted drug delivery. The inability of current treatment strategies such as surgery, chemotherapy, radiotherapy, and immunotherapy failed to target the drug at the required site, brought into light the protein-based carrier for targeted delivery (Zhang et al., 2006) (Fig. 6.3).

Targeted delivery of Magnetic nanoparticles

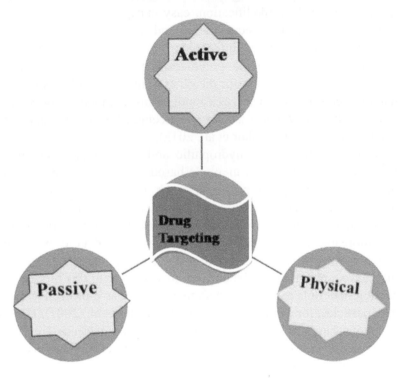

FIGURE 6.3   Targeted delivery of MNPs.

6.3.2.1   PASSIVE TARGETING

The passive carrier of nanocarrier systems is the technique that involves enhanced permeation and retention and due to the grouping of tumor blood vessels that are leaky, and the false particle screening through the intravenously delivered drugs dispense uniformly through the body. Nanoparticle size, charge, pace of penetration and intracellular internalization affect their enhanced permeability and retention (Albanese et al., 2012). It has been observed that PEGylated nanoparticles below 100 nm size suppressed the absorption of plasma proteins, thereby reducing hepatic filtration

(Allexis et al., 2008). Negatively charged nanoparticles are retained by the blood, unlike the positively charged ones that are readily absorbed due to the negative surface charge of the affected organs. Molecular simulations were performed on a large scale to clarify the influence of PEGylated nanoparticles cellular uptake, to understand the nanoparticles of various shapes with the similar surface area, PEG implanting density, and the strength of ligand–receptor combination (Li and Liu 2015).

Enhanced permeation and retention effect is not that successful since most of the nanoparticles that are targeted passively failed to reach the tumor on intravenous administration. Environmental and external stimuli trigger at the target site of the affected tissue or organ greatly improve the targeting efficiency (Bae and Park 2011). Carriers such as liposomes (doxorubicin and daunorubicin), micelles, polymer–drug conjugates possess macromolecules that dissolve, entrap, and conjugate with the drug surface. The carriers progress toward the uptake of nanoparticles both actively and passively when their surfaces are functionalized with the cell-penetrating peptides with positive charge (Saha et al., 2010; Zorko and Langel 2005).

6.3.2.2   ACTIVE TARGETING

The uptake selectivity of nanoparticles is upgraded by the active targeting of the ligands bound to the surface of nanoparticles with high affinity. High binding affinity of the ligands with target cells increases the delivery efficiency and protects the nanoparticles from enzyme destruction. In the active targeting, the nanoparticle is functionalized with a ligand that binds to molecules on cells. While in the case of tumor cells, they express the same molecule as that of the healthy cells defying the nanoparticles from reaching its target. This is circumvented by using variety or a manifold of ligands. In vivo screening of phage and aptamer helps in identifying the potential tumor receptors, that is, cytoplasmic proteins such as annexin, p32 protein, plectin, angiogenic vessel proteins like integrins (Fogal et al., 2008; Ruoslahti et al., 2010).

Antibodies can be considered as ligands due to their easy availability and high specificity, but cannot be used due to their inefficiency to conjugate to the nanoparticles. A promising alternative to antibodies are peptides that are stable, easy to produce and are handy. Arginylglycil aspartic acid is the commonly used ligand for integrin receptors (Low et al., 2008). Therapeutic efficiency is limited due to the binding incompatibility of ligands and receptors, which is overcome by the different charges and multiple ligands. Tumor penetration is reduced hindering selectivity by the stronger binding. Active targeting differs from the passive targeting by modifying the pattern distribution of the carrier allocating it to a specific organ (Cheng et al., 2012).

6.3.2.3 PHYSICAL TARGETING

When external stimulation such as radiation, magnetic effect, and photothermal therapy is applied for navigating drugs to the target site, the concept is known as physical targeting. Gold nanoparticles are predominantly used and even tested in animals for their low toxicity (Huang et al., 2008). Photothermal therapy with few adverse effects is frequently applied to the accurate delivery of nanoparticles to the cancer cells that are killed by the heat that is generated by the conversion of light energy. Apart from gold, other photothermal materials were explored that instigate drug release by their pH and heat. Polymer-conjugated graphene oxide causes cell death due to its pH-sensitive nature. Carbon nanotubes with strong absorbance demonstrate photothermal hyperthermia by inhibiting the G2 to M phase in the cell cycle (Zhang et al., 2013). However, there exist certain limitations with photothermal therapy for the reason that cancer cells can mostly tolerate the stress created by the environment.

MNPs generate heat energy known as magnetic hyperthermia by their continuous oscillation in the magnetic field. Therapy correlates with the heat energy produced that varies based on the magnetic field sensitivity and strength and on the space between the target and MNPs. Both photothermal therapy and magnetic hyperthermia can be performed in vivo and in vitro, and a number of clinical trials have been carried out. Combination of active and physical targeting promote the internalization of nanoparticle successfully (Sreedhar and Csermely 2004).

6.4   ROUTE OF ADMINISTRATION

The route of administration of the nanoparticles determines the bioavailability and biocompatibility of the loaded drug at the target site. Nanoparticles are administered through oral, intravenous, and intraperitoneal routes. They are targeted at the site of interest based on their administration route and offer a significant improvement over traditional oral and intravenous methods of administration in terms of efficiency and effectiveness. Subcutaneous or intradermal injections may be desirable for lymphatic targeting, whereas intraperitoneal injection may be more efficacious for brain targeting. Only a limited number of polymers can be used for the formulation of nanoparticles designed to deliver drugs in vivo. Indeed, a suitable polymer must be quickly eliminated from the body to allow repeated administrations while avoiding accumulation. The application method can influence the response to the treatment: intravenous administration of the nanoparticles was not effective, while the intra-arterial application gave good results despite the potential thrombotic risk. Indeed, the huge number of possible polymer structures and the large variety of preparation techniques were allowed for tuning the nanoparticle delivery system to the specific therapeutic application, administration route, and type of protein drug (Jurgons et al., 2006) (Fig. 6.4).

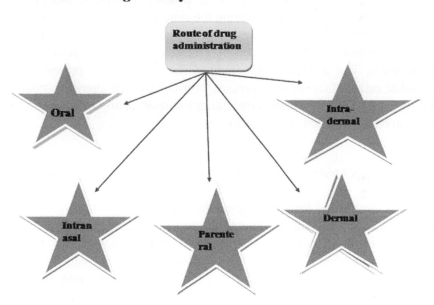

FIGURE 6.4 Route of drug delivery.

6.4.1 INTRODUCTION OF BASIC ROUTES OF ADMINISTRATION

Successful drug therapy depends on the drug bioavailability, which in turn depends on the route of administration, absorption, and metabolism of the drug. Variations in the pharmacological effects of the drug-loaded nanoparticles occur based on the administration route. Different drug delivery routes include nasal, parenteral, ophthalmic, pulmonary, dermal, and transdermal. Among which, the oral route is the recurrent form of usage for the administration of nanoparticles loaded with the drug.

6.4.1.1   ORAL DRUG DELIVERY

The oral route is advantageous over other routes of delivery as it provides convenience, painless administration, and self-application leading to high compliance and is hence widely used. Large surface area for absorption, solubility, permeability, and blood supply are made available by the gastro-intestinal tract. However, all the drugs do not possess the required bioavailability, which is circumvented by the development of polymeric nanoparticles encapsulated with the drugs enhancing the drug solubility and drug delivery (Bernkop-Schnurch 2013).The uptake of the nanoparticle is facilitated and the drug permeability is enhanced by chitosan and its derivatives by two mechanisms. The anionic components of the glycoprotein present on the epithelial cell surface which interacts with the surface charge is positive, that is, mucoadhesion and the paracellular transport is improved by the tight junction opening (Chen et al., 2013).

6.4.1.2   INTRANASAL DRUG DELIVERY

The nasal route has fascinated great interest as an alternative route for the administration of various agents. In general, this route is used for the administration of mucolytics, decongestants, and antibiotics. However, the progression of nanotechnology has enabled studies involving anticancer drugs, central nervous system drugs, peptides, analgesics, and diagnostic agents. The nasal mucosa due to its epithelial microvilli, rich vasculature, highly porous endothelial membrane, and a large surface area presents many beneficial characteristics for the systemic absorption of drugs thereby facilitating drug permeation, such that drugs can be absorbed directly into the systemic circulation. These characteristics lead to a fast onset of action, quickly reaching therapeutic plasma levels of the drug (Pires et al., 2009). All these factors reduce side effects, permit dose reduction, and increase patient compliance to healing. Additionally, the intranasal administration presents a noninvasive and painless alternative to the intravenous and oral routes, thus maximizing patient comfort.

6.4.1.3 PARENTERAL DRUG DELIVERY

Generally, the therapeutic results achieved with orally administered drugs needed for the application of higher doses than those that would be essential to achieve the same effect using the parenteral route. This is mainly because this parenteral pathway allows for the rapid and absolute absorption of drugs, distinct to what happens in the oral route where a large fraction of the dose is lost due to first pass liver metabolism. In addition to which the parenteral route allows the use of minor doses and has extended therapeutic effects when compared to other routes (Merisko et al., 2011). It is also projected to lower the incidence of side effects compared to that of the oral administration. Drugs with poor dissolution properties are generally promising candidates to have their effect extended by parenteral administration. These molecules are often difficult to formulate using conventional approaches and are associated with formulation-related performance issues, for example, lack of dose proportionality, poor bioavailability, slow onset of action, and other attributes leading to poor patient compliance. Other advantages of parenteral administration include the continuous infusion of drugs with a short half-life as well as drug administration to unconscious and comatose patients (Rowland 1972; Zhang 2013).

6.4.1.4 DERMAL AND TRANSDERMAL DRUG DELIVERY

Nanoparticle application has been primarily paying attention to parenteral and oral applications. In the present days, besides these uses considering the advantages of nanoparticles for dermal application such as the protection of incorporated active compounds against chemical degradation and flexibility in modulating the release of the compound, nanosystems applied to the skin are drawing more and more attention of researchers (Pawar and Babu, 2010). Nanoparticles applied to the skin can have one of two desired effects: local activity within the skin (dermal drug delivery) or systemic activity after nanoparticle permeation through the skin (transdermal drug delivery). The stratum corneum is the main barrier of the skin in both the dermal and transdermal routes, which has to be overcome for suitable drug delivery (Neubert 2011).

6.4.2 INNOVATIONAPPROACHFORUSINGSTEMCELLSINMNP

Stem cell therapy is an upcoming field as it increases the cell survival through theranostic approaches in the field of regenerative medicine. MNP-labeled mesenchymal stem cells (MSCs) had lower cytotoxicity and higher detection limits. Cytotoxicity assessment in vitro and visualization of rhodamine B/RITC-labeled, multimodal, and silica-coated nanoparticles in vivo was made possible for the labeling of MSCs derived from the human cord blood. Perfect healing is achieved by the destruction of cancer stem cells, and hence many approaches have attempted to attack cancer stem cells. Several factors are related to the difficulty in abolishing cancer stem cells that include cell protection through cell cycle arrest, elimination of drugs by adenosine triphosphate-binding cassette transporters, and reducing systems to protect from oxidative stress. Microneedle array and iontophoresis technologies has been employed for the targeted delivery of encapsulated drugs to hair follicle stem cells (Zhang et al., 2013).

Bone marrow-derived MSCs that are undifferentiated were cocultured with bone morphogenetic protein-2 loaded with other heparin-conjugated poly(lactic-co-glycolic acid) (PLGA) nanoparticles. More bone formation than control group is induced significantly using heparin-conjugated PLGA nanoparticles loaded with bone morphogenetic protein-2 (BMP-2) when tested both in vitro and in vivo (Kim et al., 2008). Dendrimers possessing highly branched dendritic branches have the advantage of actively entrapping the drug by electrostatic interactions within the branches. This makes these dendrimers attractive in bone tissue engineering for the drug delivery systems, for example in the transfection of MSCs using the folate–poly(amidoamine) (PAMAM). This dendrimer was used to carry the human BMP-2 gene-containing plasmid. The study reported the possibility of PAMAM dendrimers used for inducing in vitro differentiation of MSCs to osteoblast phenotype (Santos et al., 2009).

The human MSCs culture with the drug-loaded nanoparticles was proliferated and differentiated osteogenically in a much improvised manner when compared to the scaffolds lacking the drugs (Nguyen et al., 2012). The osteoblastic differentiation of MSCs is promoted and proliferated in the presence of BMP-2-loaded nanotube (Hu et al., 2012). Molecular layer deposition (MLD), a monomolecular step synthesis process where the molecules are interconnected in designated sequences tailor the organic materials. This MLD technique is anticipated to offer improved ways to carry drugs to the targeted cancer cells and cancer stem cells without attacking normal cells (Yoshimura et al., 2012).

Cell therapy has emerged due to the minimum capacity of the adult's heart in mice for endogenous regeneration. After acute myocardial infarction and heart failure, cell therapy can be performed for the regeneration of damaged vascular and cardiac tissue. Relatively, large cells such as the

MSCs, approximately 20–25 μm in diameter were widely used for the intracoronary (IC) injection. Cardiac progenitor cell-enrichment strategies often have not been fruitful due to nonavailability of well-characterized antibodies for a cardiac-specific phenotype. Furthermore, circumventing the major cell manipulation in cell cultures and improving the enrichment with biocompatible engineered SPION tagging in a single step, has the prospective for application in cell therapy. Hence, the primary strategic advance is to evaluate the migration, homing, and function of the stem cells, which will in the course of time assist in maximizing the effectiveness of these novel therapies. Owing to the higher spatial resolution in shaping, the outcome of transplanted stem cells and the accessibility of well-defined pathological and anatomical information about the surrounding tissue, MRI has gained significant importance. SPIONS have dual ability to internalizing into cells and accessible to the external magnetic field, consequently has made them useful tools for theranostic purposes (Vinod Kumar et al., 2015). Cell size is an important factor for the microembolism after intra-arterial cell injection. After intravascular delivery, larger stem cells such as the MSCs, are simply entrapped within the microcirculation. IC injection permits the relatively homogeneous distribution of donor stem cells to the target area with less myocardial injury in a physiological manner when compared to the intramyocardial injection (Fukushima et al., 2007). Nanoprobe-sorted cells were treated with 5-azacitidine to induce a single-step directed differentiation into cardiomyocytes. There was a noteworthy change in the labeled cells morphology after 7 days of induction with 5-azacitidine and the emergence of myogenic-like cells. Cardiomyogenic-specific morphology was more prominent after around 21 days post-induction. The resultant cardiomyocyte-like cells were characterized using cardiac cell-specific surface markers cluster of differentiation (CD105, signal regulatory protein alpha, and kinase domain receptor) by flow cytometry. Ultrasound is rapidly used for the echocardiography and the imaging of the stem cells that are delivered to the target site.

MNPs are mostly used in the gene/nucleic acid transfer technology. Such nucleic acid transfer is an important tool that is routinely used in current life science research, such as for the control of target gene expression and cell labeling. Recently, gene transfer technology has been lucratively used for the induced pluripotent stem cell production (Okita et al., 2010) and in caspase nuclease ribonucleic acid (RNA)-guided genome editing or transcription activator-like effector nuclease (Jinek et al., 2012) editing, thus suggesting a further increase in its importance in the near future. Currently, cell transplantation therapy is being carried out by many research groups worldwide using the somatic stem cells (Kami et al., 2013).

6.5   IMAGING/IN VIVO TRACKING OF MNPS

There is an immediate requirement for the visualization of cell bio-distribution and development of cell therapy in a noninvasive way. Information regarding the persistence and motility of transferred cells, most favorable route of delivery, and the remedial doses for the subjects are offered by the imaging of cell trafficking. Cell tracking is a noninvasive technique emerged for the visualization of cells in their native environment. MRI is a potential tool developed that uses MNPs. It has been a challenge to the clinical researchers in verifying cell trafficking instantly after inoculation and its migration in the course of time. This deliberately slow immune cell therapy is overcome by the in vivo cell tracking which provides prospective information thereby beat the regulatory barriers. MNPs have wide application in different types of imaging modalities.

SPIONs emerged as MR imaging contrast agents followed by multimodal agents that visualize the MNPs across different platforms. The use of MRI in tandem with contrast agents for cellular labeling has been gaining importance as an experimental research in the field of stem cells and tissue engineering, where the success of the treatment modality is staggered with the lack of knowledge on the fate of the posttransplanted cells. Thus, the ability to labeling cells with para-MNPs and tracking the cellular migration, homing, engraftment, and eventually determining its fate provides valuable information for improvement and also in determining its translational efficacy.

Experimental studies in cellular labeling with nanoparticles have been long ongoing. SPIONs in particular, have been of interest to both researchers and clinicians because of its size, cellular uptake, superparamagnetic properties, biocompatibility, relatively low toxicity, and safer incorporation of the degraded iron into the hemoglobin of the circulatory system, in comparison to other contrast agents. Moreover, the feasibility, safety, and efficacy of stem cell labeling have been well investigated in MSCs. This enables the clinicians in obtaining pathologic information using image capabilities and a contrast agent (Brans et al., 2006; Hamoudeh et al., 2008; Rashi et al., 2016). MRI helps in obtaining metabolic and functional information with high resolution using magnetic fields that help in aligning nuclear magnetization of body's hydrogen atoms in the presence of radio frequency pulse. MNPs and paramagnetic chelates are the contrast agents for the MRI (Weissleder and Pittet 2008) offering a spatial resolution and in shortening the relaxation times.

Optical imaging helps in monitoring the photons emitted from the fluorescent agents in the near infrared range based on its ability to effectively pass through the biological tissue due to these MNPs. It is used for the molecular screening of surface-based diseases and in disease resection facilitation. The fluorescent molecules absorb light of particular wavelength and emit light of higher wavelength. Contrast agents are used in fluorescence reflectance imaging and in topography systems (Jobsis 1977) for the continuous imaging. However, optical imaging lacks mere penetration of light and minimum resolution. Positron emission tomography (PET) imaging uses high-energy gamma rays produced by the decay of positron-emitting radionuclides (tracers). They are then introduced in the body for the imaging with respect to the nuclear medicine. Tracers include isotopes that can be incorporated into the metabolites. This technique is dependent on the concentration of the tracer and generates three-dimensional images for the illumination of biological processes. PET is preferable for its extraordinary sensitivity though it offers limited spatial resolution (Debbage and Jaschke 2008).

MNPs for their intrinsic magnetic properties are used in the above imaging technologies and in overcoming the limitations. Detectability of these particles depends on the size, material coating, and crystallinity, and multi-model imaging agents are created by their modification with the reporters. Magnetic sensitivity depends on the acquisition time, the strength of the magnetic field, and characteristics of the gradient (Trehin et al., 2006). Yang et al. targeted the tumors in breast cancer by the MRI and thus demonstrated using the multifunctional magneto-polymeric nanohybrids made of doxorubicin a drug and magnetic nanocrystals which were further encapsulated using a copolymer shell (Yang et al., 2007). MNPs are highly functional and efficient in the MRI and the targeted drug delivery.

6.5.1 IN-VIVO ANIMAL EXPERIMENT FOR IMAGING MNPS

The use of MNPs for drug delivery began late in the 1970s and reported by individual researchers with respect to the tumors. Earlier Widder et al. delivered antitumor drugs by employing magnetically responsive microspheres (Widder et al., 1978and 1979). With regard to these studies, other research groups initiated to carryout animal studies using MNPs (Alexiou, 2000). Liu et al. targeted MNPs to the brain of the animals using accurate ultrasound and magnetic targeting in synergy. External magnetic field and ultrasound enhanced the concentration of MNPs; however, the external magnetic field has taken away the tumor cells 15-fold higher when compared to the epirubicin. No MNP debris was observed 6 h after their administration in the control animals and further confirmed by Prussian blue staining and confocal microscopy (Liu et al., 2010; Vinod Kumar et al., 2015). The capability of polycyanoacrylate nanospheres possessing a magnetic core was investigated for the antitumor drug 5-florouracil that improves the therapeutic efficacy diminishing the underlying toxic effects (Arias et al., 2008). Chertok et al. delivered successfully polyethylene imine MNPs (GPEI) in the rats with brain tumors resulting in high cellular association and low toxicity. Magnetic targeting in conjugation with intracarotid administration resulted in tumor entrapment of GPEI in comparison to that of intravenous administration (Chertok et al., 2010). Esophageal cancer in rabbits was targeted by the oral administration under the effect of the magnetic field (Ito et al., 1990). Further evaluation of its potential application was carried out in an in vivo setting in C57BL/6 mice. This was achieved using an MR-based experiment for its traceability and signal to noise optimization. Thus, the improvised protocol reported by Rashi et al. helps in further characterization, evaluation, and extrapolation for the dual purpose of cell therapy while tracking the cells for the intended purpose (Rashi et al., 2016).

Magnetic targeting was successful leading to few clinical trials to date. Lubbe et al. performed the phase I clinical trial for the drug delivery using MNPs complexed to epirubicin by the electrostatic interactions between the phosphate group of the particle surface and amino sugars present in the drug (Lubbe et al., 1996). This clinical study added to the previous studies in mice and rats for their treatment by using low nanoparticle concentration and high ferrofluid and magnetic-targeted delivery and tumor occlusion. This study was performed in Germany by recruiting 14 patients with liver cancer, out of them only 6 patients witnessed the targeted delivery of the epirubicin drug at the tumor site. However, nanoparticles have not targeted the tumor at the liver in other in vivo studies.

Phase II clinical trials were performed in 32 patients with hepatocellular carcinoma utilizing MNPs-coupled doxorubicin hydrochloride drug. The drug was delivered through the hepatic artery. The drug targeted the tumors present in most of the patients, in the presence of external magnetic field and localized using the MRI (Koda et al., 2002). Resistant and nonresistant cancers related to a wide range of cell lines is alleviated using the MNPs coated with polycyanoacrylate nanospheres for the phase I and phase II clinical trials (Merle et al., 2006). Microspheres even eradicate the small subcutaneous B lymphomas in the mice using the ^{90}Y, a γ-emitting radioisotope, that is, for the radionuclide therapy (Häfeli et al., 2009). A third phase clinical trial, an extension of phase II, was performed in four patients of the same study population, wherein the nanoparticles were targeted to the tumor site in the presence of rare earth magnets. This complex has increased the focus at the tumor site (Wilson et al., 2004).

Drug release in vivo was determined in different animals like Sprague-Dawley male rats weighing 280–300 g which are healthy (Hua Xu et al., 2005), 8–9 week old mice, and in sheep (Katja Schulze et al., 2005) as the experimental animals. In humans, the drug delivery to the musculoskeletal system (Zavisova et al., 2007), to the lungs (Gonda et al., 2000), and in xenograft tumor models for the cancer treatment (Li et al., 2012) using the magnetic delivery system without any side effects came into practice. Progress in current technologies and the expansion of MNPs as drug delivery systems to carry drugs to tumor hypoxic zones have fast-tracked in the past decade and led to the development of various magnetic nanoformulations such as liposomes, metallic/nonmetallic, and polymeric nanoparticles (Table 6.1).

TABLE 6.1 Magnetic Nanoparticle Imaging Through Animal Studies.

Sr. no.	Clinical trial	Study subjects and disorder	Drug used with nanoparticles	Study
1.	Phase I	14 patients with liver cancer	Epirubicin	Lubbe et al., 1996
2.	Phase II	32 patients with hepatocellular carcinoma	Doxorubicin hydrochloride	Koda et al., 2002; Merle et al., 2006
3.	Phase III	4 patients of Phase II trial	Rare earth magnets	Wilson et al., 2004

6.6 FUTURE DIRECTIONS

MNPs are in clinical use since many years especially as contrast enrichment agent for MRI (Pankhurst et al., 2003). DNA delivery using MNPs was demonstrated initially using the green fluorescent protein-encoded adeno-associated virus coating on the nanoparticle at the University of Florida. This coating enhanced the transduction efficiency of the cultured cells in vivo and in vitro in mice (Mah et al., 2000). Polyethyleneimine, a cationic polymer, is one of the first studied transfection agents, the secondary amine groups of which condense the DNA and attach it to the particle surface in the presence of electrostatic interactions (Abdallah et al., 1996). Polyethyleneimine is beneficial for the lysosomal rupture and its contents release after internalization maintaining the pH of the lysosome. This highlights the entry of DNA complex through clathrin-dependent pits by endocytosis (Akinc et al., 2005; Schillinger et al., 2005). The initial report on polyethyleneimine-coated nanoparticles was generated for the gene delivery, which is nonviral in the presence of MNPs in vitro. Significant transfection is achieved by the sedimentation of the coating particle at the target site. SPIONS coated with DNA vectors reducing the duration of gene delivery, their conjugation with adenoviral vectors led to the transduction of cell lines without any viral receptors (Scherer et al., 2002).

Brain tumors were targeted using polyethyleneimine-modified MNPs resulting in low toxicity and high association within the cells. These nanoparticles play a crucial role in treating the brain tumors, as they do not accumulate successfully at the tumor site in the absence of magnetic field (Chertok et al., 2010). MNPs of iron oxide was developed by Liu et al. by encapsulating them within the polymer on the surface of which anticancer agent epirubicin was immobilized. This magnetic drug delivery system targets the brain using a synergistic ultrasound and magnetic delivery system (Liu et al., 2010). Alternative approach for synthesizing polyethyleneimine-coated MNPs has come into practice where dextran silica particles were used (Mc Bain et al., 2007). Transfection efficiency was improved by applying dynamic magnetic fields obtained from rare earth magnets (Mc Bain et al., 2008). Magnetofectin is utilized to transfect lung epithelial cells, blood endothelial cells, and to deliver antisense oligonucleotides and small interfering RNA to downregulate gene expression (Gersting et al., 2004, Krotz and Sohn et al., 2003; Krotz, de Wit et al., 2003). Polyethylene amine increases cellular uptake of magnetofectins leading to endocytosis and further internalization in the presence of magnetic field (Huth et al., 2004). Schilinger et al., recently, reported the decreased expression of luciferase in HeLa (first human cell lines) using MNP-associated SiRNA (Schilinger et al., 2005).

Magnetic drug delivery system using doxorubicin was developed by Chen et al. where it was bound to the magnetite nanoparticles and embedded in PEG functionalized silica shell that is porous. Silica shell blocks the doxorubicin discharge from the nanoparticle apart from protecting the drug molecules (Chen et al., 2010). Cai et al. recently reported a remarkable advancement using the nanotube spearing for the effective gene delivery using the nanoparticles. In this procedure, DNA-coated nickel-entrenched carbon nanotubes are introduced into the cells oriented toward the magnetic field. These nanotubes are pulled toward the cells aligning in the magnetic flux, allowing the nanotubes to pierce and pass through the membrane sending the target DNA and other cell types such as B cell lymphoma and primary neurons retaining the cell viability (Cai et al., 2005). Esophageal cancer was first targeted in rabbits under magnetic field influenced by oral administration (Ito et al., 1990). Magnetic microspheres approximately 100 nm in size were nontoxic and used in the clinical trials on cancer therapy in 14 patients of Germany (Lubbe et al., 1996). Arias et al. investigated the capabilities of polycyanoacrylate nanospheres with a magnetite core as delivery systems for the antitumor drug 5-flurouracil (Arias et al., 2008).

Recent advances in the nanoparticle-based drug and gene delivery are promising and accomplished clinical success by reaching the market-place, though there are still certain obstacles to overcome. Nanoparticles are designed using new materials such that they attain elevated magnetic moments for them to capture easily. There is an enduring progress in permanent magnet technology and novel, implantable materials capable of creating high magnetic fields and high gradients look set to, significantly enhance the ability to target sites deeper in the tissue. Patients possessing tumor sites near the surface of the body and in the liver are likely to be the first to benefit from this technique. Nevertheless, the potential for clinical application of magnetic targeting to a range of tumors and other diseases makes this an extremely fertile field of applied research. As we move toward our future, we are in immediate need of better characterization tools in order to evaluate and as well understand the behavior of novel MNPs in the body.

6.7 CONCLUSION

This chapter covers the physical and chemical properties, applications, their fabrication, enhanced targeting through passive, active, and physical magnetic targeting mechanisms, route of their administration through oral, parenteral, intranasal, dermal, and transdermal routes, imaging capabilities, innovative approaches for using stem cells and their in vivo tracking. MNPs possess the ability to overcome the in vivo barriers due to their physico-chemical properties. Active targeting, in particular, offers high sensitivity due to the ability to direct MNP localization, but attention has to be paid to the targeting agent that is being used, and the method of MNP attachment employed. Finest method for the manufacture of MNPs with desired size range is based on the physicochemical characteristics of a drug and its entrap-ment efficiency. The nanoparticle is fabricated in the presence of less toxic reagents, by economic scale-up through procedural simplification and by optimization to improve yield and entrapment efficiency. Basic knowledge is being gained by the nanoparticle engineers through controlled studies of individual physicochemical parameters leading to the more elaborate and functional MNPs. Simultaneously, questions arise about the removal and long-term toxicity of these particles to clinical entry. Once we address the issues, MNPs will move closer to clinical application, improving the diagnosis, treatment, and monitoring of our most unmanageable diseases.

KEYWORDS

- **magnetic nanoparticles**
- **drug targeting**
- **superparamagnetic iron oxide nanoparticles**
- **magnetic resonance imaging**
- **disease therapy**
- **administration route**

REFERENCES

Abdallah, B.; et al. A Powerful Nonviral Vector for in Vivo Gene Transfer into the Adult Mammalian Brain: Polyethylenimine. *Hum Gene Ther.* **1996,** *7*, 1947–1954.

Abolfazl, A.; Mohamad, S.; Soodabeh, D. Magnetic Nanoparticles: Preparation, Physical Properties, and Applications in Biomedicine. *Nanoscale Res. Lett.* **2012,** *7*, 144.

Akinc, A.; et al. Exploring Polyethyleniminemediated DNA Transfection and the Proton Sponge Hypothesis. *J. Genet. Med.* **2005,** *7*, 657.

Albanese, A.; Tang, P. S.; Chan, W. C. The Effect of Nanoparticle Size, Shape, and Surface Chemistry on Biological Systems. *Annu. Rev. Biomed. Eng.* **2012,** *14*, 1–16.

Alexiou, C.; et al. Locoregional Cancer Treatment with Magnetic Drug Targeting. *Cancer Res.* **2000,** *60*, 6641–6648.

Alexis, F.; et al. Factors Affecting the Clearance and Biodistribution of Polymeric Nanoparticles. *Mol. Pharm.* **2008,** *5*(4), 505–515.

Ameur, S.; et al. Sensitive Immunodetection Through Impedance Measurements Onto Gold Functionalized Electrodes. *Appl. Biochem. Biotechnol.* **2000,** *89*, 161–170.

Arias, J. L.; et al. Magnetite/Poly(alkylcyanoacrylate) (core/shell) Nanoparticles as 5-Fluorouracil Delivery Systems for Active Targeting. *Eur. J. Pharm. Biopharm.* **2008,** *69*, 54–63.

Bae, Y. H.; Park, K. Targeted Drug Delivery to Tumors: Myths, Reality and Possibility. *J. Controlled Release* **2011,** *153*(3), 198–205.

Bernkop-Schnurch A. Nanocarrier Systems for Oral Drug Delivery: Do We Really Need Them? *Eur. J. Pharm. Sci.* **2013,** *49*, 272–277.

Brans, B.; et al. Clinical Applications of Newer Radionuclide Therapies. *Eur. J. Cancer* **2006,** *42*, 994–1003.

Briley-Saebo, K.; et al. Hepatic Cellular Distribution and Degradation of Iron Oxide Nanoparticles Following Single Intravenous Injection in Rats: Implications for Magnetic Resonance Imaging. *Cell Tissue Res.* **2004,** *316*(3), 315–323.

Cai, D.; et al. Highly Efficient Molecular Delivery into Mammalian Cells Using Carbon Nanotube Spearing. *Nat. Methods.* **2005**, *2*, 449–454.

Chen, F. H.; et al. Synthesis of a Novel Magnetic Drug Delivery System Composed of Doxorubicin-Conjugated Fe3O4 Nanoparticle Cores and a PEG-Functionalized Porous Silica Shell. *Chem. Commun.* **2010**, *46*, 8633–8635.

Chen M-C, Mi F-L, Liao Z-X, et al. Recent advances in chitosan-based nanoparticles for oral delivery of macromolecules. *Adv Drug Deliv Rev.* **2013**, *65*, 865–879.

Cheng, Z.; et al. Multifunctional Nanoparticles: Cost Versus Benefit of Adding Targeting and Imaging Capabilities. *Science* **2012**, *338*(6109), 903–910.

Chertok, B.; David, A. E.; Yang, V. C. Polyethyleneimine-Modified Iron Oxide Nanoparticles for Brain Tumor Drug Delivery Using Magnetic Targeting and Intra-Carotid Administration. *Biomaterials* **2010**, *31*, 6317–6324.

Dabbousi, B. O. (CdSe) ZnS Core-Shell Quantum Dots: Synthesis and Characterization of a Size Series of Highly Luminescent Nanocrystallites. J. *Phys.* Chem. B 1997, *101*(46), 9463–9475.

Debbage, P.; Jaschke, W. Molecular Imaging with Nanoparticles: Giant roles for Dwarf Actors. *Histochem. Cell Biol.* **2008**, *130*, 845–875.

Duran, J. D. G.; et al. Magnetic Colloids as Drug Vehicles. *J. Pharm. Sci.* **2008**, *97*, 2948–2983.

Ersoy, H.; Rybicki, F. J. Biochemical Safety Profiles of Gadolinium Based Extracellular Contrast Agents and Nephrogenic Systemic Fibrosis. *J. Magn. Reson. Imaging* **2007**, *26*(5), 1190–1197.

Ferrari, M. Cancer Nanotechnology: Opportunities and Challenges. *Nat. Rev. Cancer* **2005**, *5*, 161–171.

Fogal, V.; et al. Mitochondrial/Cell-Surface Protein p32/gC1qR as a Molecular Target in Tumor Cells and Tumor Stroma. *Cancer Res.* **2008**, *68*(17), 7210–7218.

Fukushima, S.; et al. Direct Intramyocardial but not Intracoronary Injection of Bone Marrow Cells Induces Ventricular Arrhythmias in a Rat Chronic Ischemic Heart Failure Model. *Circulation* **2007**, *115*, 2254–2261.

Gao, X. H.; et al. In Vivo Cancer Targeting and Imaging with Semiconductor Quantum Dots. *Nat. Biotechnol.* **2004**, *22*(8), 969–976.

Gersting, S. W.; et al. Gene Delivery to Respiratory Epithelial Cells by Magnetofection. *J. Gene Med.* **2004**, *6*, 913–922.

Gonda, I. The Ascent of Pulmonary Drug Delivery. *J. Pharm. Sci.* **2000**, *89*, 940–945.

Green, M. Organometallic Based Strategies for Metal Nanocrystals Synthesis. *Chem. Commun.* **2005**, *24*, 3002–3011.

Häfeli, U.; Gubin, S. P. Magnetic Nanoparticles. Wiley-VCH Verlag GmbH and Co, KGaA: Weinheim, 2009.

Hamoudeh, M.; et al. Radionuclides Delivery Systems for Nuclear Imaging and Radiotherapy of Cancer. *Adv. Drug Deliv. Rev.* **2008**, *60*, 1329–1346.

Hewakuruppu, Y. L.; et al. "Plasmonic" Pump–Probe "Method to Study Semi-Transparent Nanofluids." *Appl. Opt.* **2013**, *52*(24), 6041–6050.

Hines, M. A.; Guyot-Sionnest, P. Synthesis and Characterization of Strongly Luminescing ZnS-Capped CdSe Nanocrystals. J. Phys. Chem. **1996**, *100*(2), 468–471.

Hu, Y.; et al. TiO2 Nanotubes as Drug Nanoreservoirs for the Regulation of Mobility and Differentiation of Mesenchymal Stem Cells. *Acta Biomater.* **2012**, *8*, 439–448.

Huang, X.; et al. Plasmonic Photothermal Therapy (PPTT) Using Gold Nanoparticles. *Lasers Med. Sci.* **2008**, *23*(3), 217–228.

Huth, S.; et al. Insights into the Mechanism of Magnetofection Using PEI-Based Magnetofectins for Gene Transfer. *J. Gene Med.* **2004**, *6*, 923–936.

Issa, B.; et al. Magnetic Nanoparticles: Surface Effects and Properties Related to Biomedicine Applications. *Int. J. Mol. Sci.* **2013**, *14*(11), 21266–21305.

Ito, R.; et al. Magnetic Granules: a Novel System for Specific Drug Delivery to Esophageal Mucosa in Oral Administration. *Int. J. Pharm.* **1990**, *61*, 109–117.

Jeong, B.; Bae, Y. H.; Kim, S. W. Drug Release from Biodegradable Injectable Thermosensitive Hydrogel of PEG-PLGA-PEG Triblock Copolymer. *J. Controlled Release* **2000**, *63*, 155–163.

Jinek, M.; et al. A Programmable Dual-RNA-Guided DNA Endonuclease in Adaptive Bacterial Immunity. *Science* **2012**, *337*(6096), 816–821.

Jobsis, F. F. Noninvasive, Infrared Monitoring Of Cerebral and Myocardial Oxygen Sufficiency and Circulatory Parameters. *Science* **1977**, *198*, 1264–1267.

Jurgons, R.; et al. Drug Loaded Magnetic Nanoparticles for Cancer Therapy. *J. Phys. Condens. Matter* **2006**, *18*, S2893–S2902.

Kami, D.; Watakabe, K.; Yamazaki-Inoue, M.; Minami, K.; Kitani,T.; Itakura, Y.; Toyoda, M.; Sakurai, T.; Umezawa, A; Gojo, S. Large-scale cell production of stem cells for clinical application using the automated cell processing machine. *BMC Biotechnol.* **2013**, *13*, 102.

Katza, S.; et al. Intraarticular Application of Super Paramagnetic Nanoparticles and Their Uptake by Synovial Membrane-an Experimental Study in Sheep. *J. Magn. Magn. Mater.* **2005**, *293*, 419–432.

Kim, J.; et al. Designed Fabrication of Multifunctional Magnetic Gold Nanoshells and Their Application to Magnetic Resonance Imaging and Photothermal Therapy. *Angew. Chem. Int. Ed.* **2006**, *45*(46), 7754–7758.

Kim, S. E.; et al. Enhancement of Ectopic bone Formation by Bone Morphogenetic Protein-2 Delivery Using Heparin-Conjugated PLGA Nanoparticles with Transplantation of Bone Marrow-Derived Mesenchymal Stem Cells. *J. Biomed. Sci.* **2008**, *15*, 771–777.

Koda, J.; et al. Phase I/II Trial of Hepatic Intraarterial Delivery of Doxorubicin Hydrochloride Adsorbed to Magnetic Targeted Carriers in Patients with Hepatocarcinoma. *Eur. J. Cancer* **2002**, *38*(Suppl 7), S18.

Krotz, F.; et al. Magnetofection Potentiates Gene Delivery to Cultured Endothelial Cells. *J. Vasc. Res.* **2003**, *40*, 425–434.

Krotz, F.; et al. Magnetofection-a Highly Efficient tool for Antisense Oligonucleotide Delivery in Vitro and in Vivo. *Mol. Ther.* **2003**, *7*, 700–710.

Laurent, S. D.; et al. Magnetic Iron Oxide Nanoparticles: Synthesis, Stabilization, Vectorization, Physicochemical Characterizations, and Biological Applications. *Chemical Rev.* **2008**, *108*, 2064–2110.

Lazaro, F. J.; et al. Magnetic Characterization of Rat Muscle Tissues After Subcutaneous Iron Dextran Injection. *Biochem. Biophys. Acta* **2005**, *1740*, 434–445.

Lecommandoux, S.; et al. Smart Hybrid Magnetic Self-Assembled Micelles and Hollow Capsules. *Prog. Solid State Chem.* **2006**, *34*, 171–179.

Lee, J.; et al. Simple Synthesis of Functionalized Super Paramagnetic Magnetite/Silica Core/ Shell Nanoparticles and Their Application as Magnetically Separable High Performance Biocatalysts. *Small* **2008**, *4*(1), 143–152.

Lewinski, N.; Colvin V.; Drezek, R. Cytotoxicity of Nanoparticles. *Small* **2008**, *4*, 26–49.

Li, C.; Li, L.; Keate, A. C. Targeting Cancer Gene Therapy with Magnetic Nanoparticles. *Oncotarget* **2012**, *3*, 365–370.

Li, Y.; Kröger, M.; Liu, W. K. Shape Effect in Cellular Uptake of PEGylated Nanoparticles: Comparison Between Sphere, Rod, Cube and Disk. *Nanoscale* **2015**, *7*(40), 16631–16646.

Liu, C.; et al. Reduction of Sintering During Annealing of FePt Nanoparticles Coated with Iron Oxide. *Chem. Mater.* **2005**, *17*(3), 620–625.

Liu, H. L.; et al. Magnetic Resonance Monitoring of Focused Ultrasound/Magnetic Nanoparticle Targeting Delivery of Therapeutic Agents to the Brain. *Proc. Natl. Acad. Sci.* **2010**, *107*, 15205–15210.

Low, P. S.; Henne, W. A.; Doorneweerd, D. D. Discovery and Development of Folic-Acid-Based Receptor Targeting for Imaging and Therapy of Cancer and Inflammatory Diseases. *Acc. Chem. Res.* **2008**, *41*(1), 120–129.

Lu, Y.; et al. Modifying the Surface Properties of Super Paramagnetic Iron Oxide Nanoparticles Through a Sol-Gel Approach. *Nano Lett.* **2002**, *2*(3), 183–186.

Lubbe, A. S.; et al. Clinical Experiences with Magnetic Drug Targeting: A Phase I Study with 4'-Epidoxorubicin in 14 Patients with Advanced Solid Tumors. *Cancer Res.* **1996**, 56, 4686–4693.

Mah, C.; et al. Microsphere-Mediated Delivery of Recombinant AAV Vectors in Vitro and in Vivo. *Mol. Ther.* **2000**, *1*, S239.

Martina, M. S.; et al. Generation of Superparamagnetic Liposomes Revealed as Highly Efficient MRI Contrast Agents for in Vivo Imaging. [*J. Am. Chem. Soc.* **2005**, *127*, 10676–10685.

McBain, S. C.; et al. Magnetic Nanoparticles as Gene Delivery Agents: Enhanced Transfection in the Presence of Oscillating Magnet Arrays. *Nanotechnology* **2008**, *19*, 405102.

McBain, S. C.; Yiu, H. H. P.; El Haj, A. Polyethyleneimine Functionalized Iron Oxide Nanoparticles as Agents for DNA Delivery and Transfection. *J. Mater. Chem.* **2007**, *17*, 2561–2565.

McBain, S. C.; H. P. Yiu, H. H. P.; Dobson, J. Magnetic Nanoparticles for Gene and Drug Delivery. *Int. J. Nanomedicine.* **2008**, *3*(2), 169–180.

Merisko-Liversidge, E.; Liversidge, G. G. Nanosizing for Oral and Parenteral Drug Delivery: A Perspective on Formulating Poorly-Water Soluble Compounds Using Wet Media Milling Technology. *Adv. Drug Delivery Rev.* **2011**, *63*, 427–440.

Merle, P.; et al. Phase 1 Study of Intra-Arterial Hepatic (IAH) Delivery of Doxorubicintransdrug (DT) for Patients with Advanced Hepatocellular Carcinoma (HCC). *J. Clin. Virol.* **2006**, *36*(2), 179.

Meyers, D. *Surfactant Science and Technology*, 3rd ed.; John Wiley and Sons Inc; Hoboken: New Jersey, 2005.

Morcos, S. K. Nephrogenic Systemic Fibrosis Following the Administration of Extracellular Gadolinium Based Contrast Agents: Is the Stability of the Contrast Agent Molecule an Important Factor in the Pathogenesis of this Condition? *Br. J. Radiol.* **2007**, *80*(950), 73–76.

Mulder, W. J. M.; et al. Lipidbased Nanoparticles for Contrast-Enhanced MRI and Molecular Imaging. *NMR Biomed.* **2006**, *19*, 142–164.

Nagavarma, B. V. N.; et al. Different Techniques for Preparation of Polymeric Nanoparticles—A Review. *Asian J. Pharm. Clin. Res.* **2012**, *5*(Suppl 3), 35–49.

Nasongkla, N.; et al. Multifunctional Polymeric Micelles as Cancer-Targeted, MRI-Ultrasensitive Drug Delivery Systems. *Nano Lett.* **2006**, *6*, 2427–2430.

Neubert, R. H. H. Potentials of New Nanocarriers for Dermal and Transdermal Drug Delivery. *Eur. J. Pharm. Biopharm.* **2011**, *77*(1), 1–2.

Nguyen, L. T.; et al. Electrospun Poly(L-lactic acid) Nanofibres Loaded with Dexamethasone to Induce Osteogenic Differentiation of Human Mesenchymal Stem Cells. *J. Biomater. Sci. Polym.* **2012,** *23*(14), 1771–1791.

Nitin, N.; et al. Functionalization and Peptide-Based Delivery of Magnetic Nanoparticles as an Intracellular MRI Contrast Agent. *J. Biol. Inorg. Chem.* **2004,** *9*(6), 706–712.

Nurmi, J. T.; et al. Characterization and Properties of Metallic iron Nanoparticles: Spectroscopy, Electrochemistry, and Kinetics. *Environ. Sci. Technol.* **2005,** *39*, 1221–1230.

Okita, K.; et al. Generation of Mouse-Induced Pluripotent Stem Cells with Plasmid Vectors. *Nat. Protoc.* **2010,** *5*(3), 418–428.

Omid, V.; Jonathan, W. G.; Miqin, Z. Design and Fabrication of Magnetic Nanoparticles for Targeted Drug Delivery and Imaging. *Adv. Drug Deliv. Rev.* **2010,** *62*(3), 284–304.

Owens III, D. E.; Peppas, N. A. Opsonization, Biodistribution, and Pharmacokinetics of Polymeric Nanoparticles. *Int. J. Pharm.* **2006,** *307*(1), 93–102.

Pankhurst, Q. A.; et al. Applications of Magnetic Nanoparticles in Biomedicine. *J. Phys. D.* **2003,** *36*, R167–R181.

Pawar, K. R.; Babu, R. J. Polymeric and Lipid-Based Materials for Topical Nanoparticle Delivery Systems. *Crit. Rev. Ther. Drug.* **2010,** *27*(5), 419–459.

Pellegrino, T.; et al. Hydrophobic Nanocrystals Coated with an Amphiphilic Polymer Shell: A General Route to Water Soluble Nanocrystals. *Nano Lett.* **2004,** *4*(4), 703–707.

Peng, X. G.; et al. Epitaxial Growth of Highly Luminescent CdSe/CdS Core/Shell Nanocrystals with Photostability and Electronic Accessibility. J. Am. Chem. Soc. **1997,** *119*(30), 7019–7029.

Piao, Y.; et al. Wrapbake-Peel Process for Nanostructural Transformation from Beta-FeOOH Nanorods to Biocompatible Iron Oxide Nanocapsules. *Nat. Mater.* **2008,** *7*(3), 242–247.

Pires, A.; et al. Intranasal Drug Delivery: How, Why and What for? *Pharm. Sci.* **2009,** *12*(3), 288–311.

Qu, L. H.; Peng, Z. A.; Peng, X. G. Alternative Routes Toward High Quality CdSe Nanocrystals. Nano Lett. **2001,** *1*(6), 333–337.

Quellec, P.; et al. Protein Encapsulation within Polyethylene Glycol-Coated Nanospheres. I. Physicochemical Characterization. *J. Biomed. Mater. Res.* **1998,** *42*, 45–54.

Rashi, J.; et al. A Modified Approach to Image Guided Cell Based Therapy for Cardiovascular Diseases Using Cardiac Precursor Nanoprobe– GloTrack. *Protocol Exchange* **2016,** DOI: 10.1038/protex.2016.004.

Rockenberger, J.; Scher, E. C.; Alivisatos, A. P. A New Nonhydrolytic Single-Precursor Approach to Surfactantcapped Nanocrystals of Transition Metal Oxides. J. Am. Chem. Soc. 1999, *121*(49), 11595–11596.

Rowland, M. Influence of Route of Administration on Drug Availability. *J. Pharm. Sci.* **1972,** *61*(1), 70–74.

Ruoslahti, E.; Bhatia, S. N.;Sailor, M. J. Targeting of Drugs and Nanoparticles to Tumors. *J. Cell Biol.* **2010,** *188*(6), 759–768.

Saha, R. N.; et al. Nanoparticulate Drug Delivery Systems for Cancer Chemotherapy. *Mol. Membr Biol.* **2010,** *27*(7), 215–231.

Santos, J. L.; et al. Osteogenic Differentiation of Mesenchymal Stem Cells Using PAMAM Dendrimers as Gene Delivery Vectors. *J. Control. Release* **2009,** *134*(2), 141–148.

Santra, S.; et al. Synthesis and Characterization of Silica-Coated Iron Oxide Nanoparticles in Microemulsion: The Effect of Nonionic Surfactants. *Langmuir* **2001,** *17*, 2900–2906.

Sanvicens, N.; Marco, M. P. Multifunctional Nanoparticles–Properties and Prospects for Their Use in Human Medicine. *Trends Biotechnol.* **2008,** *26*(8), 425–433.

Savva, M.; Duda, E.; Huang, L. A Genetically Modified Recombinant Tumor Necrosis Factor-Alpha Conjugated to the Distal Terminals of Liposomal Surface Grafted Poly-Ethyleneglycol Hains. *Int. J. Pharm.* **1999**, *184*, 45–51.

Scherer, F.; et al. Magnetofection: Enhancing and Targeting Gene Delivery by Magnetic Force in Vitro and in Vivo. *Gene Ther.* **2002**, *9*, 102–109.

Schillinger, U.; et al. Advances in Magnetofection Magnetically Guided Nucleic Acid Delivery. *J. Magn. Magn. Mater.* **2005**, *293*, 501–508.

Schulze, K.; et al. Uptake and Biocompatibility of Functionalized Poly(vinylalcohol) Coated Superparamagnetic Maghemite Nanoparticles by Synoviocytes in Vitro. *J. Nanosci. Nanotechnol.* **2006**, *6*(9–10), 2829–2840.

Schütt, W.; Teller, J.; Zborowski, M. *Scientific and Clinical Applications of Magnetic Carriers;* Plenum Publishing Corp: NY, USA, 2005; Vol. 4, pp 145–160.

Seyda, B.; Banu, Y.; Ali Demir, S. Magnetic Nanoparticles: Synthesis, Surface Modifications and Application in Drug Delivery. In *Recent Advances in. Novel Drug Carrier Systems*; Sezer A. D. Ed.; InTech: 2012. DOI: 10.5772/52115. https://www.intechopen.com/books/recent-advances-in-novel-drug-carrier-systems/magnetic-nanoparticles-synthesis-surface-modifications-and-application-in-drug-delivery.

Shenoy, D. B.; Amiji, M. M. Poly(ethylene oxide)-Modified poly([-caprolactone) Nanoparticles for Targeted Delivery of Tamoxifen in Breast Cancer. *Int. J. Pharm.* **2005**, *293*(1), 261–270.

Solanki, A.; Kim, J. D.; Lee K. B. Nanotechnology for Regenerative Medicine: Nanomaterials for Stem Cell Imaging. *Nanomedicine* **2008**, *3*(4), 567–578.

Sosnovik, D. E.; Nahrendorf, M.; Weissleder, R. Molecular Magnetic Resonance Imaging in Cardiovascular Medicine. *Circulation* **2007**, *115*(15), 2076–2086.

Sreedhar, A. S.; Csermely, P. Heat Shock Proteins in the Regulation of Apoptosis: New Strategies in Tumor Therapy: A Comprehensive Review. *Pharmacol. Ther.* **2004**, *101*(3), 227–257.

Srivastava, C.; et al. Size Effect Ordering in [FePt](100-x) Cr-x Nanoparticles. *J. Appl. Phys.* **2006**, *99*, 054304.

Subramani, K. Applications of Nanotechnology in Drug Delivery Systems for the Treatment of Cancer and Diabetes. *Int. J. Nanotech.* **2006**, *3*, 557–580.

Sun, E. Y.; Josephson, L.; Weissleder, R. "Clickable" Nanoparticles for Targeted Imaging. *Mol. Imaging.* **2006**, *5*(2), 122–128.

Sun, S.; Zeng, H. Size-Controlled Synthesis of Magnetite Nanoparticies. *J. Am. Chem. Soc.* **2002**, *124*, 8204–8205.

Thakur, S.; et al. The Effect of Polyethylene Glycol Spacer Chain Length on the Tumortargeting Potential of Folate-Modified PPI Dendrimers. *J. Nanopart. Res.* **2013**, *15*, 1625.

Torchilin, V. P. Recent Advances with Liposomes as Pharmaceutical Carriers. *Nat. Rev. Drug Discov.* **2005**, *4*, 145–160.

Tratnyek, P. G.; Johnson, R. L. Nanotechnologies for Environmental Cleanup. *Nano Today* **2006**, *1*(2), 44–48.

Trehin, R.; et al. Fluorescent Nanoparticle Uptake for Brain Tumor Visualization. *Neoplasia.* **2006**, *8*(4), 302–311.

Velge-Roussel, F.; et al. Immunochemical Characterization of Antibody-Coated Nanoparticles. *Experientia.* **1996**, *52*(8), 803–806.

Vinod, K. V.; et al. Fluorescent Magnetic Iron Oxide Nanoparticles for Cardiac Precursor Cell Selection from Stromal Vascular Fraction and Optimization for Magnetic Resonance Imaging. Int. J. Nanomedicine **2015**, *10*, 711–726.

Wang, L. Y.; et al. Iron Oxide-Gold Coreshell Nanoparticles and Thin Film Assembly. *J. Mater. Chem.* **2005**, *15*(18), 1821–1832.

Weissleder, R.; Pittet, M. J. Imaging in the Era of Molecular Oncology. *Nature.* **2008**, *452*, 580–589.

Widder, K. J.; Senyei, A. E.; Ranney, D. F. Magnetically Responsive Microspheres and Other Carriers for the Biophysical Targeting of Antitumor Agents. *Adv. Pharmacol. Chemother.* **1979**, *16*, 213–271.

Widder, K. J.; Senyel, A. E.; Scarpelli, G. D. Magnetic Microspheres: a Model System of Site Specifi c Drug Delivery in Vivo. *Proc. Soc. Exp. Biol. Med.* **1978**, *158*(2), 141–146.

Wilson, M. W.; et al. Regional Therapy with a Magnetic Targeted Carrier Bound to Doxorubicin in a Dual MR Imaging/Conventional Angiography Suite–Initial Experience with Four Patients. *Radiology.* **2004**, *230*(1), 287–293.

Xu, H.; et al. Site-Directed Research of Magnetic Nanoparticles in Magnetic Drug Targeting. *J. Magn. Magn. Mater.* **2005**, *293*(1), 514–519.

Yadollahpour, A. Magnetic Nanoparticles: A Review of Chemical and Physical Characteristics Important in Medical Applications. *Orient. J. Chem.* **2015**, *31*(1), 25–30.

Yang, J.; et al. Multifunctional Magneto-Polymeric Nanohybrids for Targeted Detection and Synergistic Therapeutic Effects on Breast Cancer. *Angew. Chem.* **2007**, *46*(46), 8836–8839.

Yoshimura, T.; et al. Cancer Therapy Utilizing Molecular Layer Deposition and Self-Organized Lightwave Network: Proposal and Theoretical Prediction. *IEEE J. Sel. Top. Quantum Electron.* **2012**, *18*, 1192–1199.

Zavisova, V.; et al. Encapsulation of Indomethacin in Magnetic Biodegradable Polymer Nanoparticles. *J. Magn. Magn. Mater.* **2007**, *311*(1), 379–382.

Zhang, H.; et al. Targeting and Hyperthermia of Doxorubicin by the Delivery of Single-Walled Carbon Nanotubes to EC-109 Cells. *J. Drug Target.* **2013**, *21*(3), 312–319.

Zhang, J. I. N.; et al. Design of Nanoparticles as Drug Carriers for Cancer Therapy. *Cancer Genomics Proteomics* **2006**, *3*(3–4), 147–157.

Zhang, L.; Zhang, N. How Nanotechnology Can Enhance Docetaxel Therapy. *Int. J. Nanomed* **2013**, *8*, 2927–2941.

Zhang, Z.; et al. Polymeric Nanoparticles-Based Topical Delivery Systems for the Treatment of Dermatological Diseases. *Wires Nanomed. Nanobi.* **2013**, *5*(3), 205–218.

Zorko, M.; Langel, U. Cell-Penetrating Peptides: Mechanism and Kinetics of Cargo Delivery. *Adv. Drug Delivery Rev.* **2005**, *57*(4), 529–545.

CHAPTER 7

OVERVIEW OF APPLICATIONS OF GOLD NANOPARTICLES IN THERAPEUTICS

JULIANA PALMA ABRIATA[1], MARCELA TAVARES LUIZ[1], GIOVANNI LOUREIRO RASPANTINI[1], JOSIMAR O. ELOY[2], RAQUEL PETRILLI[3], JULIANA MALDONADO MARCHETTI[1], and ROBERT LEE[4,*]

[1]School of Pharmaceutical Sciences of RIbeirao Preto, University of Sao Paulo, Ribeirao Preto, Sao Paulo, Brazil

[2]School of Pharmaceutical Sciences, Sao Paulo State University, Araraquara, Sao Paulo, Brazil

[3]University of Western Sao Paulo (UNOESTE), Presidente Prudente, São Paulo, Brazil

[4]Ohio State University, Columbus, Ohio, USA

*Corresponding author. E-mail: lee.1339@osu.edu

ABSTRACT

Gold nanoparticles (GNPs) have been applied in the therapy due to their advantageous optical, chemical, and catalytic characteristics. They can be synthesized by various methods, resulting in different final products. GNPs have been used to promote drug delivery through passive and active targeting. Many ligands, such as antibodies, small molecules, and peptides, have been conjugated to the surface of GNPs aiming at rapid cellular uptake and improved efficacy and selectivity. For this propose, those ligands can be attached to GNP surface by physical adsorption or covalent coupling. Besides, GNPs can be designed to respond to different stimuli, both intrinsic and extrinsic, and thus deliver the drug to the target site. Then, these NPs will be activated by one or more specific triggers from inner or outer body, releasing its content in the target. Therefore, GNPs are versatile drug delivery systems. In this context, the present chapter aims to present and discuss some aspects of GNP development and its use in the drug delivery field. For this, we will first describe different types of GNPs, their production methods, and characterization. Then, the different functionalization options, using small molecules or macromolecules will be discussed. Finally, the stimuli-responsive GNPs will be discussed.

7.1   INTRODUCTION

The application of metallic nanoparticles (NPs) in biomedicine has been studied for over 40 years. A variety of metallic NPs has been developed and, among them, gold nanoparticles (GNPs) are the most employed systems due to their advantageous optical, chemical, and catalytic characteristics (de Araújo et al., 2017; Pena-pereira et al., 2017). As a consequence, GNPs are considered a multifunctional platform for imaging and diagnosis (Cho et al., 2017; Wang et al., 2005), theranostics (Heo et al., 2012; Zhao et al., 2016), targeted delivery of therapeutic molecules (Fernandes et al., 2017; Saber et al., 2017), cellular and tissue sensitization (Hainfeld et al., 2004, 2010), and for stimuli–response approaches (Gamal-Eldeen et al., 2016; Mocan et al., 2017).

GNPs are known to be inert and stable against oxidation. They can be synthesized by various methods, resulting in different particle architecture. In this context, the first method was developed by Turkevich, and after that, Frens and Brust developed different methods for GNP production, which are used since then (Priyadarshini and Pradhan, 2016; Shah et al., 2014; Turkevich et al., 1951).

Upon passive accumulation, GNPs accumulate at tumor sites or inflamed tissues due to the leaky architecture of the blood vessels, which are responsible for oxygen and nutrients supply (Dreaden et al., 2012). This phenomenon is named as enhanced permeability and retention (EPR) effect and was first described by Matsumura and Maeda (Maeda, 2010; Matsumura and Maeda, 1986). Another possibility is the coating of GNP surface using polyethylene glycol (PEG). The use of PEG coating can protect and reduce GNP clearance by the reticuloendothelial system (RES) (Ajnai et al., 2014). Another approach involves active targeting. It refers to the functionalization of NPs surface with different ligands that can be recognized by overexpressed receptors at the target cells. The use of active targeting can promote rapid cellular uptake and improve efficacy and selectivity (Petrilli et al., 2014). For instance, GNPs surface can be modified with antibodies, small molecules, peptides, among others, and thus they can be targeted to selective disease sites (Ajnai et al., 2014). Briefly, these moieties can be attached to GNP surface by physical adsorption or by covalent coupling. The physical adsorption is a rapid and simple process; however, changes in physical parameters can produce the detachment of the ligand. On the contrary, the covalent attachment is currently the most widely used technique, and it is frequently based on a strong bond between gold and thiol groups (Priyadarshini and Pradhan, 2016). The covalent bond increases the stability, which is necessary for GNP administration (Dreaden et al., 2012).

GNPs can also be designed to respond to different stimuli, both intrinsic and extrinsic and thus deliver the drug to the target site. Intrinsic stimuli correspond to microenvironmental changes that occur either pathologically or physiologically, whereas extrinsic stimuli correspond to those externally applied, such as lasers or ultrasound (Tian et al., 2016). GNPs can be designed to respond to alterations in the pH, temperature, light irradiation, ultrasound, magnetic field, electric current, and also external application of heat (Eloy et al., 2015). Thus, following administration of GNPs, intravenously or through a different route, it will accumulate at the target site, both passively by the EPR effect of actively based on different targeting moieties. Then, these NPs can be activated by one or more specific triggers from internal or external stimuli, releasing its content in the target (Yao et al., 2016).

Therefore, GNPs are versatile drug delivery systems. Herein, our aim is to present and discuss some key aspects of GNP development and use them in the drug delivery field. For this, we will first describe different types of GNPs, their production methods, and characterization. Then, the different functionalization methods, using small molecules or macromolecules will be discussed. Finally, the stimuli-responsive GNPs will be presented and discussed.

7.2 GOLD NANOPARTICLE (GNP) SYNTHESIS

GNPs may be produced by two main different approaches: "top-down" and "bottom-up" manufacturing. Bottom-up methods involve the manipulation of substances at atomic level aiming the generation of bigger structures, using chemically modulated tools. On the other hand, top-down methods use physical and chemical processes to reduce the particle size. Both approaches aim to design structures with specific sizes, shapes, and physicochemical properties (Alex and Tiwari, 2015). An extensive amount of techniques using both approaches have been described through the last decades. The present section intends to introduce and discuss their capabilities, limitations, and synthesis methods.

7.2.1   COMMON SYNTHESIS METHODS

7.2.1.1   CITRATE REDUCTION METHOD

Back in 1951, the first GNP manufacturing method was described (Turkevich et al., 1951). Turkevich et al. used hydrogen tetrachloroaurate (HAuCl$_4$) and citric acid in boiling water to start the nucleation process that generates GNPs. Citric acid plays a dual role in the synthesis method, first by reducing Au^{3+} to Au^0 and then by generating dicarboxy acetone, which acts by stabilizing the structure and preventing aggregation. Later on, Frens improved the method by controlling the gold–citrate ratio to modify the particle size (Frens, 1973). Further studies revealed about other factors that can influence the size and shape of synthesized GNPs, such as temperature, pH, additives, and several stabilizers types and concentrations.

7.2.1.2   TWO-PHASE SYNTHESIS FOLLOWED BY THIOL STABILIZATION

The two-phase synthesis followed by thiol stabilization, also known as Brust–Schiffrin method was published in 1994 and was described as an easy method for the synthesis of highly stable GNPs. The strategy dwells on growing the nuclei concomitantly with the attachment of thiol monolayers at the particle surface using two distinct phases. The obtained NPs had sizes ranging from 1.0 to 3.0 nm and with 2.0–2.5 nm maximum distribution (Brust et al., 1994). This method gained a lot of popularity within a short time period as the outer thiol groups enabled solubilization in organic solvents without permanent aggregation and simplified subsequent surface modification/functionalization steps.

7.2.2   OTHER METHODS

GNPs have distinct applications and benefits on therapeutics due to their immense versatility, in which localized surface plasmon resonance may differ according to its shape and size (Jain et al., 2008). The synthesis methods described previously specified techniques that can generate GNPs mainly from spherical shapes. More complex shapes, such as rods, wires, belts, cages, stars, and dendrites, of which, some are illustrated in Figure 7.1, are also called anisotropic GNPs and require higher complexity methods. Such shape control is much more difficult as specific reactants need to be applied in a controlled manner, time, and with different physical aid methods involved.

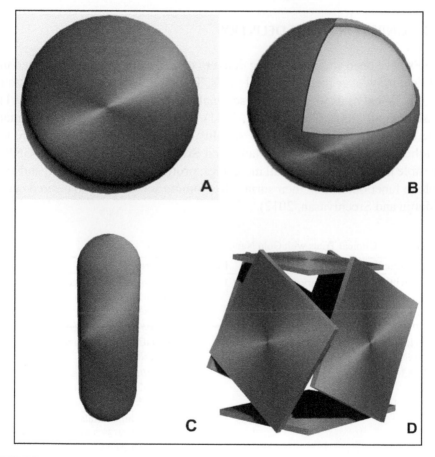

FIGURE 7.1 Nanospheres (A), nanocapsules (B), nanorods (C), and nanocages (D).

Among the popular physical methods used in GNPs manufacturing, photochemistry, sonochemistry, radiolysis, and thermolysis are the most used (Daniel and Astruc, 2003; Frazier et al., 2015; Gormley et al., 2012). Many shape-controlled synthesis methods involve a photoreduction process by ultraviolet irradiation in the presence of polymer capping agents, such as polyvinyl alcohol, polyvinylpyrrolidone, and PEG (Zhou et al., 1999). Liang described a well-established method, seed-mediated method, for gold nanorods (GNRs), which is relatively simple and has a high yield (Xu et al., 2014). Tsai's silver-induced approach produces nanostars with high applicability for photothermal therapy (PTT), especially in cancer treatment (Cheng et al., 2012).

7.3   GNPS FOR DRUG DELIVERY

GNPs have been studied for drug delivery systems and have been studied in clinical trials over the last several years. There are some studies in phases I and II clinical trials for some diseases, including cancer treatment (Table 7.1) (Astete et al., 2007; Thakor et al., 2011). This system has a great potential due to its low cytotoxicity, high stability, biocompatibility, and is easy to synthesize and functionalize (Manju and Sreenivasan, 2012). Hence, GNPs are able to load drugs, as small molecules, proteins, DNA, and RNA, and are able to functionalize on their surface to promote active target to desired cells (Manju and Sreenivasan, 2012).

TABLE 7.1   Clinical Trials with Gold Nanoparticles (GNPs).

Description	Disease	Phase	References
GNPs with iron oxide–silica shells versus stenting	Coronary artery disease, atherosclerosis	1	https://clinicaltrials.gov/ct2/show/NCT01436123?term=gold+nanoparticles&rank=1
NU-0129 spherical nucleic acid GNP	Gliosarcoma, recurrent glioblastoma	1	https://clinicaltrials.gov/ct2/show/NCT03020017?term=gold+nanoparticles&rank=2
C19-A3 GNP	Type 1 diabetes	1	https://clinicaltrials.gov/ct2/show/NCT02837094?term=gold+nanoparticles&rank=3
GNP	Atherosclerotic lesions	1 and 2	https://clinicaltrials.gov/ct2/show/NCT01270139?term=gold+nanoparticles&rank=4
GNP	Pulmonary hypertension	RP	https://clinicaltrials.gov/ct2/show/NCT02782026?term=gold+nanoparticles&rank=5
CNM-Au8 GNP	Stomach diseases	1	https://clinicaltrials.gov/ct2/show/NCT02755870?term=gold+nanoparticles&rank=7

RP: Recruiting participants.

7.3.1 GNP SURFACE MODIFICATION

Several surface functionalization methods have been studied and developed for biological application of GNPs. Thus, the GNP surface is very suitable for attachment of a large number of molecules and it provides a sustained and controlled release (Brown et al., 2010; Rana et al., 2013).

These moieties can be attached to GNP surface by physical adsorption or by covalent coupling. The first one, called noncovalent interactions, occurs by electrostatic or hydrophobic interaction (Fig. 7.2). It is a rapid and simple process as it avoids the synthesis step. However, changes in physical parameters, such as pH or ionic strength can produce the detachment of the ligand. In addition, this type of interaction can allow the direct ligand release, which may not be able to support sustained release (Jazayeri et al., 2016; Rana et al., 2013). On the contrary, the covalent attachment is currently the most widely used technique, and it is frequently based on a strong bond between gold and thiol groups (Priyadarshini and Pradhan, 2016). The covalent bond increases the stability, which is necessary for GNP administration (Dreaden et al., 2012; Jazayeri et al., 2016; Yeh et al., 2012).

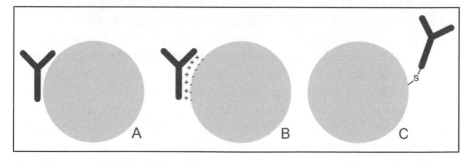

FIGURE 7.2 Types of GNP conjugation. hydrophobic interaction (A), electrostatic interaction (B), and covalent binding (C).

The biological response can be modified through the attachment of molecule on GNP surfaces. GNP surfaces have several interesting properties, such as strong attraction for thiol, proteins, carboxylic acid, disulfides, and other molecules (Khan et al., 2014).

As mentioned before, covalent bonds are used for the attachment of many drugs onto the GNPs. Safwat and coworkers performed modifications on GNP surface with thioglycolic acid (TGA) and glutathione (GSH) through thiol linkers, loading it with 5-fluorouracil (5-FU). This system showed a higher anticancer action against colon cancer cells when compared with free 5-FU (Safwat et al., 2016).

In another study, pyrimidine-capped GNPs were synthesized in methanol media through the reduction of tetrachloroauric acid by sodium borohydride, along with the respective pyrimidine ligand. GNPs coupled with this ligand showed an important reduction of bacterial resistance (Zhao et al., 2010).

Banu et al. (2015) synthesized folate-conjugated and doxorubicin (DOX)-loaded GNP bioconjugate. First, they synthesized the GNPs and then conjugated them with folate PEG to GNP surface and then, the surface was modified with poly(sodium-4-styrenesulfonate) (PSS) followed by the conjugation of DOX drug. This system was developed to overcome the limitations of DOX treatment of breast cancer and improve the therapy with PTT (Banu et al., 2015).

GNPs can be used for both drug delivery and photoacoustic imaging. Manivasagan et al. (2016) developed an effective system based on paclitaxel-loaded chitosan oligosaccharide-stabilized GNPs against cancer cells and it was possible to use these NPs as the contrast agents for tissue and cell imaging (Manivasagan et al., 2016). To ensure GNPs' safe delivery in tumors, NPs need to become "invisible" in the RES and they can be modified by polymers such as PEG. This polymer is attached to GNPs by thiol linkers at the end of the polymer (Trouiller et al., 2015). In this way, GNPs covered with thiolated PEG are used to prevent nonspecific interactions with the immune system. On the other hand, carboxyl group terminated into PEG is able to provide a suitable and stable covalent binding group. This group can be modified with (1-ethyl-3-(3-dimethylaminopropyl)carbodiimide hydrochloride)–N-hydroxysuccinimide (EDC/NHS) reagents to attach antibody onto GNP through NHS ester formers (Jazayeri et al., 2016).

Additionally, noncovalent bindings have been extensively studied. Previous reports indicated the efficient delivery to cancer cells through GNPs covered with carboxymethyl chitosan. The GNP surface was attached with DOX amino group bindings (electrostatic and hydrogen bonds). In this study, researchers demonstrated the activity of GNPs loaded with DOX promoting an efficient uptake (Madhusudhan et al., 2014). Another type of noncovalent binding, the ionic interaction, was used in a study conducted by Kim et al. (2016) wherein GNPs were incorporated into liposomes formulations containing DOX and the anticancer activity was evaluated. DOX was loaded through ionic interaction and GNPs covered with Pluronic F68 loaded with DOX were incorporated onto vehicle (Kim et al., 2016).

7.4   TARGETING

The GNP targeting in tumor site can be promoted through two ways, passive and active targeting. Both mechanisms aim to increase the drug concentration on tumor cells and consequently making the therapy more specific and efficient. Therefore, this chapter intends to explain both ways of targeting and suggests many ligands that may be used for this purpose.

7.4.1   PASSIVE TARGETING

The accumulation of GNPs in tumors through passive targeting can be explained by the EPR effect which occurred due to solid tumor inherent characteristics. In this microenvironment, the rapid and defected angiogenesis process usually forms leaky capillary fenestrations, which permit the accumulation of NPs into tumor interstitium. Moreover, the retention of these particles occurs due to the dysfunctional lymphatic drainage on tumor site (Huang et al., 2010; Wang and Thanou, 2010). For this purpose, the size and charge of NPs need to be controlled. The ideal NP size for EPR effect should be between 10 and 100 nm for an efficient extravasation in fenestrated vasculature, avoiding the renal filtration and the capture by the liver. Besides, particles with neutral or anionic charge should prevent renal elimination (Danhier et al., 2010).

Moreover, studies have shown that NPs coated with an inert surface, like PEG, may promote an increase of EPR effect due to the prevention of plasma clearance by immune system cells. It occurred due to the accumulation of serum protein binding on GNP surface (Sykes et al., 2014). Therefore, PEG-coated GNP is a way to improve therapeutic efficacy due to an increase in drug blood circulation and a decrease of clearance and immunogenicity (Liu et al., 2014a).

7.4.2   ACTIVE TARGETING

The active targeting of nanocarriers to specific cells has been extensively investigated for increasing drug concentration on target site, enhancing therapeutic efficacy and reducing the subsequent side effects. The challenge of active targeting systems is conjugating ligands on NP surfaces complementary with overexpressed receptor in therapeutic target cell and with no significant expression in normal cells (Wang and Thanou, 2010). Among these ligands, there are several studies on peptides, proteins, aptamers, small molecules, and antibodies conjugated with GNPs, which are aimed for therapy and diagnostic applications.

The complex of these ligands formed by the covalent bond with specific cell overexpressed receptor can internalize a high drug concentration through receptor-mediated endocytosis process (Vigderman and Zubarev, 2013). GNPs conjugated with these ligands have been evaluated by several researchers for chemotherapy and PTT on target site. Thus, the studies have reported promising results of the accumulation of this type of nanocarrier at the target site and subsequent increased therapeutic action (Dykman and Khlebtsov, 2012).

7.4.2.1 PROTEINS

Antibodies or immunoglobulins (Ig), common ligands easily conjugated with GNPs, are proteins produced by the immune system which have high affinities for specific antigens. Among the many classes of antibodies (IgA, IgE, IgM, and IgG), the immunoglobulin IgG is the most widely used as a targeting ligand coupled with drug delivery systems. These antibodies are composed of two heavy and two light-chain proteins held together by disulfide bonds. The region where the antigen is recognized and bonded is located in the upper portion of the two Fab fragments, called V_L and V_H domains (Crivianu-Gaita and Thompson, 2016; Dykman and Khlebtsov, 2012).

Antibodies have been applied for cancer therapy by targeting GNPs with antitumor drugs or using this kind of nanocarries for PTT on tumor cells. Different receptors expressed on tumor cells are being investigated for targeting GNPs in the tumor microenvironment, such as the epidermal growth factor receptor 1 (EGFR-1 or ErbB1), human epidermal growth factor receptor 2 (HER-2 or ErbB-2), vascular endothelial growth factor receptor 2 (VEGFR-2), CD20 receptor on B cells, CD44 on pancreatic cells, Mucin-7, and trophoblast cell surface antigen 2 (TROP-2) (Bergeron et al., 2015; Bisker et al., 2012; Chen et al., 2015; Das et al., 2011; Yang, 2014).

In recent years, epidermal growth factor receptor (EGFR), a cell surface receptor tyrosine kinase has become a main receptor for active targeting of drug delivery systems, given that this receptor is extensively expressed in many types of malignant tumors, such as lung, breast, ovarian, pancreatic, cervical, renal and prostate cancer (Patra et al., 2008). The receptor overexpression occurs due to the high capacity of tumor cells to further cell proliferation, angiogenesis, tumor metastasis and block apoptosis, promoting multiple tumorigenic processes (Kao et al., 2014). So, functionalizing GNPs with anti-EGRF antibodies represents a strategy to actively target these NPs and promote several antitumor drugs accumulation and improved PTT on tumor microenvironment. For example, Zhang et al. (2014) have shown that combination of EGFR monoclonal antibody conjugated with GNRs and near infra-red (NIR) irradiation could induce Hep-2 cells apoptosis both in vitro and in vivo, being PTT mediated by this kind of nanocarrier considered better than the traditional chemotherapy and radiotherapy (Zhang et al., 2014). In another study, the researchers evaluated the effectiveness of gold nanoshells functionalized with anti-EGFR to release DOX on pulmonary cancer cell membranes. When irradiated with NIR that system showed an increased accumulation of DOX on the cell surface by EGFR targeting.

Another member of EGFR family is a cell surface receptor tyrosine kinase called human epidermal growth factor receptor 2 (HER2 or ErbB2). Several tumor cells present overexpression of the HER2 receptor, like ovarian, lung, and gastric carcinomas, where HER2 receptor is overexpressed in 20–30% of breast cancer cases (Qureshi et al., 2015). Trastuzumab (Herceptin®), an FDA-approved anti-HER2 monoclonal antibody, has been widely used as a therapeutic agent against tumor cells and for active targeting of NPs (Chatto-padhyay et al., 2012). According to the results obtained by Stuchinskaya et al (2011), the functionalization of GNPs with anti-HER2 monoclonal antibody demonstrated an in vitro selective targeting of these nanocarriers on breast cancer cells, which express HER2 receptor with consequent enhanced PTT (Stuchinskaya et al., 2011).

Many other receptors have been investigated for targeting GNPs mediated by antibodies. The Mucin-7 receptor, a transmembrane protein expressed on urothelial cancer cell surface, has been researched for targeting GNPs using specific monoclonal antibodies aiming to kill these tumor cells by PTT. The in vitro results showed that active targeting of GNPs using Mucin-7 was able to kill selectively many lines of cancer cells (MBT2, T24, 9202, and 8301) and preserve normal cells (Chen et al., 2015). Furthermore, Liu et al. (2014c) obtained promising in vitro results using GNPs functionalized with monoclonal antibodies (anti-TROP2) for specific trophoblast cell surface antigen 2 (TROP2), a transmembrane glycoprotein present in high levels on cervical cancer cell surface. This research showed selective apoptosis and DNA damage caused by photothermal ablation (Yang, 2014).

Additionally, targeting GNPs to vascular endothelium through vascular endothelial growth factor receptor 2 (VEGFR-2) represents an important route for antiangiogenic cancer therapy. In the tumor microenvironment, the angiogenesis process is stimulated by the tumor cell upregulation of vascular endothelial growth factor (VEGF) and the high expression of VEGFR-2 in tumor vasculature (Byrne et al., 2008; Hida et al., 2016). Thus, GNPs functionalized with anti-VEGRF-2 have been evaluated due to the inhibition of VEGF binding on receptor, which causes an endocytic pathway for GNP internalization, promoting tumor vasculature suppression. In 2011, Das and coworkers evaluated the in vitro effect of GNPs conjugated with anti-VEGFR containing DOX for potential antiangiogenic therapy and the results showed that this system was selective and promoted large GNPs uptake in cell lines that overexpressed VEGFR, in comparison with another cell line that is negative for this receptor, becoming a promising drug delivery method for antiangiogenic cancer therapy (Das et al., 2011).

In addition to antibodies, drug delivery systems can be used by conjugation with the natural ligands for active targeting, such as transferrin and epidermal growth factor (EGF) proteins. The transferrin receptor (TfR), a glycoprotein which is important for DNA synthesis, may be able to bind on nanocarriers surface functionalized with transferrin and can promote active targeting for different types of malignant cells that more abundantly express these receptors, such as in brain, breast, and lung (Amreddy et al., 2015; Mohamed et al., 2014). Li et al. (2009) investigated the in vitro selective targeting of GNPs conjugated with transferrin on tumor cell lines (3T3) and normal cell lines. The results showed that the cellular uptake of this conjugate by the tumor cells was four times higher than in normal lines and sixfold when compared with GNPs without transferrin, becoming a potential tool for imaging and cancer therapy (Li et al., 2009). Furthermore, Dixit et al. (2015) developed the GNPs with transferrin and EGF for PTT in glioblastomas. They showed that these systems were five to six times more effective to promote the uptake of GNPs in glioma cell lines (U87 and LN229) than untargeted GNPs. Besides, this functionalization was able to traverse the blood–brain barrier more efficiently than GNPs without these proteins being conjugated (Dixit et al., 2015).

Furthermore, peptides like the Fab fragments can be used as ligands on nanocarriers to promote targeted delivery of some drugs. Moreover, a smaller Fab fragment named single-chain antibody (scFv), consisting of only two domains, can also be used as ligand on GNPs. This fragment does not contain Fc portion, a constant region, but only scFv portion which has a high antigen specificity and affinity (Crivianu-Gaita and Thompson, 2016). Thus, these kinds of ligands conjugated with GNPs have greater potential for application, mainly for diagnostics of various diseases, increasing the sensitivity of several immunoassays; also, they are promising ligands for drug delivery systems (Kirui et al., 2010; Liu et al., 2009).

7.4.2.2   PEPTIDES

The application of peptide sequences which specifically bind to receptors on target cell has also been the subject of a large number of researches. The functionalization with these target moieties, for instance, Arg–Gly–Asp (RGD), Asn–Gly–Arg (NGR), octreotide, somatostatin, goserelin, and bombesin peptides, to GNPs has a large potential for active targeting of these nanocarriers to tumor microenvironment for cancer therapy (Corti et al., 2013; Gormley et al., 2011a; Heidari et al., 2015; Huang et al., 2010; Jaskula-Sztul et al., 2012).

The functionalization of GNPs with RGD peptide, an RGD sequence, has been extensively investigated for active targeting to tumor tissues with overexpression of $\alpha_v\beta_3$ integrin receptor. During the tumor development, $\alpha_v\beta_3$ integrin receptor is responsible for angiogenesis process, promoting tumor development and metastasis (Arosio et al., 2011; Kumar et al., 2012). One research group investigated the influence of GNPs conjugated with and without RGD peptide on cellular uptake and demonstrated that the GNPs with surface modification showed high biorecognition with $\alpha_v\beta_3$ integrin present on endothelial cells, becoming a potential nanocarrier for cancer therapy by photothermal ablation (Gormley et al., 2011a).

Bombesin is another peptide largely used for conjugation with GNPs for active targeting on cancer lines, such as prostate and breast and lung carcinoma, which present overexpression of the gastrin-releasing peptide receptor, a transmembrane G-protein. Suresh et al. (2014) demonstrated that the uptake of GNPs functionalized with bombesin is mediated by receptor endocytosis (clathrin) (Suresh et al., 2014). In addition, Heidari et al. (2015) evaluated the uptake of this nanocarrier toward breast cancer cells (T47D cell line) and the effect of PTT in vivo using a mouse model. They obtained a high specificity and uptake of the GNPs conjugated with bombesin in vitro and an increased accumulation of this nanocarrier in the tumor tissue, resulting in the complete disappearance of breast tumor in the in vivo model when PTT was applied (Heidari et al., 2015).

7.4.2.3   APTAMERS

The aptamers are short single-chain oligopeptides which have broadly been applied for actively targeted nanocarriers. The aptamers development begins with the selection of sequences of RNA or DNA chains according to the receptor present in target cell. Then, these sequences are isolated, amplified, and sent for the enrichment process by an in vitro process called systematic evolution of ligands by exponential enrichment (SELEX). Generally, the SELEX process is able to develop aptamers with around 70–80 nucleotides (Crivianu-Gaita and Thompson, 2016). Thus, sequences of DNA or RNA selected promote the formation of aptamers with high affinity and specificity for the desired target. Another advantage of these oligopeptides is their lack of immunogenicity and toxicity compared to others ligands, for example, the monoclonal antibodies. In addition, aptamers are usually thermally stable and easily synthesized and modified by chemical reactions (Alibolandi et al., 2015; Ghasemi et al., 2015).

Recently, promising results were obtained by functionalization of drug delivery systems with aptamers. Researchers have shown that the combinations of the oligopeptides to GNPs led to higher accumulation of these systems to desired target and thus they provide greater therapeutic effect. Among these, aptamers which are targeted to overexpressed receptors on tumor cells have been extensively evaluated, for instance, aptamers specific to nucleolin receptor, prostate-specific membrane antigen (PSMA) receptor, and EGFR (Tiwari et al., 2011).

According to Kim et al. (2010), GNPs functionalized with RNA aptamer coupling to the PSMA resulted in a higher accumulation of DOX in tumor cells that had an overexpression of PSMA protein receptor when compared to cells that do not express this type of protein (Kim et al., 2010). In addition, the studies show that using PTT with GNPs functionalized with AS1411 aptamer have shown higher uptake and serious photodamage in breast cancer line MCF7 when compared with the normal epithelium cell (Shieh et al., 2010). Besides, another study conducted by Choi et al. (2014) demonstrated that GNPs conjugated with anti-EGFR aptamer exhibited an excellent in vitro active targeting when compared to cell line with high expression of EGFR (A431 line) and cell line without this receptor (MCF7 line). Furthermore, they showed that PTT using GNPs conjugated with anti-EGFR aptamer promoted efficient in vivo photothermal ablation (Choi et al., 2014).

7.4.2.4 SMALL MOLECULES

In recent years, small molecules have been conjugated with GNPs in order to increase the accumulation of these particles at the target site due to specific receptor overexpression on cancer cells. The advantages of the small molecules are their non-immunogenic and nontoxic characteristics (Zhong et al., 2014). A large number of small molecules such as folate, methotrexate, carbohydrates, anisamide, and SV119 have been studied as targeting ligands for cancer therapy (Chen et al., 2007; Geng et al., 2011; Huang et al., 2016; Sun et al., 2014).

Folic acid (FA) (vitamin B$_9$), a small molecule with weight molecular of 441 Da, is required by eukaryotic cells for pyrimidine and purine syntheses, the essential DNA components. The folate receptor (FR), a highly specific single-chain glycoprotein, is expressed in eukaryotic cells for their cellular development. Researchers showed that this receptor is overexpressed about 100–300-fold in tumor cells as compared to normal cells, such as lung, ovarian, cervical, and colon carcinomas (Samadian et al., 2016; Saxena et al., 2012). Hence, folate has been attached to drug delivery systems to enhance therapeutic efficacy against tumor cell through active targeting in tumor microenvironment (Kuo and Lee, 2015; Tao et al., 2012).

In the last decade, GNPs functionalized with FA have been extensively used for targeting and internalization of these nanocarriers with promising results. Thus, studies using these small molecules conjugated with GNPs for targeting of antitumor drugs and application of PTT demonstrated that these systems promoted an increased cellular uptake through endocytosis when compared to nonfunctionalized particles, increasing in vitro efficacy (Huang et al., 2016; Manju and Sreenivasan, 2012).

In addition to FRs, other cell receptors for small molecules targeting, such as glucose, asialoglycoprotein, and sigma receptors, have been reported. The glucose receptor is also overexpressed on several tumor cells because these cells have an exacerbated cell growth and therefore require higher nutritional support (Han et al., 2016). According to Geng et al. (2011) thiol-glucose-bound GNPs (Glu-GNPs) exhibit a higher uptake (about 31%) by ovarian cancer lines (SKOV3) when compared with nontargeted GNP; moreover, the large intracellular uptake of Glu-GNPs promoted an increased inhibition of cell proliferation when radiation was applied (Geng et al., 2011). Another carbohydrate used to target GNPs into tumor cells is galactose. This carbohydrate binds specifically to extracellular asialoglycoprotein receptor (ASGP-R), a hepatocyte-specific receptor, and promotes GNP uptake (Lai et al., 2010). Zhu et al. (2015) demonstrated that GNPs functionalized with galactose could improve the in vitro efficacy of radiotherapy using HepG2 cell line (Zhu et al., 2015).

7.5 STIMULI-RESPONSIVE GNPS

Stimuli-responsive NPs are designed to respond to specific intrinsic and extrinsic stimuli and deliver drugs to the target site. The former usually refers to pH and temperature, while the latter is commonly associated with light irradiation, ultrasound, magnetic field, electric current, and external application of heat or cold (Eloy et al., 2015).

pH-responsive NPs are frequently employed for enhanced drug delivery at the tissue level, for example, exploiting the pH gradient in the tumor microenvironment, and at the intracellular level, designed to escape acidic endo-lysosomal compartment, with consequent cytoplasmic drug release. Briefly, NPs respond to pH stimuli through physiochemical change in material structure and surface, for instance, swelling, dissociation, and charge switching (Sperling et al., 2010).

The design of pH-sensitive NPs can involve chemical bonds, present either in the backbone or side chain of materials, which remain stable in physiological pH but are unstable in acidic environment. Among these chemical bonds, hydrazone is the most commonly employed linker for acid-sensitive release, being cleavable at pH lower than 6.8 (Liu et al., 2014b). For instance, DOX has been linked to angiopep-2-functionalized GNPs through the hydrazone bond, allowing the enhanced drug release at lower pH values, 5.0 and 6.0, compared with 6.8 and 7.4. The outcome was better intracellular delivery to glioma cells and longest survival of glioma-bearing mice (Ruan et al., 2015).

Other chemical bonds might also be useful for pH sensitivity. GNPs with surface modified with thiol containing ligands, TGA, and glutathione, have been conjugated to 5-fluorouracil, which was better released at pH 3.0, with improved cytotoxic effect in colon carcinoma cell lines (Safwat et al., 2016). Finally, GNPs have also been modified with azide bond through the polymer poly(4-vinylpyridine) (P4VP), causing pH-responsive release at pH 5.0 (Zhang et al., 2009).

Amino acids such as tryptophan have been employed for the synthesis and stabilization of GNPs. Within this context, Das and coworkers synthesized GNPs by the reduction of chloroauric acid by tryptophan, which demonstrated pH sensitivity properties, useful for NP accumulation at nonpolar liquid–liquid interface in tumor tissue environment (Das et al., 2015). For peptide delivery, insulin has been linked to GNPs through carboxyl groups of TGA. Interestingly, in low pH, carboxylic acid of TGA is protonated, leading to stronger hydrogen-bonding interaction with protonated amine and carboxylic groups of insulin, which is released upon higher pH values. Furthermore, insulin-loaded GNPs were incorporated into polypyrrole nanobiocomposites, which respond to electrical current, enhancing insulin release (Shamaeli and Alizadeh, 2015).

GNPs have also been loaded into pH-sensitive materials, such as microgel particles composed by copolymer of N-[3-(dimethylamino)propyl]methacrylamide (DMAPMA) and N-isopropylacrylamide (NIPAM), which present an increase in mobility as a result of protonation of the amine groups containing the microgel particle (Bradley and Garcia-risueño, 2011). Moreover, O'Neal and co-workers prepared hydrogen-bonding polymer nanocomposites of poly(ethylene oxide) (PEO) and poly(methacrylic acid) (PMAA) for intercalation of GNPs and demonstrated that GNPs were released in a pH-dependent fashion from polymeric network (Neal et al., 2017).

Mild hyperthermia, with a generation of heat around 41–43°C, has been associated with radiation and chemotherapy for cancer treatment, with synergistic effect (Heinfield et al., 2017). Furthermore, thermosensitive drug delivery systems are reported to increase therapeutic effectiveness due to the promotion of tumor vascular permeability, characteristic of the EPR effect, allowing for better NP accumulation and enhanced release from temperature-sensitive NPs (Ta and Porter, 2013). For instance, recently, GNPs have been attached with thermosensitive polymers, such as poly(N-isopropylacrylamide) (PNIPAM) and poly(N,N-dimethylacrylamide) (PDMAM), and with methacrylate monomers with azobenzene side chains (Huebner et al., 2016; Luo et al., 2016). Furthermore, GNPs have also been loaded into thermosensitive liposomes based on 1,2-dipalmitoyl-sn-glycero-3-phosphocholine (DPPC), 1-palmitoyl-2-hydroxy-sn-glycero-3-phosphocholine (MPPC) for enhanced hydrophilic drug release (Mathiyazhakan et al., 2015).

Light can be absorbed by gold nanostructures with a wavelength that matches their surface plasmon resonance optical properties. Therefore, upon near-infrared (NIR) light irradiation, photothermal conversion occurs, leading to hyperthermia, also known as plasmonic PTT (Gormley et al., 2011b). Different gold nanostructures have shown this property, including nanoshells, nanorods, and nanocages (Frazier and Ghandehari, 2015). Polydispersity of GNRs and nanospheres is an important factor for plasmonic heat generation and it has been shown that the former produces more heat (Baffou et al., 2009; Qin et al., 2016). PEGylated GNRs have been prepared for in vivo application (Niidome et al., 2006).

GNRs have been demonstrated to enhance macromolecular delivery upon laser irradiation and they have improved the accumulation of N-(2-hydroxypropyl)-methacrylamide (HPMA) copolymers accumulation in tumor tissue, which enhanced the in vivo efficacy of radiotherapeutic 90Y-labeled-HPMA copolymer in prostate cancer-bearing mice. Moreover, the participation of targeting to heat-shock protein which increase site-specific delivery has been reported (Frazier and Ghandehari, 2015; Gormley et al., 2012). Recently, GNRs have been embedded with responsive copolymer micelle for phothermal effect associated with targeting for cancer therapy (Parida et al., 2017).

Light-sensitive release from GNP-containing formulations has been shown and it depends on materials, such as lipids, membrane-anchored probes, and metallic particles, which react upon light exposure following different mechanisms, including photoisomerization, photocleavage, and photopolymerization (Eloy et al., 2015). Within this context, on-demand drug release has been proposed for imaging-guided antitumor therapy with GNPs. Han and co-workers reported GNPs anchored with DOX through a thioketal linker, sensitive to reactive oxygen species (ROS). In their study, they employed protoporphyrin IX, which under light irradiation generates ROS, which cleaves the thioketal bond, causing DOX release. In another study, GNRs were used to enhance NIR absorption, for coumarin derivative chromophore moiety cleavage, enabling DOX release from β-cyclodextrins (Liang et al., 2016).

7.6 CONCLUSION

GNPs have been successfully prepared, using the two-phase synthesis followed by thiol stabilization, for instance, and extensively characterized over the last decades, becoming promising drug delivery systems due to their several advantages described herein. Noteworthy, the photothermal properties of GNPs, particularly with nanorods, allow their use in cancer treatment, and many studies have reported their successful in vivo performance. Furthermore, GNP surface has been covalently decorated with several ligands for targeted delivery, including folate, peptides, antibodies, and aptamers, the latter generated by the SELEX technology, holding great promise for therapeutic application, particularly in the cancer treatment field. Overall, studies showed better efficacy with great potential for clinical application, and some clinical trials are ongoing, in phase 1 or 2. Furthermore, the application of GNPs with potential clinical use is the stimuli-responsive materials sensitive to light due to the plasmonic properties of GNPs, pH, based on chemical sensitive bonds, or temperature, with thermosensitive polymers, for instance, for triggered delivery. However, some aspects need to be further elucidated before the translation of GNP drug delivery systems from bench to bedside, such as the scale-up of preparation methods and toxicological aspects. Furthermore, additional clinical studies are clearly needed to establish their safety and efficacy.

KEYWORDS

- **GNP**
- **drug delivery system**
- **synthesis method**
- **surface modification**
- **active targeting**
- **stimuli-responsive GNP**

REFERENCES

Ajnai, G.; Chiu, A.; Kan, T.; Cheng, C. C.; Tsai, T. H.; Chang, J. Trends of Gold Nanoparticle-Based Drug Delivery System in Cancer Therapy. *J. Exp. Clin. Med.* **2014,** *6*(6), 172–178.

Alex, S.; Tiwari, A. Functionalized Gold Nanoparticles: Synthesis, Properties and Applications—A Review. *J. Nanosci. Nanotechnol.* **2015,** *15*(3), 1869–1894.

Alibolandi, M.; Ramezani, M.; Sadeghi, F.; Abnous, K.; Hadizadeh, F. Epithelial Cell Adhesion Molecule Aptamer Conjugated PEG-PLGA Nanopolymersomes for Targeted Delivery of Doxorubicin to Human Breast Adenocarcinoma Cell Line in Vitro. *Int. J. Pharm.* **2015,** *479*(1), 241–251.

Amreddy, N.; Muralidharan, R.; Babu, A.; Mehta, M.; Johnson, E. V.; Zhao, Y. D.; Munshi, A.; Ramesh, R. Tumor-targeted and pH-controlled Delivery of Doxorubicin Using Gold Nanorods for Lung Cancer Therapy. *Int. J. Nanomed.* **2015,** *10*, 6773–6788.

Arosio, D.; Manzoni, L.; Araldi, E. M. V.; Scolastico, C. Cyclic RGD Functionalized Gold Nanoparticles for Tumor Targeting. *Bioconjug. Chem.* **2011,** *22*(4), 664–672.

Astete, C. E.; Kumar, C. S. S. R.; Sabliov, C. M. Size Control of Poly(d,l-Lactide-Co-Glycolide) and Poly(d,l-Lactide-Co-Glycolide)-Magnetite Nanoparticles Synthesized by Emulsion Evaporation Technique. *Colloids Surf. A Physicochem. Eng. Asp.* **2007,** *299*(1–3), 209–216.

Baffou, G.; Quidant, R.; Girard, C. Heat Generation in Plasmonic Nanostructures: Influence of Morphology. *Appl. Phys. Lett.* **2009,** *94*(15), 1–3.

Banu, H.; Sethi, D. K.; Edgar, A.; Sheriff, A.; Rayees, N.; Renuka, N.; Faheem, S. M.; Premkumar, K.; Vasanthakumar, G. Doxorubicin Loaded Polymeric Gold Nanoparticles Targeted to Human Folate Receptor upon Laser Photothermal Therapy Potentiates Chemotherapy in Breast Cancer Cell Lines. *J. Photochem. Photobiol.* B **2015,** *149*, 116–128.

Bergeron, E.; Boutopoulos, C.; Martel, R.; Torres, A.; Rodriguez, C.; Niskanen, J.; Lebrun, J.-J.; Winnik, F. M.; Sapieha, P.; Meunier, M. Cell-Specific Optoporation with Near-Infrared Ultrafast Laser and Functionalized Gold Nanoparticles. *Nanoscale* **2015,** *7*(42), 17836–17847.

Bisker, G.; Yeheskely-Hayon, D.; Minai, L.; Yelin, D. Controlled Release of Rituximab from Gold Nanoparticles for Phototherapy of Malignant Cells. *J. Control Release* **2012,** *162*(2), 303–309.

Bradley, M.; Garcia-risueño, B. S. Journal of Colloid and Interface Science Symmetric and Asymmetric Adsorption of pH-responsive Gold Nanoparticles onto Microgel Particles and Dispersion Characterisation. *J. Colloid Interface Sci.* **2011,** *355*(2), 321–327.

Brown, S. D.; Nativo, P.; Smith, J.-A.; Stirling, D.; Edwards, P. R.; Venugopal, B.; Flint, D. J.; Plumb, J. A.; Graham, D.; Wheate, N. J. Gold Nanoparticles for the Improved Anticancer Drug Delivery of the Active Component of Oxaliplatin. *J. Am. Chem. Soc.* **2010,** *132*(13), 4678–4684.

Brust, M.; Walker, M.; Bethell, D.; Schiffrin, D. J.; Whyman, R. Synthesis of Thiol-Derivatised Gold Nanoparticles in a Two-Phase Liquid-Liquid System. *J. Chem. Soc. Chem. Commun.* **1994,** *0*(7), 801–802.

Byrne, J. D.; Betancourt, T.; Brannon-Peppas, L. Active Targeting Schemes for Nanoparticle Systems in Cancer Therapeutics. *Adv. Drug Deliv. Rev.* **2008,** *60*(15), 1615–1626.

Chattopadhyay, N.; Fonge, H.; Cai, Z.; Scollard, D.; Lechtman, E.; Done, S. J.; Pignol, J. P.; Reilly, R. M. Role of Antibody-Mediated Tumor Targeting and Route of Administration in Nanoparticle Tumor Accumulation in Vivo. *Mol. Pharm.* **2012,** *9*(8), 2168–2179.

Chen, Y. H.; Tsai, C. Y.; Huang, P. Y.; Chang, M. Y.; Cheng, P. C.; Chou, C. H.; Chen, D. H.; Wang, C. R.; Shiau, A. L.; Wu, C. L. Methotrexate Conjugated to Gold Nanoparticles Inhibits Tumor Growth in a Syngeneic Lung Tumor Model. *Mol. Pharm.* **2007,** *4*(5), 713–722.

Chen, C. H.; Wu, Y. J.; Chen, J. J. Gold Nanotheranostics: Photothermal Therapy and Imaging of Mucin 7 Conjugated Antibody Nanoparticles for Urothelial Cancer. *Biomed. Res. Int.* **2015,** *2015*(2015). Article ID 813632.

Cheng, L.-C.; Huang, J.-H.; Chen, H. M.; Lai, T.-C.; Yang, K.-Y.; Liu, R.-S.; Hsiao, M.; Chen, C.-H.; Her, L.-J.; Tsai, D. P. Seedless, Silver-Induced Synthesis of Star-Shaped Gold/Silver Bimetallic Nanoparticles as High Efficiency Photothermal Therapy Reagent. *J. Mater. Chem.* **2012,** *22*(5), 2244–2253.

Cho, J.; Kim, A.; Kim, S.; Lee, S.; Chung, H.; Yoon, M. Development of a Novel Imaging Agent Using Peptide-Coated Gold Nanoparticles Toward Brain Glioma Stem Cell Marker CD133. *Acta Biomater.* **2017,** *47*, 182–192.

Choi, J.; Park, Y.; Choi, E. B.; Kim, H.-O.; Kim, D. J.; Hong, Y.; Ryu, S.-H.; Lee, J. H.; Suh, J.-S.; Yang, J.; et al. Aptamer-Conjugated Gold Nanorod for Photothermal Ablation of Epidermal Growth Factor Receptor-Overexpressed Epithelial Cancer. *J. Biomed. Opt.* **2014,** *19*(5), 51203.

Corti, A.; Curnis, F.; Rossoni, G.; Marcucci, F.; Gregorc, V. Peptide-Mediated Targeting of Cytokines to Tumor Vasculature: The NGR-hTNF Example. *BioDrugs* **2013,** *27*(6), 591–603.

Crivianu-Gaita, V.; Thompson, M. Aptamers, Antibody scFv, and Antibody Fab' Fragments: An Overview and Comparison of Three of the Most Versatile Biosensor Biorecognition Elements. *Biosens. Bioelectron.* **2016,***85*, 32–45.

Danhier, F.; Feron, O.; Préat, V. To Exploit the Tumor Microenvironment: Passive and Active Tumor Targeting of Nanocarriers for Anti-Cancer Drug Delivery. *J. Control Release* **2010,** *148*(2), 135–146.

Daniel, M.-C.; Astruc, D. Gold Nanoparticles: Assembly, Supramolecular Chemistry, Quantum-Size-Related Properties, and Applications toward Biology, Catalysis, and Nanotechnology. *Chem. Rev.* **2003,** *104*(1), 293–346.

Das, A.; Soehnlen, E.; Woods, S.; Hegde, R.; Henry, A.; Gericke, A.; Basu, S. VEGFR-2 Targeted Cellular Delivery of Doxorubicin by Gold Nanoparticles for Potential Antiangiogenic Therapy. *J. Nanoparticle Res.* **2011,** *13*(12), 6283–6290.

Das, A.; Chadha, R.; Maiti, N.; Kapoor, S. Synthesis of pH Sensitive Gold Nanoparticles for Potential Application in Radiosensitization. *Mater. Sci. Eng. C Mater. Biol. Appl.* **2015**, *55*, 34–41.

de Araújo, R. F. Jr.; de Araújo, A. A.; Pessoa, J. B.; Freire Neto, F. P.; da Silva, G. R.; Leitão Oliveira, A. L.; de Carvalho, T. G.; Silva, H. F.; Eugênio, M.; Sant'Anna, C.; Gasparotto, L. H. Anti-Inflammatory, Analgesic and Anti-Tumor Properties of Gold Nanoparticles. *Pharmacol. Rep.* **2017**, *69*(1), 119–129.

Dixit, S.; Novak, T.; Miller, K.; Zhu, Y.; Kenney, M. E.; Broome, A.-M. Transferrin Receptor-Targeted Theranostic Gold Nanoparticles for Photosensitizer Delivery in Brain Tumors. *Nanoscale* **2015**, *7*(5), 1782–1790.

Dreaden, E. C.; Austin, L. A.; Mackey, M. A.; El-Sayed, M. A. Size Matters: Gold Nanoparticles in Targeted Cancer Drug Delivery. *Ther. Deliv.* **2012**, *3*(4), 457–478.

Dykman, L.; Khlebtsov, N. Gold Nanoparticles in Biomedical Applications: Recent Advances and Perspectives. *Chem. Soc. Rev.* **2012**, *41*(6), 2256–2282.

Eloy, J. O.; Petrilli, R.; Lopez, R. F. V.; Lee, R. J. Stimuli-Responsive Nanoparticles for siRNA Delivery. *Curr. Pharm. Des.* **2015**, *21*(29), 4131–4144.

Fernandes, A. R.; Jesus, J.; Martins, P.; Figueiredo, S.; Rosa, D.; Martins, L. M. R. D. R. S.; Corvo, M. L.; Carvalheiro, M. C.; Costa, P. M.; Baptista, P. V. Multifunctional Gold-Nanoparticles: A Nanovectorization Tool for the Targeted Delivery of Novel Chemotherapeutic Agents. *J. Control Release* **2017**, *245*, 52–61.

Frazier, N.; Ghandehari, H. Hyperthermia Approaches for Enhanced Delivery of Nanomedicines to Solid Tumors. *Biotechnol. Bioeng.* **2015**, *112*(10), 1967–1983.

Frazier, N.; Robinson, R.; Ray, A.; Ghandehari, H. Effects of Heating Temperature and Duration by Gold Nanorod Mediated Plasmonic Photothermal Therapy on Copolymer Accumulation in Tumor Tissue. *Mol. Pharm.* **2015**, *12*(5), 1605–1614.

Frens, G. Controlled Nucleation for the Regulation of the Particle Size in Monodisperse Gold Suspensions. *Nature* **1973**, *241*(105), 20–22.

Gamal-Eldeen, A. M.; Moustafa, D.; El-Daly, S. M.; El-Hussieny, E. A.; Saleh, S.; Bacon, M. K. K. L.; Gupta, S.; Katti, K.; Shukla, R.; Katti, K. V. Photothermal Therapy Mediated by Gum Arabic-Conjugated Gold Nanoparticles Suppresses Liver Preneoplastic Lesions in Mice. *J. Photochem. Photobiol. B* **2016**, *163*, 47–56.

Geng, F.; Song, K.; Xing, J. Z.; Yuan, C.; Yan, S.; Yang, Q.; Chen, J.; Kong, B. Thio-Glucose Bound Gold Nanoparticles Enhance Radio-Cytotoxic Targeting of Ovarian Cancer. *Nanotechnology* **2011**, *22*(28), 285101.

Ghasemi, Z.; Dinarvand, R.; Mottaghitalab, F.; Esfandyari-Manesh, M.; Sayari, E.; Atyabi, F. Aptamer Decorated Hyaluronan/Chitosan Nanoparticles for Targeted Delivery of 5-Fluorouracil to MUC1 Overexpressing Adenocarcinomas. *Carbohydr. Polym.* **2015**, *121*, 190–198.

Gormley, A. J.; Malugin, A.; Ray, A.; Robinson, R.; Ghandehari, H. Biological Evaluation of RGDfK-Gold Nanorod Conjugates for Prostate Cancer Treatment. *J. Drug Target.* **2011a**, *19*(10), 915–924.

Gormley, A. J.; Greish, K.; Ray, A.; Robinson, R.; Gustafson, J. A.; Ghandehari, H. Gold Nanorod Mediated Plasmonic Photothermal Therapy: a Tool to Enhance Macromolecular Delivery. *Int. J. Pharm.* **2011b**, *415*(1–2), 315–318.

Gormley, A. J.; Larson, N.; Sadekar, S.; Robinson, R.; Ray, A.; Ghandehari, H. Guided Delivery of Polymer Therapeutics Using Plasmonic Photothermal Therapy. *Nano Today* **2012**, *7*(3), 158–167.

Hainfeld, J. F.; Slatkin, D. N.; Smilowitz, H. M. The Use of Gold Nanoparticles to Enhance Radiotherapy in Mice. *Phys. Med. Biol.* **2004,** *49*(18), N309–N315.

Hainfeld, J. F.; Dilmanian, F. A.; Zhong, Z. Gold Nanoparticles Enhance the Radiation Therapy of a Murine Squamous Cell Carcinoma. *Phys. Med. Biol.* **2010,** *55*, 3045–3059.

Han, J.; Zhang, J.; Yang, M.; Cui, D.; Fuente, J. M. Correction: Glucose-Functionalized Au Nanoprisms for Optoacoustic Imaging and Near-Infrared Photothermal Therapy. *Nanoscale* **2016,** *8*(3), 492–499.

Heidari, Z.; Salouti, M.; Sariri, R. Breast Cancer Photothermal Therapy Based on Gold Nanorods Targeted by Covalently-Coupled Bombesin Peptide. *Nanotechnology* **2015,** *26*(19), 195101.

Heo, D. N.; Yang, D. H.; Moon, H. J.; Lee, J. B.; Bae, M. S.; Lee, S. C.; Lee, W. J.; Sun, I. C.; Kwon, I. K. Gold Nanoparticles Surface-Functionalized with Paclitaxel Drug and Biotin Receptor as Theranostic Agents for Cancer Therapy. *Biomaterials* **2012,** *33*(3), 856–866.

Hida, K.; Maishi, N.; Sakurai, Y.; Hida, Y.; Harashima, H. Heterogeneity of Tumor Endothelial Cells and Drug Delivery. *Adv. Drug Deliv. Rev.* **2016,** *99*, 140–147.

Huang, X.; Peng, X.; Wang, Y.; Wang, Y.; Shin, D. M.; El-Sayed, M. A.; Nie, S. A Reexamination of Active and Passive Tumor Targeting by Using Rod-shaped Gold Nanocrystals and Covalently Conjugated Peptide Ligands. *ACS Nano* **2010,** *4*(10), 5887–5896.

Huang, S.; Duan, S.; Wang, J.; Bao, S.; Qiu, X.; Li, C.; Liu, Y.; Yan, L.; Zhang, Z.; Hu, Y. Folic-Acid-Mediated Functionalized Gold Nanocages for Targeted Delivery of Anti-miR-181b in Combination of Gene Therapy and Photothermal Therapy against Hepatocellular Carcinoma. *Adv. Funct. Mater.* **2016,** *26*(15), 2532–2544.

Huebner, D.; Rossner, C.; Vana, P. Light-Induced Self-Assembly of Gold Nanoparticles with a Photoresponsive Polymer Shell. *Polymer (Guildf)* **2016,** *107*, 503–508.

Jain, P. K.; Huang, X.; El-Sayed, I. H.; El-Sayed, M. A. Noble Metals on the Nanoscale: Optical and Photothermal Properties and Some Applications in Imaging, Sensing, Biology, and Medicine. *Acc. Chem. Res.* **2008,** *41*(12), 1578–1586.

Jaskula-Sztul, R.; Xiao, Y.; Javadi, A.; Eide, J.; Xu, W.; Kunnimalaiyaan, M.; Gong, S.; Chen, H. Abstract 1953: Co-Delivery of Doxorubicin and siRNA Using Octreotide-Conjugated Gold Nanorods for Targeted Neuroendocrine Cancer Therapy. *Cancer Res.* **2012,** *72*(Suppl. 8), 1953–1953.

Jazayeri, M. H.; Amani, H.; Pourfatollah, A. A.; Pazoki-Toroudi, H.; Sedighimoghaddam, B. Various Methods of Gold Nanoparticles (GNPs) Conjugation to Antibodies. *Sens. Bio-Sensing Res.* **2016,** *9*, 17–22.

Kao, H.-W.; Lin, Y.-Y.; Chen, C.-C.; Chi, K.-H.; Tien, D.-C.; Hsia, C.-C.; Lin, W.-J.; Chen, F.-D.; Lin, M.-H.; Wang, H.-E. Biological Characterization of Cetuximab-Conjugated Gold Nanoparticles in a Tumor Animal Model. *Nanotechnology* **2014,** *25*(29), 295102.

Khan, A.; Rashid, R.; Murtaza, G.; Zahra, A. Gold Nanoparticles: Synthesis and Applications in Drug. *Trop. J. Pharma. Res.* **2014,** *13*(7), 1169–1177.

Kim, D.; Jeong, Y. Y.; Jon, S. A Drug Loaded Aptamer-Gold Nanoparticle Bioconjugate for Combined CT Imaging and Therapy of Prostate Cancer. *ACS Nano.* **2010,** *4*(7), 10–12.

Kim, K.; Oh, K. S.; Park, D. Y.; Lee, J. Y.; Lee, B. S.; Kim, I. S.; Kim, K.; Kwon, I. C.; Sang, Y. K.; Yuk, S. H. Doxorubicin/gold-Loaded Core/shell Nanoparticles for Combination Therapy to Treat Cancer through the Enhanced Tumor Targeting. *J. Control Release* **2016,** *228*, 141–149.

Kirui, D. K.; Rey, D. A.; Batt, C. A. Gold Hybrid Nanoparticles for Targeted Phototherapy and Cancer Imaging. *Nanotechnology* **2010,** *21*(10), 105105–105114.

Kumar, A.; Ma, H.; Zhang, X.; Huang, K.; Jin, S.; Liu, J.; Wei, T.; Cao, W.; Zou, G.; Liang, X. J. Gold Nanoparticles Functionalized with Therapeutic and Targeted Peptides for Cancer Treatment. *Biomaterials* **2012,** *33*(4), 1180–1189.

Kuo, Y. C.; Lee, C. H. Inhibition against Growth of Glioblastoma Multiforme in Vitro Using Etoposide-Loaded Solid Lipid Nanoparticles with P-Aminophenyl-α-D-Manno-Pyranoside and Folic Acid. *J. Pharm. Sci.* **2015,** *104*(5), 1804–1814.

Lai, C. H.; Lin, C. Y.; Wu, H. T.; Chan, H. S.; Chuang, Y. J.; Chen, C. T.; Lin, C. C. Galactose Encapsulated Multifunctional Nanoparticle for HepG2 Cell Internalization. *Adv. Funct. Mater.* **2010,** *20*(22), 3948–3958.

Li, J.-L.; Wang, L.; Liu, X.-Y.; Zhang, Z.-P.; Guo, H.-C.; Liu, W.-M.; Tang, S.-H. In Vitro Cancer Cell Imaging and Therapy Using Transferrin-Conjugated Gold Nanoparticles. *Cancer Lett.* **2009,** *274*(2), 319–326.

Liang, Y.; Gao, W.; Peng, X.; Deng, X.; Sun, C.; Wu, H.; He, B. Biomaterials Near Infrared Light Responsive Hybrid Nanoparticles for Synergistic Therapy. *Biomaterials* **2016,** *100*, 76–90.

Liu, Y.; Liu, Y.; Mernaugh, R. L.; Zeng, X. Single Chain Fragment Variable Recombinant Antibody Functionalized Gold Nanoparticles for a Highly Sensitive Colorimetric Immunoassay. *Biosens. Bioelectron.* **2009,** *24*(9), 2853–2857.

Liu, H.; Doane, T. L.; Cheng, Y.; Lu, F.; Srinivasan, S.; Zhu, J. J.; Burda, C. Control of Surface Ligand Density on PEGylated Gold Nanoparticles for Optimized Cancer Cell Uptake. *Part. Part. Syst. Charact.* **2014a,** *32*(2), 197–204.

Liu, J.; Huang, Y.; Kumar, A.; Tan, A.; Jin, S.; Mozhi, A.; Liang, X. pH-sensitive Nano-Systems for Drug Delivery in Cancer Therapy. *Biotechnol. Adv.* **2014b,** *32*(4), 693–710.

Liu, T.; Tian, J.; Chen, Z.; Liang, Y.; Liu, J.; Liu, S.; Li, H.; Zhan, J.; Yang, X. Anti-TROP2 Conjugated Hollow Gold Nanospheres as a Novel Nanostructure for Targeted Photothermal Destruction of Cervical Cancer Cells. *Nanotechnology* **2014c,** *25*(34), 345103.

Luo, C.; Dong, Q.; Qian, M.; Zhang, H. Thermosensitive Polymer-Modified Gold Nanoparticles with Sensitive Fluorescent Properties. *Chem. Phys. Lett.* **2016,** *664*, 89–95.

Madhusudhan, A.; Reddy, G. B.; Venkatesham, M.; Veerabhadram, G.; Kumar, A. D.; Natarajan, S.; Yang, M. Y.; Hu, A.; Singh, S. S. Efficient pH Dependent Drug Delivery to Target Cancer Cells by Gold Nanoparticles Capped with Carboxymethyl Chitosan. *Int. J. Mol. Sci.* **2014,** *15*(5), 8216–8234.

Maeda, H. Tumor-Selective Delivery of Macromolecular Drugs via the EPR Effect: Background and Future Prospects. *Bioconjug. Chem.* **2010,** *21*(5), 797–802.

Manivasagan, P.; Bharathiraja, S.; Bui, N. Q.; Lim, I. G.; Oh, J. Paclitaxel-Loaded Chitosan Oligosaccharide-Stabilized Gold Nanoparticles as Novel Agents for Drug Delivery and Photoacoustic Imaging of Cancer Cells. *Int. J. Pharm.* **2016,** *511*(1), 367–379.

Manju, S.; Sreenivasan, K. Gold Nanoparticles Generated and Stabilized by Water Soluble Curcumin-Polymer Conjugate: Blood Compatibility Evaluation and Targeted Drug Delivery onto Cancer Cells. *J. Colloid Interface Sci.* **2012,** *368*(1), 144–151.

Mathiyazhakan, M.; Yang, Y.; Liu, Y.; Zhu, C.; Liu, Q. Non-Invasive Controlled Release from Gold Nanoparticle Integrated Photo-Responsive Liposomes through Pulse Laser Induced Microbubble Cavitation. *Colloids Surf. B* **2015,** *126*, 569–574.

Matsumura, Y.; Maeda, H. A New Concept for Macromolecular Therapeutics in Cancer Chemotherapy: Mechanism of Tumoritropic Accumulation of Proteins and the Antitumor Agents Smancs. *Cancer Res.* **1986,** *46*, 6387–6392.

Mocan, L.; Matea, C.; Tabaran, F. A.; Mosteanu, O.; Pop, T.; Puia, C.; Agoston-coldea, L.; Zaharie, G.; Mocan, T.; Buzoianu, A. D.; Lancu, C. Selective ex Vivo Photothermal

Nano-Therapy of Solid Liver Tumors Mediated by Albumin Conjugated Gold Nanoparticles. *Biomaterials* **2017**, *119*, 33–42.

Mohamed, M. S.; Veeranarayanan, S.; Poulose, A. C.; Nagaoka, Y.; Minegishi, H.; Yoshida, Y.; Maekawa, T.; Kumar, D. S. Type 1 Ribotoxin-Curcin Conjugated Biogenic Gold Nanoparticles for a Multimodal Therapeutic Approach towards Brain Cancer. *Biochim. Biophys. Acta.* **2014**, *1840*(6), 1657–1669.

Neal, J. T. O.; Bolen, M. J.; Dai, E. Y.; Lutkenhaus, J. L. Hydrogen-Bonded Polymer Nanocomposites Containing Discrete Layers of Gold Nanoparticles. *J. Colloid Interface Sci.* **2017**, *485*, 260–268.

Niidome, T.; Yamagata, M.; Okamoto, Y.; Akiyama, Y.; Takahashi, H.; Kawano, T.; Katayama, Y.; Niidome, T. PEG-Modified Gold Nanorods with a Stealth Character for in Vivo Applications. *J. Control. Release* **2006**, *114*(3), 343–347.

Parida, S.; Maiti, C.; Rajesh, Y.; Dey, K. K.; Pal, I.; Parekh, A.; Patra, R.; Dhara, D.; Kumar, P.; Mandal, M. Gold Nanorod Embedded Reduction Responsive Block Copolymer Micelle-Triggered Drug Delivery Combined with Photothermal Ablation for Targeted Cancer Therapy. *Biochim. Biophys. Acta* **2017**, *1861*(1), 3039–3052.

Patra, C. R.; Bhattacharya, R.; Wang, E.; Katarya, A.; Lau, J. S.; Dutta, S.; Muders, M.; Wang, S.; Buhrow, S. A.; Safgren, S. L.; et al. Targeted Delivery of Gemcitabine to Pancreatic Adenocarcinoma Using Cetuximab as a Targeting Agent Targeted Delivery of Gemcitabine to Pancreatic Adenocarcinoma Using Cetuximab as a Targeting Agent. *Cancer Res.* **2008**, *15*(6), 1970–1978.

Pena-pereira, F.; Lavilla, I.; Bendicho, C. Chemical Unmodified Gold Nanoparticles for in-Drop Plasmonic-Based Sensing of Iodide. *Sens. Actuators, B* **2017**, *242*, 940–948.

Petrilli, R.; Eloy, J. O.; Marchetti, J. M.; Lopez, R. F. V.; Lee, R. J. Targeted Lipid Nanoparticles for Antisense Oligonucleotide Delivery. *Curr. Pharm. Biotechnol.* **2014**, *15*(9), 847–855.

Priyadarshini, E.; Pradhan, N. Gold Nanoparticles as Efficient Sensors in Colorimetric Detection of Toxic Metal Ions: a Review. *Sens. Actuators, B* **2016**, *238*, 888–902.

Qin, Z.; Wang, Y.; Randrianalisoa, J.; Raeesi, V.; Chan, W. C. W.; Lipi, W.; Bischof, J. C. Quantitative Comparison of Photothermal Heat Generation between Gold Nanospheres and Nanorods. *Sci. Rep.* **2016**, *6*,1–13, Article number: 29836.

Qureshi, A.; Gurbuz, Y.; Niazi, J. H. Label-Free Capacitance Based Aptasensor Platform for the Detection of HER2/ErbB2 Cancer Biomarker in Serum. *Sens. Actuators B* **2015**, *220*, 1145–1151.

Rana, S.; Bajaj, A.; Mout, R.; Rotello, V. M. Monolayer Coated Gold Nanoparticles for Delivery Applications. *Adv. Drug Deliv. Rev.* **2013**, *64*(2), 200–216.

Ruan, S.; Yuan, M.; Zhang, L.; Hu, G.; Chen, J.; Cun, X.; Zhang, Q.; Yang, Y.; He, Q.; Gao, H. Biomaterials Tumor Microenvironment Sensitive Doxorubicin Delivery and Release to Glioma Using Angiopep-2 Decorated Gold Nanoparticles. *Biomaterials* **2015**, *37*, 425–435.

Saber, M. M.; Bahrainian, S.; Dinarvand, R.; Atyabi, F. Targeted Drug Delivery of Sunitinib Malate to Tumor Blood Vessels by cRGD-Chiotosan-Gold Nanoparticles. *Int. J. Pharm.* **2017**, *517*(1–2), 269–278.

Safwat, M. A.; Soliman, G. M.; Sayed, D.; Attia, M. A. Gold Nanoparticles Enhance 5-Fluorouracil Anticancer Efficacy against Colorectal Cancer Cells. *Int. J. Pharm.* **2016**, *513* (1–2), 648–658.

Samadian, H.; Hosseini-Nami, S.; Kamrava, S. K.; Ghaznavi, H.; Shakeri-Zadeh, A. Folate-Conjugated Gold Nanoparticle as a New Nanoplatform for Targeted Cancer Therapy. *J. Cancer Res. Clin. Oncol.* **2016**, *142*(11), 2217–2229.

Saxena, V.; Naguib, Y.; Hussain, M. D. Folate Receptor Targeted 17-Allylamino-17-Demethoxygeldanamycin (17-AAG) Loaded Polymeric Nanoparticles for Breast Cancer. *Colloids Surf. B* **2012,** *94*, 274–280.

Shah, M.; Badwaik, V.; Kherde, Y.; Waghwani, H. K.; Modi, T.; Aguilar, Z. P.; Rodgers, H.; Hamilton, W.; Marutharaj, T.; Webb, C.; et al. Gold Nanoparticles: Various Methods of Synthesis and Antibacterial Applications Monic. *Front. Biosci.* **2014,** *19*, 1320–1344.

Shamaeli, E.; Alizadeh, N. Functionalized Gold Nanoparticle-Polypyrrole Nanobiocomposite with High Effective Surface Area for Electrochemical/pH Dual Stimuli-responsive Smart Release of Insulin. *Colloids Surf. B* **2015,** *126*, 502–509.

Shieh, Y. A.; Yang, S. J.; Wei, M. F.; Shieh, M. J. Aptamer-Based Tumor-Targeted Drug Delivery for Photodynamic Therapy. *ACS Nano* **2010,** *4*(3), 1433–1442.

Sperling, R. A.; Parak, W. J. Surface Modification, Functionalization and Bioconjugation of Colloidal Inorganic Nanoparticles. *Philos. Trans. A. Math. Phys. Eng. Sci.* **2010,** *368*(1915), 1333–1383.

Stuchinskaya, T.; Moreno, M.; Cook, M. J.; Edwards, D. R.; Russell, D. A. Targeted Photodynamic Therapy of Breast Cancer Cells Using Antibody-Phthalocyanine-Gold Nanoparticle Conjugates. *Photochem. Photobiol. Sci.* **2011,** *10*(5), 822–831.

Sun, T.; Wang, Y.; Wang, Y.; Xu, J.; Zhao, X.; Vangveravong, S.; Mach, R. H.; Xia, Y. Using SV119-Gold Nanocage Conjugates to Eradicate Cancer Stem Cells through a Combination of Photothermal and Chemo Therapies. *Adv. Healthc. Mater.* **2014,** *3*(8), 1283–1291.

Suresh, D.; Zambre, A.; Chanda, N.; Hoffman, T. J.; Smith, C. J.; Robertson, J. D.; Kannan, R. Bombesin Peptide Conjugated Gold Nanocages Internalize via Clathrin Mediated Endocytosis. *Bioconjug. Chem.* **2014,** *25*(8), 1565–1579.

Sykes, E. A.; Chen, J.; Zheng, G. Investigating the Impact of Nanoparticle Size on Active and Passive Tumor Targeting Efficiency. *ACS Nano.* **2014,** *8*(6), 5696–5706.

Ta, T.; Porter, T. M. Thermosensitive Liposomes for Localized Delivery and Triggered Release of Chemotherapy. *J. Control. Release* **2013,** *169*(1–2), 112–125.

Tao, Y.; Han, J.; Dou, H. Surface Modification of Paclitaxel-Loaded Polymeric Nanoparticles: Evaluation of in Vitro Cellular Behavior and in Vivo Pharmacokinetic. *Polym. (United Kingdom)* **2012,** *53*(22), 5078–5085.

Thakor, A. S.; Jokerst, J.; Zavaleta, C.; Massoud, T. F.; Gambhir, S. S. Gold Nanoparticles: A Revival in Precious Metal Administration to Patients. *Nano Lett.* **2011,** *11*(10), 4029–4036.

Tian, L.; Lu, L.; Qiao, Y.; Ravi, S.; Salatan, F.; Melancon, M. P. Stimuli-Responsive Gold Nanoparticles for Cancer Diagnosis and Therapy. *J. Funct. Biomater.* **2016,** *7*(3), 19–50.

Tiwari, P. M.; Vig, K.; Dennis, V. A.; Singh, S. R. Functionalized Gold Nanoparticles and Their Biomedical Applications. *Nanomaterials* **2011,** *1*(1), 31–63.

Trouiller, A. J.; Hebie, S.; El Bahhaj, F.; Napporn, T. W.; Bertrand, P. Chemistry for Oncotheranostic Gold Nanoparticles. *Eur. J. Med. Chem.* **2015,** *99*(1), 92–112.

Turkevich, J.; Stevenson, P. C.; Hillier, J. A Study of the Nucleation and Growth Processes in the Synthesis of Colloidal Gold. *Discuss. Faraday Soc.* **1951,** *11*, 55–75.

Vigderman, L.; Zubarev, E. R. Therapeutic Platforms Based on Gold Nanoparticles and Their Covalent Conjugates with Drug Molecules. *Adv. Drug Deliv. Rev.* **2013,** *65*(5), 663–676.

Wang, M.; Thanou, M. Targeting Nanoparticles to Cancer. *Pharmacol. Res.* **2010,** *62*(2), 90–99.

Wang, H.; Huff, T. B.; Zweifel, D. A.; He, W.; Low, P. S.; Wei, A.; Cheng, J.-X. In Vitro and in Vivo Two-Photon Luminescence Imaging of Single Gold Nanorods. *Proc. Natl. Acad. Sci. U. S. A.* **2005,** *102*(44), 15752–15756.

Xu, X.; Zhao, Y.; Xue, X.; Huo, S.; Chen, F.; Zou, G.; Liang, X.-J.; Atwater, H. A.; Polman, A.; Zhang, Z. J.; et al. Seedless Synthesis of High Aspect Ratio Gold Nanorods with High Yield. *J. Mater. Chem. A* **2014**, *2*(10), 3528–3535.

Yao, J.; Feng, J.; Chen, J. External-Stimuli Responsive Systems for Cancer Theranostic. *Asian J. Pharm. Sci.* **2016**, *11*(5), 585–595.

Yeh, Y.-C.; Creran, B.; Rotello, V. M.; Zhang, T.; Chen, P.; Sun, Y.; Xing, Y.; Yang, Y.; Dong, Y.; Xu, L.; et al. Gold Nanoparticles: Preparation, Properties, and Applications in Bionanotechnology. *Nanoscale* **2012**, *4*(6), 1871–1880.

Zhang, T.; Wu, Y.; Pan, X.; Zheng, Z.; Ding, X.; Peng, Y. An Approach for the Surface Functionalized Gold Nanoparticles with pH-Responsive Polymer by Combination of RAFT and Click Chemistry. *Eur. Polym. J.* **2009**, *45*(6), 1625–1633.

Zhang, S.; Li, Y.; He, X.; Dong, S.; Huang, Y.; Li, X.; Li, Y.; Jin, C.; Zhang, Y.; Wang, Y. Photothermolysis Mediated by Gold Nanorods Modified with EGFR Monoclonal Antibody Induces Hep-2 Cells Apoptosis in Vitro and in Vivo. *Int. J. Nanomedicine* **2014**, *9*(1), 1931–1946.

Zhao, Y.; Tian, Y.; Cui, Y.; Liu, W.; Ma, W.; Jiang, X. Small-Molecule Capped Gold Nanoparticles as Potent Antibacterial Agents That Target Gram-Negative Bacteria. *J. Am. Chem. Soc.* **2010**, *132*(35), 12349–12356.

Zhao, L.; Kim, T.; Kim, H.; Ahn, J.; Yeon, S. Enhanced Cellular Uptake and Phototoxicity of Vertepor Fi N-Conjugated Gold Nanoparticles as Theranostic Nanocarriers for Targeted Photodynamic Therapy and Imaging of Cancers. *Mater. Sci. Eng. C* **2016**, *67*, 611–622.

Zhong, Y.; Meng, F.; Deng, C.; Zhong, Z. Ligand-Directed Active Tumor-Targeting Polymeric Nanoparticles for Cancer Chemotherapy. *Biomacromolecules* **2014**, *15*(6) 1955–1969.

Zhou, Y.; Wang, C. Y.; Zhu, Y. R.; Chen, Z. Y. A Novel Ultraviolet Irradiation Technique for Shape-Controlled Synthesis of Gold Nanoparticles at Room Temperature. *Chem. Mater.* **1999**, 11, 2310–2312.

Zhu, C.; Zheng, Q.; Wang, L.; Xu, H.-F.; Tong, J.; Zhang, Q.; Wan, Y.; Wu, J. Synthesis of Novel Galactose Functionalized Gold Nanoparticles and Its Radiosensitizing Mechanism. *J. Nanobiotechnol.* **2015**, *13*(1), 1–11.

CHAPTER 8

SUPERPARAMAGNETIC IRON OXIDE NANOPARTICLES: APPLICATION IN DIAGNOSIS AND THERAPY OF CANCER

LJILJANA DJEKIC

Faculty of Pharmacy, University of Belgrade, Belgrade, Serbia

Corresponding author. E-mail: ljiljanadjek@gmail.com

ABSTRACT

Superparamagnetic iron oxide nanoparticles (SPIONs) are products of nanotechnology which have been intensively experimentally considered as drug delivery carriers, hyperthermia, and contrast agents with the particular focus on cancer treatment. This chapter provides an overview of the main research aspects in this field, including common approaches for magnetite core synthesis; main strategies for coating and functionalization of SPIONs surface with targeting ligands, imaging, or therapeutic moieties, drug release mechanisms; the principles of their usage in magnetic fluid hyperthermia and magnetic resonance imaging; safety considerations. The potential for enhancement of diagnostic and therapy of cancer is described in detail. The multifunctionality of SPIONs is pointed as their specific feature enables integration of cancer diagnosis and therapy (theranostic).

8.1   INTRODUCTION

The difficulties in diagnosis and treatment of cancer strongly encouraged the development of novel noninvasive or minimally invasive strategies as an alternative to the approaches usually used such as surgery, radiotherapy, and chemotherapy associated with harmful side effects. Development of nano-pharmaceutics is already accepted as a potent platform for enhancement of cancer therapy (Cole et al., 2011; Lima-Tenorio et al., 2015). The usefulness of nanomaterials as drug delivery carriers is mainly based on their small size and large surface area available for functionalization for the purpose of targeting therapy and imaging. Iron oxide nanoparticles have been the most extensively investigated magnetic nanoparticles for biomedical applications due to their biocompatibility, biodegradability to nontoxic iron and oxygen component, ease of synthesis, and multifunctionality (Mahmoudi et al., 2011; Sun et al., 2008b; Veiseh et al., 2010). They are typically produced as homogenous nanosuspensions (ferrofluids) of inorganic core comprising nanocrystalline magnetite (Fe_3O_4) or maghemite (γFe_2O_3) protected with a biocompatible coating. Superparamagnetic iron oxide nanoparticles (SPIONs) have unique physical and chemical properties different from the atom and the bulk counterparts (Gupta and Gupta, 2005). SPIONs are superparamagnetic particles which magnetization arises under the external magnetic field from electron exchange between the Fe^{2+} and Fe^{3+} ions; however, they do not retain any magnetism after removal of the magnetic field. They can interact with an external magnetic field focused on a specific body area, thus enabling magnetic resonance imaging (MRI) or cancer therapy (Kim et al., 2001). During the last two decades several contrast agents for MRI, drug carriers, and magnetic fluid hyperthermia (MFH) based on iron oxide nanoparticles are commercialized. Recent advances in nanotechnology have enabled the research on the targeted delivery of SPIONs by functionalization of the nanoparticles with highly specific targeting peptides, antibodies, and small molecules as well as with permeation agents and fluorophores (Tietze et al., 2015). This chapter reviews applications of SPIONs as MRI contrast agents and as carriers for drug delivery including the recent developments in this area.

8.2 METHODS OF PREPARATION OF SUPERPARAMAGNETIC IRON OXIDE NANOPARTICLES (SPIONS)

8.2.1 METHODS OF MAGNETITE CORE SYNTHESIS

SPIONs are nanoparticles of composite morphology that typically consists of a magnetite core, a coating, and multiple functional molecules (ligands) on their surface (i.e., or core–shell structure). The common coating materials are biocompatible synthetic polymers, polysaccharides, lipids, proteins, silica, and gold. Coatings stabilize magnetite nanoparticles in physiologic fluids and enable additional chemical modifications (Cole et al., 2011). For magnetite nanoparticles synthesis several techniques have been developed.

Design and preparation methods of the SPIONs provide significant differences in magnetic properties of the nanoparticles (Sun et al., 2008b). Magnetite particles obtained under different synthetic conditions may also display large differences regarding their magnetic properties. These differences are attributed to changes in structural disorder (Taylor et al., 2001), creation of antiphase boundaries (Zhou et al., 2001), or the existence of a magnetically dead layer at the particle surface (Kim et al., 2001).

SPIONs for biomedical application are predominately synthesized by an aqueous coprecipitation process in the presence of the coating material (i.e., bottom-up or chemical synthesis approaches) (Molday and Mackenzie, 1982; Shen et al., 1993). For example, Kim et al. performed the synthesis of the nonionic surfactant (polyoxyethylene (10) oleyl ether) coated SPIONs with narrow size distribution by chemical coprecipitation method with a superparamagnetic property that is detectable in an MRI scanner and with the small size of the nanoparticles suitable for targeting the nanosized intercellular space in the living brain (Kim et al., 2001). The method is based on a coprecipitation of Fe^{3+} and Fe^{2+} ions in an aqueous solution of their salts (e.g., chlorides, sulphates, and nitrates), in an alkaline medium and under a stream of inert gas, at room temperature or with heating. The size, shape, and composition of the nanoparticles depends largely on the choice of salt, Fe^{2+}/Fe^{3+} ratio, temperature, pH of the solution, ionic strength, and the presence of the coating material (Ha et al., 2008; Lu et al., 2007; Storm et al., 1995). A complete precipitation of Fe_3O_4 is expected between pH 9 and 14 at a molar ratio Fe^{3+}/Fe^{2+} of 2:1 (Gupta and Gupta, 2005). Although such simple methods are suitable to produce nanoparticles with uniform composition and size (Willad et al., 2004), the average particle diameter is usually larger than 10 nm. Moreover, the drawback of the method is the incorporation of impurities disrupting the crystal structure (Gupta and Gupta, 2005) and compromising their magnetic properties (Gupta and Wells, 2004). To help improve the uniformity and prevent oxidation in air and agglomeration stability of Fe_3O_4 nanoparticles, their core is often coated with polymers that may act as "in situ" surface coatings present during the precipitation process (e.g., poly(acrylic acid)) (Si et al., 2004) or graft poly(ethylene glycol) (PEG)-g-poly(glycerol monoacrylate) polymers (Wan et al., 2006) or they may be introduced after the nucleation and growth processes are completed (post-synthesis addition). The polymer coating can also affect morphology, limit the crystallinity, and the magnetic susceptibility of the formed SPIONs. For example, Lee et al. found that crystallinity decreased with increasing concentration of poly(vinyl alcohol) (PVA) during the coprecipitation synthesis of SPIONs (Lee et al., 1996). Water-in-oil microemulsions are suitable as the reaction media for obtaining a uniform particle size by copre-cipitation method in a controlled manner. Mixing of the two water-in-oil microemulsions, containing soluble salts ($FeCl_2x6H_2O$ and $FeCl_3x6H_2O$) and the precipitation agent (e.g., NH_4OH), results in nanoprecipitation within water droplets. The nanodroplets of aqueous phase of the microemul-sion continuously collide, coalesce, and break again (Djekic and Primorac, 2008; Djekic et al., 2011) allowing the formation of the precipitate as a result of interdroplet exchange and nuclei aggregation (Gupta and Gupta, 2005). The nanoparticles are usually stabilized by coating with the oleic acid and subsequently with an additional layer of sodium dodecyl sulfate. When ethanol or acetone added, the precipitate forms, and after that separate from the mixture by filtration or centrifugation (Fig. 8.1). Generally, the yield is very low and the method is not suitable for the industrial scale production of the nanoparticles (Ha et al., 2008; Lu et al., 2007). The microemulsion reac-tion media are also suitable for preparation of spherical, oval, or cylindrical nanoparticles (Fig. 8.2). A water-in-oil microemulsion stabilized with three different nonionic surfactants (Triton X-100, Brij-97 and Igepal CO-520) has been used for the preparation of silica-coated iron oxide nanoparticles (Bulte et al., 1992) as small as 1–2 nm and of very uniform size (standard deviation less than 10%). A uniform silica coating as thin as 1 nm encapsu-lating the bare nanoparticles is formed by the base-catalysed hydrolysis and the polymerization reaction of tetraethyl orthosilicate in the microemulsion. Reversed microemulsion method is applied for gold coating of SPIONs with enhanced chemical inertness and ability to form self-assembled monolayers on their surface using alkanethiols (Lu et al., 2007; Yigit et al., 2007; Yigit et al., 2008). Gomez-Lopera et al. (2001) have described a simple and reproducible double emulsion method for preparing colloidal particles with a magnetite nucleus and a biodegradable poly(DL-lactide) polymer coating to be loaded with therapeutic drugs and used as drug delivery systems.

Small (2–20 nm) monodisperse nanocrystals of magnetite can be obtained by thermal decomposing the organometallic compounds in an organic solvent, during boiling surfactant was added as a stabilizer (i.e., top-down approach). The organometallic precursors are metal acetylacetonates [M (acac) n] (M=Fe, n=2 or 3, acac=acetylacetonate), metal cupferonates [MxCupx] (M=metal ion; Cup=N-Nitroso-N-phenylhydroxylamine, $C_6H_5N(NO)O^-$), or carbonyl compounds. For the stabilization of the nanoparticles, oleic acid and hexadecylamine are usually used. The parameters which determine the size and crystallinity of magnetic nanoparticles are quantitative ratio of reagents (e.g., organometallic compound, a surfactant, and a solvent), temperature, reaction time, and the period of crystal growth (Ha et al., 2008;

Lu et al., 2007). A hydrophobic coating on the nanoparticle surface requires additional modification to achieve hydrophilicity and compatibility with aqueous media as well as biomedical applicability (Lee et al., 2007; Sun et al., 2004; Xu and Sun, 2007).

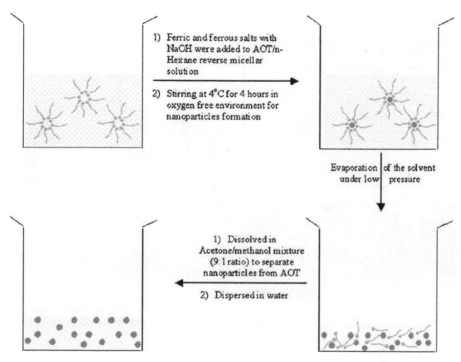

FIGURE 8.1 Strategy of preparing highly monodispersed iron oxide nanoparticles inside the without microemulsion droplets. Iron salts were dissolved inside the aqueous cores of reverse micelles and precipitated using alkali solutions to get the particles of the desired size.

Source: Reprinted from Gupta, A. K.; Gupta, M. Synthesis and Surface Engineering of Iron Oxide Nanoparticles for Biomedical Applications. *Biomaterials* **2005**, *26*, 3995–4021. © 2005, with permission from Elsevier.

Solgel is a method applicable for the synthesis of SPION with potential application in cancer therapy by magnetic fluid hyperthermia, however not suitable for production of SPIONs to be used as contrast agents (Saldivar-Ramirez et al., 2014; Sánchez et al., 2014). Solgel synthesis is based on the hydrolysis and polycondensation of metal precursors, metal, or metalloid element which form colloidal liquid system (sol) in the presence of various reactive ligands. The obtained sol is then dried by solvent removal or by chemical reaction at room temperature leading to form gel which consists of a network containing a liquid phase (Kalia et al., 2014). Crystalline nanoparticles are obtained from gel by heating. The method is suitable for synthesis of monodisperse and large size nanoparticles (tens to hundreds of nanometers), even in a large scale. The important advantages of the solgel method are good size and structure control of the nanoparticles, and the possibility to obtain pure amorphous particles (Salunkhe et al., 2014).

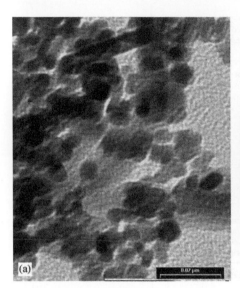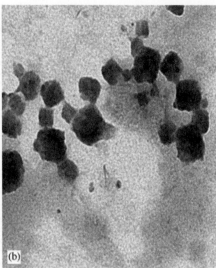

FIGURE 8.2 Transmission electron microscopy pictures of magnetic particles prepared in (a) bulk solutions and (b) in w/o microemulsions.

Source: Reprinted from Gupta, A. K.; Gupta, M. Synthesis and Surface Engineering of Iron Oxide Nanoparticles for Biomedical Applications. *Biomaterials* **2005**, 26, 3995–4021. © 2005, with permission from Elsevier.

8.2.2 SURFACE MODIFICATIONS OF SPIONS WITH BIOMEDICAL RELEVANCE

Size and surface properties (such as charge and hydrophobicity) of the SPIONs have great relevance for their stability, physicochemical, and magnetic properties (Jiles, 1991; Schulze et al., 2005), ease of ligand coupling, and pharmacokinetic performance (Cole, 2011). The magnetite nanoparticle surfaces could be modified through the creation of few atomic layers of organic polymer or inorganic metallic or oxide coatings, suitable for further functionalization by the attachment of various bioactive ligands (Berry and Curtis, 2003). The main goals of the SPIONs coating are to:

1. Provide surfaces available for the conjugation of drug molecules, targeting ligands, and reporter moieties
2. Protect particles against agglomeration provoked by their high surface energy and partial neutralization of surface charges on the nanoparticles in biological fluids due to the presence of salts or other electrolytes
3. Prevent nonspecific interactions in the biological milleau (e.g., adsorption of plasma protein to the surfaces of SPIONs [opsonization] upon intravenous injection, as the first step in their clearance by the reticuloendothelial system [RES]) (Berry and Curtis, 2003; Veiseh et al., 2010).

In the absence of any surface coating, the hydrophobic surface of the nanoparticles of iron oxide and a large specific surface area (a large surface area-to-volume ratio) are favorable for establishing a strong hydrophobic interactions and formation of large clusters and aggregates with strong magnetic dipole–dipole attractions and ferromagnetic behavior (Hamley, 2003) then they are suspended in an aqueous vehicle or in the body after administration. Thus, surface modification of SPIONs is necessary (Gupta and Gupta, 2005). The prevention of aggregation is achieved by adding nonimmunogenic and nonantigenic stabilizing coating materials (small molecules and polymers) with a high affinity for the iron oxide core (McCarthy and Weissleder, 2008). Stabilizers are physically adsorbed onto the surface of the magnetite cores and prevent their aggregation by electrostatic repulsion or steric effects. The concentration of the stabilizer should not be too high in order to avoid an increase of the diameter above 100 nm, since in this way the risk of opsonization after (intravenous) i.v. injection is increased (Arruebo et al., 2007; Chomoucka et al., 2010; Kim et al., 2010; Lu et al., 2007). Colloidal suspensions of magnetic particles (Fe_3O_4 or Fe_2O_3), forming magnetizable fluids that remain liquid in the most intense magnetic fields (*ferrofluids*) find widespread applications (Gupta and Gupta, 2005; McCarthy and Weissleder, 2008).

Coatings typically have included a variety of synthetic polymers (macrogols [PEGs], polyvinylpyrrolidone, PVA, poly(lactic-co-glycolic acid) [PLGA], carbomers, poloxamers, polyethyleneimine [PEI]), semisynthetic polymers (sodium carboxymethyl cellulose), natural polymers (dextrans, starch, chitosan, gelatin, and pullulan), lipids (fatty acids, phospholipids), surfactants (sodium oleate, sodium lauryl sulfate), and inorganic materials (silica, gold) (Gupta and Gupta, 2005; Sun et al., 2008b). The most widely utilized polymer coatings for in vivo applications are the polysaccharide dextran and PEG.

The coating can be achieved through a number of approaches at the time of precipitation to prevent aggregation (in situ coating) or after precipitation (post-synthesis adsorption or postsynthesis end grafting). In situ and post-synthesis modification with polysaccharides and copolymers lead to coatings that uniformly encapsulate cores. Lee et al. (1996) have modified nanoparticle's surface with PVA by precipitation of iron salts in PVA aqueous solution to form stable dispersion. They found that the crystallinity of the particles decreased with increasing PVA concentration, while the morphology and particle size remained almost unchanged. Alternatively, end-grafted polymers (e.g. PEG) are anchored to the nanoparticle surface by the polymer end groups, forming brush-like extensions. Liposome and micelle-forming molecules create a shell around the SPION core. These structures retain hydrophobic regions that can be used for drug encapsulation. Each technique retains specific advantages and disadvantages depending upon the polymer employed (e.g., ease of coating, number of functional groups, and so forth). Coating materials and immobilization strategies each influence the magnetic properties of magnetic nanoparticles (MNPs) in different ways. Several studies have revealed that the coating thickness, and hydrophobicity can drastically affect the magnetic properties of MNPs (Duan et al., 2008; LaConte et al., 2007).

8.2.2.1   POLYMERIC COATINGS

Polymeric coatings have been engineered to enhance SPIONs pharmacokinetics, endosomal release, and tailored drug loading and release behavior (Sun et al., 2008b). Coating of SPIONs with hydrophilic polymers such as PEG is a common approach to minimize or eliminate the protein adsorption and thus increase the circulation time of nanoparticles in the bloodstream. PEG-coated (PEGylated) SPIONs are commonly regarded to as "stealth" nanoparticles regarding the recognition by the RES (Harris and Chess, 2003). The attached polymer molecules may form loops, tails, brushes, or shells depending on the molecular weight and geometric orientation of the polymer on the surface of the particles, therefore, participating in their effective hydrodynamic size (Fig. 8.3). For instance, PEI is water-soluble cationic polymer that can take both linear and branched forms (Kircheis et al., 2001). Dextran and other forms of dextran polymers (e.g., carboxy-dextran, carboxymethyl dextran), have been used to coat SPIONs with varying hydrodynamic sizes (Laurent et al., 2008). Conventional dextran coatings are based on hydrogen bonding, making the polymer susceptible to detachment. Most clinical preparations (Ferridex®, Combidex®, Resovist®, and AMI-228/ferumoxytrol) have been employed on dextran or similar carbohydrate coatings (Jung, 1995; McCarthy and Weissleder, 2008; Weissleder et al., 1990), and one recent preparation has described the use of citrate-stabilized particles (Taupitz et al., 2004). Dextran-coated SPIONs have become an important part of clinical cancer imaging, and have been shown to increase the accuracy of cancer nodal staging (Ferrari, 2005; Harisinghani et al., 2003; Harisinghani et al., 2004). These particles have also been utilized to better delineate primary tumors (Enochs et al., 1999), image angiogenesis (Tang et al., 2005), and detect metasteses (Harisinghani et al., 2001; Saini et al., 2000). One of the main drawbacks is that the dextran coating is in equilibrium with the surrounding medium, as it is not strongly associated with the iron oxide core (McCarthy and Weissleder, 2008).

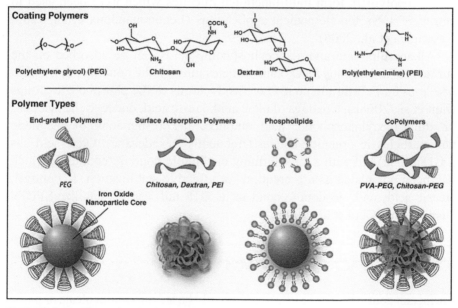

FIGURE 8.3 Illustration depicting the assembly of polymers onto the surface of magnetic nanoparticle cores.

Source: Reprinted from Veiseh, O.; Gunn, J. W.; Zhang, M. Design and Fabrication of Magnetic Nanoparticles for Targeted Drug Delivery and Imaging. *Adv. Drug Delivery Rev.* **2010,** *62,* 284–304. © 2010, with permission from Elsevier.

Chitosan is a cationic, hydrophilic polymer suitable for functionalization. The situ coating is not acceptable due to its poor solubility at pH values necessary for precipitation of SPIONs (Kumar et al., 2004). Therefore, chitosan is physically adsorbed onto oleic acid-coated SPIONs (Kim et al., 2005). The cationic nature of the polymer allows complexation with genetic material through electrostatic interactions (Bhattarai et al., 2008). These

SPIONs were used to enhance gene transfection. In addition, chitosan-based coatings are suitable for functionalization with other ligands through both amino and hydroxyl functional groups.

Copolymers (e.g., PEI and PEG polymers [Jain et al., 2005], PEG-g-chitosan-g-PEI [Yang et al., 2006], a triblock PEG-poly(methacrylic acid)-poly(glycerol monomethacrylate) copolymer) have been developed to combine the distinct functionalities of their constituents. They can be attached to the SPION surface by layer-by-layer deposition based on elec-trostatic interactions (Ito et al., 2004), hydrophilic/hydrophobic interactions (Grief and Richardson, 2005), and covalently grafting polymer layers to base coatings (McFerrin and Sontheimer, 2006). Additionally, amphiphilic block copolymers form multifunctional micelles which have been used to entrap SPIONs for biomedical applications (Lecommandoux et al., 2005; Nasongkla et al., 2006).

Alkanesulphonic and alkanephosphonic acids can be adsorbed on the surfaces of amorphous Fe_2O_3 nanoparticles through the bonding of phospho-nate ions on Fe^{3+} through one O or two O atoms of the phosphonate groups (Sun et al., 2008b). Coatings of oleic acid, lauric acid, dodecyl-, hexadecyl-, or dihexadecylphosphonic acid stabilize the dispersions of magnetic nanoparticles in organic vehicles (frequently hexadecane). Oleic acid has a C18 (oleic) tail with a kink-forming cis-double bond in the middle which has been postulated as a prerequest for effective stabilization. In contrast, stearic acid, with no double bond in its C18 tail, cannot stabilize SPION suspensions (Choi et al., 2006).

8.2.2.2   SURFACTANT COATINGS

Surfactant coatings of the SPIONs are usually based on sodium oleate or phos-pholipids. Hydrophilic SPIONs can be coated by spherical aggregates (micelles or vesicles) post-synthesis or by particle synthesis within their aqueous core (Martina et al., 2005; Yang et al., 2007). For example, small (15 nm) uniform SPIONs were obtained by precipitation into the liposomal core (Decuyper and Joniau, 1988). Martina et al. (2005) developed long circulating magnetic fluid-loaded liposomes by encapsulating maghemite nanocrystals within unilamellar vesicles of egg phosphatidylcholine and distearoylphosphatidylethanolamine-poly(ethylene glycol) 20Amphiphilic coatings of SPIONs such as micelles and liposomes are also suitable for simultaneous encapsulation of the active substance molecules (Mornet et al., 2004).

8.2.2.3 SURFACE MODIFICATIONS WITH INORGANIC MOLECULES

Silica, gold, or gadolinium, are currently frequently used in biocompatible inorganic coatings for SPIONs providing protection to the core from oxidation and suitability for functionalization with biologically active agents (Bulte et al., 1992; Cerdan et al., 1989; Farokhzad et al., 2004; Herr et al., 2006; Weissleder et al., 2005; Yigit et al., 2007). Functionalization of silanol groups of silica with alcohols and silane coupling agents (Herr et al., 2006) enable to produce stable dispersions in nonaqueous solvents as well to covalently attached specific ligands. Silica or gold-encapsulated SPIONs represent attractive approach for developing MRI contrast agents. In the study of Ma et al. (2006), the synthesis of multifunctional SPIONs comprising iron oxide cores (with the diameter of approximately 10 nm) and SiO_2 shells (10–15 nm thick) including a luminescent organic dye ruthenium inside the second silica shell was demonstrated. In this study, an organic dye, tris (2,2'-bipyridine) ruthenium, was doped inside a second silica. Therefore, the designed nanoparticles exhibited both superparamagnetic and luminescent properties and were proposed for use in biomedical imaging applications.

8.2.2.4 SURFACE FUNCTIONALIZATION WITH TARGETING LIGANDS

SPIONs are most often functionalized with more than one type of ligand including targeting agents, permeation enhancers, imaging dyes, and therapeutic agents (e.g., peptide-targeted magnetic nanoparticles can include fluorescent dyes or radionuclides and drug molecules) (McCarthy and Weissleder, 2008). The common ligands are peptides as well as small hydrophilic molecules (e.g., reactive amines, alcohols, carboxylic acids, sulfhydryls, or anhydrides), preferable regarding steric constraints that significantly enhance biomedical advances of the SPIONs. Molecules such as 1-ethyl-3-(3-dimethylaminopropyl) carbodiimide hydrochloride, N-succinimidyl 3-(2-pyridyldithio) propionate, N-hydroxysuccinimide, or N,N'-methylenebisacrylamide are usually used for attachment of the hydrophilic coating to a protein coating specific for cell surface (Roberts et al., 2002). Functional ligands can be conjugated on the MNP surfaces or incorporated within these nanostructures (Fig. 8.4). The conjugation of the ligand molecules with nanoparticle surfaces occurs through covalent linkage strategies (e.g., direct nanoparticle conjugation, click chemistry, and covalent linker chemistry) and physical interactions (electrostatic, hydrophilic/hydrophobic, and affinity interactions), in accordance with the chemical properties and functional groups on the SPIONs coating and a ligand. Strong covalent linkages can be formed between functional groups, typically amino, carboxylic acid, and thiol groups, which are available on the coating polymer molecules (e.g., chitosan, PEI, dextran, and PEG) as well as on the targeting, therapeutic, and imaging agent, through amide or ester bonds. Physical interactions include electrostatic, hydrophilic/hydrophobic, and affinity interactions. Electrostatic interactions have proved useful in the binding of plasmid DNA (Gomez-Lopera et al., 2001; Voit et al., 2001) and cationic proteins (Boussif et al., 1995). Hydrophobic/hydrophilic interactions are weak and useful mainly in adsorption of hydrophobic drugs onto hydrophobic coatings of SPIONs, and thus, suitable for triggered release intracellularly when the coating degrades. Affinity interactions are very strong, insensitive to variations in environmental conditions such as pH, salinity, or hydrophilicity, and effective for bioconjugation of targeting ligands (McCarthy and Weissleder, 2008; Veiseh et al., 2010).

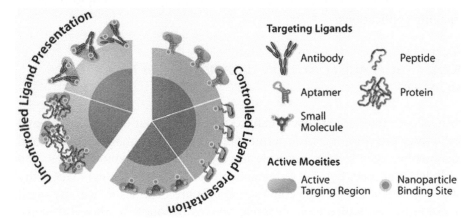

FIGURE 8.4 Illustration of the supermolecular assembly and presentation of targeting antibodies, proteins, peptides, aptamers and small molecules on the surfaces of SPIONs. Note that protein and antibody assembly is difficult to control. Small organic molecules do assemble well but their small size may cause their active targeting regions to be sterically blocked by polymeric coatings. Peptides and aptamers assembly can be controlled through their engineering, and can be modified to assemble in a manner that ensures their active sites are available for interaction with targets on cell surfaces.

Source: Reprinted from Veiseh, O.; Gunn, J. W.; Zhang, M. Design and Fabrication of Magnetic Nanoparticles for Targeted Drug Delivery and Imaging. *Adv. Drug Delivery Rev.* **2010**, *62*, 284–304. © 2010, with permission from Elsevier.

8.3   APPLICATION OF SPIONS IN DIAGNOSIS OF CANCER

Magnetic nanoparticles are used in in vivo diagnosis for about 40 years. Over the last two decades on the market were placed several contrast agents for MRI of malignant diseases. The main benefits of MRI are a high three-dimensional spatial resolution and high contrast differentiation between soft tissues. On the other hand, MRI requires placement of the patient in the MRI apparatus, preferably without motion, during relatively long acquisition times (minutes to hours). SPIONs that have been developed as imaging agents for MRI which possess proper design features regarding size, coating, and molecular functionalization (e.g., PEG coating, liver targeting molecules) (Lima-Tenório et al., 2015). Li et al. (2015) prepared tumor-targeted magnetic micelles by self-assembling of SPIONs-loaded amphiphilic copolymeric micelles based on 2,2,3,4,4,4-hexafluorobutyl methacrylate, methacryloxyethyl trimethylammonium chloride, and methoxy PEG monomethacrylate (methoxy PEG monomethacrylate), functionalized with folate-conjugated bovine serum albumin (FA–BSA) through electrostatic interactions. Wang et al. (2015), described a simple process to prepare PEG–SPIONs and PEG/PEI–SPIONs, by the thermal decomposition of iron (III) acetylacetonate in the mixture of PEG and PEI. The designed SPIONs are intravenously injected for in vivo MRI of the mouse brains and they showed good contrast effect. Xie et al. (2015) demonstrated that SPIONs exhibited vascular imaging effects of the mouse brains after 24 h of intravenous injection.

SPIONs have traditionally been used for imaging through passive targeting but also attractive options are multimodal imaging, cellular-specific targeting, and theranostics (Lima-Tenório et al., 2015). Multimodal imaging SPIONs incorporate multiple imaging moieties for the use in integrated imaging systems including magnetic resonance (MR), optical, or nuclear imaging systems providing a variety of information regarding the pathological process using the unique imaging capabilities of each system (Weissleder and Pittet, 2008; Weissleder and Mahmood, 2001). The average particle size of the first generation of SPIONs-based contrast agents is larger than 50 nm. Ferumoxides AMI-25 (Endorem® and Feridex IV®) for i.v. application contains SPIONs with average size of 150 nm with a small core of magnetite (5.6 nm) and a coating of dextran and suspended in an isotonic glucose solution. Resovist® i.v. injection contains magnetite particles (about 4.2 nm) and coated with carboxydextran up to the overall diameter of about 60 nm. After i.v. injection, SPIONs are engulfed fast by the RES macrophages and thus enable diagnosis of tumors or metastases in liver and spleen. SPIONS rapidly accumulates in tissues rich in Kupffer cells, while accumulation in the malignant tissue does not occur. Under the influence of an external magnetic field, the nanoparticles magnetize and the pathological tissue become "visible" on the scanner by creating a contrast between healthy and pathological tissues thus enabling discrimination between different tissue types (Sun et al., 2008b). The second generation of the SPION-based contrast agents has been developed with the aim to prolong the time of their retention in the circulation and enable diagnosis of metastasis in lymph nodes. Powder for solution for infusion Sinerem® is intravascular contrast agent for MRI diagnostic with the particle size about 30 nm. This is contrast agent specific for the RES (liver, spleen, bone marrow, and lymph nodes) and was used to evaluate the primary expansion of the tumors in the area of the pelvis. Maximum portion of the applied dose of Sinerem®, that is, distributed in the liver and lymph nodes after administration is up to 2 and 43%, respectively, and when applied Endorem® is 6 and 1%, respectively. However, the analysis among a large group of patients showed that Sinerem® did not provide the expected improvements in the diagnosis of malignancy in the lymph nodes. Upon the request of the manufacturer, Sinerem® was withdrawn from the European market in 2007.

The next generation of the active targeting SPIONS has the potential for significant improvement of tumor detection and localization (Koo et al., 2006). For example, the literature demonstrated the specific accumulation of chlorotoxin (CTX)-targeted SPIONs in 9L glioma flank xenografts enhancing the contrast of tumors and a surrounding healthy tissue in comparison to nontargeted control SPIONs (Sun et al., 2008c). Different contrast agents (T1 contrast agent gadolinium-diethyleneamine penta-acetic acid; T2 contrast agents Feridex®; unmodified magnetic micelles; and, FA-BSA-modified magnetic micelles) were evaluated in nude BALB/c mice with tumor growth to about 400 mm^3 after a subcutaneous injection of human hepatoma Bel-7402 cells to the lower back. It was observed that the signal intensity in the group treated with the FA-BSA-modified magnetic micelles decreases most significantly between 1 and 24 h, while the tumors with Feridex® and unmodified magnetic micelles show minimal contrast change in the same period. The distribution of the four contrast agents in the normal liver at 24-h postinjection was measure to investigate magnetic micelle deposition and revealed the high selectivity and sensitivity to hepatoma. FA-BSA modified magnetic micelles showed to prolong the circulation time and enable the particles to be delivered to the tumor tissue. It was also observed that the folate-mediated active targeting could improve the uptake of magnetic micelles by the tumor (Li et al., 2015).

8.4 APPLICATION OF SPIONS IN CANCER THERAPY

The utilization of SPIONs as carriers for the site-specific delivery of anticancer drugs and/or magnetic hyperthermia agents represents the most promising area of their biomedical application. Chemotherapeutic drugs frequently combined with SPIONs are paclitaxel, doxorubicin (DOX), methotrexate (MTX), and etoposide (Hu et al., 2008; Jain et al., 2008b; Kohler et al., 2005; Kohler et al., 2006; Liong et al., 2008; Schulze et al., 2005). Recently, Fang et al. (2014) described dual (DOX and curcumin)-loaded nanocapsule assembled from PVA and polyacrylic acid and iron oxide nanoparticles as an interesting strategy for magnetic targeting in brain tumor-bearing mice with more efficient suppression of cancer growth in vivo than does the delivery of either drug individually. Additionally, magnetic nanoparticles could be an attractive option for cancer treatment by macromolecules, for example, survivin small interfering ribonucleic acid (siRNA), BIRC5 (a cancer-specific protein previously known as *survivin*) siRNA (Kumar et al., 2010; Medarova et al., 2007), EGFRvIII (a 145-kDa golgi membrane mutated regulatory protein expressed in tumor cells) antibody (Bhojani et al., 2010; Hadjipanayis et al., 2010), CTX (Veiseh et al., 2009; Veiseh et al., 2010), dendrimers and anti-EGFRvIII siRNA (Agrawal et al., 2009). Drug molecules can be chemically linked to nanoparticles either directly or through molecular cross-linkers, physically adsorbed by electrostatic or hydrophobic interactions, or encapsulated in the coating (Cole et al., 2011). The high surface area-to-volume ratios of SPIONs allow a large number of therapeutic molecules to be attached to individual nanoparticles (Sun et al., 2008b). The drug is usually bound to the surface of the SPION nanospheres or encapsulated in liposomes or microspheres comprising SPIONs (Mahmoudi et al., 2011). The biocompatible ferrofluids are injected intravascularly (typically intravenously) and SPIONs are accumulated at the target site by high-gradient magnetic fields. In a recent study of Licciardi et al. (2013), it was reported that the external magnetic field drastically modified biodistribution of the 10 nm iron oxide nanocarriers coated with the FA-functionalized amphiphilic copolymer α,β-poly(N-2-hydroxyethyl)–d, l-aspartamide–co–(N-2-ethylen-isobutirrate)-graft–poly(butyl methacrylate) (PHEA–IB–p(BMA) copolymer), in rats, thus demonstrating the properties of efficient targeted anticancer drug delivery system. In early studies for enhancement of magnetic targeting, suggested an intra-arterial DOX-loaded SPIONs injection proximal to the tumor site and achieved the 200 times more targeting yield in comparison with an intravenous injection (Widder et al., 1981; Widder et al., 1983).

SPIONs can accumulate in many tumor tissues passively through the enhanced permeability and retention effect (Maeda, 2010) (i.e., because the leaky vasculature of a solid tumor and the poor lymphatic drainage found in some types of tumors) instead of the normal tissue vessels where endothelial cells are closely packed and form a barrier for nanoparticle penetration. The most important factors that determine their pharmacokinetics and effectiveness of the therapy are the average hydrodynamic radius and surface characteristics (Yoo et al., 2010). Optimization of the blood circulation time and bioavailability requires the comprehensive design and fitting of the particle size, charge, and surface chemistry (Chatterjee et al., 2003). Following systemic (intravenous) administration, particles with diameters greater than 200 nm and with hydrophobic surface, adsorb to nanoparticle surfaces circulating opsonin proteins and rapidly sequestered by the RES (mainly by the Kupffer cells in the liver, spleen, and the GIT), resulting in decrease of the blood circulation time. Nevertheless, such particles may be useful as MRI agents. In contrast, particles with diameters of less than 10 nm are rapidly removed through extravasations and renal clearance. Hydrophilic particles ranging from 10 to 00 nm are small enough to be unrecognizable by RES and thus demonstrate the prolonged blood circulation times and may offer the most effective distribution in diseased tissues (Gupta and Wells, 2004; Medarova et al., 2007). Sufficient hydrophilicity of the SPIONs to minimize or avoid the plasma protein adsorption and recognition by macrophage cells is usually provided by surface coverage with amphiphilic polymers such as macrogols (PEGs), PEG derivatives, poloxamers, and poloxamines. Concentration of the coating material would have to be balanced against the minimum amount required to ensure particle stability and provide sufficient chemical functionality for desired modifications (Cole et al., 2011). Hydrophilic polymer chains attached to a nanoparticle surface provide steric resistance to opsonization and macrophage uptake processes. Moreover, hydrophilicity facilitates the intracellular uptake of the SPIONs bearing cancer therapy by specific cancer cells (Zhang et al., 2002). Several studies (Gomez-Lopera et al., 2001) have pointed the benefits of biodegradable polymers (e.g., poly(D, L-lactide)) as almost ideal coatings of the magnetic core of the SPIONs, which are nontoxic, non-immunogenic, magnetic field responsive, and useful for specific imaging (outside RES) and drug delivery with enhanced efficacy and specificity (Chilkoti et al., 2002; Najafi et al., 2003; Otsuka et al., 2003). The SPIONs shells usually consist of pH- and thermo-sensitive polymers (e.g., block copolymers and copolymer hydrogels). The drug is localized within the stimulus sensitive shell and it will be released upon reaching the targeted cancerous tissue, under the corresponding stimulus. The thermo-sensitive polymeric shell change their structure or conformation breaking of covalent or non-covalent chemical bonds with the drug when temperature of magnetic core is increased (e.g., by an alternating current magnetic field) (Berry et al., 2003; Portet et al., 2001). Similarly, pH-sensitive coating polymers can be utilized to cancer targeting and they may change and release the drug under the pH differences between the caner and the healthy tissue (Ballauff et al., 2007; Dimitrov et al., 2007; Jeong et al., 2002; Klouda and Mikos, 2008; Kono, 2001; Ruel-Gariepy and Leroux, 2004). Additionally, a design of the target-specific SPIONs by introduction of the tumor cell-specific ligands at their surface can increase the selectivity of the drug delivery at the diseased tissue and thus decrease side effects (Koo et al., 2005; Mishra et al., 2010). Moreover, the drug-loaded SPIONs localization under the external magnetic field is a promising strategy for additional enhancement of the therapy (Veiseh et al., 2010). Drug-targeting and magnetic hyperthermia therapy depend on the accuracy during the exposition to the magnetic field and several other parameters such as physicochemical and magnetic properties of SPIONs, strength, and geometry of the magnetic field, depth of the target tissue, blood vascular supply, and flow rate (Cole et al., 2011; Dobson, 2006; Neuberger et al., 2005). SPIONs reach to the healthy tissues and these gets situated between the magnetic sources and pathological tissues are inevitably exposed to the influence of the magnetic field. Therefore, the geometry of the magnetic field is of great importance for the quality of treatment. The risk of agglomeration of nanoparticles in situ after the termination of the influence of the external magnetic field and ability of embolization of the blood vessels should be considered. The main disadvantage of this strategy is that the intensity of the magnetic field weakens with increasing distance of the source and the target tissue, thus the high intensity is necessary to ensure localization of nanoparticles, drug release, or to achieve hyperthermia. Therefore, a sufficient efficacy could only be achieved for targets close to the body surface (Grief and Richardson, 2005). For instance, brain tumors deep within the skull are particularly hard to target magnetically because they are at distances at which the magnetic flux density significantly decreases (Liu et al., 2010). Strategies to overcome magnet barriers under current investigation include magnetic implants (Polyak et al., 2008) to produce localized fields. Kubo et al. (2000) implanted permanent magnets at solid osteosarcoma sites in hamsters and achieved four-fold increase in delivery of the cytotoxic compounds through magnetic liposomes in comparison with typical intravenous (nonmagnetic) delivery.

After that the internalization of the nanoparticles and/or a drug payload into the targeted cells, likely by receptor-mediated endocytosis and internalization by caveolae structures, with the subsequent release of the drug takes place. Nanoparticles smaller than 50 nm or those decorated with the nonspecific permeation enhancers such as highly cationic, cell-penetrating peptides (e.g., trans-activator of transcription gene of human immunodeficiency virus—HIV-TAT, model amphiphatic peptide, and low-molecular-weight protamine), or aminated synthetic polymers (e.g. PEI), are suitable for facilitated diffusion through cell, when targeting ligands are not used or are used but do not aid in cellular uptake (Cole et al., 2011). Introduction of cleavable linkers responsive to pH, osmolarity, or enzymatic activity (Tung et al., 2000; Weissleder et al., 1999) is a common strategy to achieve the intracellular drug release to targeted subcellular organelles, such as the nucleus or mitochondria. The drug uploaded within the SPIONs can be released through enzymatic activity or changes in physiological conditions such as pH, osmolality, or temperature, and may be internalized by the endothelial cells of the target tissue or be taken up by the tumor cells (Mahmoudi et al., 2011). For SPIONs with the drug loaded on their surface, a majority of the payload is quickly released upon injection (i.e., burst effect) and therefore, a small fraction of the drug reach to the specific site after magnetic drug targeting. However, in cases where the drug, such as methotrexate has an affinity for the target cell, grafting of the drug to the surface of the SPIONs is an advantage (Kohler et al., 2005; Kohler et al., 2006; Sun et al., 2008c). Mahmoudi et al. (2009) demonstrated the reduced burst effect for tamoxifen in the therapy of breast cancer for 21% by preparation of the SPIONs with a cross-linked PEG-co-fumarate coating in comparison with the noncross-linked tamoxifen-loaded particles. In another case, magnetic liposomes and PLGA microspheres filled with drugs and SPIONs and used for targeted delivery applications (Gonzales and Krishan, 2005; Kubo et al., 2001) are recognized as promising strategy for lowering the burst effect. Yang et al. demonstrated the sustained release of a hydrophobic cisplatin from poly(ethyl-2-cyanoacrylate)-coated magnetite nanoparticles and more rapid release for the hydrophilic gemcitabine (Yang et al., 2006). Kohler et al. observed a controlled release of covalently attached MTX through amide bonds to amine-functionalized SPIONs in breast and brain tumor cells (Kohler et al., 2005; Kohler et al., 2006). The most of polymeric coatings in SPIONs form hydrogel shells suitable for optimization of their physical and chemical properties (permeability, temperature sensitivity, pH sensitivity, osmolarity sensitivity, surface functionality, swelling, biodegradability, and surface biorecognition sites) for achievement of the drug release kinetic (Peppas et al., 2006). For example, the release of chemotherapeutic drugs (examined in both in vitro and in vivo) can be controlled by local heating using an alternating current magnetic or electromagnetic field of the thermo-sensitive hydrogels on the surface of multifunctional SPIONs (e.g., for simultaneous imaging, hyperthermia, and drug delivery applications) (Babcinova et al., 2002; Viroonchatapan et al., 1998). Local heating of the membrane of the magnetoliposomes with SPIONs (8 nm) induced the DOX release upon magnetic field irradiation at a temperature higher than a phase transition temperature of 42°C (Babcinova et al., 2002).

Chemicell GmbH Company (http://www.chemicell.com/home/index.html) has developed and patented the SPIONS for antineoplastics drug delivery. Target MAG-DOX (Chemicell GmbH) is a ferrofluid comprising nanoparticles of magnetite coated with a cross-linked starch with cationic terminal groups which are substituted with DOX at a concentration of 3 mg/ml. The particle size was 50 nm. The drug substance is released by diffusion or under the influence of enzymes, or changes in pH or temperature. Other ferrofluids from the same company that can further be used for magnetic drug targeting applications and MRI are fluid MAG-nanoparticles comprising of SPIONs with diameters of 50, 100, and 200 nm in aqueous dispersions. The particles are covered with very hydrophilic polymers while the terminal functional groups such as ion-exchange groups or reactive groups for covalent immobilization can be used for binding to biomolecules. Current investigations are going in the direction to develop the third generation of SPIONs, which are extremely small and have a cancer ligands bound to the surface specific for moieties that are overexpressed or uniquely present on tumor cell plasma membranes to achieve the active target delivery and reduce the toxicity and side effects. Their surface is typically modified with different proteins, peptides (e.g., C-telopeptide CTX, arginylglycylaspartic acid, lung cancer-targeting peptide, tumor homing peptide (Cys–Arg–Glu–Lys–Ala), bombesin, F3, A54, luteinizing hormone-releasing hormone), antibodies (e.g., anti-HER2, a humanized monoclonal *antibody* that targets the human epidermal growth factor receptor-2, and anti-EGFR/EGFRvIII), and small molecules (e.g., folate) (Arruebo et al., 2007; Chomoucka et al., 2010; Cole et al., 2011; Lu et al., 2007; Kim et al., 2010). Small molecules as targeting agents are generally more robust than proteins or peptides thereby reducing the possibility of loss of functionality through the synthesis of magnetic nanoparticles. SPIONs with the folic acid whose receptor is overexpressed on the surface of many human tumor cells, including ovarian, lung, breast, endometrial, renal, and colon cancers demonstrated the highly selective binding to a variety of tumor cells to improve their detectability by MRI. Macromolecular ligands are attractive; however, their synthesis and combination with magnetic nanoparticles is expensive (Cole et al., 2011; Sun et al., 2008b).

8.5   THERANOSTICS (THERAGNOSTICS)

Superparamagnetic nanoparticles are multifunctional nanomaterials suitable for simultaneous therapy and diagnosis (i.e., theranostic or theragnostic), that is, they could target the tumor site for diagnosis (by MRI), for chemotherapeutic drug delivery and/or hyperthermia, and for monitoring the therapy response and efficacy (Schleich et al., 2013). Such integrative theranostic capacity is significant to reduce the related side effects associated with the traditional methods as well as their cost (Ahmed et al., 2012). Barick et al. (2014) prepared carboxyl decorated iron oxide nanoparticles by a faciles of t-chemical approach for MRI and hyperthermia applications. Sun et al. (2008a), developed PEG-coated SPIONs for active brain targeting with attached MTX and CTX which may facilitate delivery to brain tumor cells. MR images obtained in a 3-day study from mice with flank 9L brain tumors injected with SPIONs-MTX or SPIONs-MTX-CTX showed the selective contrast enhancement and indicated the accumulation of SPIONs comprising CTX, which also exhibits the ability to inhibit tumor invasion, which is particularly useful in the treatment of highly invasive brain tumors such as gliomas (Deshane et al., 2003). Yang et al. (2007) simultaneous targeted drug and MRI agents to breast cancer tumors through the use of multifunctional magneto-polymeric nanohybrids (MMPNs) composed of magnetic nanocrystals and DOX which were simultaneously encapsulated within an amphiphilic block copolymer shell and additionally functionalized with the breast cancer targeting/therapeutic ligand, anti-Herceptin antibody. In vivo evaluations of this nanoparticle system were performed in nude mice bearing NIH3T6.7 breast cancer tumors and MR images acquired at various time points prior to and postinjection with either HER-MMPNs, anti-herceptin antibody-conjugated multifunctional magneto-polymeric nanohybrids or a control irrelevant IgG human antibody conjugated MMPNs—IRR-MMPNs. The quantitative evaluation of MR images revealed the preferential accumulation of the targeted MNPs compared to the control MNPs. Further evaluation revealed that the HER-MMPNs which were decorated with targeting ligands and loaded with DOX were most effective in inhibiting tumor growth. Zhou et al. (2014) developed a new triple functional photo-thermal therapeutic agent—PEGylated Fe-Fe$_3$O$_4$ with capacity of magnetic targeting drug delivery, near-infrared photothermal therapy, and MRI for efficient ablation of tumors.

8.6 FUTURE RESEARCH DIRECTIONS

Generally, ferrofluids for parenteral administration not only should release a chemotherapeutic drug but also should fulfill the strict requirements for sterility, non-immunogenicity, and nontoxicity as any other parenteral formulations (Mahmoudi et al., 2011). Fulfilling of these requirements is the important current challenge for clinical level evaluation of SPIONs. Biodistribution and toxicity profile of SPIONs designed for drug delivery and diagnostics depend on numerous aspects including the nanoparticle synthesis procedure, size, and shape of the particles, the overall chemical composition (coatings and ligands), present impurities, surface charac-teristics of the particles, type, charge and solubility characteristics of the coating, toxicity of any drug attached cargo, route of administration, the administered dose, the extent of tissue distribution, pharmacokinetic profile, and biodegradability. In the most cases, the majority of the administered SPIONs distribute and accumulate in tissues of the RES, primarily liver (Chertok et al., 2010; Gupta et al., 2007; Jain et al., 2008a; Kunzmann et al., 2011; Lee et al., 2010; Medarova et al., 2007; Tran and Webster, 2010), and this raises concerns regarding their long-term fate in vivo. Many developed SPIONs are well tolerated. No considerable cytotoxicity of SPIONs was found at concentrations of several hundred times of the applied dosage when tested through various methods such as the cell life cycle assay, MTT assay, comet assay, TUNEL assay (i.e., for apoptosis detection), whereas other magnetic particles have shown considerable toxicity in the same dosage (Laurent et al., 2011). However, iron oxide nanoparticles per se are nontoxic and biodegradable (eventually broken down to form blood hemoglobin) (Lin et al., 2001), other constituents of the SPIONs biodegraded individually by the body (Veiseh et al., 2010). Concerns may arise regarding inorganic and biopersistent coatings such as gold (Lin et al., 2001). The metabolism, clearance, and toxicity profiles associated with gold-coated SPIONs will be drastically different from that of an iron oxide nanoparticles coated with biodegradable molecules (e.g., phospholipids from the liposomes). Therefore, there is obvious necessity for individual evaluation of every SPION-based formulation (Sun et al., 2008b). The critical points for evaluation of the safety aspect are difficult to establish (Garnett and Kallinteri, 2006). The currently recommended preclinical assessments included evaluation of biocompatibility, immunogenicity, pharmacokinetic profile (distribution, metabolism and biodegradability, bioavailability, elimination), and toxicity. Toxicity evaluation may include organ-specific toxicity (e.g., liver functionality tests), cytokine detection, histology, lipid hydroperoxide (a biomarker for tissue oxidative stress) levels and blood counts, acute, subacute, and chronical toxicity of the formulations, individual ingredients and degradation products, in in vitro and in vivo animal models (Lewinski et al., 2008; Lima-Tenório et al., 2015; Veiseh et al., 2010).

8.7 CONCLUSION

SPIONs for utilization in cancer treatment as chemotherapeutic drugs carriers, MHF, or MRI agents, under the external magnetic field, need special surface characteristics (coatings and ligands) and possess biocompatibility, nontoxicity, high magnetic susceptibility, and target ability in a specific body area. The size, shape, composition, and particularly the nature of surfaces, play a significant role in biokinetics and biodistribution of the nanoparticles in the body and for their internalization into cells. Therefore, the increase of the potential use of the SPIONs for cancer treatment relies on the development of new preparation methods and design of the particle with well-tailored magnetic properties, surface characteristics, and safety which will be an acceptable candidate for regulatory approval for use in humans. Furthermore, the generation of dual function SPIONs may lead to expansion of their clinical utility from the MRI agents to the targeted theranostic treatment of malignant diseases. The subject of the ongoing studies, beside careful optimization of SPIONs physicochemical characteristics, is also the enhancement of the tumor therapy and diagnostic based on a good control of the magnetic field parameters (e.g., the strength and frequency of the applied magnetic field) and the depth and concentration of the SPIONs in the tissue.

KEYWORDS

- chemotherapeutics
- enhanced permeability and retention effect
- passive drug delivery
- active drug delivery
- targeted drug delivery
- ferrofluid, magnetic fluid hyperthermia (MFH)
- magnetic resonance imaging
- nanotechnology
- nanopharmaceutics

REFERENCES

Agrawal, A.; Min, D.-H.; Singh, N.; Zhu, H.; Birjiniuk, A.; von Maltzahn, G.; Harris, T. J.; Xing, D.; Woolfenden, S.; Charest, P. A.; Bhatia, S. Functional Delivery of Sirna in Mice Using Dendriworms. *ACS Nano* **2009**, *3*, 2495–2504.

Ahmed, N.; Fessi, H.; Elaissari, A. Theranostic Applications of Nanoparticles in Cancer. *Drug Discovery Today* **2012**, *17*, 928–934.

Arruebo, M.; Fernández-Pacheco, R.; Ricardo Ibarra, M.; Santamaría, J. Magnetic Nanoparticles for Drug Delivery. *Nano Today* **2007**, *2*, 22–32.

Babincova, M.; Cicmanec, P.; Altanerova, V.; Altaner, C.; Babinec, P. Ac Magnetic Field Controlled Drug Release from Magnetoliposomes: Design of a Method for Site-Specific Chemotherapy. *Bioelectrochemistry* **2002**, *55*, 17–19.

Ballauff, M.; Lu, Y. "Smart" Nanoparticles: Preparation, Characterization and Applications. *Polymer* **2007**, *48*, 1815–1823.

Barick, K. C.; Singh, S.; Bahadur, D.; Lawande, M. A.; Patkar, D. P.; Hassan, P. A. Carboxyl Decorated Fe3O4 Nanoparticles for Mri Diagnosis and Localized Hyperthermia. *J. Colloid Interface Sci.* **2014**, *418*, 120–125.

Berry, C. C.; Curtis, A. S. G. Functionalisation of Magnetic Nanoparticles for Applications in Biomedicine. *J. Phys. D Appl. Phys.* **2003**, *36*, R198–R206.

Berry, C. C.; Wells, S.; Charles, S.; Curtis, A. S. G. Dextran and Albumin Derivatised Iron Oxide Nanoparticles: Influence on Fibroblasts in Vitro. *Biomaterials* **2003**, *24*(25), 4551–4557.

Bhattarai, S. R.; Kim, S. Y.; Jang, K. Y.; Lee, K. C.; Yi, H. K.; Lee, D. Y.; Kim, H. Y.; Hwang, P. H. Laboratory Formulated Magnetic Nanoparticles for Enhancement of Viral Gene Expression in Suspension Cell Line. *J. Virol. Methods* **2008**, *147*, 213–218.

Bhojani, M. S.; Van Dort, M.; Rehemtulla, A.; Ross, B. D. Targeted Imaging and Therapy of Brain Cancer Using Theranostic Nanoparticles. *Mol. Pharmacol* **2010**, *7*, 1921–1929.

Boussif, O.; Lezoualch, F.; Zanta, M. A.; Mergny, M. D.; Scherman, D.; Demeneix, B.; Behr, J. P. A Versatile Vector for Gene and Oligonucleotide Transfer Into Cells in Culture and In-Vivo—Polyethylenimine. *Proc. Natl. Acad. Sci. U. S. A.* **1995,** *92*(16), 7297–7301.

Bulte, J. W.; Hoekstra, Y.; Kamman, R. L.; Magin, R. L.; Webb, A. G.; Briggs, R. W.; Go, K. G.; Hulstaert, C. E.; Miltenyi, S.; The, T. H.; et al. Specific Mr Imaging of Human Lymphocytes by Monoclonal Antibody-Guided Dextran-Magnetite Particles. *Magn. Reson. Med.* **1992,** *25*, 148–157.

Cerdan, S.; Lotscher, H. R.; Kunnecke, B.; Seelig, J. Monoclonal Antibody-Coated Magnetite Particles as Contrast Agents in Magnetic Resonance Imaging of Tumors. *Magn. Reson. Med.* **1989,** *12*, 151–163.

Chatterjee, J.; Haik, Y.; Chen, C.-J. Size Dependent Magnetic Properties of Iron Oxide Nanoparticles. *J. Magn. Magn. Mater.* **2003,** *257*(1), 113–118.

Chertok, B.; Cole, A. J.; David, A. E.; Yang, V. C. Comparison of Electron Spin Resonance Spectroscopy and Inductively-Coupled Plasma Optical Emission Spectroscopy for Biodistribution Analysis of Iron-Oxide Nanoparticles. *Mol. Pharm.* **2010,** *7*, 375–385.

Chilkoti, A.; Dreher, M. R.; Meyer, D. E. Design of Thermally Responsive, Recombinant Polypeptide Carriers for Targeted Drug Delivery. *Adv. Drug Delivery Rev.* **2002,** *54*(8), 1093–1111.

Choi, J. S.; Jun, Y. W.; Yeon, S. I.; Kim, H. C.; Shin, J. S.; Cheon, J. Biocompatible Heterostructured Nanoparticles for Multimodal Biological Detection. *J. Am. Chem. Soc.* **2006,** *128*, 15982–15983.

Chomoucka, J.; Drbohlavova, J.; Huska, J.; Adam, A.; Kizek, R.; Hubalek, J. Magnetic Nanoparticles and Targeted Drug Delivering. *Pharm. Res.* **2010,** *62*, 144–149.

Cole, A. J.; Yang, V. C.; David, A. E. Cancer Theranostics: the Rise of Targeted Magnetic Nanoparticles. *Trends Biotechnol.* **2011,** *29*(7), 232–332.

Decuyper, M.; Joniau, M. Magnetoliposomes-Formation and Structural Characterisation. *Eur. Biophys. J.* **1988,** *15*, 311–319.

Deshane, J.; Garner, C. C.; Sontheimer, H. Chlorotoxin Inhibits Glioma Cell Invasion Via Matrix Metalloproteinase-2. *J. Biol. Chem.* **2003,** *278*, 4135–4144.

Dimitrov, I.; Trzebicka, B.; Muller, A. H. E.; Dworak, A.; Tsvetanov, C. B. Thermosensitive Water-Soluble Copolymers with Doubly Responsive Reversibly Interacting Entities. *Prog. Polym. Sci.* **2007,** *32*, 1275–1343.

Djekic, L.; Primorac, M.; Jockovic, J. Phase Behaviour, Microstructure and Ibuprofen Solubilization Capacity of Pseudo-Ternary Nonionic Microemulsions. *J. Mol. Liq.* **2011,** *160*, 81–87.

Djekic, L.; Primorac, M. The Influence of Cosurfactants and Oils on the Formation of Pharmaceutical Microemulsions Based on Peg-8 Caprylic/Capric Glycerides. *Int. J. Pharm.* **2008,** *352*, 231–239.

Dobson, J. Magnetic Nanoparticles for Drug Delivery. *Drug Dev. Res.* **2006,** *67*, 55–60.

Duan, H. W.; Kuang, M.; Wang, X. X.; Wang, Y. A.; Mao, H.; Nie, S. M. Reexamining the Effects of Particle Size and Surface Chemistry on the Magnetic Properties of Iron Oxide Nanocrystals: New Insights Into Spin Disorder and Proton Relaxivity. *J. Phys. Chem. C* **2008,** *112*, 8127–8131.

Enochs, W. S.; Harsh, G.; Hochberg, F.; Weissleder, R. Improved Delineation of Human Brain Tumors on Mr Images Using a Long-Circulating, Superparamagnetic Iron Oxide Agent. *J. Magn. Reson. Imaging* **1999,** *9*, 228–232.

Fang, J.-H.; Lai, Y.-H.; Chiu, T.-L.; Chen, Y.-Y.; Hu, S.-H.; Chen, S.-Y. Magnetic Core–Shell Nanocapsules with Dual-Targeting Capabilities and Co-Delivery of Multiple Drugs to Treat Brain Gliomas. *Adv. Healthcare Mater.* **2014,** *3,* 1250–1260.

Farokhzad, O. C.; Jon, S.; Khademhosseini, A.; Tran, T. N.; Lavan, D. A.; Langer, R. Nanoparticle–Aptamer Bioconjugates: a New Approach for Targeting Prostate Cancer Cells. *Cancer Res.* **2004,** *64,* 7668–7672.

Ferrari, M. Cancer Nanotechnology: Opportunities and Challenges. *Nat. Rev. Cancer* **2005,** *5,* 161–171.

Garnett, M. C.; Kallinteri, P. Nanomedicines and Nanotoxicology: Some Physiological Principles. *Occup. Med.* **2006,** *56,* 307–311.

Gomez-Lopera, S. A.; Plaza, R. C.; Delgado, A. V. Synthesis and Characterization of Spherical Magnetite/Biodegradable Polymer Composite Particles. *J. Colloid Interface Sci.* **2001,** *240*(1), 40–47.

Gonzales, M.; Krishnan, K. M. Synthesis of Magnetoliposomes with Monodisperse Iron Oxide Nanocrystal Cores for Hyperthermia. *J. Magn. Magn. Mater.* **2005,** *293,* 265–270.

Goya, G. F.; Berquo, T. S.; Fonseca, F. C. Static and Dynamic Magnetic Properties of Spherical Magnetite Nanoparticles. *J. Appl. Phys.* **2003,** *94*(5), 3520–3528.

Grief, A. D.; Richardson, G. Mathematical Modelling of Magnetically Targeted Drug Delivery. *J. Magn. Magn. Mater.* **2005,** *293,* 455–463.

Gupta, A. K.; Gupta, M. Synthesis and Surface Engineering of Iron Oxide Nanoparticles for Biomedical Applications. *Biomaterials* **2005,** *26,* 3995–4021.

Gupta, A. K.; Naregalkar, R. R.; Vaidya, V. D.; Gupta, M. Recent Advances on Surface Engineering of Magnetic Iron Oxide Nanoparticles and Their Biomedical Applications. *Nanomed.* **2007,** *2,* 23–39.

Gupta, A. K.; Wells, S. Surface-Modified Superparamagnetic Nanoparticles for Drug Delivery: Preparation, Characterization, and Cytotoxicity Studies. *IEEE Trans. Nanobiosci.* **2004,** *3,* 66–73.

Hadjipanayis, C. G.; Machaidze, R.; Kaluzova, M.; Wang, L.; Schuette, A. J.; Chen, H., Wu., X.; Mao, H. Egfrviii Antibody-Conjugated Iron Oxide Nanoparticles for Magnetic Resonance Imaging-Guided Convection-Enhanced Delivery and Targeted Therapy of Glioblastoma. *Cancer Res.* **2010,** *70,* 6303–6312.

Hamley, I. W. Nanotechnology with Soft Materials. *Angew. Chem. Int. Ed.* **2003,** *42,* 1692–1712.

Ha, N. T.; Hai, N. H.; Luong, N. H.; Chau, N.; Chinh, H. D. Effects of the Conditions of the Microemulsion Preparation on the Properties of Fe3O4 Nanoparticles. *VNU J. Sci. Nat. Sci. Technol.* **2008,** *24,* 9–15.

Harisinghani, M. G.; Barentsz, J.; Hahn, P. F.; Deserno, W. M.; Tabatabaei, S.; van de Kaa, C. H.; de la Rosette, J.; Weissleder, R. Noninvasive Detection of Clinically Occult Lymph-Node Metastases in Prostate Cancer. *N. Engl J. Med.* **2003,** *348,* 2491–2499.

Harisinghani, M. G.; Weissleder, R. Sensitive, Noninvasive Detection of Lymph Node Metastases. *PLoS Med.* **2004,** *1,* e66.

Harisinghani, M. G.; Saini, S.; Weissleder, R.; Rubin, D.; deLange, E.; Harms, S.; Weinreb, J.; Small, W.; Sukerkar, A.; Brown, J. J.; Zelch, J.; Lucas, M.; Morris, M.; Hahn, P. F. Splenic Imaging with Ultrasmall Superparamagnetic Iron Oxide Ferumoxtran-10 (Ami-7227): Preliminary Observations. *J. Comput. Assist. Tomogr.* **2001,** *25,* 770–776.

Harris, J. M.; Chess, R. B. Effect of Pegylation on Pharmaceuticals. *Nat. Rev. Drug Discovery* **2003,** *2,* 214–221.

Herr, J. K.; Smith, J. E.; Medley, C. D.; Shangguan, D.; Tan, W. Aptamer-Conjugated Nanoparticles for Selective Collection and Detection of Cancer Cells. *Anal. Chem.* **2006**, *78*, 2918–2924.

Hu, S.-H.; Tsai, C.-H.; Liao, C.-F.; Liu, D.-M.; Chen, S.-Y. Controlled Rupture of Magnetic Polyelectrolyte Microcapsules for Drug Delivery. *Langmuir* **2008**, *24*, 11811–11818.

Ito, A.; Kuga, Y.; Honda, H.; Kikkawa, H.; Horiuchi, A.; Watanabe, Y.; Kobayashi, T. Magnetite Nanoparticle-Loaded Anti-Her2 Immunoliposomes for Combination of Antibody Therapy with Hyperthermia. *Cancer Lett.* **2004**, *212*, 167–175.

Jain, T. K.; Morales, M. A.; Sahoo, S. K.; Leslie-Pelecky, D. L.; Labhasetwar, V. Iron Oxide Nanoparticles for Sustained Delivery of Anticancer Agents. *Mol. Pharmacol* **2005**, *2*, 194–205.

Jain, T. K.; Reddy, M. K.; Morales, M. A.; Leslie-Pelecky, D. L.; Labhasetwar, V. Biodistribution, Clearance, and Biocompatibility of Iron Oxide Magnetic Nanoparticles in Rats. *Mol. Pharmacol.* **2008a**, *5*, 316–327

Jain, T. K.; Richey, J.; Strand, M.; Leslie-Pelecky, D. L.; Flask, C. A.; Labhasetwar, V. Magnetic Nanoparticles with Dual Functional Properties: Drug Delivery and Magnetic Resonance Imaging. *Biomaterials* **2008b**, *29*, 4012–4021

Jeong, B.; Kim, S. W.; Bae, Y. H. Thermosensitive Sol–Gel Reversible Hydrogels. *Adv. Drug Delivery Rev.* **2002**, *54*, 37–51.

Jiles, D. Introduction to Magnetism and Magnetic Materials, 1st ed.; Chapman and Hall: New York, 1991.

Jung, C. W. Surface Properties of Superparamagnetic Iron Oxide Mr Contrast Agents: Ferumoxides, Ferumoxtran, Ferumoxsil. *J. Magn. Reson. Imaging* **1995**, *13*, 675–691.

Kalia, S.; Kango, S.; Kumar, A.; Haldorai, Y.; Kumari, B.; Kumar, R. Magnetic Polymer Nanocomposites for Environmental and Biomedical Applications. *Colloid Polym. Sci.* **2014**, *292*, 2025–2052.

Kim, B. Y. S.; Rutka, J. T.; Chan, W. C. W. Nanomedicine. *N. Engl. J. Med.* **2010**, *363*, 2434–2443.

Kim, D. K.; Zhang, Y.; Voit, W.; Rao, K. V.; Kehr, J.; Bjelke, B.; Muhammed, M. Superparamagnetic Iron Oxide Nanoparticles for Biomedical Applications. *Scr. Mater.* **2001**, *44*, 1713–1717.

Kim, E. H.; Lee, H. S.; Kwak, B. K.; Kim, B. K. Synthesis of Ferrofluid with Magnetic Nanoparticles by Sonochemical Method for Mri Contrast Agent. *J. Magn. Magn. Mater.* **2005**, *289*, 328–330.

Kircheis, R.; Wightman, L.; Wagner, E. Design and Gene Delivery Activity of Modified Polyethylenimines. *Adv. Drug Delivery Rev.* **2001**, *53*, 341–358.

Klouda, L.; Mikos, A. G. Thermoresponsive Hydrogels in Biomedical Applications. *Eur. J. Pharm. Biopharm.* **2008**, *68*, 34–45.

Kono, K. Thermosensitive Polymer-Modified Liposomes. *Adv. Drug Delivery Rev.* **2001**, *53*, 307–319.

Kohler, N.; Sun, C.; Fichtenholtz, A.; Gunn, J.; Fang, C.; Zhang, M. Q. Methotrexate Immobilized Poly(Ethylene Glycol) Magnetic Nanoparticles for Mr Imaging and Drug Delivery. *Small* **2006**, *2*, 785–792.

Kohler, N.; Sun, C.; Wang, J.; Zhang, M. Q. Methotrexate-Modified Superparamagnetic Nanoparticles and Their Intracellular Uptake Into Human Cancer Cells. *Langmuir* **2005**, *21*, 8858–8864.

Koo, O. M.; Rubinstein, I.; Onyuksel, H. Role of Nanotechnology in Targeted Drug Delivery and Imaging: a Concise Review. *Nanomedicine* **2005**, *1*, 193–212.

Koo, Y. E.; Reddy, G. R.; Bhojani, M.; Schneider, R.; Philbert, M. A.; Rehemtulla, A.; Ross, B. D.; Kopelman, R. Brain Cancer Diagnosis and Therapy with Nanoplatforms. *Adv. Drug Delivery Rev.* **2006**, *58*, 1556–1577.

Kubo, T.; Sugita, T.; Shimose, S.; Nitta, Y.; Ikuta, Y.; Murakami, T. Targeted Delivery of Anticancer Drugs with Intravenously Administered Magnetic Liposomes in Osteosarcoma-Bearing Hamsters. *Int. J. Oncol.* **2000**, *17*(2), 309–315.

Kubo, T.; Sugita, T.; Shimose, S.; Nitta, Y.; Ikuta, Y.; Murakami, T. Targeted Systemic Chemotherapy Using Magnetic Liposomes with Incorporated Adriamycin for Osteosarcoma in Hamsters. *Int. J. Oncol.* **2001**, *18*(1), 121–125.

Kumar, M.; Muzzarelli, R. A. A.; Muzzarelli, C.; Sashiwa, H.; Domb, A. J. Chitosan Chemistry and Pharmaceutical Perspectives. *Chem. Rev.* **2004**, *104*, 6017–6084.

Kumar, M.; Yigit, M.; Dai, G.; Moore, A.; Medarova, Z. Image-Guided Breast Tumor Therapy Using a Small Interfering RNA Nanodrug. *Cancer Res.* **2010**, *70*, 7553–7561.

Kunzmann, A.; Andersson, B.; Thurnherr, T.; Krug, H.; Scheynius, A.; Fadeel, B. Toxicology of Engineered Nanomaterials: Focus on Biocompatibility, Biodistribution and Biodegradation. *Biochim. Biophys. Acta* **2011**, *1810*, 361–373.

LaConte, L. E. W.; Nitin, N.; Zurkiya, O.; Caruntu, D.; O'Connor CJ; Hu XP; Bao G Coating Thickness of Magnetic Iron Oxide Nanoparticles Affects R-2 Relaxivity. *J. Magn. Reson. Imaging* **2007**, *26*, 1634–1641.

Laurent, S.; Dutz, S.; Häfeli, U. O.; Mahmoudi, M. Magnetic Fluid Hyperthermia: Focus on Superparamagnetic Iron Oxide Nanoparticles. *Adv. Colloid Interface Sci.* **2011**, *166*, 8–23.

Laurent, S.; Forge, D.; Port, M.; Roch, A.; Robic, C.; Elst, L. V.; Muller, R. N. Magnetic Iron Oxide Nanoparticles: Synthesis, Stabilization, Vectorization, Physicochemical Characterizations, and Biological Applications. *Chem. Rev.* **2008**, *108*, 2064–2110.

Lecommandoux, S.; Sandre, O.; Checot, F.; Perzynski, R. Smart Hybrid Magnetic Selfassembled Micelles and Hollow Capsules. *Prog. Solid State Chem.* **2006**, *34*, 171–179.

Lee, J. H.; Huh, Y. M.; Jun, Y. W.; Seo, J. W.; Jang, J. T.; Song, H. T.; Kim, S.; Cho, E. J.; Yoon, H. G.; Suh, J. S.; Cheon, J. Artificially Engineered Magnetic Nanoparticles for Ultrasensitive Molecular Imaging. *Natural Med.* **2007**, *13*, 95–99.

Lee, J.; Isobe, T.; Senna, M. Preparation of Ultrafine Fe3o4 Particles by Precipitation in the Presence of PVA At High pH. *J. Colloid Interface Sci.* **1996**, *177*, 490–494.

Lee, M. J.-E.; Veiseh, O.; Bhattarai, N.; Sun, C.; Hansen, S. J.; Ditzler, S.; Knoblaugh, S.; Lee, D.; Ellenbogen, R.; Zhang, M.; Olson, J. M. Rapid Pharmacokinetic and Biodistribution Studies Using Chlorotoxin-Conjugated Iron Oxide Nanoparticles: A Novel Non-Radioactive Method. *PLoS One* **2010**, *5*, e9536.

Lewinski, N.; Colvin, V.; Drezek, R. Cytotoxicity of Nanoparticles. *Small* **2008**, *4*, 26–49.

Licciardi, M.; Scialabba, C.; Fiorica, C.; Cavallaro, G.; Cassata, G.; Giammona, G. Polymeric Nanocarriers for Magnetic Targeted Drug Delivery: Preparation, Char- Acterization, and in Vitro and in Vivo Evaluation. *Mol. Pharm.* **2013**, *10*, 4397–4407.

Li, H.; Yan, K.; Shang, Y.; Shrestha, L.; Liao, R.; Liu, F.; Li, P.; Xu, H.; Xu, Z.; Chu, P. K. Folate-Bovine Serum Albumin Functionalized Polymeric Micelles Loaded with Superparamagnetic Iron Oxide Nanoparticles for Tumor Targeting and Magnetic Resonance Imaging. *Acta Biomater.* **2015**, *15*, 117–126.

Lima-Tenório, M. K.; Gómez Pineda, E. A.; Ahmad, N. M.; Fessi, H.; Elaissari, A. Magnetic Nanoparticles: in Vivo Cancer Diagnosis and Therapy. *Int. J. Pharm.* **2015**, *493*, 313–327.

Lin, J.; Zhou, W.; Kumbhar, A.; Fang, J.; Carpenter, E. E.; O'Connor, C. J. Gold-Coated Iron (Fe@Au) Nanoparticles: Synthesis, Characterization, and Magnetic Field-Induced Self-Assembly. *J. Solid State Chem.* **2001**, *159*, 26–31.

Liong, M.; Lu, J.; Kovochich, M.; Xia, T.; Ruehm, S. G.; Nel, A. E.; Tamanoi, F.; Zink, J. I. Multifunctional Inorganic Nanoparticles for Imaging, Targeting, and Drug Delivery. *ACS Nano* **2008**, *2*, 889–896.

Liu, H. L.; Hua, M. Y.; Yang, H. W.; Huang, C. Y.; Chu, P. C.; Wu, J. S.; Tseng, I. C.; Wang, J. J.; Yen, T. C.; Chen, P. Y.; Wei, K. C. Magnetic Resonance Monitoring of Focused Ultrasound/Magnetic Nanoparticle Targeting Delivery of Therapeutic Agents to the Brain. *Proc. Natl. Acad. Sci. U. S. A.* **2010**, *107*, 15205–15210.

Lu, A.-H.; Salabas, E. L.; Schüth, F. Magnetic Nanoparticles: Synthesis, Protection, Functionalization, and Application. *Angew. Chem. Int. Ed.* **2007**, *46*, 1222–1244.

Maeda, H. Tumor-Selective Delivery of Macromolecular Drugs Via the Epr Effect: Background and Future Prospects. *Bioconjugate Chem.* **2010**, *21*, 797–802.

Ma, D. L.; Guan, J. W.; Normandin, F.; Denommee, S.; Enright, G.; Veres, T.; Simard, B. Multifunctional Nano-Architecture for Biomedical Applications. *Chem. Mater.* **2006**, *18*, 1920–1927.

Mahmoudi, M.; Sant, S.; Wang, B.; Laurent, S.; Sen, T. Superparamagnetic Iron Oxide Nanoparticles (Spions): Development, Surface Modification and Applications in Chemotherapy. *Adv. Drug Delivery Rev.* **2011**, *63*, 24–46.

Mahmoudi, M.; Simchi, A.; Imani, M.; Hafeli, U. O. Superparamagnetic Iron Oxide Nanoparticles with Rigid Cross-Linked Polyethylene Glycol Fumarate Coating for Application in Imaging and Drug Delivery. *J. Phys. Chem. C.* **2009**, *113*(19), 8124–8131.

Martina, M. S.; Fortin, J. P.; Menager, C.; Clement, O.; Barratt, G.; Grabielle-Madelmont, C.; Gazeau, F.; Cabuil, V.; Lesieur, S. Generation of Superparamagnetic Liposomes Revealed as Highly Efficient Mri Contrast Agents for in Vivo Imaging. *J. Am. Chem. Soc.* **2005**, *127*, 10676–10685.

McCarthy, J. R.; Weissleder, R. Multifunctional Magnetic Nanoparticles for Targeted Imaging and Therapy. *Adv. Drug Delivery Rev.* **2008**, *60*, 1241–1251.

McFerrin, M. B.; Sontheimer, H. A Role for Ion Channels in Glioma Cell Invasion. *Neuron Glia Biol.* **2006**, *2*, 39–49.

Medarova, Z.; Pham, W,; Farra, C.; Petkova, V.; Moore, A. In Vivo Imaging of Sirna Delivery and Silencing in Tumors. *Nat. Med.* **2007**, *13*, 372–377.

Mishra, B.; Patel, B. B.; Tiwari, S. Colloidal Nanocarriers: a Review on Formulation Technology, Types and Applications Toward Targeted Drug Delivery. *Nanomed.* **2010**, *6*, 9–24.

Molday, R. S.; Mackenzie, D. Immunospecific Ferromagnetic Iron-Dextran Reagents for the Labeling and Magnetic Separation of Cells. *J. Immunol. Methods* **1982**, *52*, 353–367.

Mornet, S.; Vasseur, S.; Grasset, F.; Duguet, E. Magnetic Nanoparticle Design for Medical Diagnosis and Therapy. *J. Mater. Chem.* **2004**, *14*, 2161–2175.

Najafi, F.; Sarbolouki, M. N. Biodegradable Micelles/Polymersomes from Fumaric/Sebacic Acids and Poly(Ethylene Glycol). *Biomaterials* **2003**, *24*(7), 1175–1182.

Nasongkla, N.; Bey, E.; Ren, J. M.; Ai, H.; Khemtong, C.; Guthi, J. S.; Chin, S. F.; Sherry, A. D.; Boothman, D. A.; Gao, J. M. Multifunctional Polymeric Micelles as Cancer-Targeted, Mri-Ultrasensitive Drug Delivery Systems. *Nano Lett.* **2006**, *6*, 2427–2430.

Neuberger, T.; Schopf, B.; Hofmann, H.; Hofmann, M.; von Rechenberg, B. Superparamagnetic Nanoparticles for Biomedical Applications: Possibilities and Limitations of a New Drug Delivery System. *J. Magn. Magn. Mater.* **2005**, *293*, 483–496.

Otsuka, H.; Nagasaki, Y.; Kataoka, K. Pegylated Nanoparticles for Biological and Pharmaceutical Applications. *Adv. Drug Delivery Rev.* **2003**, *54*(8), 403–419.

Peppas, N. A.; Hilt, J. Z.; Khademhosseini, A.; Langer, R. Hydrogels in Biology and Medicine: from Molecular Principles to Bionanotechnology. *Adv. Mater.* **2006**, *18*, 1345–1360.

Polyak, B.; Fishbein, I.; Chorny, M.; Alferiev, I.; Williams, D.; Yellen, B.; Friedman, G.; Levy, R. J. High Field Gradient Targeting of Magnetic Nanoparticle-Loaded Endothelial Cells to the Surfaces of Steed Stents. *Proc. Natl. Acad Sci. U. S. A.* **2008**, *105*, 698–703.

Portet, D.; Denizot, B.; Rump, E.; Lejeune, J. J.; Jallet, P. Nonpolymeric Coatings of Iron Oxide Colloids for Biological Use as Magnetic Resonance Imaging Contrast Agents. *J. Colloid Interface Sci.* **2001**, *238*(1), 37–42.

Prime, K. L.; Whitesides, G. M. Self-Assembled Organic Monolayers—Model Systems for Studying Adsorption of Proteins At Surfaces. *Science* **1991**, *252*, 1164–1167.

Roberts, M. J.; Bentley, M. D.; Harris, J. M. Chemistry for Peptide and Protein Pegylation. *Adv. Drug Delivery Rev.* **2002**, *54*(4), 459–476.

Ruel-Gariepy, E.; Leroux, J. C. In Situ-Forming Hydrogels—Review of Temperaturesensitive Systems. *Eur. J. Pharm. Biopharm.* **2004**, *58*, 409–426.

Saini, S.; Sharma, R.; Baron, R. L.; Turner, D. A.; Ros, P. R.; Hahn, P. F.; Small, W. C.; Delange, E. E.; Stillman, A. E.; Edelman, R. R.; Runge, V. M.; Outwater, E. K. Multicentre Dose Ranging Study on the Efficacy of Uspio Ferumoxtran-10 for Liver Mr Imaging. *Clin. Radiol.* **2000**, *55*, 690–695.

Saldivar-Ramirez, M. M. G.; Sanchez-Torres, C. G.; Cortes-Hernandez, D. A.; Escobedo-Bocardo, J. C.; Almanza-Robles, J. M.; Larson, A.; Resendiz-Hernandez, P. J.; Acuna-Gutierrez, I. O. Study on the Efficiency of Nanosized Magnetite and Mixed Ferrites in Magnetic Hyperthermia. *J. Mater. Sci. Mater. Med.* **2014**, *25*, 2229–2236.

Salunkh, A. B.; Khot, V. M.; Pawar, S. H. Magnetic Hyperthermia with Magnetic Nanoparticles: a Status Review. *Curr. Top Med. Chem.* **2014**, *14*, 572–594.

Sánchez, J.; Cortés-Hernández, D. A.; Escobedo-Bocardo, J. C.; Jasso-Terán, R. A.; Zugasti-Cruz, A. Bioactive Magnetic Nanoparticles of Fe–Ga Synthesized by Sol–Gel for Their Potential Use in Hyperthermia Treatment. *J. Mater. Sci. Mater. Med.* **2014**, *25*, 2237–2242.

Schleich, N.; Sibret, P.; Danhier, P.; Ucakar, B.; Laurent, S.; Muller, R. N.; Jérôme, C.; Gallez, B.; Préat, V.; Danhier, F. Dual Anticancer Drug/Superparamagnetic Iron Oxide-Loaded Plga-Based Nanoparticles for Cancer Therapy and Magnetic Resonance Imaging. *Int. J. Pharm.* **2013**, *447*, 94–101.

Schulze, K.; Koch, A.; Schopf, B.; Petri, A.; Steitz, B.; Chastellain, M.; Hofmann, M.; Hofmann, H.; von Rechenberg, B. Intraarticular Application of Superparamagnetic Nanoparticles and Their Uptake by Synovial Membrane—an Experimental Study in Sheep. *J. Magn. Magn. Mater.* **2005**, *293*, 419–432.

Shen, T.; Weissleder, R.; Papisov, M.; Bogdanov, A. Jr; Brady, T. J. Monocrystalline Iron Oxide Nanocompounds (Mion): Physicochemical Properties. *Magn. Reson. Med.* **1993**, *29*, 599–604.

Si, S.; Kotal, A.; Mandal, T. K.; Giri, S.; Nakamura, H.; Kohara, T. Size-Controlled Synthesis of Magnetite Nanoparticles in the Presence of Polyelectrolytes. *Chem. Mater.* **2004**, *16*, 3489–3496.

Storm, G.; Belliot, S. O.; Daemen, T.; Lasic, D. D. Surface Modification of Nanoparticles to Oppose Uptake by the Mononuclear Phagocyte System. *Adv. Drug Delivery Rev.* **1995**, *17*, 31–48.

Sun, C.; Fang, C.; Stephen, Z.; Veiseh, O.; Hansen, S.; Lee, D.; Ellenbogen, R. G.; Olson, J.; Zhang, M. Q. Tumor-Targeted Drug Delivery and MRI Contrast Enhancement by Chlorotoxin-Conjugated Iron Oxide Nanoparticles. *Nanomed.* **2008a**, *3*, 495–505

Sun, C.; Lee, J. S. H.; Zhang, M. Magnetic Nanoparticles in MR Imaging and Drug Delivery. *Adv. Drug Delivery Rev.* **2008b**, *60*, 1252–1265

Sun, C.; Veiseh, O.; Gunn, J.; Fang, C.; Hansen, S.; Lee, D.; Sze, R.; Ellenbogen, R. G.; Olson, J.; Zhang, M. In vivo MRI Detection of Gliomas by Chlorotoxin-Conjugated Superparamagnetic Nanoprobes. *Small* **2008c,** *4,* 372–379

Sun, S. H.; Zeng, H.; Robinson, D. B.; Raoux, S.; Rice, P. M.; Wang, S. X.; Li, G. X. Monodisperse Mfe2o4 (M=Fe, Co, Mn) Nanoparticles. *J. Am. Chem. Soc.* **2004,** *126,* 273–279.

Tang, Y.; Kim, M.; Carrasco, D.; Kung, A. L.; Chin, L.; Weissleder, R. In Vivo Assessment of Ras-Dependent Maintenance of Tumor Angiogenesis by Real-Time Magnetic Resonance Imaging. *Cancer Res.* **2005,** *65,* 8324–8330.

Taupitz, M.; Wagner, S.; Schnorr, J.; Kravec, I.; Pilgrimm, H.; Bergmann-Fritsch, H.; Hamm, B. Phase I Clinical Evaluation of Citrate-Coated Monocrystalline Very Small Superparamagnetic Iron Oxide Particles as a New Contrast Medium for Magnetic Resonance Imaging. *Invest. Radiol.* **2004,** *39,* 394–405.

Tietze, R.; Zaloga, J.; Unterweger, H.; Lyer, S.; Friedrich, R.; Janko, C.; Pöttler, M.; Dürr, S.; Alexiou, C. Magnetic Nanoparticle-Based Drug Delivery for Cancer Therapy. *Biochem. Biophys. Res. Commun.* **2015,** *468,* 463–470.

Taylor, A. P.; Barry, J. C.; Webb, R. I. Structural and Morphological Anomalies in Magnetosomes: Possible Biogenicorigin for Magnetite in Alh840. *J. Microsc.* **2001,** *201,* 84–106.

Tran, N.; Webster, T. J. Magnetic Nanoparticles: Biomedical Applications and Challenges. *J. Mater. Chem.* **2010,** *20,* 8760–8767.

Tung, C. H.; Mahmood, U.; Bredow, S.; Weissleder, R. In Vivo Imaging of Proteolytic Enzyme Activity Using a Novel Molecular Reporter. *Cancer Res.* **2000,** *60,* 4953–4958.

Veiseh, O.; Gunn, J. W.; Zhang, M. Design and Fabrication of Magnetic Nanoparticles for Targeted Drug Delivery and Imaging. *Adv. Drug Delivery Rev.* **2010,** *62,* 284–304.

Veiseh, O.; Kievit, F. M.; Fang, C.; Mu, N.; Jana, S.; Leung, M. C.; Mok, H.; Ellenbogen, R. G.; Park, J. O.; Zhang, M. Chlorotoxin Bound Magnetic Nanovector Tailored for Cancer Cell Targeting, Imaging, and Sirna Delivery. *Biomaterials* **2010,** *31,* 8032–8042.

Veiseh, O.; Sun, C.; Fang, C.; Bhattarai, N.; Gunn, J.; Kievit, F.; Du, K.; Pullar, B.; Lee, D.; Ellennbogen, R. G.; Olson, J.; Zhang, M. Specific Targeting of Brain Tumors with an Optical/Magnetic Resonance Imaging Nanoprobe Across the Blood–Brain Barrier. *Cancer Res.* **2009,** *69,* 6200–6207.

Viroonchatapan, E.; Sato, H.; Ueno, M.; Adachi, I.; Murata, J.; Saiki, I. Microdialysis Assessment of 5-Fluorouracil Release from Thermosensitive Magnetoliposomes Induced by an Electromagnetic Field in Tumor-Bearing Mice. *J. Drug Target* **1998,** *5,* 379–390.

Voit, W.; Kim, D. K.; Zapka, W.; Muhammed, M.; Rao, K. V. *Magnetic Behavior of Coated Superparamagnetic Iron Oxide Nanoparticles in Ferrofluids.* Materials Research Society Symposium Proceedings on the Symposium Synthesis, Functional Properties and Applications of Nanostructures, San Francisco (CA), United States, April 17–20, 2001; 676:Y7.8. 1–6.

Wang, J.; Zhang, B.; Wang, L.; Wang, M.; Gao, F. One-Pot Synthesis of Water- Soluble Superparamagnetic Iron Oxide Nanoparticles and Their MRI Contrast Effects in the Mouse Brains. *Mater. Sci. Eng. C Mater. Biol. Appl.* **2015,** *48,* 416–423.

Wan, S. R.; Huang, J. S.; Yan, H. S.; Liu, K. L. Size-Controlled Preparation of Magnetite Nanoparticles in the Presence of Graft Copolymers. *J. Mater. Chem.* **2006,** *16,* 298–303.

Weissleder, R.; Elizondo, G.; Wittenberg, J.; Lee, A. S.; Josephson, L.; Brady, T. J. Ultrasmall Superparamagnetic Iron Oxide: An Intravenous Contrast Agent for Assessing Lymph Nodes with Mr Imaging. *Radiology* **1990,** *175*(2), 494–498.

Weissleder, R.; Kelly, K.; Sun, E. Y.; Shtatland, T.; Josephson, L. Cell-Specific Targeting of Nanoparticles by Multivalent Attachment of Small Molecules. *Nat. Biotechnol.* **2005,** *23,* 1418–1423.

Weissleder, R.; Mahmood, U. Molecular imaging. *Radiology* **2001**, *219(2)*, 316–333.

Weissleder, R.; Pittet, M.J. Imaging in the era of molecular oncology. *Nature* **2008**, *452*, 580–589.

Weissleder, R.; Tung, C. H.; Mahmood, U.; Bogdanov, A. Jr. In Vivo Imaging of Tumors with Protease-Activated Near-Infrared Fluorescent Probes. *Nat. Biotechnol.* **1999**, *17*, 375–378.

Widder, K. J.; Morris, R. M.; Howard, D. P.; Senyei, A. E. Tumor Remission in Yoshida Sarcoma-Bearing Rats by Selective Targeting of Magnetic Albumin Microspheres Containing Doxorubicin. *Proc. Natl. Acad Sci. U. S. A.* **1981**, *78*, 579–581.

Widder, K. J.; Morris, R. M.; Poore, G. A.; Howard, D. P.; Senyei, A. E. H. Selective Targeting of Magnetic Albumin Microspheres Containing Lowdose Doxorubicin: Total Remission in Yoshida Sarcoma-Bearing Rats. *Eur. J. Cancer Clin. Oncol.* **1983**, *19*, 135–139.

Willard, M. A.; Kurihara, L. K.; Carpenter, E. E.; Calvin, S.; Harris, V. G. Chemically Prepared Magnetic Nanoparticles. *Int. Mater. Rev.* **2004**, *49*, 125–170.

Xie, S.; Zhang, B.; Wang, L.; Wang, J.; Li, X.; Yang, G.; Gao, F. Superparamagnetic Iron Oxide Nanoparticles Coated with Different Polymers and Their Mri Contrast Effects in the Mouse Brains. *Appl. Surf. Sci.* **2015**, *326*, 32–38.

Xu, C. J.; Sun, S. H. Monodisperse Magnetic Nanoparticles for Biomedical Applications. *Polym. Int.* **2007**, *56*, 821–826.

Yang, J.; Lee, C. H.; Ko, H. J.; Suh, J. S.; Yoon, H. G.; Lee, K.; Huh, Y. M.; Haam, S. Multifunctional Magneto-Polymeric Nanohybrids for Targeted Detection and Synergistic Therapeutic Effects on Breast Cancer. *Angew. Chem. Int. Ed.* **2007**, *46*, 8836–8839.

Yang, J.; Lee, H.; Hyung, W.; Park, S. B.; Haam, S. Magnetic Peca Nanoparticles as Drug Carriers for Targeted Delivery: Synthesis and Release Characteristics. *J. Microencapsulation* **2006**, *23*, 203–212.

Yang, J.; Lee, T. I.; Lee, J.; Lim, E. K.; Hyung, W.; Lee, C. H.; Song, Y. J.; Suh, J. S.; Yoon, H. G.; Huh, Y. M.; Haam, S. Synthesis of Ultrasensitive Magnetic Resonance Contrast Agents for Cancer Imaging Using Peg-Fatty Acid. *Chem. Mater.* **2007**, *19*, 3870–3876.

Yigit, M. V.; Mazumdar, D.; Kim, H. K.; Lee, J. H.; Dintsov, B.; Lu, Y. Smart "Turn-On" Magnetic Resonance Contrast Agents Based on Aptamer-Functionalized Superparamagnetic Iron Oxide Nanoparticles. *Chembiochem.* **2007**, *8*, 1675–1678.

Yigit, M. V.; Mazumdar, D.; Lu, Y. Mri Detection of Thrombin with Aptamer Functionalized Superparamagnetic Iron Oxide Nanoparticles. *Bioconjugate Chem.* **2008**, *9*, 412–417.

Yoo, J. W.; Chambers, E.; Mitragotri, S. Factors That Control the Circulation Time of Nanoparticles in Blood: Challenges, Solutions and Future Prospects. *Curr. Pharm. Des.* **2010**, *16*, 2298–2307.

Zhang, Y.; Kohler, N.; Zhang, M. Surface Modification of Superparamagnetic Magnetite Nanoparticles and Their Intracellular Uptake. *Biomaterials.* **2002**, *23(7)*, 1553–1561.

Zhou, W. L.; Wang, K.-Y.; O'Connor, C. J.; Tang, J. Granular Growth of Fe3o4 Thin Films and Its Antiphase Boundaries Prepared by Pulsed Laser Deposition. *J. Appl. Phys.* **2001**, *89*(11), 7398–7400.

Zhou, Z. G.; Sun, Y. A.; Shen, J. C.; Wei, J.; Yu, C.; Kong, B.; Liu, W.; Yang, H.; Yang, S. P.; Wang, W. Iron/Iron Oxide Core/Shell Nanoparticles for Magnetic Targeting Mri and Near-Infrared Photothermal Therapy. *Biomaterials* **2014**, *35*, 7470–7478.

PART III
Lipid-Based Nanoparticulates

CHAPTER 9

SOLID LIPID NANOPARTICLES: GENERAL ASPECTS, PREPARATION METHODS, AND APPLICATIONS IN DRUG DELIVERY

MARCOS LUCIANO BRUSCHI*, HÉLEN CÁSSIA ROSSETO, and LUCAS DE ALCÂNTARA SICA DE TOLEDO

Department of Pharmacy, Laboratory of Research and Development of Drug Delivery Systems, State University of Maringá, Maringá, PR, Brazil

Corresponding author. E-mail: mlbruschi@uem.br; mlbruschi@gmail.com

ABSTRACT

The advances in the pharmaceutical therapy on the aspects of efficacy, safety, and quality of the medicines available for the patients require the constant search for novel release systems. In this sense, there are the lipid nanoparticles (LN) of first and second generation: the solid lipid nanoparticles (SLN) and the nanostructured lipid carriers (NLCs), which are alternative systems that combine advantages of several colloidal carrying agents. Besides, the LN are prepared with ingredients considered as safe for the use in biological systems as they are physiologically compatible. This chapter brings a combination of information from the beginning of the development of these systems since the 1990s until modern studies, approaching the main preparation methods and also the LN evaluation as drug delivery systems through different application routes. In this approach, the great ease of production of these systems, using simple and easily scaling-up methods, such as the high-pressure homogenization (HPH), is observed. Thus, as they are nanostructured systems, it is really important to evaluate the intrinsic toxicity of the tiny size of these particles, especially when smaller than 100 nm for parenteral administration. In general, the LN are considered low toxicity formulation with a broad range of application where they may be hydrophilic or lipophilic; thus, they are being considered as highly versatile systems and promising to the future of medical therapy.

9.1 INTRODUCTION

The progress in the pharmaceutical therapy does not mean only the development of new drugs. It also includes difficulties related to plasmatic concentration fluctuation, ineffective concentration (drugs with low absorption and/ or rapid metabolism), low stability, solubility issues. Besides, the distribution of the active ingredient to places that are not the target and the high toxicity of determined drugs lead the pharmaceutical researchers to design the drug delivery systems different from the conventional ones (Bakan, 2001; Mehnert and Mäder, 2001; Moreton and Collett, 2005).

In this sense, the focus is turned to the enhancement of dosage forms, transforming them in real modified and controlled delivery systems of the active agents (Bakan, 2001; Mehnert and Mäder, 2001).

The modified release of a drug depends on a set of techniques that allow the release of the active agent on the affected site in the needed amount and time to create the desired effect. Thus, it may reduce the toxic side effects, besides providing greater therapeutic efficacy (Carvalho et al., 2010; Pezzini et al., 2007).

The preparation of nanoparticulate systems allows converting liquids into solids, mask odor and taste, as well as change colloids' properties and surface. It is interesting to show that they may also protect drugs from the gastric environment, enhance drug solubility, and even modulate the release characteristics and site-specific delivery decreasing side effects and toxicity, and enhancing the security (Bakan, 2001; Nadian and Lindblom, 2002; Sandor et al., 2001; Silva et al., 2015).

The nanomedicine applied for the drug release and targeting has come up in the last decade as a high-technology science, which is being gradually improved (Etheridge et al., 2013).

Moreover, several drug delivery systems are developed and improved, such as liposomes, osmotic pumps, polymeric matrixes, and emulsions (Lopes et al., 2005; Manadas et a l., 2002). Thus, the remaining doubt is: "Which system is the most suitable for the desired effect?" (Mehnert and Mäder, 2001).

Initially, in the 1970s, the preparation of particulate polymeric systems was performed with the aim to transport vaccines and anticancer drugs and to enhance the drug absorption by the cancer cell. First, capsules, in nanometric scale, were obtained with hydrophilic substances (polymers) (Üner and Yener, 2007).

At the beginning of the 1990s, the advantages of polymeric particles, emulsions, and liposomes were combined with the development of solid lipid nanoparticles (SLN) (Müller et al., 2002). SLN were developed as a transporting system to be an alternative to the existing ones. In the production of this system, a solid lipid or a mixture of solid fatty substances substitutes the liquid lipid of an oil-in-water emulsion. Thus, the particle matrix is solid and lipid in room temperature as well as in body temperature (Pardeike et al., 2009).

9.2 ADVANCES ON SOLID LIPID NANOPARTICLES (SLN)

From the beginning of the 1990s, new researches have arisen aiming to study and develop the SLN (Müller et al., 2011). They present themselves as really promising once they are considered as alternative carriers for the therapeutic agents, combining the advantages of different colloidal as emulsions, liposomes, and polymeric micro/nanoparticles, besides avoiding some issues related to the stability and scale-up (Manjunath et al., 2005; Müller et al., 19962002; Olbrich et al., 2001).

Expanding the explanations of the combined advantages from other carriers, the SLN were developed from the need of improving the emulsions. This is because what hinders the drug administration with this system is their physical instability, and also, the available oils used to prepare the formulations present low solubility. Besides, the SLN may be an alternative to liposomes because one of their greatest disadvantages is the cost of the final product, and to the polymeric particles due to their cytotoxicity and tough scale-up (Müller et al., 2000; Pardeike et al., 2009).

The first research groups that described the SLN were led by Rainer H. Müller and Maria R. Gasco. From that period onward, these carriers were studied to transport several types of actives, such as peptides, proteins, and therapeutic antigens (Almeida and Souto, 2007; Gaspar et al., 2017)

In 1991, the first patents were registered and the research and development of SLN have completed more than 20 years, with many research teams studying this field (Müller et al., 2011).

The SLN are composed of solid lipids and their average diameter is between 50 and 1000 nm according to the photon correlation spectroscopy, while the main range is between 150 and 300 nm (Müller et al., 2000, 2011).

9.3   COMPOSITION OF SLN

SLN have a solid matrix, constituted by a solid lipid or a solid lipids mixture, which distinguishes them from nanoemulsions and liposomes (Müller et al., 2011; Pardeike et al., 2009).

The use of these solid lipids, rather than liquid ones, increases the release kinetic control and enhances the stability of chemically sensitive compounds (Helgason et al., 2009).

The notable excellence of this system consists of the use of lipids compatible to the human organisms, which are easily and completely degradable, thus considered "nano-safe carriers" (Puglia and Bonina, 2012; Puglia et al., 2013). The used lipids are natural glycerides that are rapidly degraded by biological processes, such as the enzymatic decomposition, synthesizing nontoxic products such as fatty acids and glycerol that are naturally present in the human body (Müller et al., 2000; Olbrich et al., 2002; Pardeike et al., 2009).

Therefore, the most employed lipids are the triglycerides (tristearin), partial glycerides (glyceryl monostearate), docosanoic acid, waxes (cetyl palmitate), and sterols (cholesterol) (Jores et al., 2004; Mehnert and Mäder, 2001).

The solid lipids added to the water and the emulsifier are the main constituents of the SLN. All of these are generally recognized as safe (GRAS), which is another advantage of the LN (Müller et al., 2002; Sánchez-López et al., 2017).

The emulsifiers are necessary in these formulations, to ensure the system stability, which are located between lipid and aqueous phase. They influence the in vivo biodegradation of the lipid matrix, as well as the average particle size; hence, the selection of the emulsifying agent must take in consideration the intended administration route. In general lines, a great variety of emulsifiers (ionic and nonionic) may be used individually or in combination. The most used types are the poloxamer, polysorbates, lecithin, pure phospholipids, and bile acids (Mäder and Mehnert, 2004; Sánchez-López et al., 2017).

Other inherent advantages to this system are the increased drug stability, incorporation of both hydrophilic and lipophilic drugs, organic solvents free, low-cost material, broad-spectrum application, efficient production, and sterilization on the large scale (Arana et al., 2015; Goto et al., 2017; Jores et al., 2004; Kakadia and Conway, 2014; Mehnert and Mäder, 2001; Müller et al., 1996).

These systems show a vast potential application spectrum (Jores et al., 2004). This is due to the possibility of controlling several parameters of the composition and production, modifying the release kinetics depending on the used method, the biocompatibility level, the stability, and the entrapment efficiency (Goto et al., 2017; Kalaycioglu and Aydogan, 2016; Wissing et al., 2004). Moreover, SLN dispersions may be stable for up to 3 years when destabilization factors, such as light and temperature (≥8 C), are avoided (Manjunath et al., 2005; Mehnert and Mäder, 2001).

Besides all the advantages, the SLN, in determined situations, may present some flaws, such as the drug degradation induced by the high pressure, particle growth, gelling tendency, and polymorphic transitions (Mehnert and Mäder, 2001; Westesen et al., 1997).

As solid lipids form the SLN, the lipid crystallization becomes an important aspect for the formulation stability. During storage, they may converge to a more stable structure (least energy consumption), resulting in a β-modification, with a high-order degree, with no remaining imperfections on the crystal mesh, causing the drug ejection (Mäder and Mehnert, 2004; Müller et al., 2002).

For example, when working with SLN consisting of pure triglycerides, the drug incorporation capacity is possibly limited, apart from the drug ejection from the crystal web (Krasodomska et al., 2016; Westesen et al., 1997). Westesen and Bunjes (1995) observed that the crystal mesh formed by solid lipids present a lower incorporation ability compared to a matrix with amorphous liquid droplets.

In this sense, to increase the SLN loading capacity and guarantee the modified release, Jenning et al. (2000a) altered the SLN, making it possible to incorporate liquid lipids on the solid core. Therefore, the nanostructured lipid carrier (NLC) was formulated, enabling the formation of a less rigid structure (Müller et al., 2002).

The NLCs are considered as a category of SLN. It is the term used for the second-generation SLN, which is constituted by the association of solid and liquid lipids (Kovács et al., 2017).

This mixture decreases the melting point of the lipid phase of the formulation in comparison with pure solid lipids, but they remain solid at body temperature (Gainza et al., 2014; Müller et al., 2002; Olbrich et al., 2002; Pardeike et al., 2009; Souto et al., 2004). The NLC shares the mentioned

SLN advantages, and it also emerged as promising nanocarriers which overcame the SLN drawbacks (Ghate et al., 2016). In this sense, these systems
complement each other depending on the potential use, becoming industri-
ally viable for the large-scale production, nontoxic and easy active release
(Aditya et al., 2014).

The LN may be obtained by many methods and they are adjustable
and modifiable depending on the chosen technique and lipid composition
(Krasodomska et al., 2016; Müller et al., 2002).

9.4   METHODS OF PREPARATION

There are several methodologies to prepare SLN and NLC, which present
their own advantages and drawbacks, and are used depending on the
proposed objective. Among them, there is the high-pressure homogenization
(HPH), high-shear homogenization (HSH) and ultrasound, microemulsion,
solvent emulsification/evaporation, and solvent injection method (Mäder
and Mehnert, 2004; Mehnert and Mäder, 2001; Müller et al., 2000).

9.4.1   HIGH-PRESSURE HOMOGENIZATION

This method is already used for the preparation of emulsions on the large
scale in the pharmaceutical industry, highlighting it as one of the main
methods which is safe to prepare LN (Müller et al., 2000; Silva et al., 2011).
The technique is based on the possibility of homogenators to the forced pass
of a liquid under high pressure (10–200 Pa) through a micrometric orifice
(Mäder and Mehnert, 2004).

To perform this methodology, there are two approaches (hot or cold
homogenization) before the fluid is accelerated (over 100 km/h) and cavitation forces disturb the particles until they reach the narrow space (Mehnert
and Mäder, 2001; Müller et al., 1993, 1996). In both cases, the initial step
is the active incorporation in the fused lipid matrix by dissolution or dispersion, depending on the drug solubility (Müller et al., 2000).

In hot homogenization, first, the lipid or lipid mixture is heated at 5–10°C
above its melting point, and then it is added to surfactant dispersion in water.
This premixture, with the help of a stirrer, forms a pre-emulsion, which goes
to the HPH, at a temperature higher than the lipid melting point, generating
an oil-in-water nanoemulsion that is cooled to room temperature, resulting
in the SLN (Silva et al., 2011).

When using the cold homogenization, the lipid dispersion is frozen in liquid nitrogen or dry ice; in other words, it is a homogenization performed with solid lipid. This mixture is then ground in power mill to particle size reduction (50–100 µm), added to an aqueous dispersion with surfactant, and then passed through HPH at room temperature or below it (Almeida et al., 1997; Mehnert and Mäder, 2001; Müller et al., 1995).

The HPH is considered as a promising method as it may be used as hot or cold (Severino et al., 2012), allowing the manipulation of thermolabile drugs. It is well known by industries due to an easy scaling-up and no need of using the organic solvents; also, it is possible to obtain particles of small diameters and low polydispersity index (Kovačević et al., 2014; Mehnert and Mäder, 2001).

The cold homogenization overcomes some drawbacks of the hot homogenization at critical points, for example, when drug suffers degradation at high temperatures (sensitive compounds), it escapes to the aqueous phase during the homogenization (hydrophilic actives) and also the cooling process of the hot homogenization may cause stability problems. However, the particles are greater and with higher polydispersity index (Mäder and Mehnert, 2004; Müller et al., 2000; Westesen and Bunjes, 1995).

9.4.2 HIGH-SHEAR HOMOGENIZATION AND ULTRASOUND

In view of the production cost of novel systems, such as the LN, studies have developed easy-to-handle methods and more economic preparation, such as the HSH and ultrasound (Puglia et al., 2013).

Although the HPH is more widely used, the HSH is simple and accessible for the LN production (López-García and Ganem-Rondero, 2015; Pardeike et al., 2009). In their study, Righeschi et al. (2016) used the HSH–ultrasound method, based on the homogenization with Ultra-Turrax®, followed by the sonication over a short time period.

The application of the ultrasound step based on the generation of cavitational bubbles, led to the decrease in particle size (Avvaru et al., 2006; Puglia et al., 2013). However, this technique presents some disadvantages. The HSH methods may lead to the formation of microparticles and polydisperse systems (Mäder and Mehnert, 2004). Besides, there is a potential risk of contamination by metals due to the insertion of the ultrasound probe directly into the lipid dispersion (Wissing et al., 2004).

Studies have shown that the sequential use of HSH and HPH enable better control over the LN distribution and size, thus modulating the particle production according to the administration route (Severino et al., 2012).

9.4.3 MICROEMULSION

The microemulsion method to obtain LN was developed by Gasco (1993), and is based on the heating of the lipid and aqueous phases, containing one or more surfactants and cosurfactants, at the same temperature. The water dispersion is poured over the oily one with a smooth agitation, obtaining a microemulsion, which is added into water (2–10°C), under stirring conditions for the cooling and formation of the dispersed lipid microparticles.

This is also a methodology that allows the scale-up. Industries use tanks, rigidly controlling the temperature to obtain the microemulsion and the particle precipitation step; besides, the pH control is an essential parameter for the final quality of the formulation (Mäder and Mehnert, 2004; Müller et al., 2000).

The precipitation stage is considered a dilution of the systems because the formed microemulsion is added into water. Still, it is possible to prepare a more concentrated LN dispersion using equal amounts of cold water and microemulsion, as stated by Kakkar and Kaur (2011), which overcame the necessity of concentrating the final SLN dispersion. Many researchers chose this method, as Arana et al. (2015), because it is a procedure which is organic solvent free and enables rapid lipid crystallization and blow aggregation.

9.4.4 SOLVENT EMULSIFICATION/EVAPORATION

In this technique, the lipid is dissolved in a nonpolar organic solvent and the mixture is emulsified in an aqueous phase under reduced pressure. After that, the solvent is evaporated and the SLN are precipitated. As seen, this method does not involve heating steps; thus, it is suitable for thermolabile drugs (Wissing et al., 2004). It is an alternative to the preparation of LN. However, it contradicts one of the most important advantages of the SLN, which is the nonusage of organic solvent.

9.4.5 SOLVENT INJECTION

To prepare the LN, a solution of solid lipid is rapidly injected into solvents miscible in water or a mixture of polar solvents, such as acetone, ethanol, isopropanol, and methanol (Schubert and Müller-Goymann, 2005). This, in fact, is a method developed initially to prepare polymeric nanoparticles (de Labouret et al., 1995) and its main flaw is the use of organic solvent like the solvent/emulsification method, limiting the application potential, besides the low LN concentration. The advantage is restricted to the nonapplication of thermal stress (Mäder and Mehnert, 2004).

9.4.6 ACTIVES DISTRIBUTION IN THE LIQUID NANOPARTICLE DURING THE PREPARATION PROCESS

It is really important to highlight that the knowledge of the drug incorporation phenomenon on the lipid system allows to suggest how the active will be controlled during the delivery and, consequently, alter the parameters that induce the drug inclusion way on the matrix, enabling the modification of the drug release according to the needed activity (Mehnert et al., 1997; Wissing et al., 2004; zur Mühlen and Mehnert, 1998)

The lipids that constitute the SLN are concentrated and conceive a solid matrix, where the drug is entrapped. This matrix may vary in the preparation method. Therefore, the active may be incorporated in two different models: solid solution model and core–shell model (drug-enriched shell and drug-enriched core) (Üner and Yener, 2007).

When the particles are prepared by the cold homogenization, without using surfactant or solubilizing agent, the model called solid homogenous matrix is obtained. In contrast, when the hot homogenization is followed by a cooling step, a solid lipid core is formed when it reaches the recrystallization temperature (core–shell model). By reducing the dispersion temperature, the drug is concentrated on the peripheral layers (drug-enriched shell). However, when the precipitation of the active occurs before the lipid crystallization, the drug is found in the particle core, with a lipid involucre layer (drug-enriched core) (Müller et al., 2000).

Knowing this variation in the formation of the lipid matrix during the SLN preparation, it may be possible to predict how the drug would be released. It may be observed that the drug is totally encrusted in a solid matrix (Fig. 9.1a) and due to this, it is expected that the drug release delivery will be slow—by diffusion—once the drug shows a limited wettability (Müller and Heinemann, 1994).

In contrast, the inverse capsule, drug-enriched shell (Fig. 9.1b) may result in a burst release, which is interesting for therapeutics that need an attack dose, while the drug located in the core may result in a sustained release due to the lipid–drug interactions (Fig. 9.1c) (Küchler et al., 2010).

In this sense, it is possible to modify the drug release choosing different production methods to have the drug incorporated in one of the three models, or even making an association of them, allowing that the drug presents a rapid initial release and later a sustained release. Therefore, the LN may be carrier systems which can be applied to many administration routes (Wissing et al., 2004).

FIGURE 9.1   Schematic illustrations of the three types of drug (white dots) localization in solid lipid nanoparticles: (a) drug is totally encrusted in a solid matrix; (b) drug-enriched shell; (c) drug located in the core.

When studying carriers, depending on the type of prepared system (imperfect type, amorphous type, and multiple type), the drug accommodates in different ways into the lipid matrix (Üner and Yener, 2007).

Accordingly, the modified drug release, through different administration routes, using LN was already demonstrated in several studies (Esposito et al., 2012; Gohla et al., 2000; Strasdat and Bunjes, 2013; Wissing et al., 2004; zur Mühlen et al., 1998).

9.5   SLN AS DRUG DELIVERY SYSTEM

It is really important to analyze the administration route of the SLN as it is decisive to the behavior of the particles and drugs in vivo. Other than that, the interactions with the biological processes that occur inside the human body, such as the distribution phase and enzymatic reactions are also relevant when considering the SLN formulations effects and the kinetics profile in vivo (Mehnert and Mäder, 2001). Therefore, in Table 9.1, some different formulations are exposed alongside to the suggested and/or tested route of administration.

The SLN are really attractive for topical use due to their lipid characteristics, which shows high interest for skin application because they may form an adhesive film on the skin with occlusive characteristics (Müller, 1998). Many drugs have been incorporated in different SLN formulations for topical utilization. Other than that, the SLN dispersions may be added into dosage forms that are really advantageous for topical application, such as hydrogels, creams, and lotions (Jenning et al., 2000b).

TABLE 9.1 Different Routes of Administration for Lipid Nanoparticles (Solid Lipid Nanoparticles [SLN] and Nanostructured Lipid Carrier [NLC]).

Route of administration	Type of formulation	Reference(s)
Topical	Imidazole SLN and NLC	Souto and Müller (2006)
	Triptolide SLN	Küchler et al. (2010)
	Isotretinoin SLN	Liu et al. (2007)
	Flurbiprofen SLN	Jain et al. (2005)
Peroral	Insulin SLN	Sarmento et al. (2007)
	Cyclosporine SLN	Müller et al. (2006)
	Camptothecin SLN	Yang et al. (1999)
	Zidovudine SLN	Purvin et al. (2014)
Rectal	Metoclopramide SLN	Mohamed et al. (2013)
	Flurbiprofen SLN	Din et al. (2015)
	Diazepam SLN	Abdelbary and Fahmy (2009)
Nasal	Budesonide SLN	Chavan et al. (2013)
	Alprazolam SLN	Singh et al. (2012)
Ophthalmic	Tobramycin SLN	Cavalli et al. (2002)
	Estradiol, hydrocortisone, or pilocarpine SLN	Friedrich et al. (2005)
	Calendula officinalis extract SLN	Arana et al. (2015)
Parenteral	Doxorubicin SLN	Fundarò et al. (2000)

For example, Küchler et al. (2010) prepared SLN with morphine to be applied on the skin, for pain reduction and healing improvement. The formulations consist of glyceryl behenate, poloxamer 188, and the drug was obtained by HPH. They showed wound closure acceleration, prolonged morphine release, and low cytotoxicity and irritation, presenting an attractive dosage form for the wound management.

Sarmento et al. (2007) prepared SLN with cetyl palmitate and poloxamer 407 containing insulin so as to reduce the plasma glucose levels. The obtained formulation was more efficient than the common oral insulin solution for the reduction of these levels when tested on diabetic male Wistar rats. Besides, the preparation could protect the drug from degradation in the gastrointestinal tract, aiding the intestinal absorption.

A camptothecin SLN was administered by gavage in C57BL/6J mice and was evaluated according to the in vivo distribution. The researchers showed that in comparison with a camptothecin solution, the lipid formulation presented a higher concentration in the plasma and also on the evaluated organs (liver, heart, brain, and lung) (Yang et al., 1999).

Another application of the SLN is through the rectal route as a suppository, mainly for water-insoluble drugs such as diazepam. In this context, the main route for the epileptic cases, Abdelbary and Fahmy (2009) successfully incorporated diazepam into SLN made of poloxamer 188, glyceryl behenate, and glyceryl monostearate.

In contrast, an interesting route for the administration of drugs with quick absorption rates and fast action, and also that avoids the first-pass effect is the nasal one (Üner and Yener, 2007).

Moreover, the nasal route may be used as the main route to transfer the drug directly to the brain, helping in the treatment of central nervous system infections. In this sense, an SLN formulation prepared with alprazolam was evaluated when used through the intranasal membrane. This preparation enhanced the rate and the transportation degree of alprazolam SLN to the brain when compared to alprazolam solution. Therefore, it may be possible to reduce the number of needed doses as well as the concentration, thus increasing the therapeutic index and patient compliance (Singh et al., 2012).

Colloidal systems, such as the SLN dispersions, are known to enhance the ocular bioavailability (Barbault-Foucher et al., 2002; De Campos et al., 2001; Ludwig, 2005; Seyfoddin et al., 2010). Therefore, Cavalli et al. (2002) prepared SLN composed of tobramycin, an aminoglycosides that are usually not absorbed by this route, for ophthalmic application. They showed that this is a promising vehicle for a drug employed for the treatment of "resistant" *Pseudomonas keratitis*, or even for prophylaxis against bacterial endophthalmitis, before cataract surgery.

This route is very interesting because it is cheap to prepare drug delivery systems where even small industries, which may produce formulations with low cost and high technological attractiveness (Seyfoddin et al., 2010). This route was even used to reduce the inflammation and pain found in the ocular tissue after cataract surgery, employing a celecoxib-loaded SLN (Sharma et al., 2016).

The LN come up as the very promising ophthalmic drug delivery system because of their, usually spherical shape, while particles with edges or pointy angled are not well tolerated (Seyfoddin et al., 2010). Thus, these fatty nanoparticles present small chance of irritation by this sensitive surface.

The parenteral route is interesting for drugs that should not go through the first-pass effect and present short half-lives (some minutes), such as peptides and proteins (Üner and Yener, 2007). An extensive review of the parenteral application of SLN has been carried out by Wissing et al. (2004).

As it was already said, the SLN lipids are GRAS and physiologically or physiologically similar; thus; they are well tolerated, which is a characteristic extremely important for drugs and formulations that are administered intravenously, intra-arterially, intramuscularly, and by the other parenteral routes (Mehnert and Mäder, 2001; Müller et al., 2002; Sánchez-López et al., 2017; Üner and Yener, 2007).

All these studies show that the SLN are a versatile drug delivery system, which may be prepared with a diverse number of drugs, as well as be used on different administration routes.

9.6 TOXICITY

The nanoparticles have contributed to a huge revolution in the industrial sector, in areas such as electronics, biotechnology, engineering, and pharmacy/medicine. In the healthcare field, they have been used as new dosage forms for delivering drugs, proteins, DNA, and so forth. These kinds of formulations have been obtained from metal and nonmetal, polymeric materials, lipids, among other materials (Bahadar et al., 2016; De Jong and Borm, 2008; Dreher, 2003; Lewinski et al., 2008).

These dosage forms are attractive for medical purposes due to some unique features such as the huge surface area of contact (even on low weights), the carrying ability; also, due to their small size, they find low or no difficulty in entering the organisms and crossing biological barriers, reaching organs, previously, almost inaccessible. Furthermore, they display some characteristics that differ from the bulk material as the thermal, mechanical, and electrical properties. In the beginning, these nanoproducts were seen, without any doubts, as positive in every application (Bahadar et al., 2016; De Jong and Borm, 2008; Dreher, 2003).

However, these tiny particles, present some drawbacks—they are toxic to living beings, even inert materials may show this flaw. Regarding this aspect, the research over the possible toxic effects of the nanoparticles due to their scale is relatively new (Colvin, 2003; Dagani, 2003; Lam et al., 2004; Vishwakarma et al., 2010; Warheit et al., 2004). Some researchers even consider that nanoparticles with size less than 10 nm behave more gas-like, which could pass through skin and some sensitive tissues such as the lung cells' membranes.

As the SLN are usually prepared with physiologically or physiologically similar lipids, they tend to be one of the least toxic nanoparticles (Müller et al., 2011). Moreover, Weyhers et al. (2006) evaluated the toxicity of SLN prepared with a wax, cetyl palmitate, which is not physiological. Although this wax presents some side effects on men, it is observed that these nanoparticles are suitable for intravenous administration, being well-tolerated and nottoxic, even in parenteral application.

In contrast to most nanoparticles, the SLN do not show some properties, for example, they do not suffer endocytosis, thus not being taken up by the body cells. Moreover, another point is the "biopersistency" (Müller et al., 2011), which is the capacity of a material to remain in organisms (or in the environment). In the human body, the nanoproducts may or may not suffer metabolism, and those that stay forever and are not metabolized, come up as high-risk substances/compounds. In this sense, the human body can easily handle the SLN (again due to their GRAS composition) because the organism possesses mechanisms to lead with their components.

The first group to analyze the in vitro toxicity was Müller et al. (1996). They compared the viability of human granulocytes when tested against SLN containing magnetite and a well-established polymeric (polylactic acid—PLA) nanoparticle preparation. The comparison parameter was the concentration that reduced the viability to 50% (ED50%), the low and high-molecular-weight PLA nanoparticles presented ED50% of 0.30 and 0.38%, respectively. The SLN presented a value higher than 10%, showing that the prepared SLN are less toxic than the PLA nanoparticles.

This shows that the SLN are well tolerated, being even better than particles made of polymers greatly accepted by the scientific community as the PLA. The LN are shown to be more than 10 times less toxic than the other particle types.

9.7 CONCLUSION

After more than 20 years of research since its appearance (1990s), the LN (SLN and NLC) may be employed as alternative solution to carry and release different drugs by diverse routes of administration, such as topical, peroral, parenteral, nasal, rectal, and even ophthalmic.

It is possible to work with thermolabile substances as there are methods which are likely to prepare them at room temperature. Besides, the methodologies for their preparation are easy to handle and to scale up, especially the HPH technique. In addition, they allow having an idea of how the drug release may happen due to the way the lipid recrystallization occurs.

Working with the second generation of LN, the NLC, present the main advantage, the lower risk of drug expulsion, as they do not suffer polymorphic alterations through time and may then be considered when it is needed to solve this aspect in an SLN formulation, considering that the lipid constituents are determined to modify that leads to the formation of the "perfect lipid crystal" and consequent active ejection.

Independent of the used LN generation, the preparation of a nanoparticulate lipid product is a very interesting perspective that depletes the acceptance of these systems by the pharmaceutical and cosmetic markets as the use of lipids and surfactants is considered safe (GRAS), which are present in oral or dermal products available. This corroborates with the cost reduction for the development of novel formulations as it does not require toxicity studies of the used excipients, except for the products for the parenteral application.

Thus, when compared to other nanosystems, the LN present lower toxicity as they are prepared with physiologically compatible constituents. Before performing the toxicity test of the determined formulation, it is assumed that LN smaller than 100 nm possess greater chances of causing any toxicity reaction as they may be absorbed by any cells in the body. However, observing the already accomplished studies, most of the therapeutic or cosmetic applications use bigger and biodegradable particles, which present very small risk of toxicity.

Therefore, good perspectives of the nanoparticulate lipid products reach the pharmaceutical market, which may be verified by the several researches of these systems for many different applications. Besides, according to Müller et al. (2011), this group is responsible for great part of the development of this type of system. Many pharmaceutical industries are already developing SLN or NLC products in preclinic phase, which strongly stands out the wide horizon of these formulations as drug delivery systems.

KEYWORDS

- **SLN**
- **NLC**
- **drug delivery**
- **nanotoxicity**
- **GRAS composition**
- **high-pressure homogenization**

REFERENCES

Abdelbary, G.; Fahmy, R. H. Diazepam-Loaded Solid Lipid Nanoparticles: Design and Characterization. *AAPS PharmSciTech* **2009**, *10*(1), 211–219.

Aditya, N. P.; Macedo, A. S.; Doktorovova, S.; Souto, E. B.; Kim, S.; Chang, P.-S.; Ko, S. Development and Evaluation of Lipid Nanocarriers for Quercetin Delivery: a Comparative Study of Solid Lipid Nanoparticles (SLN), Nanostructured Lipid Carriers (NLC), and Lipid Nanoemulsions (LNE). *LWT – Food Sci. Technol.* **2014**, *59*(1), 115–121.

Almeida, A. J.; Souto, E. Solid Lipid Nanoparticles as a Drug Delivery System for Peptides and Proteins. *Adv. Drug Delivery Rev.* **2007**, *59*(6), 478–490.

Almeida, A. J.; Runge, S.; Müller, R. H. Peptide-Loaded Solid Lipid Nanoparticles (SLN): Influence of Production Parameters. *Int. J. Pharm.* **1997**, *149*(2), 255–265.

Arana, L.; Salado, C.; Vega, S.; Aizpurua-Olaizola, O.; de la Arada, I.; Suarez, T.; Usobiaga, A.; Arrondo, J. L. R.; Alonso, A.; Goñi, F. M.; Alkorta, I. Solid Lipid Nanoparticles for Delivery of Calendula Officinalis Extract. *Colloids Surf. B* **2015**, *135*, 18–26.

Avvaru, B.; Patil, M. N.; Gogate, P. R.; Pandit, A. B. Ultrasonic Atomization: Effect of Liquid Phase Properties. *Ultrasonics* **2006**, *44*(2), 146–158.

Bahadar, H.; Maqbool, F.; Niaz, K.; Abdollahi, M. Toxicity of Nanoparticles and an Overview of Current Experimental Models. *Iran. Biomed. J.* **2016**, *20*(1), 1–11.

Bakan, J. A. Microencapsulação. In *Teoria e Prática na Indústria Farmacêutica;* Lachman, L., Lieberman, H., Kanig, J. L., Eds.; Fundação Calouste Gulbenkian: Lisbon, 2001; pp 707–735.

Barbault-Foucher, S.; Gref, R.; Russo, P.; Guechot, J.; Bochot, A. Design of Poly-ε-Caprolactone Nanospheres Coated with Bioadhesive Hyaluronic Acid for Ocular Delivery. *J. Controlled Release* **2002**, *83*(3), 365–375.Carvalho, F. C.; Bruschi, M. L.; Evangelista, R. C.; Palmira, M.; Gremião, D. Mucoadhesive Drug Delivery Systems. *Braz. J. Pharm. Sci.* **2010**, *46*(1), 1–18.

Cavalli, R.; Gasco, M. R.; Chetoni, P.; Burgalassi, S.; Saettone, M. F. Solid Lipid Nanoparticles (SLN) as Ocular Delivery System for Tobramycin. *Int. J. Pharm.* **2002**, *238*, 241–245.

Chavan, S. S.; Ingle, S. G.; Vavia, P. R. Preparation and characterization of solid lipid nanoparticle-based nasal spray of budesonide. *Drug Deliv. Transl. Res.* 2013, *3*(5), 402–408.

Colvin, V. L. The Potential Environmental Impact of Engineered Nanomaterials. *Nat. Biotechnol.* **2003**, *21*(10), 1166–1171.

Dagani, R. O. N. Nanomaterials: Safe or Unsafe. *Sci. Technol.* **2003**, *28*, 30–33.

De Campos, A. M.; Sánchez, A.; Alonso, M. J. Chitosan Nanoparticles: a New Vehicle for the Improvement of the Delivery of Drugs to the Ocular Surface. Application to Cyclosporin A. *Int. J. Pharm.* **2001**, *224*(1), 159–168.

De Jong, W. H.; Borm, P. J. Drug Delivery and Nanoparticles: Applications and Hazards. *Int. J. Nanomed.* **2008**, *3*(2), 133–149.

De Labouret, A.; Thioune, O.; Fessi, H.; Devissaguet, J. P.; Puisieux, F. Application of an Original Process for Obtaining Colloidal Dispersions of Some Coating Polymers. Preparation, Characterization, Industrial Scale-Up. *Drug Dev. Ind. Pharm.* **1995**, *21*(2), 229–241.

Din, F. U.; Mustapha, O.; Kim, D. W.; Rashid, R.; Park, J. H.; Choi, J. Y.; Ku, S. K.; Yong, C. S.; Kim, J. U.; Choi, H. G. Novel Dual-Reverse Thermosensitive Solid Lipid Nanoparticle-Loaded Hydrogel for Rectal Administration Of Flurbiprofen with Improved Bioavailability and Reduced Initial Burst Effect. *Eur. J. Pharm. Biopharm.* **2015**, *94,* 64–72.Dreher, K. L. Health and Environmental Impact of Nanotechnology: Toxicological Assessment of Manufactured Nanoparticles. *Toxicol. Sci.* **2003**, *77*(1), 3–5.

Esposito, E.; Mariani, P.; Ravani, L.; Contado, C.; Volta, M.; Bido, S.; Drechsler, M.; Mazzoni, S.; Menegatti, E.; Morari, M.; Cortesi, R. Nanoparticulate Lipid Dispersions for Bromocriptine Delivery : Characterization and in Vivo Study. *Eur. J. Pharm. Biopharm.* **2012,** *80*(2), 306–314.

Etheridge, M. L.; Campbell, S. A.; Erdman, A. G.; Haynes, C. L.; Wolf, S. M.; McCullough, J. The Big Picture on Nanomedicine: the State of Investigational and Approved Nanomedicine Products. *Nanomed. Nanotechnol. Biol. Med.* **2013,** *9*(1), 1–14.Friedrich, I.; Reichl, S.; Müller-Goymann, C. C. Drug Release and Permeation Studies of Nanosuspensions Based on Solidified Reverse Micellar Solutions (SRMS). *Int. J. Pharm.* **2005,** *305*(1–2), 167–175.

Fundarò, A.; Cavalli, R.; Bargoni, A.; Vighetto, T.; Zara, G. P.; Gasco, M. R. Non-stealth and stealth solid lipid nanoparticles (SLN) carrying doxorubicin: pharmacoknetics and tissue distribution after i.v. administration to rats. *Pharmacol. Res.* 2000, 42(4), 337–343.

Gainza, G.; Pastor, M.; Aguirre, J. J.; Villullas, S.; Pedraz, J. L.; Maria, R.; Igartua, M. A Novel Strategy for the Treatment of Chronic Wounds Based on the Topical Administration of rhEGF-Loaded Lipid Nanoparticles : in Vitro Bioactivity and in Vivo Effectiveness in Healing-Impaired Db/Db Mice. *J. Controlled Release* **2014,** *185,* 51–61.

Gasco, M. R. Method for Producing Solid Lipid Microspheres Having a Narrow Size Distribution. U.S. Patent 5,250,236, 1993.

Gaspar, D. P.; Serra, C.; Lino, P. R.; Gonçalves, L.; Taboada, P.; Remuñán-López, C.; Almeida, A. J. Microencapsulated SLN: an Innovative Strategy for Pulmonary Protein Delivery. *Int. J. Pharm.* **2017,** *516*(1–2), 231–246.

Ghate, V. M.; Lewis, S. A.; Prabhu, P.; Dubey, A.; Patel, N. Nanostructured Lipid Carriers for the Topical Delivery of Tretinoin. *Eur. J. Pharm. Biopharm.* **2016,** *108,* 253–261.

Gohla, S. H.; Jenning, V.; Gysler, A.; Schäfer-Korting, M. Vitamin A Loaded Solid Lipid Nanoparticles for Topical Use : Occlusive Properties and Drug Targeting to the Upper Skin. *Eur. J. Pharm. Biopharm.* **2000,** *49*(3), 211–218.

Goto, P. L.; Siqueira-Moura, M. P.; Tedesco, A. C. Application of Aluminum Chloride Phthalocyanine-Loaded Solid Lipid Nanoparticles for Photodynamic Inactivation of Melanoma Cells. *Int. J. Pharm.* **2017,** *518*(1–2), 228–241.

Helgason, T.; Awad, T. S.; Kristbergsson, K.; McClements, D. J.; Weiss, J. Effect of Surfactant Surface Coverage on Formation of Solid Lipid Nanoparticles (SLN). *J. Colloid Interface Sci.* **2009,** *334*(1), 75–81.

Jain, S. K.; Chourasia, M. K.; Masuriha, R.; Soni, V.; Jain, A.; Jain, N. K.; Gupta, Y. Solid Lipid Nanoparticles Bearing Flurbiprofen For Transdermal Delivery. *Drug Deliv.* **2005,** *12*(4), 207–215.

Jenning, V.; Thünemann, A. F.; Gohla, S. H. Characterisation of a Novel Solid Lipid Nanoparticle Carrier System Based on Binary Mixtures of Liquid and Solid Lipids. *Int. J. Pharm.* **2000a,** *199*(2), 167–177.

Jenning, V.; Schäfer-Korting, M.; Gohla, S. Vitamin A-Loaded Solid Lipid Nanoparticles for Topical Use: Drug Release Properties. *J. Controlled Release* **2000b,** *66*(2–3), 115–126.

Jores, K.; Mehnert, W.; Drechsler, M.; Bunjes, H.; Johann, C.; Ma, K.; Mäder, K. Investigations on the Structure of Solid Lipid Nanoparticles (SLN) and Oil-Loaded Solid Lipid Nanoparticles by Photon Correlation Spectroscopy, Field-Flow Fractionation and Transmission Electron Microscopy. *J. Controlled Release* 2004, 95(2), 217–227.

Kakadia, P. G.; Conway, B. R. Solid Lipid Nanoparticles: a Potential Approach for Dermal Drug Delivery. *Am. J. Pharmacol. Sci.* **2014,** *2*(5A), 1–7.

Kakkar, V.; Kaur, I. P. Evaluating Potential of Curcumin Loaded Solid Lipid Nanoparticles in Aluminium Induced Behavioural , Biochemical and Histopathological Alterations in Mice Brain. *Food Chem. Toxicol.* **2011,** *49*(11), 2906–2913.

Kalaycioglu, G. D.; Aydogan, N. Preparation and Investigation of Solid Lipid Nanoparticles for Drug Delivery. *Colloids Surf. A* **2016,** *510,* 77–86.

Kovács, A.; Berkó, S.; Csányi, E.; Csóka, I. Development of Nanostructured Lipid Carriers Containing Salicyclic Acid for Dermal Use Based on the Quality by Design Method. *Eur. J. Pharm. Sci.* **2017,** *99,* 246–257.

Kovačević, A. B.; Müller, R. H.; Savić, S. D.; Vuleta, G. M.; Keck, C. M. Solid Lipid Nanoparticles (SLN) Stabilized with Polyhydroxy Surfactants: Preparation, Characterization and Physical Stability Investigation. *Colloids Surf. A* **2014,** *444,* 15–25.Krasodomska, O.; Paolicelli, P.; Cesa, S.; Casadei, M. A.; Jungnickel, C. Protection and Viability of Fruit Seeds Oils by Nanostructured Lipid Carrier (NLC) Nanosuspensions. *J. Colloid Interface Sci.* **2016,** *479,* 25–33.

Küchler, S.; Wolf, N. B.; Heilmann, S.; Weindl, G.; Helfmann, J.; Yahya, M. M.; Stein, C.; Schäfer-korting, M. 3D-Wound Healing Model : Influence of Morphine and Solid Lipid Nanoparticles. *J. Biotechnol.* **2010,** *148,* 24–30.

Lam, C.; James, J. T.; Mccluskey, R.; Hunter, R. L. Pulmonary Toxicity of Single-Wall Carbon Nanotubes in Mice 7 and 90 Days After Intratracheal Instillation. *Toxicol. Sci.* **2004,** *77,* 126–134.

Lewinski, N.; Colvin, V.; Drezek, R. Cytotoxicity of Nanopartides. *Small* **2008,** *4*(1), 26–49.

Liu, J.; Hu, W.; Chen, H.; Ni, Q.; Xu, H.; Yang, X. Isotretinoin-Loaded Solid Lipid Nanoparticles with Skin Targeting for Topical Delivery. *Int. J. Pharm.* **2007,** *328*(2), 191–195.

López-García, R.; Ganem-Rondero, A. Solid Lipid Nanoparticles (SLN) and Nanostructured Lipid Carriers (NLC): Occlusive Effect and Penetration Enhancement Ability. *J. Cosmet. Dermatol. Sci. Appl.* **2015,** *5,* 62–72.

Lopes, C. M.; Lobo, J. M. S.; Costa, P. Formas Farmacêuticas de Liberação Modificada: Polímeros Hidrifílicos. *Rev. Bras. Ciênc. Farm.* **2005,** *41*(2), 143–154.

Ludwig, A. The Use of Mucoadhesive Polymers in Ocular Drug Delivery. *Adv. Drug Delivery Rev.* **2005,** *57*(11), 1595–1639.

Mäder, K.; Mehnert, W. Solid Lipid Nanoparticles – Concepts, Procedures, and Physicochemical Aspects. In *Lipospheres in Drug Targets and Delivery;* Nastruzzi, C., Ed.; CRC Press: Boca Raton, 2004; pp 1–22.

Manadas, R.; Pina, M. E.; Veiga, F. A Dissolução in Vitro Na Previsão Da Absorção Oral de Fármacos Em Formas Farmacêuticas de Liberação Modificada. *Rev. Bras. Ciênc. Farm.* **2002,** *38*(4), 375–399.

Manjunath, K.; Reddy, J. S.; Venkateswarlu, V. Solid Lipid Nanoparticles as Drug Delivery Systems. *Methods Find. Exp. Clin. Pharmacol.* **2005,** *27*(2), 1–20.

Mehnert, W.; Mäder, K. Solid Lipid Nanoparticles: Production, Characterization and Applications. *Adv. Drug Delivery Rev.* **2001,** *47,* 165–196.

Mehnert, W.; zur Mühlen, A.; Dingler, A.; Weyhers, H.; Müller, R. H. Solid Lipid Nanoparticles (SLN) – Ein Neuartiger Wirkstoff-Carrier Für Kosmetika Und Pharmazeutika. II. Wirkstoff-Inkorporation, Freisetzung Und Sterilizierbarkeit. *Pharm. Ind.* **1997,** *59*(6), 511–514.

Mohamed, R. A.; Abass, H. A.; Attia, M. A.; Heikal, O. A. Formulation and Evaluation of Metoclopramide Solid Lipid Nanoparticles for Rectal Suppository. *J. Pharm. Pharmacol.* **2013,** *65*(11), 10607–16021.

Moreton, C.; Collett, J. Formas Perorais de Liberação Modificada. In *Delineamento de Formas Farmacêuticas;* Aulton, M. E., Ed.; ArtMed: Porto Alegre, 2005; pp 298–313.

Müller, B. W. Topische Mikroemulsionen Als Neue Wirkstoff-Traègersysteme. In *Pharmazeutische Technologie: Moderne Arzneiformen;* Müller, R. H., Hildebrand, G. E., Eds.; Wissenschaftliche Verlagsgesellschaft: Stuttgart, 1998; pp 161–168.

Müller, R. H.; Heinemann, S. Fat Emulsions for Parenteral Nutrition. III: Lipofundin MCT/LCT Regimens for Total Parenteral Nutrition (TPN) with Low Electrolyte Load. *Int. J. Pharm.* **1994,** *101*(3), 175–189.

Müller, R. H.; Mäder, K.; Gohla, S. Solid Lipid Nanoparticles (SLN) for Controlled Drug Delivery – a Review of the State of the Art. *Eur. J. Pharm. Biopharm.* **2000,** *50*(1), 161–177.

Müller, R. H.; Radtke, M.; Wissing, S. A. Solid Lipid Nanoparticles (SLN) and Nanostructured Lipid Carriers (NLC) in Cosmetic and Dermatological Preparations. *Adv. Drug Delivery Rev.* **2002,** *54,* S131–S155.

Müller, R.; Runge, S.; Ravelli, V.; Mehnert, W.; Thünemann, A.; Souto, E. Oral Bioavailability of Cyclosporine: Solid Lipid Nanoparticles (SLN®) Versus Drug Nanocrystals. Int J Pharm. **2006,** 317, 82–89.

Müller, R. H.; Shegokar, R.; Keck, C. M. 20 Years of Lipid Nanoparticles (SLN and NLC): Present State of Development and Industrial Applications. *Curr. Drug Discovery Technol.* **2011,** *8,* 207–227.

Olbrich, C.; Bakowsky, U.; Lehr, C.-M.; Müller, R. H.; Kneuer, C. Cationic Solid-Lipid Nanoparticles can Efficiently Bind and Transfect Plasmid DNA. *J. Controlled Release* **2001,** *77*(3), 345–355.

Olbrich, C.; Gessner, A.; Kayser, O.; Müller, R. H. Lipid-Drug-Conjugate (LDC) Nanoparticles as Novel Carrier System for the Hydrophilic Antitrypanosomal Drug Diminazenediaceturate. *J. Drug Targeting* **2002,** *10*(5), 387–396.

Pardeike, J.; Hommoss, A.; Müller, R. H. Lipid Nanoparticles (SLN, NLC) in Cosmetic and Pharmaceutical Dermal Products. *Int. J. Pharm.* **2009,** *366,* 170–184.

Pezzini, B. R.; Silva, M. A. S. S.; Ferraz, H. G. Formas Farmacêuticas Sólidas Orais de Liberação Prolongada: Sistemas Monolíticos E Multiparticulados. *Rev. Bras. Ciênc. Farm.* **2007,** *43,* 491–502.

Puglia, C.; Bonina, F. Lipid Nanoparticles as Novel Delivery Systems for Cosmetics and Dermal Pharmaceuticals. *Expert Opin. Drug Delivery* **2012,** *9*(4), 429–441.

Puglia, C.; Offerta, A.; Rizza, L.; Zingale, G.; Bonina, F.; Ronsisvalle, S. Optimization of Curcumin Loaded Lipid Nanoparticles Formulated Using High Shear Homogenization (HSH) and Ultrasonication (US) Methods. *J. Nanosci. Nanotechnol.* **2013,** *13*(10), 6888–6893.

Purvin, S.; Vunddanda, P. R.; Singh, S. K.; Jain, A.; Singh, S. Pharmacokinetic and Tissue Distribution Study of Solid Lipid Nanoparticles of Zidovudine in Rats. *J. Nanotech.* **2014,** *2014,* 1–7.

Righeschi, C.; Bergonzi, M. C.; Isacchi, B.; Bazzicalupi, C.; Gratteri, P.; Bilia, A. R. Enhanced Curcumin Permeability by SLN Formulation: the PAMPA Approach. *LWT – Food Sci. Technol.* **2016,** *66,* 475–483.

Sánchez-López, E.; Espina, M.; Doktorovova, S.; Souto, E. B.; García, M. L. Lipid Nanoparticles (SLN, NLC): Overcoming the Anatomical and Physiological Barriers of the Eye – Part II – Ocular Drug-Loaded Lipid Nanoparticles. *Eur. J. Pharm. Biopharm.* **2017,** *110,* 58–69.

Sandor, M.; Enscore, D.; Weston, P.; Mathiowitz, E. Effect of Protein Molecular Weight on Release from Micron-Sized PLGA Microspheres. *J. Controlled Release* **2001**, *76*(3), 297–311.

Sarmento, B.; Martins, S.; Ferreira, D.; Souto, E. B. Oral Insulin Delivery by Means of Solid Lipid Nanoparticles. *Int. J. Nanomed.* **2007**, *2*(4), 743–749.

Schubert, M. A.; Müller-Goymann, C. C. Characterisation of Surface-Modified Solid Lipid Nanoparticles (SLN): Influence of Lecithin and Nonionic Emulsifier. *Eur. J. Pharm. Biopharm.* **2005**, *61*(1), 77–86.

Severino, P.; Santana, M. H. A.; Souto, E. B. Optimizing SLN and NLC by 22 Full Factorial Design: Effect of Homogenization Technique. *Mater. Sci. Eng. C* **2012**, *32*(6), 1375–1379.

Seyfoddin, A.; Shaw, J.; Al-Kassas; R. Solid Lipid Nanoparticles for Ocular Drug Delivery. *Drug Delivery* **2010**, *17(7)*, 467–489.

Sharma, A. K.; Sahoo, P. K.; Majumdar, D. K.; Sharma, N.; Sharma, R. K.; Kumar, A. Fabrication and Evaluation of Lipid Nanoparticulates for Ocular Delivery of a COX-2 Inhibitor. *Drug Delivery* **2016**, *23*(9), 3364–3373.

Silva, A. C. C.; González-Mira, E.; García, M. L. L.; Egea, M. A. A.; Fonseca, J.; Silva, R.; Santos, D.; Souto, E. B. B.; Ferreira, D. Preparation, Characterization and Biocompatibility Studies on Risperidone-Loaded Solid Lipid Nanoparticles (SLN): High Pressure Homogenization versus Ultrasound. *Colloids Surf. B* **2011**, *86*(1), 158–165.

Silva, C. O.; Rijo, P.; Molpeceres, J.; Figueiredo, I. V.; Ascensão, L.; Fernandes, A. S.; Roberto, A.; Reis, C. P. Polymeric Nanoparticles Modified with Fatty Acids Encapsulating Betamethasone for Anti-Inflammatory Treatment. *Int. J. Pharm.* **2015**, *493*(1), 271–284.

Singh, A. P.; Saraf, S. K.; Saraf, S. A. SLN Approach for Nose-to-Brain Delivery of Alprazolam. *Drug Delivery Transl. Res.* **2012**, *2*, 498–507.

Souto, E.; Wissing, S.; Barbosa, C.; Müller, R. Evaluation of the Physical Stability of SLN and NLC before and after Incorporation into Hydrogel Formulations. *Eur. J. Pharm. Biopharm.* **2004**, *58*(1), 83–90.

Souto, E. B.; Müller, R. H. The Use of SLN and NLC as Topical Particulate Carriers for Imidaloze Antifungal Agents. *Pharmazie.* **2006**, *61*(5), 431–437.

Strasdat, B.; Bunjes, H. Food Hydrocolloids Incorporation of Lipid Nanoparticles into Calcium Alginate Beads and Characterization of the Encapsulated Particles by Differential Scanning Calorimetry. *Food Hydrocolloids* **2013**, *30*(2), 567–575.

Üner, M.; Yener, G. Importance of Solid Lipid Nanoparticles (SLN) in Various Administration Routes and Future Perspectives. *Int. J. Nanomed.* **2007**, *2*(3), 289–300.

Vishwakarma, V.; Samal, S. S.; Manoharan, N. Safety and Risk Associated with Nanoparticles – a Review. *J. Miner. Mater. Charact. Eng.* **2010**, *9*(5), 455–459.

Warheit, D. B.; Laurence, B. R.; Reed, K. L.; Roach, D. H.; Reynolds, G. A. M.; Webb, T. R. Comparative Pulmonary Toxicity Assessment of Single-Wall Carbon Nanotubes in Rats. *Toxicol. Sci.* **2004**, *125*, 117–125.

Westesen, K.; Bunjes, H. Do Nanoparticles Prepared from Lipids Solid at Room Temperature Always Possess a Solid Lipid Matrix? *Int. J. Pharm.* **1995**, *115*(1), 129–131.

Westesen, K.; Bunjes, H.; Koch, M. H. Physicochemical Characterization of Lipid Nanoparticles and Evaluation of Their Drug Loading Capacity and Sustained Release Potential. *J. Controlled Release* **1997**, *48*(2–3), 223–236.

Weyhers, H.; Ehlers, S.; Hahn, H.; Souto, E. B.; Müller, R. H.; Müller, R. H. Solid Lipid Nanoparticles (SLN) – Effects of Lipid Composition on in Vitro Degradation and in Vivo Toxicity. *Pharmazie* **2006**, *61*(6), 539–544.

Wissing, S. . A.; Kayser, O.; Mu, R. H.; Müller, R. H. Solid Lipid Nanoparticles for Parenteral Drug Delivery. *Adv. Drug Delivery Rev.* **2004,** *56*(9), 1257–1272.

Yang, S. C.; Lu, L. F.; Cai, Y.; Zhu, J. B.; Liang, B. W.; Yang, C. Z. Body Distribution in Mice of Intravenously Injected Camptothecin Solid Lipid Nanoparticles and Targeting Effect on Brain. *J. Controlled Release* **1999,** *59*, 299–307.

zur Mühlen, A.; Mehnert, W. Drug Release and Release Mechanism of Prednisolone Loaded Solid Lipid Nanoparticles. *Pharmazie* **1998,** *53*(8), 552–555.

zur Mühlen, A.; Schwarz, C.; Mehnert, W. Solid Lipid Nanoparticles (SLN) for Controlled Drug Delivery – Drug Release and Release Mechanism. *Eur. J. Pharm. Biopharm.* **1998,** *45,* 149–155.

CHAPTER 10

SOLID LIPID NANOPARTICLES IN DRUG DELIVERY FOR SKINCARE

SHEEFALI MAHANT[1], SUNIL KUMAR[2], RAKESH PAHWA[3],
DEEPAK KAUSHIK[1], SANJU NANDA[1], and REKHA RAO[2,*]

[1]Department of Pharmaceutical Sciences, Maharishi Dayanand University, Rohtak, Haryana 124001, India

[2]Department of Pharmaceutical Sciences, Guru Jambheshwar University of Science and Technology, Hisar, Haryana 125001, India

[3]University Institute of Pharmaceutical Sciences, Kurukshetra University, Kurukshetra, Haryana 136119, India

[*]Corresponding author. E-mail: rekhaline@gmail.com

ABSTRACT

Skin being a cardinal organ of the body needs adequate nourishment and care. Following a skin care regimen, appropriate and timely treatment can lead to a considerable decline in the prevalence of skin diseases. Considering the lipoidal nature of human skin, a number of lipid-based dermatological carriers have been explored in the past, but the invention of solid lipid nanoparticles (SLNs) in the 1990s could be considered as a landmark in the arena of dermal drug delivery. SLNs, which are colloidal carriers formulated by using lipids that are solid at room temperature, not only offers the advantage of modified drug release and epidermal targeting but are also devoid of problems associated with drug loading, stability, and scale-up. The large surface area, occlusive property, skin hydration and ultraviolet resistance of these nanocarriers make them ideal for dermal application. Ease of fabrication and characterization further augment their commercialization potential. Owing to their inherent merits, SLNs have been used for cosmetic application of several active moieties, such as CoQ_{10}, retinol and tocopheryl acetate. The present chapter provides a detailed account of SLNs as dermal carriers, covering their composition, production, characterization, release profile, cosmetic benefits and studies carried out. A list of recent patents on SLNs for drug delivery in skin care has also been included.

10.1   LIPIDIC SYSTEMS FOR DELIVERING ACTIVES TO SKIN

The high burden of skin conditions prevalent in high as well as low-income nations upholds the need to prioritize skin care and develop novel treatment strategies catering to the dermatological segment. The skin condition experts' team of the Global Burden of Disease 2010 study has emphasized the importance of prevention of skin disorders (Hollestein and Nijsten, 2014).

Skin is a cardinal organ of the human body. It performs the important functions of protection, excretion and temperature regulation in the body, apart from endocrine activity (Baroli, 2009; Bensouilah and Buck, 2006). In view of its fundamental role in maintaining the homeostasis of the body, the skin must be provided with adequate nourishment, just like any other organ. The normal attributes of the healthy skin may be disturbed by deficiency diseases, systemic diseases and endocrine disorders (Souto and Müller, 2008).

Skincare calls for products which are capable of delivering the actives to the skin without altering its structural integrity and composition (Puglia and Bonina, 2012; Souto and Müller, 2008). In this context, a thorough understanding of the skin morphology is indispensable. Briefly, human skin consists of the epidermis (composed of a matrix of connective tissue) resting upon the underlying dermis (comprising nerves, blood vessels, and lymph vessels). Hair follicles present in the dermis or subcutis emerge on the skin surface, passing through the epidermis (McGrath and Uitto, 2010).

The horny layer of the epidermis, formed by corneocytes is chiefly responsible for the barrier properties of the skin. The barrier imposed by this layer is more pronounced for hydrophilic moieties in comparison to lipophilic active constituents (Schäfer-Korting et al., 2007; Souto and Müller, 2008). This fact can be ascribed to the presence of epidermal lipids which surround the corneocytes. The epidermal lipids include ceramides, cholesterol, long-chain free fatty acids and cholesteryl sulfate. The moisture content of normal skin ranges between 10 and 20% and it is a result of the proteolysis of the cellular content of corneocytes. Osmotic activity of the amino acids, thus produced, binds water molecules, contributing to skin hydration (Baroli, 2009; Wissing et al., 2001). Aside from its barrier function, the horny layer acts as a depot for topically administered active moieties (Schäfer-Korting et al., 2007).

In light of the lipoidal nature of the stratum corneum (SC), the uppermost layer of the epidermis, it may be inferred that lipid-based drug delivery systems would be the most appropriate choice for delivering pharmacological and cosmetic actives to the skin (Souto and Müller, 2008). Correspondingly, the recent decades have witnessed an upsurge in the investigations carried out for dermal delivery of actives using lipid-based formulations (Souto and Müller, 2008). The lipid-based formulations described for topical application to the skin encompass oil-in-water (o/w) emulsions, microemulsions, multiple emulsions (water-in-oil/oil-in-water []), nanoemulsions, and liposomes (Souto and Müller, 2008; Yoon et al., 2013).

With regard to dermatological application, liposomes may be considered "blockbuster carrier systems," as the cosmetic benefits of liposomal formulations outshine all other delivery systems explored prior or subsequent to their development in the 1960s (Müller et al., 2007). For many years following the emergence of liposomes, research efforts were in progress, with the view to come up with a dermal carrier that not only offered the advantages inherent in liposomes but was also capable of circumventing the limitations of drug loading, stability, and scale-up associated with the liposomal formulation. The challenge was successfully met by the advent of solid lipid nanoparticles (SLNs) in the 1990s. The credit of developing these alternative colloidal carrier systems goes to two research groups, that is, Müller and Lucks (Germany) and Dr. Maria Gasco (Italy), who developed SLNs by high-pressure homogenization (HPH) and microemulsion technique, respectively. Notwithstanding the research efforts of these scientists, Westesen was also working on the design and development of SLNs by HPH at the same time (Müller, 2007; Pardeike et al., 2009). Thence, lipid nanoparticles (LNs) received a great deal of attention and several research publications and patents followed (Gasco, 2007; Pardeike et al., 2009).

10.2 CONCEPT OF LIPID NANOPARTICLES (LNS)

SLNs may be defined as colloidal carriers composed of lipids which are solid at room temperature. Their structural similarity with physiological lipids confers on them the merits of low toxicity and appreciable in vivo tolerance (Docktorova and Souto, 2009). Not only this, in context of skin care products, the SLNs offer other advantages, namely the protection of sensitive active ingredients against degradation, thereby resolving their stability issues. Examples include retinol and coenzyme Q10 which have been successfully encapsulated in SLNs and found to be stable for long period of time. This, further, provides opportunity for designing controlled drug delivery system of such agents for their cosmetic benefits (Farbound et al., 2011). These nanometer size particles have large surface area which makes them ideal for dermal applications (Gokce et al., 2012). Advantages of SLNs specific to skin care products are their excellent occlusive, skin hydration, and ultraviolet-resistant properties (Farbound et al., 2011; Schäfer-Korting et al., 2007). Apart from these, SLNs have been found promising for the formulation of perfumery products (Souto and Müller, 2008).

10.2.1 FORMULATION ASPECTS

SLNs for dermal administration consists of lipids and stabilizers (usually surfactants, cosurfactants, and coating substances) (Svilenov and Tzachev, 2014). The solid lipid (0.1–30% w/w) is dispersed in an aqueous system comprising of 0.5–50% (w/w) of surfactant (Puglia and Bonina, 2012). The lipids which find application in SLNs are triglycerides, waxes and partial glycerides and, fatty acids. Triglycerides include glyceryl tristearate and glyceryl tripalmitate. Examples of waxes are cetyl palmitate, beeswax and carnauba wax, while free fatty acids used are behenic acid, stearic acid, and so on. The common surfactants reported for SLNs are poloxamer188, lecithin, polyglycerol methyl glucose distearate, Polysorbate 80, and sodium cocoamphoacetate (Schäfer-Korting et al., 2007; Svilenov and Tzachev, 2014). Other excipients used in the formulation of SLNs are antioxidants, preservatives, viscosity enhancers, and electrolytes. Majority of the ingredients composing SLNs fall under generally recognized as safe category, as designated by the Food and Drug Administration. This minimizes regulatory hurdles with regard to their commercial use (Svilenov and Tzachev, 2014). The average particle size of these carriers falls in the range of 80–1000 nm. Further, with the view to enable dermal application, SLNs with lipid content 35–40% can be loaded into a suitable base, such as cream or gel. This prevents the problems of aggregation or dissolution associated with these formulations. Analytical investigations have attested to their good stability in such bases, when stored for a period of 6 months (Lippacher et al., 2004).

10.2.2 PRODUCTION METHODS

The methods for SLNs production may be classified into three categories, namely high energy approaches, low energy approaches, and those utilizing organic solvents (Svilenov and Tzachev, 2014).

Among the high energy approaches, HPH is a well-established technique and the most widely employed. This method is based upon subjecting the hot precursor emulsion to high pressure (500 bar, three cycles) and high shear stress, giving rise to cavitation forces. As a result, the emulsion droplets are broken down to submicron range. The main drawbacks are the thermodynamic and mechanical stress generated during the process. For thermolabile ingredients, however, cold HPH is used (Schäfer-Korting et al., 2007; Shidhaye et al., 2008; Yoon et al., 2013).

High shear homogenization (HSH) is another energy-intensive method, based on the use of rotor–stator homogenizer and employs high shear rates of 5000–25,000 rpm for mixing. However, the low polydispersity of the resulting nanoparticles is a disadvantage. Therefore, it is often coupled with HPH or ultrasonication. The latter utilizes powerful ultrasound waves with frequency above 20 kHz (Svilenov and Tzachev, 2014; Yoon et al., 2013). Electrospray technique is a rather new technique in this category, which is based upon the use of electrodynamic atomization to yield SLN. The SLN, so produced, are in powder form and have narrow range of size distribution (Svilenov and Tzachev, 2014).

The low energy approaches include the use of microemulsion templates, which is one of the earliest methods employed for SLN production. Briefly, a hot microemulsion is produced using molten lipid phase and aqueous phase containing surfactant mix. This is subsequently diluted with cool water (2–10°C), causing precipitation of LNs and polydispersity (Battaglia et al., 2014; Doktorovova and Souto, 2009; Yoon et al., 2013).

The double emulsion method finds application in preparing LNs with hydrophilic moieties, including peptides. First, a primary w/o emulsion is prepared for the hydrophilic moiety, which is, thereafter, emulsified with aqueous solution of a hydrophilic emulsifier to yield w/o/w emulsion. This is followed by stirring and filtration (Svilenov and Tzachev et al., 2014). This category also includes phase inversion method, coacervation method, and membrane contactor method for SLN preparation (Battaglia et al., 2014; Svilenov and Tzachev, 2014).

The third approach involves the use of organic solvents, the most common techniques being emulsification solvent evaporation, emulsification solvent diffusion, and solvent injection. These methods require low energy input and heating at high temperature is also circumvented. In the former two methods, o/w or w/o/w emulsions are prepared with the help of an organic solvent, which is either volatile in nature or partially miscible with water. Later, the solvent is either allowed to evaporate (solvent evaporation) or removed by dilution with water (solvent diffusion), leading to the formation of LNs. On the other hand, in solvent injection method, the organic solvents used are highly miscible with water (e.g. dimethyl sulphoxide, isopropanol, and ethanol). This eliminates emulsification. Rather, a solution of lipid and active compound is prepared in the organic solvent. This solution is injected into the aqueous phase containing the surface-active agent, with stirring. Owing to its high miscibility, upon contact with water, the organic solvent migrates into the water, enabling precipitation of LNs (Battaglia et al., 2014; Svilenov and Tzachev, 2014; Yoon et al., 2013).

Supercritical fluid technique uses fluids (e.g. carbon dioxide) in the supercritical range for dissolving lipids and active compound, instead of organic solvents. Since the particles produced by this method are relatively larger in size, it is often used in conjunction with some other homogenization method (Svilenov and Tzachev, 2014).

10.3   CHARACTERIZATION OF SOLID LIPID NANOPARTICLES (SLNS)

Characterization of SLNs using suitable techniques is crucial in order to ensure quality, and stability of the products and to ascertain the release kinetics (Das and Chaudhury, 2011; Yoon et al., 2013). SLNs can be characterized by various properties, such as particle size, polydispersity, zeta potential (ZP), morphology, crystallinity, and polymorphism.

10.3.1   DETERMINATION OF PARTICLE SIZE, POLYDISPERSITY INDEX, AND ZETA POTENTIAL

Photon correlation spectroscopy (PCS) and laser diffraction (LD) are the most extensively employed techniques for routine analysis of particle size (Iqbal et al., 2012; Kakkar et al., 2011; Schlupp et al., 2011; Wissing and Muller, 2003). PCS (sometimes referred to as dynamic light scattering) measures the fluctuation in the intensity of the scattered light due to movement of particles (Iqbal et al., 2012; Müller et al., 2000; Teeranachaideekul et al., 2007). This technique covers size ranging from a few nanometers to 3 μm (approximately). Therefore, PCS offers great utility in assessment of LNs. However, it is not able to determine larger micron-sized particles. They can be examined with the help of LD measurements. LD can evaluate particle sizes greater than 3 μm (Müller et al., 2006; 2008; Yoon et al., 2013). This method is based on the fact that diffraction angle is a function of particle radius. Particles with greater size tend to scatter light at small angles while the smaller ones scatter light at large angles. Although PCS is relatively precise and sensitive, it is generally recommended that both the techniques be utilized simultaneously, particularly for nonspherical particles (Das and Chaudhury, 2011; Iqbal et al., 2012).

On account of polydisperse nature of SLNs, assessment of polydispersity index (PDI) is also essential for the evaluation of particle size distribution. Polydispersity can be estimated by PCS (Baboota et al., 2009; Liu et al., 2007a; Xie et al., 2010; Wissing and Muller, 2001). Samples with PDI values less than 0.3 are generally considered to consist of homogenous SLNs (Hu et al., 2004; Jeon et al., 2013; Zhang et al., 2009).

The ZP is a measure of the overall charge of a particle in a given medium, which can be determined by PCS too (Das and Chaudhury 2011; López-García and Ganem-Rondero, 2015). Measurement of ZP enables to predict the stability of colloidal dispersions (Chen et al., 2006; Mei et al., 2005; Müller et al., 2000). In general, high ZP value (more than ±30 mV) provides greater electric repulsion to inhibit the aggregation of particles, thereby, electrically stabilizing the nanoparticles dispersion (Iqbal et al., 2012; Raza et al., 2013; Ridolfi et al., 2012). Howbeit, this principle may not be relevant for all types of colloidal dispersions, especially those containing steric stabilizers.

10.3.2 MORPHOLOGICAL ANALYSIS

Techniques such as, transmission electron microscopy (TEM), scanning electron microscopy (SEM), and atomic force microscopy (AFM) can be employed to investigate the shape and morphological appearance of SLNs (Jain et al., 2005; Omwoyo et al., 2014; Yoon et al., 2013). These techniques are also useful for the determination of particle size and its distribution. SEM is based on the principle of electron transmission from the sample surface, while electron transmission through the sample forms the basis of TEM (Iqbal et al., 2012; Yoon et al., 2013). In numerous investigations, SEM and TEM images have revealed the representative spherical shape of SLNs (Baboota et al., 2009; Jain et al., 2010; Kakkar et al., 2011;

Liu et al., 2007b; Lv et al., 2009; Shah et al., 2007; Silva et al., 2011; Varshosaz et al., 2010). Although regular SEM may not be sensitive enough for the nanosized particles, field emission SEM can be used to detect nanosized particles (Sahana et al., 2010). However, procedure of sample preparation (e.g., drying process) is likely to influence the particle's shape in this technique. Cryo-SEM may be useful in this case, where SLNs are cryo-transferred with liquid nitrogen before conducting SEM analysis and images of the frozen sample are taken (Lee et al., 2007). In addition, AFM technique has drawn the attention of scientific community globally and has been widely reported for nanoparticle characterization (Jain et al., 2011; Mehnert et al., 2001; Xie et al., 2011). AFM presents a three-dimensional profile of the particle surface. In contrast, only two-dimensional image of a sample is revealed by electron microscopy (Das and Chaudhury, 2011; Shahgaldian et al., 2003).

10.3.3 CRYSTALLINITY AND LIPID MODIFICATION

Special consideration must be paid to assess degree of crystallinity of the lipid and its modification in SLNs, as these factors strongly affect drug loading as well as release rate (Müller et al., 2000; Venkateswarlu and Manjunath, 2004). Differential scanning calorimetry (DSC) and X-ray diffractometry (XRD) are extensively utilized techniques for investigation of the crystallinity, along with polymorphic transitions of the excipients of SLNs (Baboota et al., 2009; Jain et al., 2011; Radomska-Soukharev, 2007). It is worthwhile to mention that DSC is based on the principle that melting points and melting enthalpies vary for different lipid modifications. DSC furnishes information on the crystallization behavior and melting of all the components of SLNs. Furthermore, this technique is also valuable to recognize the mixing behavior of the lipids present in the structure (Jenning et al., 2000a; Üner M, 2006).

XRD can identify precise crystalline character of the lipid matrices, depending on their crystal structure. Long and short spacing present in the lipid lattice can also be assessed by means of XRD patterns (Bunjes and Koch, 2005; Estella-Hermoso de Mendoza et al., 2009; Kalam et al., 2010; Müller et al., 2000; Yoon et al., 2013). X-ray diffraction is utilized as a technique which distinguishes the various lipid polymorphs. Diverse polymorphic forms can be unambiguously characterized by their spacings (Üner M, 2006; Westesen et al., 1993). X-ray diffraction profiles can also corroborate the polymorphism behavior established by DSC measurements (Bhalekar et al., 2009; Jenning et al., 2000b).

10.3.4 ASSESSMENT OF ADDITIONAL COLLOIDAL STRUCTURES

The magnetic resonance techniques such as nuclear magnetic resonance (NMR) and electron spin resonance (ESR) are useful for examining dynamic phenomena as well as characteristic features of nanocompartments in colloidal lipid dispersions. Supercooled melts formed due to the low line widths of the lipid protons can be detected by ¹H NMR analysis (Zimmermann et al., 2005). This technique is based on the varied time of proton relaxation observed in the liquid and semisolid/solid state (Iqbal et al., 2012). Information about the mobility and structure of both the nanoparticle system and the incorporated drug can also be explored by NMR investigations (Wissing et al., 2004; Zimmermann et al., 2005). ESR necessitates the use of a paramagnetic spin probe to elucidate SLNs. ESR technique facilitates direct, repeatable, and noninvasive evaluation of the distribution of the spin probe among the aqueous and lipid phase (Braem et al., 2007; Iqbal et al., 2012; Kuchler et al., 2010).

10.4 COSMETIC BENEFITS OF SLNS

In consideration of the vital functions of the skin, it should essentially be nourished and tended with care. Topically applied formulations containing noncytotoxic ingredients, which reinforce the functions of the skin and impart protection, are best suited to serve as skin care products (Grana et al., 2013; Kakadia and Conway, 2014). Figure 10.1 depicts the role of SLNs in repairing damaged skin.

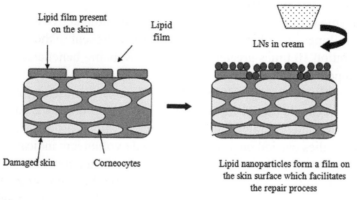

FIGURE 10.1 Lipid nanoparticles (LNs) incorporated in a suitable base (cream, gel, or lotion) help to repair the damaged skin by reinforcing the lipid film on the skin surface.

Apparently, the aesthetic features and elegance of the formulation are of paramount importance (Souto and Müller, 2008; Wissing et al., 2001). Cosmetic formulations are typically expected to facilitate personal hygiene and act superficially upon cutaneous application (Yukuyama et al., 2015). LNs have proved especially useful in skin care segment, owing to their cosmetic attributes (Lasoń and Ogonowski, 2011). Table 10.1 provides a glimpse of the cosmetic benefits of SLNs.

10.5 SLNS AS SKIN CARE PRODUCTS

The primary purpose of the dermal dosage forms is the rescue of the active molecules through SC and their pharmacological action on skin. SLNs are potential delivery systems utilized in fabrication and delivery of numerous active molecules for skin care. Representative active molecules investigated in recent years for skin care have been enlisted in Table 10.2.

For efficient dermal delivery of isotretinoin, Liu et al. encapsulated it in SLNs using hot homogenization technique. Various physicochemical characterization methods were used to evaluate these nanoformulations. In vitro studies using rat, skin were performed, showing enhanced cumulative drug uptake followed by skin targeting. In this study, isotretinoin-loaded SLNs were reported as promising lipid carrier for topical administration.

So et al. synthesized 6-methyl-3-phenethyl-3,4-dihydro-1H-quinazoline-2-thione (JSH-18) as an inhibitor of melanin synthesis for skin whitening efficiency, due to melanin inhibition. SLNs were prepared using JSH-18 and loaded in cream. The optimized JSH-18 SLN cream was investigated in hairless rats for skin whitening action and reported as appropriate carrier for such agents (So et al., 2010).

Coenzyme Q10 has been distinguished as potential active compound for heart, liver, lungs, and skin. It was reported to be present in the epidermis in a concentration 10 times greater than in dermis, being beneficial as an anti-aging compound (Shindo et al., 1994). Farboud et al. fabricated coenzyme Q10-loaded SLNs through HPH method. Various parameters in terms of physicochemical characteristics were explored to optimize the formulation and SLNs were incorporated into the cream base. In vivo skin elasticity and hydration studies carried out on healthy female volunteers indicated greater dermal penetration and pharmacological activity of the new formulation as compared to simple cream. Hence, SLNs may be the promising carrier to maintain coenzyme Q10 effectiveness for dermal delivery (Farboud et al., 2011). The effective dermal delivery of various compounds has numerous

TABLE 10.1 Cosmetic Benefits of Solid Lipid Nanoparticles-Based Formulations.

Feature	Description	Incorporated active compounds	References
Adhesiveness and occlusion	Small particle size and large surface are enabled adhesion to the skin and formation of an occlusive film on the skin surface	—	Hamishekhar et al., 2015; Souto and Müller, 2008; Puglia et al., 2012
Lubricant and emollient property	Depends upon the viscoelastic properties of the formulation and choice and concentration of lipid phase. Very pure lipids such as tripalmitin and tristearin give best results	—	Souto and Müller, 2008
Skin hydration (Moisturization)	Occlusive property of the solid lipid nanoparticles (SLNs) restores skin lipids and prevents transepidermal water loss (TEWL)	Ascorbyl palmitate	Üner et al., 2005 Müller et al., 2000
Drug targeting to upper layers of the skin	Greater penetration of the active compound into epidermis is achieved due to occlusion effect and enhanced skin hydration	Retinol and retinyl palmitate	Jenning et al., 2000c
Ultraviolet (UV) blocking property	SLNs possess the ability to reflect and scatter UV radiation on their own and show synergistic effect with molecular sunscreens	Tocopherol acetate 2-Hydroxy-4-methoxybenzophenone	Wissing et al., 2001 Wissing and Müller, 2001b
Release modulation	Release of the active compound depends upon the method of SLN production, composition of SLN, solubilization of the active compound by the surfactant used and o/w partition coefficient of the active compound	Vitamin A Fragrances Insect repellants	Jenning et al., 2000d Pardeike et al., 2009
Chemical stability of labile actives	Solid lipid matrix of the SLN protects the labile active moiety from degradation by oxygen and water. The lipid should be selected on the basis of its ability to solubilize the active compound during shelf-life of the product	Coenzyme Q10 Ascorbyl palmitate	Üner et al., 2005 Müller et al., 2000 Farboud et al., 2011

TABLE 10.1 *(Continued)*

Feature	Description	Incorporated active compounds	References
pH control	SLN dispersions with optimum pH can be produced to avoid skin irritation	—	Souto and Müller, 2008
Osmotic effect	Very hypertonic preparation can cause discomfort to the user. SLN based formulations show optimum isotonicity	—	Souto and Müller, 2008
Wrinkle smoothening effect	Application of SLN-loaded cream reduces wrinkle depth	Retinol	Shidhaye et al., 2008
Whitening effect	SLNs weaken the coloration of colored actives or those which get converted into colored intermediates during product's shelf-life	Coenzyme Q10 Vitamins	Wissing et al., 2001a

TABLE 10.2 Representative Active Molecules Encapsulated in Solid Lipid Nanoparticles (SLNs) in Recent Years.

Sr. no.	Drug molecule	Preparation technique	Remarks	References
1	Isotretinoin	Hot homogenization	Skin targeting topical delivery	Liu et al., 2007b
2	JSH 18	–	Optimum efficiency for skin whitening	So et al., 2010
3	Coenzyme Q10	High-pressure homogenization (HPH)	Prolonged released and dermal penetration	Farbound et al., 2011
4	Coenzyme Q10	Thin film hydration and high shear homogenization (HSH)	Evaluation of the influence of the nanosystem on effective delivery into the skin	Gokce et al., 2012
5	Neem oil	Double emulsification	Prolonged treatment of acne	Vijayanet al., 2013
6	Coenzyme Q10	High-speed homogenization	Enhanced dermal delivery	Karkmet al., 2013
7	Coenzyme Q10	Continental emulsification	Promote the penetration properties	Lohan et al., 2015
8	Hydroquinone	Hot melt homogenization	Enhanced stability &Dermal delivery	Ghanbarzabeh et al., 2015
9	Safranal	HSH, ultrasound, and HPH	Sunscreen and moisturizing potential evaluation.	Khameneh et al., 2015
10	rhEGF	–	Topical wound dressing	Gainza et al., 2015
11	N-6-furfuryl adenine	Hot microemulsion	Prevent Photoaging	Goindi et al., 2015
12	Lauric acid and retinoic acid	Hot homogenization	Better efficacy in combination therapy	Silva et al., 2015
13	Flutamide	Hot melt homogenization	Follicular targeting in the treatment of androgenic alopecia.	Hamishekhar et al., 2015
14	Diflucortolonevalerate	One-step production	Increased encapsulation efficiency	Abdel-Salam et al., 2015

TABLE 10.2 *(Continued)*

Sr. no.	Drug molecule	Preparation technique	Remarks	References
15	Amphotericin B	Solvent diffusion	Enhanced antifungal efficacy	Butani et al., 2015
16	Aceclofenac	Ultrasonic emulsification	Enhanced skin delivery	Raj et al., 2015
17	Ketoprofen, indomethacin, nimesulide	Microwave assisted one-pot microemulsion	Enhanced entrapment efficiency of nonsteroidal anti-inflammatory drugs	Shah et al., 2016
18	Diflunisal	Hot homogenization	Enhanced pharmacological potency	Kaur et al., 2016
19	LL37 and Serpin A1	Emulsion solvent diffusion	Synergistic antibacterial activity	Fumakia and Ho, 2016
20	Benzophenone	–	Skin protection against sunlight	Gilbert et al., 2016
21	Rhodamine B	Hot melt homogenization	Explored effect of particle size	Adib et al., 2016

benefits in the management of skin diseases. Following year, Gokce et al. designed coenzyme Q10-loaded liposomes as well as SLNs to enhance dermal delivery. Thin film hydration techniques and HSH were used for the preparation of liposomes and SLNs, respectively. The physicochemical properties such as particle size, ZP, and encapsulation efficiency were determined. In addition to DSC and TEM, the cytotoxicity studies of the formulations on the human fibroblasts cell lines were distinguished. The observations of this study showed biocompatibility, which was also found suitable for cell proliferation. In comparison to SLNs, the liposomal formulations exhibited increased cell proliferation and protective effect corresponding to reactive oxygen species accumulation. Hence, liposomal preparation was found better over SLNs for dermal delivery of coenzyme Q10 (Gokce et al., 2012). Korkmaz et al. incorporated coenzyme Q10 in SLNs using high-speed homogenization technique. The prepared SLN formulations were loaded into Carbopol hydrogels. Different physicochemical parameters along with rheological studies were explored. Dermal delivery of drug-loaded SLNs was measured using the rat abdominal skin. Results suggested coenzyme Q10 SLN-loaded gels as a good delivery system for maintaining its antioxidant potential (Korkmaz et al., 2013).

Vijayan et al. formulated and characterized neem oil-loaded SLNs. The SLNs were prepared using various concentrations of lecithin and Tween 80 through double emulsification method. Results indicated that increase in surfactant concentration decreased the average particle size of SLNs loaded with neem oil. Neem oil SLNs with high entrapment efficiency (82.10%) were obtained, which were also found stable after 3 weeks. Further, this formulation with higher lecithin content, exhibited satisfactory prolonged antibacterial action on acne microbes (Vijayan et al., 2013).

To overcome the challenges of the effective delivery of hydroquinone, Ghanbarzadeh and research group-loaded hydroquinone in SLNs through hot melt homogenization technique and carried out their physicochemical evaluation. In vitro permeation of SLN hydrogels was assessed with the help of excised rat skin. The results demonstrated better hydroquinone skin localization and retarded systemic absorption of hydroquinone SLN-loaded gel, in comparison to hydroquinone Carbopol gel. Conclusively, SLNs have been considered as potential nanocarrier for dermal delivery of hydroquinone in hyperpigmentation disorders (Ghanbarzadeh et al., 2015).

Raj et al. developed aceclofenac (ACF) SLNs and incorporated these in hydrogel for improved topical delivery. SLNs were fabricated using ultrasonic emulsification technique and optimized based on lipid content and stirring speed. The optimized nanoformulation was characterized by surface morphology, ZP, particle size, PDI. Moreover, DSC and X-ray diffraction studies were performed. In vivo evaluation of ACF-SLN-loaded hydrogel demonstrated better inhibition of edema in comparison to plain ACF hydrogel, after 24 h. On the basis of the findings, ACF SLN formulation was reported as potential topical carrier for skin targeting of this drug (Raj et al., 2015).

Khameneh et al. fabricated safranal-loaded SLNs comprising glyceryl monostearate and Tween 80, using high pressure and HSH and, ultrasound techniques. In addition to physicochemical properties, moisturizing activity and sun protection factor of the prepared formulations were evaluated in vitro through the transpore tape. The increased sun protection factor in comparison to reference 8% homosalate, indicated SLNs as a promising dermal approach for safranal, to attain sunscreen properties (Khameneh et al., 2015).

Gainza and his research group encapsulated recombinant human epidermal growth factor in LNs for topical treatment of the chronic wound. The prepared SLN and nanostructured lipid carriers (NLC) dressings were evaluated in vitro. There was no significant difference in in vitro observations of both types of the formulations. Because of their potential long-term stability, researchers concluded fibrin-based lipid nanoscaffolds to be a vital tool for dermal wound dressing (Gainza et al., 2015).

Goindi et al. fabricated kinetin (N-6-furfuryl adenine), an active natural phytochemical, loaded SLNs using hot microemulsion method. Kinetin-loaded SLN-based topical preparations were evaluated for physicochemical properties (drug entrapment efficiency, particle size, ZP, and pH), spreadability, photoprotective, and ex vivo skin permeation and stability. The permeation through mice skin exhibited three times more permeation of kinetin in comparison to conventional cream (Garnier®). The results indicated potential cosmetic and therapeutic benefits of this molecule when loaded in SLNs (Goindi et al., 2015).

Silva et al. designed SLNs loaded with lauric acid (saturated fatty acid) and retinoic acid in combination with hot melt homogenization technique. The effect of stearylamine (lipophilic) and physicochemical evaluation of prepared SLNs was examined. The in vitro studies of the formulation were performed against *Staphylococcus epidermidis, Staphylococcus aureus*, and *Propionibacterium acnes*. Still, good encapsulation stability of SLNs was found in the presence of stearylamine. The results obtained from the above studies, suggested that the combination approach in SLNs represented a boon for topical treatment for acne vulgaris (Silva et al., 2015).

Hamishehkar et al. formulated SLNs loaded with flutamide (potent anti-androgen) by hot melt homogenization technique for targeting the hair follicles to treat androgenic alopecia. The prepared SLN formulations were subjected to evaluation: in vitro (drug permeation and accumulation studies using rat skin) and in vivo (histopathological study using male hamsters) to appraise dermal drug delivery. In addition, stability studies, X-ray diffraction studies, and encapsulation efficiency were also explored. The results of all observations generated great interest in the topical delivery of the flutamide by incorporating it in SLNs (Hamishehkar et al., 2015).

Abdel-Salam et al. fabricated diflucortolone valerate-loaded SLNs through one step production method. In order to obtain semisolid formulations, high lipid concentration and lipid selection were considered promising parameters. Evaluation of average particle size, entrapment efficiency, rheology, and SEM of the prepared formulations were carried out. It was concluded that lipid-based surfactants such as labrafil and labrasol possessed promising role in enhancing the entrapment efficiency of diflucortolone valerate for topical use (Abdel-Salam et al., 2015).

To enhance antifungal potential, Butani et al. prepared amphotericin B-loaded SLNs using solvent diffusion technique. The fabricated nano-formulation was characterized by particle size, surface morphology, drug entrapment, ZP, in vitro antifungal activity, ex vivo skin irritation, permeation, and retention studies. To evaluate the interaction between lipid and drug, Fourier-transform infrared spectroscopy, powder X-ray diffraction studies, and DSC were carried out. The obtained results showed no interaction between lipid and amphotericin B and equal distribution of the drug in the matrix. The lyophilized SLNs were found potentially more stable without cryoprotectant (sucrose) as exhibited by stability studies. The optimized formulation displayed two-fold higher amphotericin B permeation with respect to plain drug dispersion and more prominent inhibition zone in fungal species (*Trichophytonrubrum*). It was concluded that SLNs can be considered as good dermal delivery system to enhance the antifungal efficacy of amphotericin B (Butani et al., 2015).

Lohan et al. developed coenzyme Q10-loaded ultrasmall LNs to improve their penetration in the skin. Their cellular uptake and distribution within the cytoplasm were explored in human keratinocyte cell lines. The results of Cell Proliferation Kit II (XTT) cell viability test showed no sign of toxicity and demonstrated greatest reduction in radical formation (Lohan et al., 2015).

Recently, stearic acid-based SLN loaded with ketoprofen, indomethacin, and nimesulide were formulated through microwave-assisted one-pot micro-emulsions process. The characterization of the prepared formulations for their physicochemical and drug release behavior was carried out. In vitro cell culture studies on human lung epithelial cells have also been executed. From the results of drug release studies, the SLNs showed biphasic drug release pattern. The concentration-dependent cytotoxicity in human A549 cells and decreased secretion of IL-8 and IL-6 in lipopolysaccharide-induced cells was observed from cell viability and cellular uptake studies, and anti-inflammatory activity studies, respectively. The obtained results signified that the SLNs produced through microwave have significant potential for encapsulation of the nonsteroidal anti-inflammatory drugs, which could be further developed for delivery through oral, topical, and/or nasal route (Shah et al., 2016).

Kaur and her research group designed diflunisal-loaded SLNs through hot homogenization technique based on microemulsification technique. The prepared formulations were evaluated for particles size, morphology, entrapment efficacy, spreadability, rheology, and stability. Preclinically, the cumulative drug amount permeated per area, permeation flux, and skin retention through mice skin was measured. Hence, the obtained results asserted efficient potential of prepared formulations with escalating pharmacological potency (Kaur et al., 2016).

To promote wound healing through combination therapy, Fumakia and Ho prepared LL37 (endogenous host defense peptide) and Serpin A1 (elastase inhibitor) SLNs using emulsion solvent diffusion technique. In vitro release profile, ex vivo skin permeation studies, degradation studies, antibacterial studies, cellular assays, in vitro wound healing assay, in vitro cellular cytotoxicity studies, and cellular monolayer integrity were carried out. The results suggested enhanced wound closure as well as increased the antibacterial activity based on the synergistic action of moieties loaded in SLNs (Fumakia and Ho, 2016).

With the view to impart skin protection from ultraviolet radiations, Gilbert and research group fabricated benzophenone-3-loaded lipid nanocarriers NLC, SLNs, nanocapsules (NC), and nanostructured polymeric lipid carriers (NPLC). Percutaneous absorption, along with cutaneous bioavailability of the fabricated formulations was compared. From the observations of aforementioned studies, NPLC and NC were found to minimize skin permeation of the drug, with highest sun protection factor. Both these lipid-based polymeric carriers remarkably reduced benzophenone-3 skin permeation (Gilbert et al., 2016). Adib et al. explored the influence of particle size on cutaneous penetration, depth and rate of rhodamine B deposition in various skin layers. The observations from this study illustrated sub-100 nm sized SLNs as an appropriate carrier for penetration into deeper layers of skin, mainly through hair follicles (Adib et al., 2016).

10.6   PATENTS ON SLNS

As discussed above, SLNs have come forth as a promising approach to overcome potential drawbacks of conventional therapeutic systems, such as low bioavailability, low drug loading, and severe toxic effects. There has been a spurt in the number of patents filed in this area, signifying its growing importance in drug delivery nowadays. Table 10.3 provides a list of most recent patents and patent applications on SLNs.

The worldwide databases of European patent office (http://ep.espacenet. com) and United States patent office (www.uspto.gov) were accessed to collect the patents and patent applications.

10.7   DRUG LOADING AND DRUG RELEASE

The distribution of active ingredient within the lipid matrix of the nanoparticle may assume different patterns, subject to the composition of the system (as determined by the lipid, active ingredient, and the surfactant) and the preparation conditions employed (Shidhaye et al., 2008). Muller et al. have proposed three distinct models for the localization of the active ingredient into SLNs (Müller et al., 2002).

The drug dissolution model or the homogeneous matrix model consists of the active ingredient assuming homogeneous distribution within the lipid matrix. This is basically a distribution pattern wherein the drug is present either as molecular dispersion or as amorphous clusters within the SLN matrix. Such a structure results from the incorporation of highly lipophilic active moiety using hot homogenization technique or when SLNs are produced by cold homogenization method.

The second model referred to as drug-enriched shell model, is a system produced by the precipitation of lipid, followed by phase separation during cooling step. In such a case, the active ingredient gets embedded in the outer shell and the core of the SLNs is devoid of the active moiety.

The third method is called drug-enriched core model. In contrast to the second model, in this case, the active ingredient begins to precipitate, leading to the formation of the SLN with very low content of the active compound. Such a model is governed by the Fick's law of diffusion, hence, permitting controlled release of the active compound. Further, it is even possible to have a system that is actually a mix of all the three models stated above (Müller et al., 2002, Pardeike et al., 2009; Shidhaye, et al. 2008). Two more core drug incorporation models have been suggested in

TABLE 10.3   Recent Patents on SLNs.

Patent no.	Composition	Drug	Method	Remark	References
Chinese Patent Application CN 105343031	Palmiticacid and polyglycerol10 laurate	Ivermectin	Freeze-drying	Achieved controlled release of the drug	Dawei et al., 2016
Chinese Patent Application CN 105311643	Stearic acid (SA-R8)	Paclitaxel	Film dispersion method	Increased absorption through intestinal mucosae	Ren, 2016
Chinese Patent application CN105560216	Soya lecithin, cholesterol	Entecavir	Supercritical fluid technology	Improved stability, higher bioavailability, and reduced toxic effect	Shanchum et al., 2016
Korean Patent application KR20150134443	Palmitic acid and cholesterol	Curcumin	Emulsification by ultrasonic homogenization	Increased stability in acidic stomach environment, improved delivery from blood brain barrier	Tag and Prakash, 2016
US Patent US 9393201	Lipid mixture of poloxamer in solid lipid myristyl alcohol	Oxaliplatin	Supercritical fluid technology	Stablity against gastric acid; improved patient compliance through avoiding the inconvenience of injection and greatly reducing medical cost	Lee et al, 2016
Chinese patent application CN 104771383	Phoshpolipids	Sqauinavir Phospholipid	Solvent diffusion technology	Increased absorption in gastrointestinal tract and improved bioavailability	Liyong et al., 2015
Chinese Patent CN104586817	Lecithin, Palmitoyl phosphatidylethanol-amine	Docetaxel	Lyophilization	Improved oral bioavailability	Qingri et al., 2015
Indian Patent Application IN26KO2014	Cholesterol, oleic acid and Phosphatidylcholine	Acylovir	Hot homogenization method	Ability to avoid the systematic penetration through the skin.	Majumdar et al., 2015

TABLE 10.3 *(Continued)*

Patent no.	Composition	Drug	Method	Remark	References
US Patent 8865129	Emulsifying wax	Doxorubicin	Complex Encapsulation	Use as contrast agents or drug delivery vehicles.	Walters, 2014
US Patent 8715736	Compritol and Miglyol	Celecoxib	Hot melt homogenization technique	Increased skin permeation of encapsulated CXB	Sachdeva and Patlolla, 2014
US Patent 8911788	Palm oil, Oleic acid and Stearic acid	Erythromycin	Hot melt homogenization technique	Enhanced stability of drug against degradation during transit in the stomach following oral absorption.	Ioualalen and Raynal, 2014
Chinese Patent CN102552156	Phospholipid and cholesterol	Nimodipine	Freeedrying	High efficiency, low toxicity and enhanced Stability	Tang et al., 2013
Chinese Patent application CN103446076	Stearic acid, Palmitic acid	Ally lisothiocyanate	"HPH"	Increased solubility and reduced irritation of drug	Weidong et al., 2013
Chinese Patent application CN 102342914	Phospholipids	Calcipotriol	Highpressure homogenization	Improved skin keratin penetrating capacity	
Chinese patent application CN102327235	Stearic acid and Lauric acid	Cefixime	Highpressure homogenization	High drug loading amount, uniformity in grain size, and better controlled-release effects	Liao, 2012
Chinese patent application, CN102512369	Phospholipids	Glycyrrhetinic acid	Freeze-drying	Reduced dosage, enhanced the curative effect and reduced toxic and side effects of the medicine	Zhou et al., 2012
US Patent 7611733	Stearic acid	Platinum complex	Hot homogenization technique	Improvement of the therapeutic index decreased local toxicity and increased rate and extent of drug distribution	Gascom et al., 2009

which the active compound particles are present attached to the surface of the SLN, either singularly around it or as clusters (Müller et al., 2002, Pardeike et al., 2009; Schäfer-Korting et al., 2007). Not only this, the active compound may also be localized in coexisting structures, that is, micelles, liposomes, and drug nanocrystals (Svilenov and Tzachev, 2014). Figure 10.2 gives a diagrammatic representation of the drug incorporation models for SLNs.

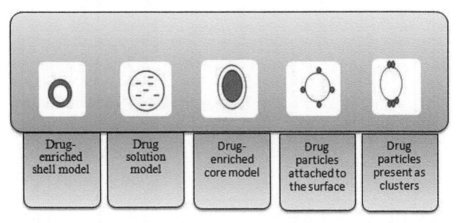

FIGURE 10.2   The three possible arrangements of drug and lipid molecules in solid LNs.

The low loading capacity of SLNs is one of its main shortcomings. The loading capacity of these lipidic carriers may be as low as 10% of the amount of lipid used, implying that the quantity of the active compound existing in the total dispersion may be limited to only 1%. The nature of the lipid core and moderate lipophilicity of active compound, account for this limiting feature of SLN (Schäfer-Korting et al., 2007).

An important point to be considered with regard to the inclusion of the active ingredient in the SLN matrix is the structural characteristics of the lipid used. API expulsion may result from the lipid matrix during crystallization (transformation to stable β form) because of unavailability of sufficient space while the crystal lattice formation takes place. This is particularly significant when the lipids used for the SLN formulation are homogeneous lipids (such as tripalmitin and tristearin). The triglycerides form perfect crystal lattice, hindering the incorporation of active moiety. On the other side, the use of mixed lipids, for example, polyacid triglycerides or mixtures of mono-, di-, and triglycerides leads to the formation of the imperfect crystal lattice. The defects present in the ensuing product allow greater amount of active compound to be incorporated in the SLN (Müller et al., 2002; Radtke and Müller, 2016; Schäfer-Korting et al., 2007).

The release of the active ingredient from SLN has been investigated by various research groups, using various model drugs. Research findings indicate that drug release is a function of production technique and composition of the nanodispersion obtained. While the former concerns the temperature conditions employed during preparation, the latter is primarily concerned with the lipid and surfactant used (Schäfer-Korting et al., 2007). Often a biphasic release profile is obtained with SLNs, showing burst release initially and prolonged release at later time points, whereas high production temperature and the use of hot homogenization method give rise to burst release profile, with the use of cold homogenization, burst release phase is absent. Likewise, high surfactant concentration also yields initial burst release of the active compound while low surfactant concentration has been found to diminish the burst. This may be expounded by the partitioning behavior of the active ingredient between the lipid and aqueous phase during heating and cooling of the hot dispersion, as a part of hot homogenization technique. As the lipid/water mix is heated, the active ingredient exhibits increased solubility in the aqueous phase, leading to the distribution of the active ingredient in favor of the aqueous phase from the molten lipid droplets. Subsequent to the homogenization step, the o/w nanodispersion is allowed to cool, as a result of which, the lipid core begins to crystallize. However, the concentration of the active moiety is relatively high at this point. As the nanodispersion is allowed to cool further, the aqueous phase gets supersaturated with the active ingredient, resulting in its redistribution into the lipidic phase. Yet, the solidification of the core in process permits the active ingredient to accumulate at the liquid outer shell alone (Müller et al., 2002).

To conclude, localization of the active ingredient in the lipid nanoparticle has an impact upon its release kinetics. Variation in production and/or formulation parameters can help in achieving modified-release products. Increasing aqueous solubility of the active ingredient yields high burst release, which is in turn favored by elevated production temperature and using surfactant in greater concentration (valid only in cases where the surfactant is responsible for solubilization of the active ingredient). Besides, the release kinetics are also affected by in vitro dissolution method employed (e.g., the active moiety may interact with the components of the dialysis bag). Moreover, the in vitro dissolution studies do not take into consideration the enzymatic degradation of the products in vivo and their interaction with cell organelles and lipid membranes. Thus, they may not provide correct picture of the in vivo release pattern (Müller et al., 2002; Svilenov and Tzachev, 2014).

10.8 CONCLUSION

Several useful aspects of SLNs have made them a promising candidate as carrier system for dermally active substances. Their occlusive properties, improved skin hydration, and penetration help in targeting the active substance to upper layers of the skin, in addition to achieving prolonged release. Ease of fabrication and industrial scale-up add to their merits, owing to which they are well-accepted by the industry. The lipids used in SLN formulation, being natural and biodegradable materials, constitute little or no health hazard. Their ability to achieve modified drug release pattern makes them distinctly suitable for delivering actives which are likely to cause skin irritation and sensitization. Perfumes and insect repellants are better delivered by incorporating into SLN based formulations. The physical stability of SLNs is a matter of concern. However, there are well-established methods for their characterization and stability assessment.

The birth of SLNs was a harbinger of a new era in the dermatological drug delivery. SLNs laid the foundation for the development of next generation of LNs, represented by nanostructured lipid carriers. Together, these nanocarriers comprise the most versatile and flexible delivery systems and have formed a niche in the skin care industry.

KEYWORDS

- lipid nanoparticles
- dermatology
- solid fatty acids
- drug
- delivery
- topical administration
- occlusion

REFERENCES

Abdel-Salam, F. S.; Elkheshen, S. A.; Mahmoud, A. A.; Ammar, H. O. Diflucortolone Valerate Loaded Solid Lipid Nanoparticles as a Semisolid Topical Delivery System. *Bull. Fac. Pharm. Cairo Univ.* **2015,** *54*(1), 1–7.

Adib, Z. M.; Ghanbarzadeh, S.; Kouhsoltani, M.; Khosroshahi, A. Y.; Hamishehkar, H. The Effect of Particle Size on the Deposition of Solid Lipid Nanoparticles in Different Skin Layers: a Histological Study. *Adv. Pharm. Bull.* **2016,** *6*(1), 31–36.

Baboota, S.; Shah, F. M.; Ali, J.; Ahuja, A. Effect of Poloxamer 188 on Lymphatic Uptake of Carvedilol-Loaded Solid Lipid Nanoparticles for Bioavailability Enhancement. *J. Drug Targeting* **2009,** *17,* 249–256.

Baroli, B. Penetration of Nanoparticles and Nanomaterials in the Skin: Fiction or Reality? *J. Pharm. Sci.* **2009,** *99,* 21–42.

Battaglia, L.; Gallarate, M.; Panciani, P. P.; Ugazio, E.; S Apino, S.; Peira, E.; Chirio, D. Techniques For The Preparation of Solid Lipid Nanoparticles. In *Application of Nanotechnology in Drug Delivery*; Sezer, A. D., ed.; *2014,* 51–70, [Online].http://www. intechopen.com/books/application-of-nanotechnology-in-drug-delivery/ techniques-for-the-preparation-of-solid-lipid-nano-and-microparticles (accessed Aug 25, 2016).

Bensouilah, J.; Buck, P. Skin Structure and Function. Aromadermatology: Aromatherapy in the Treatment and Care of Common Skin Conditions, Radcliffe Publishing House Ltd: Oxford, 2006; pp1–11.

Bhalekar, M. R.; Pokharkar, V.; Madgulkar, A.; Patil, N.; Patil, N. Preparation and Evaluation ofMiconazole Nitrate-Loaded Solid Lipid Nanoparticles for Topical Delivery. *AAPS Pharm. Sci. Tech.* **2009,** *10,* 289–296.

Braem, C.; Blaschke, T.; Panek-Minkin, G.; Herrmann, W.; Schlupp, P.; Paepenmüller, T.; Müller-Goyman, C.; Mehnert, W.; Bittl, R.; Schäfer-Korting, M.; Kramer, K. D. Interaction of Drug Molecules with Carrier Systems as Studied by Parelectric Spectroscopy and Electron Spin Resonance. *J. Controlled Release* **2007,** *119,* 128–135.

Bunjes, H.; Koch, M. H. J. Saturated Phospholipids Promote Crystallization But Slow Down Polymorphic Transitions in Triglyceride Nanoparticles. *J. Controlled Release* **2005,** *107,* 229–243.

Butani, D.; Yewale, C.; Misra, A. Topical Amphotericin B Solid Lipid Nanoparticles: Design and Development. *Colloids Surf. B* **2016,** *139,* 17–24.

Das, S.; Chaudhury, A. Recent Advances in Lipid Nanoparticle Formulations with Solid Matrix for Oral Drug Delivery. *AAPS Pharm. Sci. Tech.* **2011,** *12,* 62–76.

Dawei, G.; Dandan, D.; Liping, W. Ivermectin Solid Lipid Nanoparticle and Preparation Method Thereof. Chinese Patent 1,053,430,31, A, February 24, 2016.

Doktorovova, S.; Souto, E. B. Nanostructured Lipid Carrier-Based Hydrogel Formulations for Drug Delivery: A Comprehensive Review. *Expert Opin. Drug Delivery* **2009,** *6*(2), 165–176.

Estella-Hermoso de Mendoza, A.; Campanero, M. A.; Mollinedo, F.; Blanco-Prieto, M. J. Lipid Nanomedicines for Anticancer Drug Therapy. *J. Biomed Nanotechnol.* **2009,** *5,* 323–343.

Farboud, E. S.; Nasrollahi, S. A.; Tabbakhi, Z. Novel Formulation and Evaluation of a Q- 10 Loaded Solid Lipid Nanoparticle Cream: in Vitro and in Vivo Studies. *Int. J. Nanomed.* **2011,** *6,* 611–617.

Fumakia, M. D.; Ho, E. A. Nanoparticles Encapsulated with LL37 and SerpinA1 Promotes Wound Healing and Synergistically Enhances Antibacterial Activity. *Mol. Pharmaceutics* **2016,** *13*(7), 2318–2331.

Gainza, G.; Chu, W. S.; Guy, R. H.; Pedraz, J. L.; Hernandez, R. M.; Delgado-Charro, B.; Igartua, M. Development and in Vitro Evaluation of Lipid Nanoparticle-Based Dressings for Topical Treatment of Chronic Wounds. *Int. J. Pharm.* **2015,** *490*(1), 404–411.

Gasco, M. R. Lipid Nanoparticles: Perspectives and Challenges. *Adv. Drug Delivery Rev.* **2007,** *59,* 377–378.

Gasco, M. R.; Gasco, P.; Alberto, B. Nanoparticle Formulations of Platinum Compounds. U.S. Patent 7,611,733, B2, November 03, 2009.

Ghanbarzadeh, S.; Hariri, R.; Kouhsoltani, M.; Shokri, J.; Javadzadeh, Y.; Hamishehkar, H. Enhanced Stability and Dermal Delivery of Hydroquinone Using Solid Lipid Nanoparticles. *Colloids Surf. B* **2015,** *136,* 1004–1010.

Gilbert, E.; Roussel, L.; Serre, C.; Sandouk, R.; Salmon, D.; Kirilov, P.; Haftek, M.; Falson, F.; Pirot, F. Percutaneous Absorption of Benzophenone-3 Loaded Lipid Nanoparticles and Polymeric Nanocapsules: a Comparative Study. *Int. J. Pharm.* **2016,** *504*(1), 48–58.

Goindi, S.; Guleria, A.; Aggarwal, N. Development and Evaluation of Solid Lipid Nanoparticles of N-6-Furfuryl Adenine for Prevention of Photoaging. *J. Biomed. Nanotechnol.* **2015,** *11*(10), 1734–1746.

Grana, A.; Limpach, A.; Chauhan, H. Formulation Considerations and Applications of Solid Lipid Nanoparticles. *Am. Pharm. Rev.* **2013,** *16*(1), 19–25.

Hamishehkar, H.; Ghanbarzadeh, S.; Sepehran, S.; Javadzadeh, Y.; Adib, Z. M.; Kouhsoltani, M. Histological Assessment of Follicular Delivery of Flutamide by Solid Lipid Nanoparticles: Potential Tool for the Treatment of Androgenic Alopecia. *Drug Dev. Ind. Pharm.* **2016,** *42*(6), 846–853.

Hamishehkar, H.; Same, S.; Adibkia, K.; Zarza, K.; Shokri, J.; Taghaee, M.; Kouhsoltani, M. A. Comparative Histological Study on the Skin Occlusion Performance of a Cream Made of Solid Lipid Nanoparticles and Vaseline. *Res. Pharm. Sci.* **2015,** *10*(5), 378–387.

Hollestein, L. M.; Nijsten, T. An Insight Into Global Burden of Skin Diseases. *J. Invest. Dermatol.* **2014,** *134,* 1499–1501.

Hu, F. Q.; Hong, Y.; Yuan, H. Preparation and Characterization of Solid Lipid Nanoparticles Containing Peptide. *Int. J. Pharm.* **2004,** *273,* 29–35.

Ioualalen, K.; Raynal, R. A. Galenical System for Active Transport, Method for Preparation and Use. U.S. Patent 8,911,788, B2, December 16, 2014.

Iqbal, M. A.; Md, S.; Sahni, J. K.; Baboota, S.; Dang, S.; Ali, J. Nanostructured Lipid Carriers System: Recent Advances in Drug Delivery. *J. Drug Targeting* **2012,** *20,* 813–830.

Jain, S.; Jain, S.; Khare, P.; Gulbake, A.; Bansal, D.; Jain, S. K. Design and Development of Solid Lipid Nanoparticles for Topical Delivery of an Anti-Fungal Agent. *Drug Delivery* **2010,** *17,* 443–451.

Jain, S.; Mistry, M. A.; Swarnakar, N. K. Enhanced Dermal Delivery of Acyclovir Using Solid Lipid Nanoparticles. *Drug Delivery Transl. Res.* **2011,** *1,* 395–406.

Jain, S. K.; Chourasia, M. K.; Masuriha, R.; Soni, V.; Jain, A.; Jain, N. K.; Gupta, Y. Solid Lipid Nanoparticles Bearing Flurbiprofen for Transdermal Delivery. *Drug Delivery* **2005,** *12,* 207–215.

Jenning, V.; Mäder, K.; Gohla, S. H. Solid Lipid Nanoparticles (SLNTM) Based on Binary Mixtures of Liquid and Solid Lipids: A 1H NMR Study. *Int. J. Pharm.* **2000a,** *205,* 15–21.

Jenning, V.; Thunemann, A. F.; Gohla, S. H. Characterization of a Novel Solid Lipid Nanoparticle Carrier System Based on Binary Mixtures of Liquid and Solid Lipids. *Int. J. Pharm.* **2000b,** *199,* 167–177.

Jenning, V.; Gysler, A.; Schäfer-Korting, M.; Gohla, S. H. Vitamin A Loaded Solid Lipid Nanoparticles for Topical Use: Occlusive Properties and Drug Targeting to the Upper Skin. *Eur. J. Pharm. Biopharm.* **2000c,** *49,* 211–218.

Jenning, V.; Schäfer-Korting, M.; Gohla, S. Vitamin A Loaded Solid Lipid Nanoparticles for Topical Use: Drug Release Properties. *J. Controlled Release* **2000d,** *66,* 115–126

Jeon, H. S.; Seo, J. E.; Kim, M. S.; Kang, M. H.; Oh, D. H.; Jeon, S. O.; Jeong, S. H.; Choi, Y. W.; Lee, S. A Retinyl Palmitate-Loaded Solid Lipid Nanoparticle System: Effect of Surface Modification with Dicetyl Phosphate on Skin Permeation in Vitro and Anti-Wrinkle Effect in Vivo. *Int. J. Pharm.* **2013,** *452,* 311–320.

Kakadia, P. G.; Conway, B. R. Solid Lipid Nanoparticles: A Potential Approach for Dermal Drug Delivery. *AJPS* **2014,** *2,* 1–7.

Kakkar, V.; Singh, S.; Singla, D.; Kaur, I. P. Exploring Solid Lipid Nanoparticles to Enhance the Oral Bioavailability of Curcumin. *Mol. Nutr. Food Res.* **2011,** *55,* 495–503.

Kalam, M. A.; Sultana, Y.; Ali, A.; Aqil, M.; Mishra, A. K.; Chuttani, K. Preparation, Characterization, and Evaluation ofGatifloxacin Loaded Solid Lipid Nanoparticles as Colloidal Ocular Drug Delivery System. *J. Drug Targeting* **2010,** *18,* 191–204.

Kaur, A.; Goindi, S.; Katare, O. P. Formulation, Characterisation and in Vivo Evaluation ofLipid-Based Nanocarrier for Topical Delivery of Diflunisal. *J. Microencapsulation* **2016,** *33*(5), 475–486.

Khameneh, B.; Halimi, V.; Jaafari, M. R.; Golmohammadzadeh, S. Safranal-Loaded Solid Lipid Nanoparticles: Evaluation of Sunscreen and Moisturizing Potential for Topical Applications. *Iran. J. Basic Med. Sci.* **2015,** *18*(1), 58–63.

Korkm, E.; Gokce, E. H.; Ozer, O. Development and Evaluation of Coenzyme Q10 Loaded Solid Lipid Nanoparticle Hydrogel for Enhanced Dermal Delivery. *Acta Pharm.* **2013,** *63*(4), 517–529.

Küchler, S.; Herrmann, W.; Panek-Minkin, G.; Blaschke, T.; Zoschke, C.; Kramer, K. D.; Bittl, R.; Schäfer-Korting, M. Sln for Topical Application in Skin Diseases-Characterization of Drug-Carrier and Carrier-Target Interactions. *Int. J. Pharm.* **2010,** *390,* 225–233.

Lasoń, E.; Ogonowski, J. Solid Lipid Nanoparticles- Characteristics, Application and Obtaining. *CHEMIK* **2011,** *65*(10), 960–967.

Lee, M. K.; Lim, S. J.; Kim, C. K. Preparation, Characterization and in Vitro Cytotoxicity of Paclitaxel Loaded Sterically Stabilized Solid Lipid Nanoparticles. *Biomaterials* **2007,** *28,* 2137–2146.

Lee, S. J.; Kim, Y. H.; Lee, S. H.; Kim, K. S. OxaliplatinNanoparticles and Method for Preparing Same. U. S. Patent 9,393,201, B1, July 19, 2016.

Liao, A. Solid Cefixime Lipid Nanoparticle Preparation. Chinese Patent 102,327,235, A, January 25, 2012.

Lippacher, A.; Müller, R. H.; Mäder, K. Liquid and Semisolid Sln Dispersions for Topical Application: Rheological Characterization. *Eur. J. Pharm. Biopharm.* **2004,** *58*(3), 561–567.

Liu, J.; Gong, T.; Wang, C.; Zhong, Z.; Zhang, Z. Solid Lipid Nanoparticles Loaded with Insulin by Sodium Cholate-Phosphatidylcholine-Based Mixed Micelles: Preparation and Characterization. *Int. J. Pharm.* **2007a,** *340,* 153–162.

Liu, J.; Hu, W.; Chen, H.; Ni, Q.; Xu, H.; Yang, X. Isotretinoin-loaded Solid Lipid Nanoparticles with Skin Targeting for Topical Delivery. *Int. J. Pharm.* **2007b,** *328,* 191–195

Liyong, J.; Xiaoying, Y.; Yongzhong, G. Saquinavir Phospholipid Compound Solid Lipid Nanoparticle and Preparation Method Thereof. Chinese Patent 104, 771, 383, March 25, 2015.

Lohan, S. B.; Bauersachs, S; Ahlberg, S.; Baisaeng, N.; Keck, C. M.; Müller, R. H.; Witte, E.; Wolk, K.; Hackbarth, S.; Röder, B.; Lademann, J. Ultra-Small Lipid Nanoparticles Promote the Penetration of Coenzyme Q10 in Skin Cells and Counteract Oxidative Stress. *Eur. J. Pharm. Biopharm.* **2015,** *89,* 201–207.

López-García, R.; Ganem-Rondero, A. Solid Lipid Nanoparticles (Sln) and Nanostructured Lipid Carriers (Nlc): Occlusive Effect and Penetration Enhancement Ability. *J. Cosmet. Dermatol. Sci. Appl.* **2015**, *5*, 62–72.

Lv, Q.; Yu, A.; Xi, Y.; Li, H.; Song, Z.; Cui, J.; Cao, F.; Zhai, G. Development and Evaluation of Penciclovir-Loaded Solid Lipid Nanoparticles for Topical Delivery. *Int. J. Pharm.* **2009**, *372*, 191–198.

Majumdar, S.; Samanta, S. K.; Sarkar, S. Acyclovir Loaded Solid Lipid NanoparticulateGel Formulation for Topical Application and Method of Formulating the Same. Indian Patent Application IN26KO, 20 2014.

McGrath, A.; Uitto, J. Anatomy and Organization of Human Skin. In *Rook's Textbook of Dermatology*; Burns, T.,Breathnach, S.,Cox, N.,Griffiths, C., Ed.; Wiley- Backwell: UK, 2010; pp3.1–3.15.

Mehnert, W.; Mäder, K. Solid Lipid Nanoparticles Production, Characterization and Applications. *Adv. Drug Delivery Rev.* **2001**, *47*, 165–196.

Mei, Z.; Wu, Q.; Hu, S.; Li, X.; Yang, X. TriptolideLoaded Solid Lipid Nanoparticle Hydrogel for Topical Application. *Drug Delivery. Ind. Pharm.* **2005**, *31*, 161–168.

Müller, R. H. Lipid Naoparticles: Recent Advances. *Adv. Drug Delivery Rev.* **2007**, *59*, 375–376.

Müller, R. H.; Mader, K.; Gohla, S. Solid Lipid Nanoparticles (Sln) for Controlled Drug Delivery-A Review of the State of the Art. *Eur. J. Pharm. Biopharm.* **2000**, *50*, 161–177.

Müller, R. H.; Petersen, R. D.;Hommoss, A.;Pardeike, J. Nanostructured Lipid Carriers (Nlc) in Cosmetic Dermal Products. *Adv. Drug Delivery Rev.* **2007**, *59*(6), 522–530.

Müller, R. H.; Radtke, M.; Wissing, S. A. Solid Lipid Nanoparticles (Sln) and Nanostructured Lipid Carriers (Nlc) in Cosmetic and Dermatological Preparations. *Adv. Drug Delivery Rev.* **2002**, *54*, S131–S155.

Müller, R. H.; Runge, S.; Ravelli, V.; Mehnert, W.; Thunemann, A. F.; Souto, E. B. Oral Bioavailability of Cyclosporine: Solid Lipid Nanoparticles (Sln) Versus Drug Nanocrystals. *Int. J. Pharm.* **2006**, *317*, 82–89.

Müller, R. H.; Runge, S. A.; Ravelli, V.; Thunemann, A. F.; Mehnert, W.; Souto, E. B. Cyclosporine-Loaded Solid Lipid Nanoparticles (Sln): Drug-Lipid Physicochemical Interactions and Characterization ofDrug Incorporation. *Eur. J. Pharm. Biopharm.* **2008**, *68*, 535–544.

Omwoyo, W. N.; Ogutu, B.; Oloo, F.; Swai, H.; Kalombo, L.; Melariri, P.; Mahanga, G. M.; Gathirwa, J. W. Preparation, Characterization, and Optimization ofPrimaquine-Loaded Solid Lipid Nanoparticles. *Int. J. Nanomed.* **2014**, *9*, 3865–3874.

Pardeike, J.; Hommoss, A.; Müller, R. H. Lipid Nanoparticles (Sln, Nlc) in Cosmetic and Pharmaceutical Dermal Products. *Int. J. Pharm.* **2009**, *366*(1), 170–184.

Puglia, C.; Bonina, F. Lipid Nanoparticles as Novel Delivery Systems for Cosmetics and Dermal Pharmaceuticals. *Expert Opin. Drug Delivery* **2012**, *9*(4), 429–441.

Qingri, C.; Lili, S.; Yue, C.; Jinghao, C. Docetaxel Solid Lipid Nanoparticle Lyophilized Preparation and Preparationmethod. Chinese Patent 104586817, May 6, 2015.

Radomska-Soukharev, A. Stability of Lipid Excipients in Solid Lipid Nanoparticles. *Adv. Drug Delivery Rev.* **2007**, *59*, 411–418.

Radtke, M.; Müller, R. H. New Drugs: Nanotechnology. Nanostructured Lipid Drug Carriers. http://citeseerx.ist.psu.edu/viewdoc/download?doi=10.1.1.477.326&rep=rep 1&type=pdf (accessed Oct 24, 2016).

Raj, R.; Mongia, P.; Ram, A.; Jain, N. K. Enhanced Skin Delivery ofAceclofenac Via Hydrogel-Based Solid Lipid Nanoparticles. *Artif. Cells Nanomed. Biotechnol.* **2015**, *44*(6), 1434–1439.

Raza, K.; Singh, B.; Singal, P.; Wadhwa, S.; Katare, O. P. Systematically Optimized Biocompatible Isotretinoin-Loaded Solid Lipid Nanoparticles (Slns) for Topical Treatment of Acne. *Colloids Surf. B Biointerfaces* **2013**, *105*, 67–74.

Ren, S. Paclitaxel Lipid Nanoparticle Preparation Method. Chinese Patent 1,053,116,43, 2016.

Ridolfi, D. M.; Marcato, P. D.; Justo, G. Z.; Cordi, L.; Machado, D.; Durán, N. Chitosan-Solid Lipid Nanoparticles as Carriers for Topical Delivery ofTretinoin. *Colloids Surf. B Biointerfaces* **2012**, *93*, 36–40.

Sachdeva, M. S.; Patlolla, R. Nanoparticle Formulations for Skin Delivery. U.S. Patent 8715736, B2, May 6, 2014.

Sahana, B.; Santra, K.; Basu, S.; Mukherjee, B. Development of Biodegradable Polymer Based Tamoxifen Citrate Loaded Nanoparticles and Effect of Some Manufacturing Process Parameters on Them: A Physicochemical and in Vitro Evaluation. *Int. J. Nanomed.* **2010**, *5*, 621–630.

Schäfer-Korting, M.; Mehnert, W.; Korting, H. C. Lipid Nanoparticles for Improved Topical Application ofDrugs for Skin Diseases. *Adv. Drug Delivery Rev.* **2007**, *59*(6), 427–443.

Schlupp, P.; Blaschke, T.; Kramer, K. D.; Holtje, H. D.; Mehnert, W.; Korting Schafer, M. Drug Release and Skin Penetration from Solid Lipid Nanoparticles and a Base Cream: a Systematic Approach from a Comparison ofThree Glucocorticoids. *Skin Pharmacol. Physiol.* **2011**, *24*, 199–209.

Shah, K. A.; Date, A. A.; Joshi, M. D.; Patravale, V. B. Solid Lipid Nanoparticles (Sln) ofTretinoin: Potential in Topical Delivery. *Int. J. Pharm.* **2007**, *345*, 163–171.

Shah, R. M.; Eldridge, D. S.; Palombo, E. A.; Harding, I. H. Microwave-Assisted Formulation of Solid Lipid Nanoparticles Loaded with Non-Steroidal Anti-Inflammatory Drugs. *Int. J. Pharm.* **2016**, *515*(1), 543–554.

Shahgaldian, P.; Da Silva, E.; Coleman, A. W.; Rather, B.; Zaworotko, M. J. Para-Acyl-Calix-Arene Based Solid Lipid Nanoparticles (Slns): A Detailed Study of Preparation and Stability Parameters. *Int. J. Pharm.* **2003**, *253*, 23–38.

Shanchun, W.; Xiquan, Z.; Hongmei, G.; Fan Hengfeng, F.; Chang, T.; Shanchun, W.; Huangyan, Y. EntecavirSolid Lipid Nanoparticle and Preparation Method Thereof. Chinese Patent 105,560,216, 2016.

Shidhaye, S. S.; Vaidya, R.; Sutar, S.; Patwardhan, A.; Kadam, V. J. Solid Lipid Nanoparticles and Nanostructured Lipid Carriers- Innovative Generations of Solid Lipid Carriers. *Curr. Drug Delivery* **2008**, *5*, 324–331.

Shindo, Y.; Witt, E.; Han, D.; Epstein, W.; Packer, L. Enzymic and Non-Enzymic Antioxidants in Epidermis and Dermis ofHuman Skin. *J. Invest. Dermatol.* **1994**, *102*(1), 122–124.

Silva, A. C.; González-Mira, E.; García, M. L.; Egea, M. A.; Fonseca, J.; Silva, R.; Santos, D.; Souto, E. B.; Ferreira, D. Preparation, Characterization and Biocompatibility Studies on Risperidone-Loaded Solid Lipid Nanoparticles (Sln): High Pressure Homogenization Versus Ultrasound. *Colloids Surf. B Biointerfaces* **2011**, *86*, 158–165.

Silva, E. L.; Carneiro, G.; de Araujo, L. A.; de Jesus, M.; Trindade, V.; Yoshida, M. I.; Orefice, R. L.; de MacedoFarias, L.; de Carvalho, M. A. R.; Santos, S. G. D.; Assis, G. Solid Lipid Nanoparticles Loaded with Retinoic Acid and Lauric Acid as an Alternative for Topical Treatment ofAcne Vulgaris. *J. Nanosci. Nanotechnol.* **2015**, *15*(1), 792–799.

So, J. W.; Kim, S.; Park, J. S.; Kim, B. H.; Jung, S. H.; Shin, S. C.; Cho, C. W. Preparation and Evaluation ofSolid Lipid Nanoparticles with Jsh18 for Skin-Whitening Efficacy. *Pharm. Dev. Technol.* **2010**, *15*(4), 415–420.

Souto, E. B.; Müller, R. H. Cosmetic Features and Applications of Lipid Nanoparticles (Sln®, Nlc®). *Int. J. Cosmet. Sci.* **2008**, *30*(3), 157–165.

Svilenov, H.; Tzachev, C. Solid Lipid Nanoparticles- a Promising Drug Delivery System. In *Nanomedicine*; Sefalian, P. A., Mel, A. D., Kalaskar, D. M., Eds.; One Central Press: United Kingdom 2014; pp187–237.

Tag, K. Y.; Prakash, R. A Lipophilic Drug Loaded Solid Lipid Nanoparticle Surface Modified with N-Trimethyl Chitosan. Korean Patent 20,150,134,443, 2016.

Tang, X.; Guan, T.; Zhang, Y.; Yaxuan, W. Nimodipine Freeze Dried Solid Lipid Nanoparticle and Preparation Method Thereof. Chinese Patent 102,552,156, A, July 11, 2013.

Teeranachaideekul, V.; Souto, E. B.; Junyaprasert, V. B.; Müller, R. H. Cetyl Palmitate-Based Nlc for Topical Delivery ofCoenzyme Q(10)–Development, Physicochemical Characterization and in Vitro Release Studies. *Eur. J. Pharm. Biopharm.* **2007,** *67*, 141–148.

Üner, M. Preparation, Characterization and Physico-Chemical Properties of Solid Lipid Nanoparticles (Sln) and Nanostructured Lipid Carriers (Nlc): Their Benefits as Colloidal Drug Carrier Systems. *Pharmazie* **2006,** *61*, 375–386.

Üner, M.; Wissing, S. A.; Yener, G.; Müller, R. H. Skin Moisturizing Effect and Skin Penetration of Ascorbyl Palmitate Entrapped in Solid Lipid Nanoparticles (Sln) and Nanostructured Lipid Carriers (Nlc) Incorporated Into Hydrogel. *Pharmazie* **2005,** *60*, 751–755.

Varshosaz, J.; Minayian, M.; Moazen, E. Enhancement of Oral Bioavailability of Pentoxifylline by Solid Lipid Nanoparticles. *J. Liposome Res.* **2010,** *20*, 115–123.

Venkateswarlu, V.; Manjunath, K. Preparation, Characterization and in Vitro Release Kinetics of Clozapine Solid Lipid Nanoparticles. *J. Controlled Release* **2004,** *95*, 627–638.

Vijayan, V.; Aafreen, S.; Sakthivel, S.; Reddy, K. R. Formulation and Characterization of Solid Lipid Nanoparticles Loaded Neem Oil for Topical Treatment of Acne. *J. Acute Dis.* **2013,** *2*(4), 282–286.

Walters, M. A. Lanthanoid Complex Capsule and Particle Contrast Agents, Methods of Making and Using Thereof. U.S. Patent 8,865,129, B2, October 21, 2014.

Weidong, C.; Lei, W.; Guomi, F.; Xianwen, Z.; Tingting, Z.; Tong, L.; Qing, L.; Xiaolei, C. Allyl Isothiocyanate Solid Lipid Nanoparticle and Preparation Method Thereof. Chinese Patent CN103446076,A, December 18, 2013.

Westesen, K.; Siekmann, B.; Koch, M. H. J. Investigations on the Physical State of Lipid Nanoparticles by Synchrotron Radiation X-Ray Diffraction. *Int. J. Pharm.* **1993,** *93*, 189–199.

Wissing, S. A.; Lippacher, A.; Müller, R. H. Investigations on the Occlusive Properties ofSolid Lipid Nanoparticles (Sln). *J. Cosmet. Sci.* **2001,** *52*, 313–324.

Wissing, S. A.; Müller, R. H. A Novel Sunscreen System Based on TocopherolAcetate Incorporated into Solid Lipid Nanoparticles. *Int. J. Cosmet Sci.* **2001a,** *23*, 233–243

Wissing, S. A.; Müller, R. H. Solid Lipid Nanoparticles (SLN)- a Novel Carrier for UV Blockers. *Pharmazie* **2001b,** *53*, 783–786

Wissing, S. A.; Müller, R. H.; Manthei, L.; Mayer, C. Structural Characterization of Q10-Loaded Solid Lipid Nanoparticles by NMR Spectroscopy. *Pharm. Res.* **2004,** *21*(3), 400–405.

Wissing, S. A.; Müller, R. H. The Influence ofSolid Lipid Nanoparticles on Skin Hydration and Viscoelasticity-In Vivo Study. *Eur. J. Pharm. Biopharm.* **2003,** *56*, 67–72.

Xie, S.; Pan, B.; Wang, M.; Zhu, L.; Wang, F.; Dong, Z.; Wang, X.; Zhou, W. Formulation, Characterization and Pharmacokinetics of Praziquantel-Loaded Hydrogenated Castor Oil Solid Lipid Nanoparticles. *Nanomedicine* **2010,** *5*, 693–701.

Xie, S.; Zhu, L.; Dong, Z.; Wang, X.; Wang, Y.; Li, X.; Zhou, W. Preparation, Characterization and Pharmacokinetics of Enrofloxacin-Loaded Solid Lipid Nanoparticles: Influences of Fatty Acids. *Colloids Surf. B Biointerfaces* **2011,** *83*, 382–387.

Yoon, G.; Park, W. J.; Yoon, I. Solid Lipid Nanoparticles (Slns) and Nanostructured Lipid Carriers (Nlcs): Recent Advances in Drug Delivery. *J. Pharm. Invest.* **2013,** *43*, 353–362.

Yukuyama, M. N.; Ghisleni, D. D. M.; Pinto, T. G. A.; Chacra- Bou, N. A. Nanoemulsion: Process Selection and Application in Cosmetics- a Review. *Int. J. Cosmet. Sci.* **2015,** *38,* 13–24.

Zhang, J.; Fan, Y.; Smith, E. Experimental Design for the Optimization of Lipid Nanoparticles. *J. Pharm. Sci.* **2009,** *98,* 1813–1819.

Zhou, X.; Wang, J.; Xianming, Hu.; Xiaoju, Z.; Jiong, W.; Xianming, H. Glycyrrhetinic Acid Solid Lipid Nanoparticles and Preparation Method for Same. Chinese Patent CN102512369, A, June 27, 2012.

Zimmermann, E.; Souto, E. B.; Müller, R. H. Physicochemical Investigations on the Structure of Drug-Free and Drug Loaded Solid Lipid Nanoparticles (SLNTM) by Means of Dsc and 1H NMR. *Pharmazie* **2005,** *60,* 508–513.

PART IV
Newer Nanoarchitectures

CHAPTER 11

DENDRIMERS IN GENE DELIVERY

BHUPINDER KAUR, SURYA PRAKASH GAUTAM*, RANJIT SINGH, and NARINDER KUMAR

CT Institute of Pharmaceutical Sciences, Shahpur Campus, Jalandhar, India

Corresponding author. E-mail: suryagautam@ymail.com; gautamsuryaprakash@gmail.com

ABSTRACT

Dendrimers acclaimed its fascinating position in the nanoworld. By virtue of its unique polymeric architecture, it exhibits precise compositional and constitutional properties. The combination of a discrete number of functionalities and their high local densities make dendrimers as multifunctional platforms for amplified substrate binding. As a result of their unique architecture and construction, dendrimers possess inherently valuable physical, chemical, and biological properties. Various new methods are developed for the synthesis of dendrimers. Varieties of dendrimers are synthesized to develop dendrimer-based formulations. Especially in drug delivery, they have contributed significantly. Dendrimers-based formulations are significantly contributing in the market and many more products are in pipeline.

11.1 INTRODUCTION

Dendrimers have gained paramount importance as a drug delivery system, especially for gene delivery. Exclusive physical, chemical, and biological properties make them attractive materials for the development of nanomedicines. Key properties such as defined size, shape, molecular weight (MW), high solubility, miscibility, high ratio of multivalent surface moieties to molecular volume and their compatibility with DNA, heparin, and polyanions make these nanoscaled materials highly interesting for the development of their applications in drug delivery, gene therapy, and chemotherapy (Stiriba et al., 2002); more precisely; they are referred to as modern-day polymers and offer much extensive properties in comparison to conventional polymers. In gene delivery, dendrimer is used as synthetic (nonviral) vectors and for the development of such vectors, there is a need to make a link between dendrimer structure and the morphology and physicochemistry of the respective DNA to form excellent complexes and, furthermore, to get efficient biological performance of these nano-structure systems at the cellular and systemic level, which are also key factors for their success in gene delivery (Boas et al., 2004; Gillies et al., 2005). This chapter focuses on the basic mechanisms of gene delivery and dendrimer-based gene delivery. Currently, dendrimer-based transfection system has become routine tool for many researchers but therapeutic delivery of modified gene to nucleic acids (NAs) remains a challenge due to some barriers.

11.1.1   GENE DELIVERY

The process of safely and efficiently transferring foreign DNA into host cells using some transfection systems is not only used to treat genetic diseases but also to produce large quantities of secreted proteins for the direct therapeutic application. Nowadays, gene delivery is a fundamental goal in research with objective to overcome two major constraints (reduce compliance related to) such as safety and efficacy of gene transfer and focusing more to understand mechanism for DNA stability and enhance DNA uptake (Mulligan et al., 1993; Crystal, 1995; Anderson et al., 1998). Current methods for gene transfer in vitro and in vivo include three major groups: viral (transduction), physical (direct microinjections) and chemical methods (synthetic nonviral system).

11.1.2   GENE THERAPY

A variety of genetic mutation cause gene defect and alter cellular system of organism, which result in genetic deficiencies and abnormalities, such as cancer, acquired immune deficiency syndrome, cardiovascular diseases, and certain autoimmune disorders that remain resistant to established traditional treatments (Fig. 11.1). In spite of that, gene therapy is considerable because of its significant potential for the treatment of inherited and acquired life-threatening diseases by introducing new genetic coding, repairing of abnormal gene through selective reverse mutation or change the regulation for gene pairs for functional proteins of cells so as to normalize the cells by removing defective genes that are responsible for disease (Pearson et al., 2004; Knoell et al., 1998; Lehrman et al., 1999).

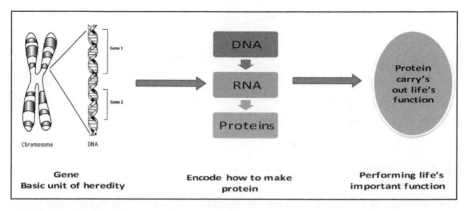

FIGURE 11.1 The role of gene in gene therapy.

11.1.3 GENE THERAPY STRATEGIES

The delivery of therapeutic NAs or DNA, normally in the form of plasmids, but increasingly also as smaller oligomers, remains one of the major obstacles currently hampering the further exploitation of genetic therapies. Specific and efficient delivery of genetic material to diseased sites and to particular cell populations is addressed using a variety of methods such as follows:

1. Viral (transduction) method
2. Physical (direct microinjection) method
3. Chemical (nanocarrier and dendrimers) method

Viral-mediated delivery has limitations such as problem of toxicity, restricted targeting, limited DNA-carrying capacity, which necessitate the development of some advanced strategies. Toxicity and immunogenicity of viral-mediated delivery hamper their use in basic research (Fig. 11.2).

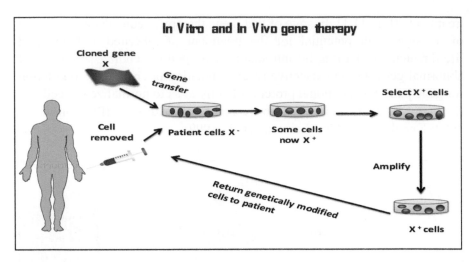

FIGURE 11.2   In vitro and in vivo gene therapy.

11.1.3.1   VIRAL METHOD

This method is further based on the insertion of modified functional gene into viral vector to penetrate in the host genome. Once the encapsulated viral vector enters the cell, the degradation of encapsulated membrane starts and this will release the viral vector that further release the modified gene into the nucleus and switch on the instructions that are necessary for the cell to synthesize the protein that was previously missed or altered as shown in Figure 11.3 (Lachmann et al., 1997). However, unfortunately, due to some adverse effects such as immunogenicity, difficulties in handling large-scale production, and limited length of the genes, this system was not successful (Gardlik et al., 2005; Lundstrom et al., 2003). Moreover, potential and real risks of some adenoviruses were also observed during the clinical trial.

11.1.3.2   PHYSICAL METHOD

Physical methods for gene transfer include biolistics, jet injection, hydrodynamic injections, ultrasound, and electroporation method. In this method, genes are directly inserted into cytosol of both small and large NA molecules, as well as any other non-permeable molecule with the help of electric impulse, fine needle, or high-pressure gas (Fig. 11.4),

11.1.4   BARRIERS TO GENE DELIVERY

There are two major barriers or constraints to get efficient gene delivery: first, target delivery of NA into cell without potential risks; second, efficient penetration of the NAs through the plasma membrane into the cell (Jin et al., 2014). The observation shows that naked plasmid DNA should normally be degraded in the systemic circulation to avoid the degradation of the packaged plasmid DNA. This packaging occurs with the help of a delivery system, which protects the NA from degradation. Furthermore, the nonviral gene delivery system should help to target the therapeutic NA to the desired site of action and facilitate efficient intracellular transfer, typically to the nucleus (Fischer et al., 1999). By using this method, the most common strategy employed for the packaging of DNA is based on the electrostatic interaction between the negative charge of NA and the positive charges of the synthetic vector, which will form complex and condense the NA into nanoparticles. Various cationic lipids or polymers are commonly used as synthetic vector and, these particles have also been termed as lipoplex, polyplex, or dendriplex when dendrimers are being used (Eichman et al., 2000; Gao and Huang, 2009; Tomalia et al., 2003). The suspension of these particles is going to be stable colloidally if the particles are charged and cationic carrier possesses charge in excess; this will help in the repulsion of particles and make suspension stable. Furthermore, this positive charge is also helpful to facilitate cell adsorption and to mediate better endosome uptake into cells. However, its nonspecific nature shows conflicts between in vivo and in vitro experiments.

11.2   DENDRIMERS

Nowadays, in comparison to the traditional polymers, dendrimers (from the Greek dendron: tree and meros: part) are gaining ground in research due to their specific and distinct properties, which differentiate them from other polymers (Karak et al., 1997). In actual, dendrimers consist of a central core molecule which acts as the root from which a number of highly branched, treelike arms originate, known as arborols (Fig. 11.6). The dendritic structure of dendrimers was originated from the class of polymer known as cascade. Furthermore, the chemistry of dendrimers is quite adaptable to facilitating the synthesis of a broad range of molecules with different functionality (Cloninger, 2002). Key properties in terms of the potential use of these materials in drug and gene delivery are due to the high density of terminal groups. These contribute to the surface modification of the molecules and offer multiple sites for attachment and complexation that may be responsible for the solubilization of poorly soluble drugs.

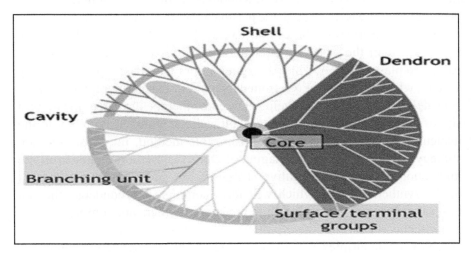

FIGURE 11.6   Structure of dendrimers.

11.2.1   STRUCTURES AND SYNTHESIS

Dendrimers are well-defined molecular architecture based on stepwise synthesis using either a divergent or a convergent method (Kojima et al., 2000; Twyman et al., 1999).

11.2.1.1   DIVERGENT APPROACH

In the divergent method, the dendrimer synthesis starts from the multi-functional core to build up one monomer generation; furthermore, the core molecule reacts with monomer molecules having one reactive group and two or more inactive groups. Then, reactive group reacts with core molecule of first-generation dendrimer. Active functional groups on the periphery undergo reaction with the repeating monomer in a systematic fashion, leading to the formation of next-generation dendrimers. Ultimately, steric effect prevents the further availability of functional groups to couple with the monomers (starburst effect).

11.2.1.2   CONVERGENT APPROACH

In the convergent method, the dendrimer is also built up layer after layer as in divergent method, but in this, it starts from the end groups and terminates at the core. The first exploration of dendrimers as molecules for gene delivery was focused on the PAMAM dendrimers (Devarakonda et al., 2004; Iwashita et al., 2012). The PAMAM dendrimers are based on an ethylenediamine or ammonia core with four and three branching points, respectively. Using a divergent approach, the molecule is built up iteratively from the core through the addition of methyl acrylate followed by amidation of the resulting ester with ethylenediamine. Each complete reaction sequence results in a new full dendrimer generation (e.g., G3, G4...) with terminal amine functionality, whereas the intermediate half generations (e.g., G2.5, G3.5...) terminate in anionic carboxylate groups (Fig. 11.7). The commercial availability and relative efficiency of PAMAM dendrimers have meant that these materials and their derivatives currently dominate the area of gene delivery with dendritic polymers.

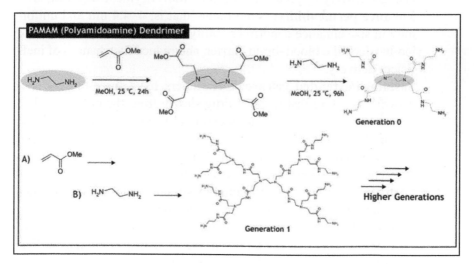

FIGURE 11.7   Synthesis of polyamidoamine (PAMAM) dendrimers.

11.2.2   CLASSIFICATION OF DENDRIMERS

The different categories of dendrimers are presented in Table 11.1.

TABLE 11.1   Classification of Dendrimers.

Chemical classification	Physical classification	Miscellaneous
1. Polyamidoamine (PAMAM) dendrimers	Simple dendrimers	Dendrophanes
2. Polypropyleneimine (PPI) dendrimers	Liquid crystalline dendrimers.	Metallodendrimers
3. polyether (PE) dendrimers	Chiral dendrimers	Polyamino phosphine
4. L-lysine-based dendrimers	Micellar dendrimers	Dendritic box
5. Phenylacetylene dendrimers	Hybrid dendrimers	Carbohydrate vaccine dendrimers

11.2.3   INTRINSIC PROPERTIES OF DENDRIMERS

1. The intrinsic properties of dendrimers are as follows: **Target delivery and reduce drug degradation:** Due to their complex, bilayer lipid structure, dendrimers form encapsulated form/structure that further helps in target delivery by protecting the drug from enzymatic degradation.
2. **Enhance permeability:** Nano range, that is, 4–4.4 nm and uniformity in size enhance dendrimer ability to cross cell membranes and bio-barriers like blood–brain barrier, and reduce the chances of inefficient delivery.
3. **Sustained/extended release:** Dendrimers have three-dimensional network structure that releases drug slowly from the polymers. The extension in circulation time is essential to produce desired therapeutic effects.
4. **High solubility:** Dendrimers' solubilizing property exerts due to ionic interaction, hydrogen bonding, and hydrophobic interaction mechanism.
5. **High stability:** Dendrimers' complex structure and surface moieties make it more stable as compared to other drugs.
6. **Drug-loading capacity:** Due to their unique structures, drug can get entrapped inside the internal cavities by encapsulation as well as electrostatically in the surface of dendrimers.
7. **Multifunctional ability:** Free surface groups can form complex with drug molecules or ligands by using cross-linking method. The surface of dendrimers offers the platform to link drug, ligand, imaging agent, dye, and may lead to the formation of smart drug delivery.
8. **Low toxicity:** Most dendrimers systems display very low cytotoxicity levels due to their ability to deliver drug at target site rather than no-specific site.

9. **Less adverse effects:** Experimental investigation has disclosed that dendrimers have low or negligible immunogenic response (Schaffert et al., 2012; Choi et al., 2004; Anderson et al., 2005; Hwang et al., 2001).

11.2.4   IN VITRO GENE DELIVERY USING DENDRIMERS

The initial in vitro observation of biological properties of PAMAM dendrimer shows nontoxicity. In cytotoxicity studies, they were compared favorably with other transfection agents, in particular, cationic polymers of higher MW such as PEI (600–1000 kDa), PLL (36.6 kDa), or DEAE–dextran (500 kDa), which are three times more toxic. Lactate dehydrogenase (LDH) release or hemolysis assays reflect that PAMAM dendrimers are less toxic in contrast to large MW PEI and PLL polymers. Nevertheless, cellular uptake of charged DNA complexes is possible due to electrostatic interactions of cationic polymer and anionic cell surfaces of dendrimers. The interaction with the membrane may cause the formation of tiny holes, and there is the possibility of erythrocyte lysis to a very small extent, which is further generation number and type specific. Size is a key factor of dendrimer cytotoxicity for both PAMAM and PPI dendrimers. Cytotoxicity of PAMAM dendrimers depends on increases in generation size, independence of surface charge, nature and density of charged groups, and so forth (Gosselin et al., 2001; Breunig et al., 2007; Dobrovolskaia et al., 2012; Shah et al., 2011). Cationic (surface) charges are in general more toxic but details depend on the specific groups involved, like, for amines, it has been proposed that primary amines are relatively more toxic than secondary or tertiary amines. NH_2 groups displayed on the outer layer of dendrimers may cause the concentration-dependent hemolysis and changes in erythrocyte morphology. Quaternization technique has previously been used to reduce the toxicity of different cationic polymers. Another technique, shielding of surface groups, has also been used to reduce toxicity, for example, covalent attachment of C12 lauroyl groups or PEG 2000 on dendrimer surface. For higher-generation dendrimers, the modification of terminal groups and higher density of nontoxic surface groups may also be beneficial to prevent a potential toxic core.

11.2.5   IN VITRO GENE DELIVERY USING DENDRIMERS

On intravenous injection of I^{125}-labeled cationic PAMAM dendrimers (G3, G4), these are rapidly eliminated from the circulation (around 99% in 1 h) and accumulate in the liver (more than 60%). Accumulation of anionic half-generation PAMAM dendrimers (2.5, 3.5, and 5.5) in the liver was reported (Daneshvar et al., 2013). Earlier observations depict that kidney retention of PAMAM dendrimers may increase with size and charge density. In general, the toxicity of cationic polymers bound to DNA decreases in in vitro assays but the particularly complex nature have a significant influence on their biodistribution, for example, their involvement with enhanced permeation and retention effect to target tumors.

11.3 DENDRIMERS AS SYNTHETIC VECTORS

11.3.1 DENDRIMER DNA COMPLEX

The complexation process of other cationic polymers with high-charge density dendrimers and dendrimer–NA complexes does not seem to differ fundamentally. Dendrimers interact with various forms of NAs, such as plasmid DNA or antisense oligonucleotides to form complexes which protect the NA from enzymatic degradation. Electrostatic interaction is responsible for forming a complex between dendrimers and NA and it lacks sequence specificity. Extended configuration of plasmid DNA is changed and a more compact configuration is achieved due to the formation of NA–dendrimer complexes (dendriplexes). The nature of the complex is dependent on number of factors such as stoichiometry and concentration of the DNA phosphates and dendrimer amines, bulk solvent properties (e.g., pH, salt concentration, and buffer strength), and even on dynamics of mixing. High ionic strength of NaCl interferes with the binding process but also appears to help to establish equilibrium (Fig. 11.8).

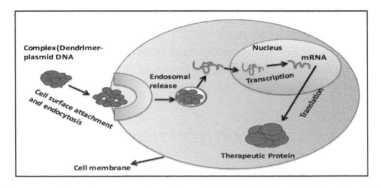

FIGURE 11.8   Schematic presentation of gene delivery process of polymer/dendrimers.

11.4 VARIOUS EXPERIMENTAL GENE THERAPIES

The ability of nonviral systems such as Superfect™ to efficiently transfect various cells in vitro has made synthetic vectors gene delivery a routine tool in research today. However, they have had little impact of genetic therapies into the clinic to date. It still remains a significant challenge for synthetic vectors to make valid predictions in in vivo studies. When one discovers the increase in complexity of the system that is being introduced by the range of possible interactions between the array of biological macromolecules and cells, this is not really surprising. Consequently, intrinsic applications of dendrimers in vivo have also focused on their use for local or in vitro administration. Despite these challenges, there are some evidence that show dendrimer-based delivery systems have a significant potential for the delivery of genetic therapies in vivo as shown in Tables 11.2 and 11.3.

11.4.1 LOCALIZED/IN VITRO ADMINISTRATION

11.4.1.1 EYE

When pilocarpine nitrate and tropicamide drugs were coadministered with PAMAM dendrimers into rabbits' eye for miotic and mydriatic activity, the result exhibited that G2 dendrimer ocular residence time is more than Carbopol® or HPMC solutions (Vandamme et al., 2005).

11.4.1.2 TUMOR

Cisplatin is a potent anticancer drug with nonspecific toxicity and poor solubility but when complexed with PAMAM dendrimers, it showed increased solubility, decreased cytotoxicity, and enhanced permeation through cell membrane and retention properties (Malik et al., 1999).

11.4.1.3 HEART

Direct injection in a murine cardiac transplant model PAMAMG5 dendrimer complexes demonstrated more widespread and prolonged expression when combined with a viral interleukin 10 gene and was able to prolong graft survival (Kukowska-Latallo et al., 2000).

TABLE 11.2 Proposed Applications of Dendrimers in Intravenous/Intraperitoneal/Intratumoral Drug Delivery Routes.

Guest molecules	Dendrimers	Interaction pattern	Goal	Administration route
Cisplatin	G3.5 PAMAM	Conjugation	Cancer chemotherapy	Intravenous
5-fluorouracil	PEG-modified G4 PAMAM	Simple encapsulation	Pharmacokinetic studies	Intravenous
Indomethacin	G4 PAMAM	Electrostatic interaction	Anti-inflammatory	Intravenous
Flurbiprofen	G4 PAMAM	Electrostatic interaction	Anti-inflammatory	Intravenous
Primaquine phosphate	Galactose-coated PPI	Simple encapsulation	Liver targeting	Intravenous
Methotrexate	Folic-acid-coated PAMAM	Conjugation	Cancer chemotherapy	Intravenous
Boron	EGF-conjugated PAMAM	Conjugation	Boron neutron capture therapy (BNCT)	Intravenous
Boron	EGF-conjugated PAMAM	Conjugation	BNCT	Intravenous and intratumoral
Boron	EGF-conjugated PAMAM	Conjugation	BNCT	Intratumoral and convection-enhanced delivery
Boron	Antibody-conjugated PAMAM	Conjugation	BNCT	Intraperitoneal
Boron	Antibody-conjugated PAMAM	Conjugation	BNCT	Intratumoral
Glucosamine	G1 PAMAM	Conjugation	Biological modulators	Intraperitoneal
Porphyrin	5-aminolevulinic acid	Conjugation	Photodynamic therapy	Intraperitoneal
DNA	G9 PAMAM	Electrostatic interaction	Lung gene transfection	Intravenous
DNA	PAMAM–cyclodextrin conjugate	Electrostatic interaction	Spleen gene transfection	Intravenous
DNA	Mannose–PAMAM–cyclodextrin conjugate	Electrostatic interaction	Kidney gene transfection	Intravenous

TABLE 11.2 (Continued)

Guest molecules	Dendrimers	Interaction pattern	Goal	Administration route
[111]In-labeled oligo-DNA	G4 PAMAM and G4–bio-tin conjugate	Electrostatic interaction	Kidney and Lung gene transfection	Intravenous
DNA	PAMAM dendrimers and avidin	Electrostatic interaction	Tumor targeting	Intraperitoneal
DNA	PPI	Electrostatic interaction	Liver gene transfection	Intravenous
DNA	PPI with methyl quaternary ammonium	Electrostatic interaction	Liver gene transfection	Intravenous
Gene medicine	G3 PPI	Electrostatic interaction	Intratumoral gene transfection	Intravenous
DNA	PEG-modified PPI	Electrostatic interaction	Transfection agents for DNA enzymes	Intravenous
DNA	PAMAM	Electrostatic interaction	Inhibition of tumor growth and angiogenesis	Intratumoral

TABLE 11.3 Proposed Applications of Dendrimers in Oral/Transdermal/Ocular Drug Delivery Routes.

Guest molecules	Dendrimers	Interaction pattern	Goal	Administration route
—	Lipidic dendrimer	—	Absorption ability by different tissues	Oral
—	Various PAMAM dendrimers	—	Intestine uptake of PAMAM dendrimers	Oral
—	Polylysine dendrimer	—	Oral uptake and translocation	Oral
Propranolol	G3 PAMAM	Conjugation	Improve oral bioavailability	Oral
5-Fluorouracil	Fatty-acid- and phospholipid-coated PAMAM	Simple encapsulation	Improve oral bioavailability	Oral
Ketoprofen	G5 PAMAM	Electrostatic interaction	Anti-inflammatory and improve oral bioavailability	Oral
Tamsulosin hydrochloride	G3 PAMAM	Simple encapsulation	Enhance transdermal delivery efficacy	Transdermal
Tamsulosin hydrochloride	G2.5 and G3 PAMAM	Simple encapsulation	Transdermal mechanism	Transdermal
Indomethacin	G4 and G4.5 PAMAM	Electrostatic interaction	Anti-inflammatory	Transdermal
Ketoprofen and diflunisal	G5 PAMAM	Electrostatic interaction	Anti-inflammatory	Transdermal
CAT reporter transgene	PAMAM	Electrostatic interaction	Skin gene transfection	Transdermal
Pilocarpine nitrate and tropicamide	PAMAM dendrimers	Simple encapsulation	Miotic activity and mydriatic activity	Ocular
Glucosamine and glucosamine-6-sulfate	G3.5 PAMAM	Conjugation	Prevent scar tissue formation	Ocular
Collagen	G2 PPI	Conjugation	Corneal tissue-engineering scaffold	Ocular
Collagen	Cell adhesion peptide-modified PPI	Conjugation	Corneal tissue-engineering scaffold	Ocular
ODN-1	Lipid–lysine dendrimers	Electrostatic interaction	Prevent ocular neovascularization	Ocular

11.4.1.4 LUNG

Intravenous injection of G9 PAMAM with DNA complex is given for gene transfection in lungs (Kihara et al., 2003).

11.4.2 SYSTEMIC ADMINISTRATION

The systemic administration of dendrimers PAMAMG3 and conjugates of α-cyclodextrin (αCD) with the terminal amines of PAMAMG3 show the improved expression for gene delivery. Previously, the number of systems developed for in vivo delivery of genes reported that the lower generations of PPI dendrimers are promising delivery systems which strike a good balance between binding/stability and toxicity, which demonstrated that these systems are also potentially useful for systemic gene therapy (Brownlie et al., 2004).

11.4.3 COMPLEX MODULATION THROUGH ADDITIVES AND/ OR CONJUGATION

The addition or modulation of the cationic transfection polymer DEAE– dextran to the dendrimer–DNA complexes appears to have an additive or possibly synergistic effect on the transfection efficiency of different PAMAM generations. Positive effects on the ability of complexes to transfect have also been reported for the combination of low MW PEI with high MW PEI and various lipidic systems (Kihara et al., 2002).

11.4.3.1 CONJUGATES

To increase the efficiency of PAMAM dendrimer, the covalent conjugates of CyD–dendrimer have also been used.

11.4.3.2 QUATERNIZATION

Quaternization technique is used for the number of purposes such as to modulate dendrimer physicochemistry and cytotoxicity and recently, this strategy is explored in conjunction with PPI dendrimers. The lower-generation PPI dendrimers (G1–G4) were modified by using methyl quaternary ammonium derivatives. The quaternization of the PPIG2 improved DNA binding and complex stability, which will also improve in vivo safety, that is, in contrast to complexes formed with the unmodified dendrimer, the formulations were now well tolerated to intravenous injection. Despite that, the modified polymer was able to facilitate transfection in the liver after systemic administration. Biocompatibility of G3 and G4 complexes also increase by about fourfold using quaternization technique (Uchegbu et al., 2001).

11.5   CONCLUSION

Dendrimers by virtue of its inherent qualities are gaining paramount importance in drug delivery. The encapsulation/complexation of drugs molecules into/with dendrimers releases them in a sustained fashion along with achieving the objectives of drug targeting. Experimental investigations as well as preclinical results are paving the way toward the clinical studies so that ultimately dendrimers-based formulation may reach the clinic. Moreover, hemolytic toxicity problem may be addressed by surface modification of dendrimers. Dendrimer nanocarriers as versatile vectors in gene delivery may become a potential tool to handle stubborn diseases such promising method for treating cancers as well as genetic disorders. In near future, DNA delivery reagent and dendrimer-based formulations will hit the market.

KEYWORDS

- dendrimers
- nanoworld
- gene delivery
- vectors
- cellular uptake
- viruses

REFERENCES

Anderson, W. F. Human Gene Therapy. *Nature* **1998,** *392*(Suppl. 6679), 25–30.

Anderson, D. G.; Akinc, A.; Hossain, N.; Langer, R. Structure/Property Studies of Polymeric Gene Delivery Using a Library of Poly(Beta-Amino Esters). *Mol. Ther.* **2005**, *3*, 426–434.

Aydin, Z.; Akbas, F.; Senel, M.; Koc, S. N. Evaluation of Jeffamine®-Cored PAMAM Dendrimers as an Efficient in Vitro Gene Delivery System. *J. Biomed. Mater. Res. A.* **2012**, *100*(10), 2623–2628.

Bhattacharya, S.; Bajaj, A. Advances in Gene Delivery Through Molecular Design of Cationic Lipids. *Chem. Commun. (Camb.)* **2009**, *31*, 4632–4656.

Boas, U.; Heegaard, P. M. Dendrimers in Drug Research. *Chem. Soc. Rev.* **2004**, *33*, 43–63.

Breunig, M.; Lungwitz, U.; Liebl, R.; Goepferich, A. Breaking up the Correlation Between Efficacy and Toxicity for Nonviral Gene Delivery. *Proc. Natl. Acad. Sci. USA.* **2007**, *104*(36), 14454–14459.

Brown, M. D.; Schatzlein, A. G.; Uchegbu, I. F. Gene Delivery with Synthetic (Non Viral) Carriers. *Int. J. Pharm.* **2001**, *229*(1–2), 1–21.

Brownlie, A.; Uchegbu, I. F.; Schatzlein, A. G. PEI-Based Vesicle-Polymer Hybrid Gene Delivery System with Improved Biocompatibility. *Int. J. Pharm.* **2004**, *274*, 41–52.

Choi, J. S.; Nam, K.; Park, J. Y.; Kim, J. B.; Lee, J. K.; Park, J. S. Enhanced Transfection Efficiency of PAMAM Dendrimer by Surface Modification with l-Arginine. *J. Controlled Release* **2004**, *3*, 445–456.

Cloninger, M. J. Biological Applications of Dendrimers. *Curr. Opin. Chem. Biol.* **2002**, *6*, 742–748.

Crystal, R. G. Transfer of Genes to Humans: Early Lessons and Obstacles to Success. *Science* **1995**, *270*(5235), 404–410.

Daneshvar, N.; Abdullah, R.; Shamsabadi, F. T.; How, C. W.; Aizat, M.; Mehrbod, P. PAMAM Dendrimer Roles in Gene Delivery Methods and Stem Cell Research. *Cell Biol. Int.* **2013**, *37*(5), 415–419.

Devarakonda, B.; Hill, R. A.; DeVilliers, M. M. The Effect of PAMAM Dendrimer Generation Size and Surface Functional Groups on the Aqueous Solubility of Nifedipine. *Int. J. Pharm.* **2004**, *284*, 133–140.

Dobrovolskaia, M. A.; Patri, A. K.; Simak, J.; Hall, J. B.; Semberova, J.; De Paoli Lacerda, S. H.; McNeil, S. E. Nanoparticle Size and Surface Charge Determine Effects of PAMAM Dendrimers on Human Platelets in Vitro. *Mol. Pharm.* **2012**, *9*(3), 382–393.

Eichman, J. D.; Bielinska, A. U.; Kukowska-Latallo, J. F.; Baker J. R. The Use of PAMAM Dendrimers in the Efficient Transfer of Genetic Material into Cells. *Pharm. Sci. Technol. Today* **2000**, *3*, 232–245.

Fischer, D.; Bieber, T.; Li, Y. X.; Elsasser, H.; Kissel, T. A Novel Non-Viral Vector for DNA Delivery Based on Low Molecular Weight, Branched Polyethylenimine: Effect of Molecular Weight on Transfection Efficiency and Cytotoxicity. *Pharm. Res.* **1999**, *16*(8), 1273–1279.

Gao, X.; Huang, L. Cationic Liposome-Mediated Gene Transfer. *Gene Ther.* **1995**, *2*(10), 710–722.

Gao, K.; Huang, L. Nonviral Methods for siRNA Delivery. *Mol. Pharm.* **2009**, *6*(3), 651–658.

Gardlik, R.; Palffy, R.; Hodosy, J.; Lukacs, J.; Turna, J.; Celec, P. Vectors and Delivery Systems in Gene Therapy. *Med. Sci. Monit.* **2005**, *11*(4), 110–121.

Gillies, E. R.; Frechet, J. M. Dendrimers and Dendritic Polymers in Drug Delivery. *Drug Discovery Today* **2005**, *10*, 35–43.

Gosselin, M. A.; Guo, W. J.; Lee, R. J. Efficient Gene Transfer Using Reversibly Cross-Linked Low Molecular Weight Polyethylenimine. *Bioconjugate Chem.* **2001**, *12*(6), 989–994.

Huang, R. Q.; Qu, Y. H.; Ke, W. L.; Zhu, J. H.; Pei, Y. Y.; Jiang, C. Efficient Gene Delivery Targeted to the Brain Using a Transferrin-Conjugated Polyethyleneglycol-Modified Polyamidoamine Dendrimer. *FASEB J.* **2007,** *21*(4), 1117–1125.

Iwashita, S.; Hiramatsu, Y.; Otani, T.; Amano, C.; Hirai, M.; Oie, K.; Yuba, E.; Kono, K.; Miyamoto, M.; Igarashi, K. Polyamidoamine Dendron-Bearing Lipid Assemblies: Their Morphologies and Gene Transfection Ability. *J. Biomater. Appl.* **2012,** *27,* 445–456.

Jin, L.; Zeng, X.; Liu, M.; Deng, Y.; He, N. Current Progress in Gene Delivery Technology Based on Chemical Methods and Nano-Carriers.*Theranostics* **2014,** *4*(3), 240–255.

Karak, N.; Maiti, S. Dendritic Polymers: a Class of Novel Material. *J. Polym. Mater.* **1997,** *14,* 105.

Khosravi, D. K.; Mozafari, M. R.; Rashidi, L.; Mohammadi, M. Calcium Based Non-Viral Gene Delivery: an Overview of Methodology and Applications. *Acta Med. Iran.* **2010,** *48*(3), 133–141.

Kihara, F.; Arima, H.; Tsutsumi, T.; Hirayama, F.; Uekama, K. Effects of Structure of Polyamidoamine Dendrimer on Gene Transfer Efficiency of the Dendrimer Conjugate with Alpha-Cyclodextrin. *Bioconjugate Chem.* **2002,** *13,* 1211–1219.

Kihara, F.; Arima, H.; Tsutsumi, T.; Hirayama, F.; Uekama, K. In Vitro and in Vivo Gene Transfer by an Optimized Alpha-Cyclodextrin Conjugate with Polyamidoamine Dendrimer. *Bioconjugate Chem.* **2003,** *14,* 342–350.

Knoell, D. M.; Yiu, I. M. Human Gene Therapy for Hereditary Diseases: a Review of Trials. *Am. J. Health Syst. Pharma.* **1998,** *55*(9), 899–904.

Kojima, C.; Kono, K.; Maruyama, K.; Takagishi, T. Synthesis of Polyamidoamine Dendrimers Having Poly (Ethylene Glycol) Grafts and their Ability to Encapsulate Anticancer Drugs. *Bioconjugate Chem.* **2000,** *11,* 910–917.

Kukowska-Latallo, J. F.; Raczka, E.; Quintana, A.; Chen, C.; Rymaszewski, M.; Baker, J. R. Intravascular and Endobronchial DNA Delivery to Murine Lung Tissue Using a Novel, Nonviral Vector. *Hum. Gene Ther.* **2000,** *11,* 1385–1395.

Lachmann, R. H.; Efstathiou, S. The Use of Herpes Simplex Virus-Based Vectors for Gene Delivery to the Nervous System. *Mol. Med. Today* **1997,** *3*(9), 404–411.

Lehrman, S. Virus Treatment Questioned after Gene Therapy Death. *Nature* **1999,** *401*(6753), 517–518.

Li, S.; Ma, Z. Review on Nonviral Gene Therapy. *Curr. Gene. Ther.* **2001,** *1*(2), 201–226.

Lundstrom, K.; Boulikas, T. Viral and Non-Viral Vectors in Gene Therapy: Technology Development and Clinical Trials. *Technol. Cancer Res. Treat.* **2003,** *2*(5), 471–486.

Luten, J.; Nostrum, C. F.; Smedt, S. C.; Hennink, W. E. Biodegradable Polymers as Non-Viral Carriers for Plasmid DNA Delivery. *J. Controlled Release* **2008,** *126*(2), 97–110.

Malik, N.; Evagorou, E. G.; Duncan, R. Dendrimer-Platinate: a Novel Approach to Cancer Chemotherapy. *Anticancer Drugs* **1999,** *10,* 767–776.

Mintzer, M. A.; Simanek, E. E. Nonviral Vectors for Gene Delivery. *Chem. Rev.* **2009,** *2,* 259–302.

Mulligan, R. C. The Basic Science of Gene Therapy. *Science* **1993,** *260*(5110), 926–932.

Parker, A. L.; Newman, C.; Briggs, S.; Seymour, L.; Sheridan, P. J. Nonviral Gene Delivery: Techniques and Implications for Molecular Medicine. *Expert Rev. Mol. Med.* **2003,** *5*(22), 1–15.

Pearson, S.; Jia, H.; Kandachi, K. China Approves First Gene Therapy. *Nat. Biotechnol.* **2004,** *22*(1), 3–4.

Rolland, A. P. From Genes to Gene Medicines: Recent Advances in Nonviral Gene Delivery. *Crit. Rev. Ther. Drug Carrier Syst.* **1998,** *15*(2), 143–198.

Schaffert, D.; Troiber, C.; Wagner, E. New Sequence-Defined Polyaminoamides with Tailored Endosomolytic Properties for Plasmid DNA Delivery. *Bioconjugate Chem.* **2012,** *6,* 1157–1165.

Shah, N.; Steptoe, R. J.; Parekh, H. S. Low-Generation Asymmetric Dendrimers Exhibit Minimal Toxicity and Effectively Complex DNA. *J. Pept. Sci.* **2011,** *17*(6), 470–478.

Stiriba, S. E.; Frey, H.; Haag, R. Dendritic Polymers in Biomedical Applications: from Potential to Clinical Use in Diagnostics and Therapy. *Angew. Chem. Int. Ed.* **2002,** *41,* 1329–1334.

Tomalia, D. A.; Majoros, I. Dendrimeric Supramolecular and Supramacromolecular Assemblies. *J. Macromol. Sci. Polym. Rev.* **2003,** *43,* 411–477.

Twyman, L. J.; Beezer, A. E.; Esfand, R.; Hardy, M. J.; Mitchell, J. C. The Synthesis of Water-Soluble Dendrimers, and Their Application as Possible Drug Delivery Systems. *Tetrahedron Lett.* **1999,** *40,* 1743–1746.

Uchegbu, I. F.; Sadiq, L.; Arastoo, M.; Gray, A. I.; Wang, W.; Waigh, R. D.; Schatzlein, A. G. Quaternary Ammonium Palmitoyl Glycol Chitosan—a New Polysoap for Drug Delivery. *Int. J. Pharm.* **2001,** *224,* 185–199.

Vandamme, T. F.; Brobeck, L. Poly(Amidoamine) Dendrimers as Ophthalmic Vehicles for Ocular Delivery of Pilocarpine Nitrate and Tropicamide. *J. Controlled Release* **2005,** *102,* 23–38.

Yamano, S.; Dai, J.; Moursi, A. M. Comparison of Transfection Efficiency of Nonviral Gene Transfer Reagents. *Mol. Biotechnol.* **2010,** *46,* 287–300.

CHAPTER 12

CARBON NANOTUBES FOR DRUG DELIVERY: FOCUS ON ANTIMICROBIAL ACTIVITY

MÁRCIA EBLING DE SOUZA[1,2], DARIANE JORNADA CLERICI[2], and ROBERTO CHRIST VIANNA SANTOS[3*]

[1]Laboratory of Microbiological Research, Centro Universitário Franciscano, Santa Maria, Rio Grande do Sul, Brazil

[2]Laboratory of Nanotechnology, Post-Graduate Program of Nanosciences, Centro Universitário Franciscano, Santa Maria, Rio Grande do Sul, Brazil

[3]Laboratory of Oral Microbiology Research, Universidade Federal de Santa Maria, Santa Maria, Rio Grande do Sul, Brazil

*Corresponding author. E-mail: robertochrist@gmail.com

ABSTRACT

Carbon nanotubes (CNTs) are a promising alternative in the delivery of various drugs because of their optical and chemical properties on their surface that promote low toxicity in biological systems and great ability to associate with drugs or molecules by adsorption or chemical bonding. On the other hand, the CNTs have a low dispersion in aqueous media, this problem can be solved through the functionalization of the surface of the nanotubes with specific chemical groups, or biological species. Several studies have been carried out with functionalized CNTs, used as vehicles for the delivery of antimicrobial drugs in a controlled and oriented way, presenting an optimization in the therapeutic activity, improving the antimicrobial action and reducing its toxicity. However, the results presented for their toxicity are still contradictory. These contradictions may be related to the various forms of production, purification, and functionalization, in addition to the different concentrations, hindering the interpretations of the results. It is necessary to develop reproducible protocols to ensure the safety of the use of CNTs as drug delivery systems.

12.1   INTRODUCTION

Throughout history, treatment with most drugs has always been limited by the inability of increasing plasmatic concentration. The time of the drug in the blood stream, low solubility in biological fluids, and in particular the deleterious side effects associated with therapy with high doses may hinder the use of the concentration required for pharmacotherapeutic success. To improve the pharmacological profiles of many classes of drug molecules, the development of new and efficient drug delivery systems is necessary (Kreuter, 2007). Currently, there are many types of administration, and within the family of nanomaterials system, carbon nanotubes (CNTs) have attracted great interest of researchers from different areas due to its unique one-dimensional structure and peculiar physical, chemical, and biological properties (Wang and Liu, 2012).

CNTs are formed by graphene sheets forming a hollow cylinder, comprising a hexagonal arrangement of carbon atoms with sp^2 hybridization (carbon–carbon distance of ~1.4 A). They are classified into two different categories: single-walled CNTs (SWCNTs) and multiwalled CNTs (MWCNTs), which consist of several concentric graphite cylinders spaced 0.34–0.36 nm from each other (Dresselhaus et al., 2004; Iijima, 1991). There are three ways for the molecule to perform the winding, depending on its direction: armchair, zigzag, and chiral. CNTs (SWCNTs and MWCNTs) are generally produced by three main techniques: arc discharge method (applying arc vaporization of two carbon rods), laser ablation method (using graphite), and chemical vapor deposition (using hydrocarbon sources: CO, methane, and acetylene). After preparation, the CNTs are subjected to purification by acid reflux, auxiliary sonication with surfactant, or air oxidation procedure to remove impurities such as amorphous carbon, fullerenes, and transition metals as catalysts introduced during synthesis (He et al., 2013).

CNTs are compounds formed exclusively of carbon atoms, they are hydrophobic and therefore insoluble in water, something that is extremely undesirable for biological and pharmacological applications (Foldvari and Bagonluri, 2008a). However, some methods have been described for chemically modifying the CNTs, through functionalization, thus allowing them to be solubilized and dispersed in water, making them less toxic

(Jain et al., 2014; Mehra et al., 2014). Furthermore, the functionalized CNTs may be bonded to a wide range of active molecules, including peptides, proteins, nucleic acids, and other therapeutic agents, facilitating handling and processing in physiological environments (Bianco et al., 2005a).

CNTs have been suggested as a promising alternative, regarding their application in the delivery of therapeutic agents due to their unique properties such as increased charging efficiency, biocompatibility, high surface area, and photoluminescence, overcoming the limitations of conventional nanocarriers (inadequate availability of chemical surface functional groups for conjugation, low entrapment/loading efficiency), make them ideal candidates in the pharmaceutical and biomedical sciences (Mehra and Jain, 2015).

CNTs have shown the main properties of an efficient delivery system of drugs, which includes its ability to perform controlled and targeted release of drugs. It happens through three methods of interaction between CNTs and the active components. The first is due to the interaction of pharmaceutical compounds with the porosity of the wall of the nanotubes, holding the assets within a mesh or bundle. The second is through the working connection of the active compound with the outer walls of CNTs. The third uses the nanotube channels as nanocatheter (Foldvari and Bagonluri, 2008b). Furthermore, the application of CNTs is increased due to its ability to penetrate efficiently the cell membrane (Bianco et al., 2005b).

Some researches are being conducted to observe the cellular uptake mechanism for nanotubes and found that the penetration of CNTs in the cytoplasm of cells can occur in the first hour of internalization. After crossing the plasma membrane, intracellular pathways of CNTs can lead to an organelle accumulation and/or elimination of nanotubes (Kang et al., 2012; Shvedova et al., 2012). The penetration of nanotubes occurs in phagocytic and non-phagocytic cells, mediated by three alternative routes which may be through the engagement of membrane as individual tubes and through direct membrane translocation of individual nanotubes and bundles of vesicular compartments (Lacerda et al., 2012). These facts are important for the intracellular trafficking, making CNTs a promising delivery system for therapeutic and diagnostic agents (Lacerda et al., 2013).

The main applications of CNTs in the biomedical field are summarized in Figure 12.1. The CNTs used as controlled and targeted delivery vehicles have shown an improvement in pharmacological activity of many bioactive molecules such as chemotherapeutics (Ali-Boucetta et al., 2008; Li et al., 2012; Liu et al 2008), antimicrobial agents (Benincasa et al., 2011; Wu et al., 2005), and anti-inflammatory (Friedrich et al., 2008; Luo et al., 2011). For more complex biological products such as vaccines based on peptides

(Pantarotto et al., 2003a) antibodies (Pantarotto et al., 2003b) and small interfering ribonucleic acids (siRNAs) (Podesta et al., 2008), the systems have to be accordingly customized. Therapeutic agents have been successfully delivered, through numerous strategies improving their action and reducing toxicity.

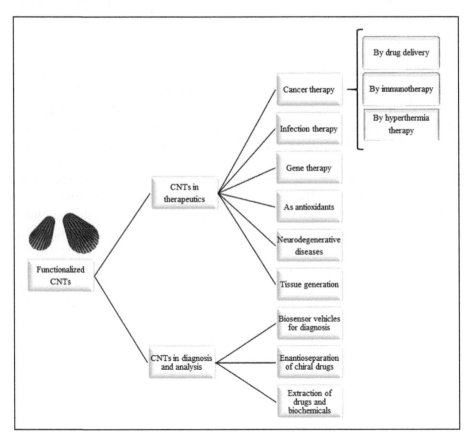

FIGURE 12.1   Carbon nanotubes applications in biomedical therapy, diagnostics, and analysis.

12.2   ANTIMICROBIAL ACTIVITY OF CARBON NANOTUBES

Antimicrobial activity of pure SWCNTs was first reported in 2007; researchers demonstrated direct evidence which showed that highly purified CNTs present strong antimicrobial activity and indicated that the likely mechanism of action is through serious damage caused to the cell membrane by direct contact with CNTs (Kang et al., 2007). In another study, experiments were performed with well-characterized SWCNTs and MWCNTs; the gene expression data showed that CNTs are much more toxic to bacteria than MWCNTs. It was reported that the size of CNTs was an important factor affecting their antibacterial activity due to its small diameter, able to easily penetrate the cell membrane (Kang et al., 2008). However, the cost of Single-walled carbon nanotubes (NTCPS) synthesis is much higher (Arias and Yang, 2009; Dunens et al., 2009; Kang et al., 2007, Nepal et al., 2008; Pasquini et al., 2012; Qi et al., 2011; Yuan et al., 2008).

Researchers have developed MWCNTs with cephalexin immobilized by a covalent bond on its surface by a polymer polyethylene glycol, and they evaluated the antimicrobial and anti-adherent activities by using Gram-negative bacteria (*Escherichia coli* and *Pseudomonas aeruginosa*) and Gram positive (*Staphylococcus aureus* and *Bacillus subtilis*). The antimicrobial activity of MWCNTs—cephalexin has been significantly improved across the Gram-positive bacteria reducing bacterial viability by 80% (twice the number found for MWCNTs alone). Besides, MWCNTs—cephalexin reduces bacterial adhesion, preventing the formation of biofilms. The authors conclude that the MWCNTs—cephalexin have the potential to be used as an effective and economical material with antibacterial properties and nonstick for environmental and medical applications (Qi et al., 2012).

Recent studies with MWCNTs evaluated the potential of various surfactants (Polysorbates and phospholipids) to obtain a better solubility and higher antimicrobial efficacy. Polysorbates showed higher solubility than phospholipids; however, the antimicrobial activity was greatest for the phospholipid, demonstrating that the antimicrobial effect is influenced not only by solubility but also largely by the type of surfactant. Furthermore, the upper antimicrobial activity was confirmed by electron microscopy images, where it was observed that significant adhesion of MWCNTs to the bacterial walls only in the presence of unsaturated phospholipids resulted in minimum inhibitory concentration values comparable to strong antimicrobial established MWCNTs, SWCNTs, and fullerenes (Lohan et al., 2016).

The mesylate pazufloxacin is an antimicrobial which belongs to the class of fluoroquinolones, which was adsorbed on the surface of MWCNTs. The in vitro release suggested that the adsorption of pazufloxacin was reversible and its release profile consisted of an initial rapid release followed by a sustained release period. Furthermore, the total concentration of pazufloxacin released from the functionalized CNTs with amino group was higher at pH 5.7 than at pH 7.0, which may be advantageous in treating infections, wherein the intracellular environment of infected cells is likely to be more acidic (Jiang et al., 2012).

Dapsone (4,4′-diaminodiphenyl sulfone) is a drug which has antimicrobial and anti-inflammatory activity used in the treatment of malaria, leprosy, pneumocystis, and toxoplasmosis associated with acquired immunodeficiency syndrome and certain conditions of dermatitis (Sago, 2002; Zhu and Stiller, 2001), as well as infections caused by *Mycobacteria* (Rastogi et al., 1993). However, it presents some undesirable effects, such as hematologic adverse reactions, that is, metahemoglobinemia and hemolytic anemia. Dapsone has been chemically functionalized on the surface of MWCNTs oxidized to form Dap-MWCNTs as a strategy to enhance the cellular uptake of dapsone and consequently its therapeutic efficacy. The complex of Dap-MWCNTs was quickly attached and phagocytosed by peritoneal macrophages in a manner similar to MWCNTs control, but its biological effect was different. Namely, its MWCNTs control-induced peritoneal macrophages apoptosis rapidly due to the induction of oxidative stress and apoptosis process, being potentiated by soluble dapsone. Meanwhile, apoptosis of peritoneal macrophages after the ingestion of Dap-MWCNTs was delayed (Vuković et al., 2010).

Amphotericin B (AMB) is a broad spectrum antimicrobial used to combat opportunistic systemic fungal infections in immunocompromised patients, being also the drug of choice for the treatment of leishmaniasis. Despite their high effectiveness and relatively low cost, the use of AMB is limited due to their undesirable side effects ranging from the infusion reaction to nephrotoxicity. MWCNTs were functionalized by a chemical process with AMB. By using MWCNTs–AMB, its leishmanicidal efficiency was significantly increased in both in vivo and in vitro tests, also presenting low cytotoxicity (Prajapati et al., 2011). In another study, MWCNTs were charged with AMB. In vitro studies were then conducted to demonstrate a controlled release of AMB at pH 4, 7.4, and 10, and the enhanced cellular uptake and increased layout of MWNTCs–AMB in organs rich in macrophages was observed, thus indicating the site-specific drug delivery (Pruthi et al., 2012).

Silver nanoparticles (AgNP) have a broad spectrum of antibacterial property with low resistance induction rates. However, AgNPS are rapidly released in practical application, resulting a decline in antimicrobial properties. A study was conducted using MWCNTs modified with polidopamine (PDA), as a carrier of the silver nanoparticles. In order to improve mechanical properties and reduce the rate of release of silver nanoparticles solutions, the coating of CNTs with PDA improved the deposition of AgNPS and increased long-term antimicrobial effect (Zhuang et al., 2016). The direct exposure of AgNP constitutes a threat to mammalian cells. Because of that CNTs were coated with biopolymers (dopamine, heparin, and chitosan) to protect the oxidized CNTs which were used as carriers of AgNP. These complexes presented excellent antimicrobial activity due to the synergistic effect of chitosan and AgNP, besides having good compatibility with endothelial cells (Nie et al., 2016). In another study, SWCNTs were functionalized with silver using a peptide with antimicrobial activity TP359. This functionalization using antimicrobial peptides reduced the dosage required for antibacterial activity of SWCNTs–Ag due to the antibacterial activity of both components, resulting in a synergistic effect while maintaining a reduced toxicity (Chaudhari et al., 2016).

Some natural products have also shown to be helpful in the dispersion of the MWCNTs. Researchers functionalized CNTs with acacia extract. MWCNTs were found dispersed throughout the 30-day trial, which suggests that the acacia extract is a good dispersion medium due to its surface-active properties. In addition to provide an alternative to MWCNTs dispersion, *Acacia* extract -MWCNTs showed excellent antimicrobial activity (Yadav et al., 2015).

Qi et al. (2011), functionalized MWCNTs with nisin (a polypeptide with important antimicrobial characteristics). This functionalization has resulted in an increase in antimicrobial activity and a reduction in adhesion of *E. coli, P. aeruginosa, B. subtilis,* and *S. aureus.* The functionalization of MWCNTs increased antimicrobial activity by seven times and antibiofilm activity by 100 times compared to pure MWCNTs.

The antimicrobial activity of CNTs was investigated in microorganisms normally found in the human digestive system, such as *Lactobacillus acidophilus, Bifidobacterium adolescentis, E. coli, Enterococcus faecalis,* and *S. aureus.* In this study, CNTs, including SWCNTs (1–3 μm), long and short S-MWCNTs (0.5–2 μm; l-MWCNTs: >50 μm), and functionalized MWCNTs (hydroxyl and carboxyl-modified, 0.5–2 μm) showed an antimicrobial action of broad spectrum. CNTs can selectively lyse the walls and membranes of microorganisms present in the human intestine, depending not only on the length and the surface of functional groups of CNTs, but also on the shapes of bacteria. The mechanism of action through membrane has shown to be associated with its diameter and length, resulting in the induction and release of DNA and intracellular components resulting in dead bacteria. Since SWCNTs are smaller and more rigid, have shown more effective action in the spherical bacteria than MWCNTs. MWCNTs may attach to the wall of the bacteria, by increasing the contact of the surface with the microorganisms (Chen et al., 2013).

Currently, CNTs are one of the fronts in a diverse field of medicine, being continually explored by researchers, scientists, and scholars. Figure 12.2 shows the increase in the number of papers indexed at PubMed U.S. National Institutes of Health's National Library of Medicine, reporting on studies of CNTs over the past decade.

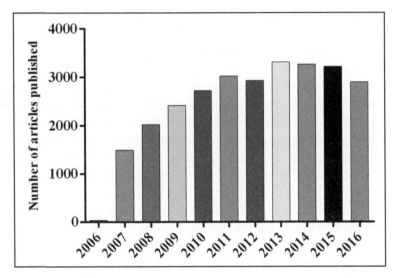

FIGURE 12.2   Increase in the number of scientific publications involving CNTs indexed at PubMed US National Institutes of Health's National Library of Medicine database in the past 10 years.

Nanotechnology is an emerging field that will have an essential role in the delivery of drugs and potentially change the way we treat diseases. Although the number of papers published in this area has been significant and increasing, there has not been any consensus on these materials for human health. Some studies indicate that these particles can cause adverse effects due to their small size and weight, allowing its easy deployment in the human body. The results of toxicological tests of the CNT in the literature seem to be contradictory. For example, some studies show that CNTs can induce cell apoptosis/necrosis (Bottini et al., 2006; Ghosh et al., 2011). While others have reported the CNTs with no sign of toxicity (Dumortier et al., 2006; Pulskamp et al., 2007). Still there are studies that affirm that CNTs present cellular toxicity, the mechanisms of induced toxicity of CNTs are distinct and disparate, mainly happening as oxidative stress (Visalli et al., 2015), DNA damage (Ghosh et al., 2011; Lan et al., 2014), and through mitochondrial impairment (Wang et al., 2012; Wei et al., 2010). The toxicity of CNTs is also related to their concentration and the dose at which the cells or organisms are exposed (Foldvari and Bagonluri, 2008b). Such contradictory results might be explained by the large number of factors, such as different assessment methods for species or different cell populations (Dumortier et al., 2006; Lan et al., 2014; Liu et al., 2013).

12.3 CONCLUSIONS

Despite several surprising results of CNTs obtained at the beginning of this research field, there are still huge opportunities to be explored and significant challenges and risks to be resolved. Research on new CNT synthesis techniques should be performed for the use of nanotubes in biomedical applications, improving productivity, and quality. A consistent and reproducible protocol must be developed to evaluate the dispersion, purity, and physical and functional properties of CNTs, so that regulatory agencies would enable the use in humans.

KEYWORDS

- **single-walled carbon nanotubesmultiwalled carbon nanotubes**
- **nanomaterials**
- **functionalization**
- **therapeutic agents**
- **drug delivery**

REFERENCES

Ali-Boucetta, H.; Al-Jamal, K. T.; McCarthy, D.; Prato, M.; Bianco, A.; Kostarelos, K. Multiwalled Carbon Nanotube-Doxorubicin Supramolecular Complexes for Cancer Therapeutics. *Chem. Commun.* **2008,** *4,* 459–461.

Arias, L. R.; Yang, L. Inactivation of Bacterial Pathogens by Carbon Nanotubes in Suspensions. *Langmuir* **2009,** *25*(5), 3003–3012.

Benincasa, M.; Pacor, S.; Wu, W.; Prato, M.; Bianco, A.; Gennaro, R. Antifungal Activity of Amphotericin B Conjugated to Carbon Nanotubes. *ACS Nano* **2011,** *5*(1), 199–208.

Bianco, A.; Kostarelos, K.; Prato, M. Applications of Carbon Nanotubes in Drug Delivery. *Curr. Opin. Chem. Biol.* **2005a,** *9*(6), 674–679.

Bianco, A.; Kostarelos, K.; Partidos, C. D.; Prato, M. Biomedical Applications of Functionalised Carbon Nanotubes. *Chem. Commun. (Camb).* **2005b,** (5), 571–577.

Bottini, M.; Bruckner, S.; Nika, K.; Bottini, N.; Bellucci, S.; Magrini, A.; Bergamaschi, A.; Mustelin, T. Multi-Walled Carbon Nanotubes Induce T Lymphocyte Apoptosis. *Toxicol. Lett.* **2006,** *160*(2), 121–126.

Chaudhari, A. A.; Ashmore, D.; Nath, S. D.; Kate, K.; Dennis, V.; Singh, S. R.; Owen, D. R.; Palazzo, C.; Arnold, R. D.; Miller, M. E.; et al. A Novel Covalent Approach to Bio-Conjugate Silver Coated Single Walled Carbon Nanotubes with Antimicrobial Peptide. *J. Nanobiotechnol.* **2016,** *14*(1), 58.

Chen, H.; Wang, B.; Gao, D.; Guan, M.; Zheng, L.; Ouyang, H.; Chai, Z.; Zhao, Y.; Feng, W. Broad-Spectrum Antibacterial Activity of Carbon Nanotubes to Human Gut Bacteria. *Small* **2013,** *9*(16), 2735–2746.

Dresselhaus, M. S.; Dresselhaus, G.; Charlier, J. C.; Hernández, E. Electronic, Thermal and Mechanical Properties of Carbon Nanotubes. *Philos. Trans. A. Math. Phys. Eng. Sci.* **2004,** *362*(1823), 2065–2098.

Dumortier, H.; Lacotte, S.; Pastorin, G.; Marega, R.; Wu, W.; Bonifazi, D.; Briand, J. P.; Prato, M.; Muller, S.; Bianco, A. Functionalized Carbon Nanotubes Are Non-Cytotoxic and Preserve the Functionality of Primary Immune Cells. *Nano Lett.* **2006,** *6*(7), 1522–1528.

Dunens, O. M.; Mackenzie, K. J.; Harris, A. T. Synthesis of Multiwalled Carbon Nanotubes on Fly Ash Derived Catalysts. *Environ. Sci. Technol.* **2009,** *43*(20), 7889–7894.

Foldvari, M.; Bagonluri, M. Carbon Nanotubes as Functional Excipients for Nanomedicines: I. Pharmaceutical Properties. *Nanotechnol. Biol Med.* **2008a,** *4*(3),173–82.

Foldvari, M.; Bagonluri, M. Carbon Nanotubes as Functional Excipients for Nanomedicines: II. Drug Delivery and Biocompatibility Issues. *Nanotechnol. Biol Med.* **2008b,** *4*(3), 183–200.

Friedrich, R. B.; Fontana, M. C.; Beck, R. C. R.; Pohlmann, A. R.; Guterres, S. S. Development and Physicochemical Characterization of Dexamethasone-Loaded Polymeric Nanocapsule Suspensions. *Quim. Nova* **2008,** *31*(5), 1131–1136.

Ghosh, M.; Chakraborty, A.; Bandyopadhyay, M.; Mukherjee, A. Multi-Walled Carbon Nanotubes (MWCNT): Induction of DNA Damage in Plant and Mammalian Cells. *J. Hazard. Mater.* **2011,** *197,* 327–336.

He, H.; Pham-Huy, L. A.; Dramou, P.; Xiao, D.; Zuo, P.; Pham-Huy, C. Carbon Nanotubes: Applications in Pharmacy and Medicine. *Biomed. Res. Int.* **2013,** 2013. Article ID 578290.

Iijima, S. Helical Microtubules of Graphitic Carbon. *Nature* **1991,** *354*(6348), 56–58.

Jain, K.; Mehra, N. K.; Jain, N. K. Potentials and Emerging Trends in Nanopharmacology. *Curr. Opin. Pharmacol.* **2014,** *15*(1), 97–106.

Jiang, L.; Liu, T.; He, H.; Pham-Huy, L. A.; Li, L.; Pham-Huy, C.; Xiao, D. Adsorption Behavior of Pazufloxacin Mesilate on Amino-Functionalized Carbon Nanotubes. *J. Nanosci. Nanotechnol.* **2012,** *12*(9), 7271–7279.

Kang, S.; Pinault, M.; Pfefferle, L. D.; Elimelech, M. Single-Walled Carbon Nanotubes Exhibit Strong Antimicrobial Activity. *Langmuir* **2007,** *23*(17), 8670–8673.

Kang, S.; Herzberg, M.; Rodrigues, D. F.; Elimelech, M. Antibacterial Effects of Carbon Nanotubes: Size Does Matter. *Langmuir* **2008,** *24*(13), 1–8.

Kang, B.; Li, J.; Chang, S.; Dai, M.; Ren, C.; Dai, Y.; Chen, D. Subcellular Tracking of Drug Release from Carbon Nanotube Vehicles in Living Cells. *Small* **2012,** *8*(5), 777–782.Kreuter, J. Nanoparticles-a Historical Perspective *Int. J. Pharm.* **2007,** *331*(1), 1–10.

Lacerda, L.; Russier, J.; Pastorin, G.; Herrero, M. A.; Venturelli, E.; Dumortier, H.; Al-Jamal, K. T.; Prato, M.; Kostarelos, K.; Bianco, A. Translocation Mechanisms of Chemically Functionalised Carbon Nanotubes across Plasma Membranes. *Biomaterials* **2012,** *33*(11), 3334–3343.

Lacerda, L.; Ali-Boucetta, H.; Kraszewski, S.; Tarek, M.; Prato, M.; Ramseyer, C.; Kostarelos, K.; Bianco, A. How do Functionalized Carbon Nanotubes Land on, Bind to and Pierce through Model and Plasma Membranes. *Nanoscale* **2013,** *5*(21), 10242–10250.

Lan, J.; Gou, N.; Gao, C.; He, M.; Gu, A. Z. Comparative and Mechanistic Genotoxicity Assessment of Nanomaterials via a Quantitative Toxicogenomics Approach across Multiple Species. *Environ. Sci. Technol.* **2014,** *48*(21), 12937–12945.

Li, J.; Yap, S. Q.; Chin, C. F.; Tian, Q.; Yoong, S. L.; Pastorin, G.; Ang, W. H. Platinum(iv) Prodrugs Entrapped within Multiwalled Carbon Nanotubes: Selective Release by Chemical Reduction and Hydrophobicity Reversal. *Chem. Sci.* **2012,** *3*(6), 2083.

Liu, Z.; Chen, K.; Davis, C.; Sherlock, S.; Cao, Q.; Chen, X.; Dai, H. Drug Delivery with Carbon Nanotubes for in Vivo Cancer Treatment. *Cancer Res.* **2008,** *68*(16), 6652–6660.

Liu, Y.; Zhao, Y.; Sun, B.; Chen, C. Understanding the Toxicity of Carbon Nanotubes. *Acc. Chem. Res.* **2013,** *46*(3), 702–713.

Lohan, S.; Raza, K.; Singla, S.; Chhibber, S.; Wadhwa, S.; Katare, O. P.; Kumar, P.; Singh, B. Studies on Enhancement of Anti-Microbial Activity of Pristine MWCNTs Against Pathogens. *AAPS Pharm. Sci. Tech.* **2016,** *17*(5), 1042–1048.

Luo, X.; Matranga, C.; Tan, S.; Alba, N.; Cui, X. T. Carbon Nanotube Nanoreservior for Controlled Release of Anti-Inflammatory Dexamethasone. *Biomaterials* **2011,** *32*(26), 6316–6323.

Mehra, N. K.; Jain, N. K. Multifunctional Hybrid-Carbon Nanotubes: New Horizon in Drug Delivery and Targeting. *J. Drug Targeting* **2015,** *24*(4), 1–15.

Mehra, N. K.; Mishra, V.; Jain, N. K. A Review of Ligand Tethered Surface Engineered Carbon Nanotubes. *Biomaterials.* **2014,** *35*(4), 1267–1283.

Nepal, D.; Balasubramanian, S.; Simonian, A. L.; Davis, V. A. Strong Antimicrobial Coatings: Single-Walled Carbon Nanotubes Armored with Biopolymers. *Nano Lett.* **2008,** *8*(7), 1896–1901.

Nie, C.; Cheng, C.; Peng, Z.; Ma, L.; He, C.; Xia, Y.; Zhao, C. Mussel-Inspired Coatings on Ag Nanoparticle-Conjugated Carbon Nanotubes: Bactericidal Activity and Mammal Cell Toxicity. *J. Mater. Chem. B* **2016,** *4*(16), 2749–2756.

Pantarotto, D.; Partidos, C. D.; Hoebeke, J.; Brown, F.; Kramer, E.; Briand, J. P.; Muller, S.; Prato, M.; Bianco, A. Immunization with Peptide-Functionalized Carbon Nanotubes Enhances Virus-Specific Neutralizing Antibody Responses. *Chem. Biol.* **2003a,** *10*(10), 961–966.

Pantarotto, D.; Partidos, C. D.; Graff, R.; Hoebeke, J.; Briand, J. P.; Prato, M.; Bianco, A. Synthesis, Structural Characterization, and Immunological Properties of Carbon Nanotubes Functionalized with Peptides. *J. Am. Chem. Soc.* **2003b,** *125*(20), 6160–6164.

Pasquini, L. M.; Hashmi, S. M.; Sommer, T. J.; Elimelech, M.; Zimmerman, J. B. Impact of Surface Functionalization on Bacterial Cytotoxicity of Single-Walled Carbon Nanotubes. *Environ. Sci. Technol.* **2012,** *46*(11), 6297–6305.

Podesta, J. E.; Al-Jamal, K. T.; Herrero, M. A.; Tian, B.; Ali-Boucetta, H.; Hegde, V.; Bianco, A.; Prato, M.; Kostarelos, K. Antitumor Activity and Prolonged Survival by Carbon Nanotube-Mediated Therapeutic siRNA Silencing in a Human Lung Xenograft Model. *Small (Weinheim an der Bergstrasse Ger.)* **2008,** *5*(10), 1176–1185.

Prajapati, V. K.; Awasthi, K.; Gautam, S.; Yadav, T. P.; Rai, M.; Srivastava, O. N.; Sundar, S. Targeted Killing of Leishmania Donovani in Vivo and in Vitro with Amphotericin B Attached to Functionalized Carbon Nanotubes. *J. Antimicrob. Chemother.* **2011,** *66*(4), 874–879.

Pruthi, J.; Mehra, N. K.; Jain, N. K. Macrophages Targeting of Amphotericin B Through Mannosylated Multiwalled Carbon Nanotubes. *J. Drug Targeting* **2012,** *20*(7), 593–604.

Pulskamp, K.; Diabaté, S.; Krug, H. F. Carbon Nanotubes Show No Sign of Acute Toxicity but Induce Intracellular Reactive Oxygen Species in Dependence on Contaminants. *Toxicol. Lett.* **2007,** *168*(1), 58–74.

Qi, X.; Poernomo, G.; Wang, K.; Chen, Y.; Chan-Park, M. B.; Xu, R.; Chang, M. W. Covalent Immobilization of Nisin on Multi-Walled Carbon Nanotubes: Superior Antimicrobial and Anti-Biofilm Properties. *Nanoscale* **2011,** *3*(4), 1874–1880.

Qi, X.; Gunawan, P.; Xu, R.; Chang, M. W. Cefalexin-Immobilized Multi-Walled Carbon Nanotubes Show Strong Antimicrobial and Anti-Adhesion Properties. *Chem. Eng. Sci.* **2012,** *84*, 552–556.Rastogi, N.; Goh, K. S.; Labrousse, V. Activity of Subinhibitory Concentrations of Dapsone Alone and in Combination with Cell-Wall Inhibitors against Mycobacterium Avium Complex Organisms. *Eur. J. Clin. Microbiol. Infect. Dis.* **1993,** *12*(12), 954–958.

Sago, J. H. R. Dapsone. *Dermatol Ther.* **2002,** *15*(4), 340–351.

Shvedova, A. A.; Pietroiusti, A.; Fadeel, B.; Kagan, V. E. Mechanisms of Carbon Nanotube-Induced Toxicity: Focus on Oxidative Stress. *Toxicol. Appl. Pharmacol.* **2012,** *261*(2), 121–133.

Visalli, G.; Bertuccio, M. P.; Iannazzo, D.; Piperno, A.; Pistone, A.; di Pietro, A. Toxicological Assessment of Multi-Walled Carbon Nanotubes on A549 Human Lung Epithelial Cells. *Toxicol. In Vitro* **2015,** *29*(2), 352–362.

Vuković, G. D.; Tomić, S. Z.; Marinković, A. D.; Radmilović, V.; Uskoković, P. S.; Čolic, M. The Response of Peritoneal Macrophages to Dapsone Covalently Attached on the Surface of Carbon Nanotubes. *Carbon* **2010,** *48*(11), 3066–3078.

Wang, X.; Liu, Z. Carbon Nanotubes in Biology and Medicine: An Overview. *Chin. Sci. Bull.* **2012,** *57*(2–3), 167–180.

Wang, X.; Guo, J.; Chen, T.; Nie, H.; Wang, H.; Zang, J.; Cui, X.; Jia, G. Multi-Walled Carbon Nanotubes Induce Apoptosis via Mitochondrial Pathway and Scavenger Receptor. *Toxicol. In Vitro* **2012,** *26*(6), 799–806.Wei, H.; Li, Z.; Hu, S.; Chen, X.; Cong, X. Apoptosis of Mesenchymal Stem Cells Induced by Hydrogen Peroxide Concerns Both Endoplasmic Reticulum Stress and Mitochondrial Death Pathway through Regulation of Caspases, p38 and JNK. *J. Cell. Biochem.* **2010,** *111*(4), 967–978.

Wu, W.; Wieckowski, S.; Pastorin, G.; Benincasa, M.; Klumpp, C.; Briand, J. P.; Gennaro, R.; Prato, M.; Bianco, A. Targeted Delivery of Amphotericin B to Cells by Using Functionalized Carbon Nanotubes. *Angew. Chemie Int. Ed. Engl.* **2005,** *44*(39), 6358–6362.

Yadav, T.; Mungray, A. A.; Mungray, A. K. Dispersion of Multiwalled Carbon Nanotubes in Acacia Extract and it's Utility as an Antimicrobial Agent. *RSC Adv.* **2015,** *5*(126), 103956–103963.

Yuan, W.; Jiang, G.; Che, J.; Qi, X.; Xu, R.; Chang, M. W.; Chen, Y.; Lim, S. Y.; Dai, J.; Chan-Park, M. B. Deposition of Silver Nanoparticles on Multiwalled Carbon Nanotubes Grafted with Hyperbranched Poly(amidoamine) and Their Antimicrobial Effects. *J. Phys. Chem. C* **2008,** *112*(48), 18754–18759.

Zhu, Y. I.; Stiller, M. J. Dapsone and Sulfones in Dermatology: Overview and Update. *J. Am. Acad. Dermatol.* **2001,** *45*(3), 420–433.

Zhuang, C.; Hao, X.; Wenchao, L.; Chen, S. Antibacterial and Release Properties of Ag Nanoparticles on Dopamine Coated Carbon Nanotubes. *Compos. Mater. An Int. J. Full* **2016,** *1*(1), 1–8.

INDEX

Printed and bound by CPI Group (UK) Ltd, Croydon, CR0 4YY

23/10/2024

01777703-0015